Asian Mathematical Conference 2016 (AMC 2016)

Journal of Physics: Conference Series Volume 893

Glasgow, United Kingdom
21 – 26 August 2016

ISBN: 978-1-5108-5061-3
ISSN: 1742-6588

Printed from e-media with permission by:

Curran Associates, Inc.
57 Morehouse Lane
Red Hook, NY 12571

Some format issues inherent in the e-media version may also appear in this print version.

Printed by Curran Associates, Inc. (2017)

Institute of Physics
Dirac House, Temple Back
Bristol BS1 6BE UK

Phone: 44 1 17 929 7481
Fax: 44 1 17 920 0979

techtracking@iop.org

Additional copies of this publication are available from:

Curran Associates, Inc.
57 Morehouse Lane
Red Hook, NY 12571 USA
Phone: 845-758-0400
Fax: 845-758-2633
Email: curran@proceedings.com
Web: www.proceedings.com

Table of contents

Volume 893

The Asian Mathematical Conference 2016 (AMC 2016)
25–29 July 2016, Bali Nusa Dua Convention Center, Bali, Indonesia

Accepted papers received: 5 September 2017
Published online: 28 October 2017

Preface

Logic and Foundations / Mathematics Education / History of Mathematics

Algebra / Lie Theory and Generalizations, Number Theory

Topology, Analysis and its Applications

Dynamical Systems and Ordinary Differential Equations

Partial Differential Equations

Probability and Statistics

Combinatorics and Graph Theory

Mathematical Aspects of Computer Science, Numerical Analysis and Scientific Computing / Mathematics in Science and Technology

Control Theory and Optimization

IOP Conf. Series: Journal of Physics: Conf. Series **893** (2017) 011001 doi:10.1088/1742-6596/893/1/011001

Preface

This volume consists of the refereed papers presented at 2016 Asian Mathematical Conference (AMC2016) held on July 25-29, 2016 at Bali Nusa Dua Convention Center (BNDCC), Nusa Dua Bali Indonesia. This conference was hosted by the Indonesian Mathematical Society (IndoMS) and the Southeast Asian Mathematical Society (SEAMS). The AMC2016 was co-organized by Institut Teknologi Bandung (ITB), Universitas Padjadjaran (Unpad), Universitas Gadjah Mada (UGM), Universitas Indonesia (UI) and Universitas Udayana (Unud).

The conference was attended by around 700 participants from all over the countries: Japan, China, Korea, India, Singapore, Malaysia, Vietnam, Thailand, Philippines, Pakistan, Australia, USA, Netherland, France, UK, Iran, Germany, Turkey, Spain and Indonesia. There were 9 plenary speakers, 3 special lecture speakers and 64 invited talks in this conference. Especially, Professor Martin Hairer (University of Warwick, 2014 Fields Medallist) gave a plenary talk.

The organizers are grateful to the Korean Mathematical Society (KMS), Mathematical Society of Japan (MSJ), Korea Institute for Advanced Study (KIAS), International Mathematical Union (IMU), Commission of Women in Mathematics IMU, Centre International de Mathematiques Pures et Appliquees (CIMPA), the Ministry of Research Technology and Higher Education, the Ministry of Tourism, Badan Keamanan Laut (BAKAMLA), Persatuan Aktuaria Indonesia (PAI), PT Komatsu Indonesia for their invaluable supports. In particular, we also thank to IOP conference series for being a technical co-sponsor for this publication.

We also thank all steering, scientific, organizing and program committee members and referees for the excellent work. Our high appreciation is to all the plenary speakers: James Maynard, Sudhir R. Ghorpade, Martin Hairer, Narutaka Ozawa, Sug Woo Shin, Hendra Gunawan, Wee Teck Gan, Van H. Vu, Zhou Xiangyu, and to all special lecture speakers, invited and contributed speakers as well as to all the participants for making this conference successful and enjoyable.

IOP Conf. Series: Journal of Physics: Conf. Series **893** (2017) 011002 doi:10.1088/1742-6596/893/1/011002

COMMITTEE

Co-Chairs	Edy Tri Baskoro (Institut Teknologi Bandung) Budi Nurani Ruchjana (Universitas Padjadjaran) Sri Wahyuni (Universitas Gadjah Mada) Kiki Ariyanti Sugeng (Universitas Indonesia)
General Secretary	Intan Muchtadi (Institut Teknologi Bandung)
Steering Committee	Edy Tri Baskoro, (President of SEAMS) - Chair Hyungju Park (International Mathematical Union) C. Herbert Clemens (International Mathematical Union) Budi Nurani Ruchjana (President of Indonesian Mathematical Society) Liqun Zhang (General Secretary of Chinese Mathematical Society) Motoko Kotani (President of Mathematical Society of Japan) Yong-Hoon Lee (President of Korean Mathematical Society) BN. Waphare (Representative of Indian Mathematical Society) Ling San (President of Singapore Mathematical Society) Le Tuan Hoa (VIASM, Vietnam)
Scientific Committee	Budi Nurani Ruchjana (Universitas Padjadjaran, Indonesia) - Chair Intan Muchtadi (Institut Teknologi Bandung, Indonesia) - Co-chair Srinivasan Kesavan (International Mathematical Union) Polly W. Sy. (International Mathematical Union) Dohan Kim (Seoul National University, Korea) Fidel Nemenzo (University of the Philippines Diliman) Wanida Hemakul (Chulalongkorn University, Thailand) Zhu Chengbo (National University of Singapore) Rosihan M. Ali (Universiti Sains Malaysia) Maslina Darus (Universiti Kebangsaan Malaysia) Leo H. Wiryanto (Institut Teknologi Bandung, Indonesia) Zulkardi (Universitas Sriwijaya, Indonesia) SEAMS Council Members
Program Committee	Intan Muchtadi (Institut Teknologi Bandung) - Chair Ikha Magdalena (Institut Teknologi Bandung) Indah Emilia Wijayanti (Universitas Gadjah Mada) Hizir Syofyan (Universitas Syiah Kuala) Rinovia Simanjuntak (Institut Teknologi Bandung) Mashadi (Universitas Riau) Irwan Endrayanto (Universitas Gadjah Mada) Slamin (Universitas Jember) Yaya S. Kusumah (Universitas Pendidikan Indonesia) Agus Suryanto (Universitas Brawijaya) Utriweni Mukhaiyar (Institut Teknologi Bandung) Marjono (Universitas Brawijaya) Hasmawati (Universitas Hasanuddin) Marwan (Universitas Syiah Kuala) Ema Carnia (Universitas Padjadjaran) Mardiyana (Universitas Sebelas Maret) Eridani (Universitas Airlangga) Dadan Kusnandar (Universitas Tanjung Pura) Ratu Ilma Indra Putri (Universitas Sriwijaya) Kristiana Wijaya (Universitas Jember) Alhadi Bustamam (Universitas Indonesia) Jalina Wijaya (Institut Teknologi Bandung)

S. Sariyasa (Universitas Pendidikan Ganesha)
Thomas Pentury (Universitas Pattimura)
Yudi Soeharyadi (Institut Teknologi Bandung)
Widowati (Universitas Diponegoro)
Salmah (Universitas Gadjah Mada)
Gustina Elfiyanti (Universitas Islam Negeri Jakarta)
Topan Eko Raharjo (Institut Teknologi Bandung)

Organizing Committee

Sri Wahyuni (Universitas Gadjah Mada) - Chair
Ch. Rini Indrati (Universitas Gadjah Mada)
Kiki Ariyanti Sugeng (Universitas Indonesia)
Edi Cahyono (Universitas Halu Oleo)
Gunardi (Universitas Gadjah Mada)
Komang Dharmawan (Universitas Udayana)
Diah Chaerani (Universitas Padjadjaran)
Wikaria Gazali (Universitas Bina Nusantara)
Tjokorda Bagus Oka (Universitas Udayana)
Ketut Budayasa (Universitas Negeri Surabaya)
Syafrizal Sy. (Universitas Andalas)
G. K. Gandhiadi (Universitas Udayana)
Saladin Uttunggadewa (Institut Teknologi Bandung)
Erma Suwastika (Institut Teknologi Bandung)
Basuki Widodo (Institut Teknologi Sepuluh Nopember)
Herni Utami (Universitas Gadjah Mada)
Anang Kurnia (Institut Pertanian Bogor)
Abdurrahman Asari (Universitas Negeri Malang)
Idha Sihwaningrum (Universitas Jenderal Sudirman)
Titin Siswantining (Universitas Indonesia)
Tulus (Universitas Sumatera Utara)
Anna Chadidjah (Universitas Padjadjaran)
Hengki Tasman (Unversitas Indonesia)
Rudi Rosadi (Universitas Padjadjaran)
Sisilia Sylviani (Universitas Padjadjaran)
Nora Hariadi (Universitas Indonesia)
Fahmi Candra Permana (Universitas Padjadjaran)

Proceeding Committee

Ikha Magdalena (Institut Teknologi Bandung)
Gustina Elfiyanti (Universitas Islam Negeri Jakarta)
Kristiana Wijaya (Universitas Jember)
Azhar Syarifuddin (Institut Teknologi Bandung)
Aditya Purwa Santika (Institut Teknologi Bandung)
Yoshua Yonatan Hamonangan (Institut Teknologi Bandung)
Nisky Imansyah Yahya (Institut Teknologi Bandung)
Rinna Samra (Institut Teknologi Bandung)

IOP Conf. Series: Journal of Physics: Conf. Series **893** (2017) 011003 doi:10.1088/1742-6596/893/1/011003

REVIEWERS

The Committees of 2016 Asian Mathematical Conference (AMC2016) would like to express gratitude to all reviewers for the volunteering support and contribution in the reviewing process:

Andry Alamsyah
Abbas Namdar Aliabadi
Syaiful Anam
Benjier Arriola
Asmiati
Hilda Assiyatun
Roger Bivand
Ema Carnia
Yus Cholily Dodi Devianto
Stanley Dewanto
Komang Dharmawan
Fengming Dong
Michiel Doorman
Orr Dunkelman
Irwan Endrayanto
Novry Erwina
Maman Fathurohman
Andrea Fenovcikova
Frans van Galen
Agnes Garciano
Agus Yodi Gunawan
Hendra Gunawan
Putu Harry Gunawan
Sugi Guritman
Rieske Hadianti
Denny Ivanal Hakim
Dewi Handayani
Eric Haryanto
Suresh Manjanath Hegde
Gatot F. Hertono
Nurdin Hinding
Ch. Rini Indrati
Sapto Wahyu Indratno
Diari Indriati
Irwansyah
Al Jupri
Kadir
Adhe Kania
Kasbawati
Berindeerjet Kaur
Hyun-Min Kim
Elvira de Lara-Tuprio
Agung Lukito
Ikha Magdalena
I Wayan Mangku
Intan Muchtadi-Alamsyah
Utriweni Mukhaiyar
Sudi Mungkasi
Dellavitha Nasution
Anak Agung Gede Ngurah
Van Nguyen
Takahiro Noi
Lienda Noviyanti
Elah Nurlaelah
Finny Oktariani
Septiadi Padmadisastra
Dila Puspita
Gantina Rachmaputri
Isnaini Rosyida
Budi Nurani Ruchjana
Farzaneh Saadati
Esengul Salturk
Aditya Purwa Santika
Suhadi Wido Saputro
Roberd Saragih
Kurnia Novita Sari
Nor Haniza Sarmin
Masitah Shahrill
Maruti Shikare
Seli Siti Sholihat
Denny Riama Silaban
Rinovia Simanjuntak
Tatag Y.E. Siswono
Slamin
Linda Smail
Kiki A. Sugeng
Suhartono
Liliek Susilowati
Taufik Edy Sutanto
Erma Suwastika
Azhar Syarifuddin
Hitoshi Tanaka
Jessada Tariboon
Kim-Chuan Toh
Binod Tripathy
Herni Utami
Sandy Vantika
Gabriel-Eduard Vilcu
Sri Wahyuni
Ian Wanless
Basuki Widodo
Widowati
Kristiana Wijaya
Leo H. Wiryanto
Lyra Yulianti
Ipung Yuwono
Jean-Claude Zambrini

IOP Conf. Series: Journal of Physics: Conf. Series **893** (2017) 011004 doi:10.1088/1742-6596/893/1/011004

Peer review statement

All papers published in this volume of *Journal of Physics: Conference Series* have been peer reviewed through processes administered by the proceedings Editors. Reviews were conducted by expert referees to the professional and scientific standards expected of a proceedings journal published by IOP Publishing.

Didactic trajectory of research in mathematics education using research-based learning

Rully Charitas Indra Prahmana[1], **Yaya S. Kusumah**[2], **Darhim**[2]

[1] Dept. of Mathematics Education, Universitas Ahmad Dahlan, Yogyakarta Indonesia

[2] Dept. of Mathematics Education, Universitas Pendidikan Indonesia, Bandung Indonesia

E-mail: rully.indra@mpmat.uad.ac.id

Abstract. This study aims to describe the role of research-based learning in design a learning trajectory of research in mathematics education to enhance research and academic writing skills for pre-service mathematics teachers. The method used is a design research with three stages, namely the preliminary design, teaching experiment, and retrospective analysis. The research subjects are pre-service mathematics teacher class of 2012 from one higher education institution in Tangerang - Indonesia. The use of research-based learning in designing learning trajectory of research in mathematics education plays a crucial role as a trigger to enhancing math department preservice teachers research and academic writing skills. Also, this study also describes the design principles and characteristics of the learning trajectory namely didactic trajectory generated by the role of research-based learning syntax.

Keywords: Design Research, Research-based Learning, Research Skill, Academic Writing Skill, Didactic Trajectory

1. Introduction

Indonesian Directorate of Higher Education Institution to encourage more research and publications by students including undergraduate students issued a policy [1]. Furthermore, the National Standard Qualification [2] requires students to publish their studies in a reputable national journal (acknowledged by KKNI level 6). On the other hand, some researchers have documented the success of research-based learning to enhance students' skills in conducting research [3, 4, 5], but most of them still focused on the students who come from the non-education department. Therefore, this study develops learning trajectory of research that focused on mathematics education.

2. Literature Review and Research Question

Research skills represent a set of abilities which are required to do a research, including the strategies and instruments to gather and evaluate information which consists of observing, formulating research question, building a hypothesis, conducting experiment, analyzing data, and drawing a conclusion [6, 7]. In accordance to that, Majelis Profesor Riset Lembaga Ilmu Pengetahuan Indonesia (The Council of Research Professor of Indonesian Institute of Science) [8] stated that research skills are the skills to conduct scientific study in attempt to find truth with the implementation of scientific methods which are based on tested scientific reasoning. Furthermore, these skills are important for preservice teachers since developing such skills would help them building a strong intellectual and practical connection between research and their own learning process [3, 9]. Thus, when students possess these skills, it would be easier for them to do a research.

Willison & O'Regan [5] developed six research skill indicators that classified into five levels of research. Furthermore, Dowse and Howie [10], conducted within the framework of design research, sought to design and develop an academic research writing intervention. Prahmana & Kusumah [11] identified that the learning process at the university did not offer enough support for students to develop their academic writing skills. Therefore, this study aims to develop a learning trajectory of research in mathematics education using design research methodology to enhance research and academic writing skills for pre-service mathematics teachers.

Based on the problems mentioned, the success of the learning activities to improve pre-service teachers skills is investigated in research question as follows, *"To what extent do the didactic trajectory on research in mathematics education in assisting pre-service mathematics teachers to enhance their research and academic writing skills?"*

3. Research Method

This study uses a design research as the research method, which is an appropriate way to answer the research questions and achieve the research objectives that start from preliminary design, teaching experiments, and retrospective analysis [12]. Design research is a methodology that has five characteristics, namely interventionist nature, process oriented, the reflective component, cyclic character, and theory-driven [13]. Design research is a cyclical process of thought experiment and instruction experiment to implementation [14]. According to Freudenthal [15], students given the opportunity to build and develop their ideas and thoughts when constructing mathematics. Teachers can select appropriate learning activities as a basis to stimulate students to think and act when creating mathematics.

The 35 pre-service mathematics teachers during eight months period from one higher education institution in Tangerang - Indonesia was collected. Initial data analysis of 35 pre-service teachers research and academic writing skills conducted in six stages and the learning trajectories on this topic were identified [11]. This study consists of three steps repeatedly done until the discovery of a didactic trajectory that a revision of the theory of learning tested.

4. Research Results and Analysis

The design principles are statements heuristics (heuristic statements) based on the input or suggestions based on research to solve a problem [16]. The statement should always develop in certain situations, so it does not guarantee success in another situation. However, the report is getting better, when validated in various situations. Therefore, all activities that are designed and implemented in this study documented and analyzed retrospectively to generate design principles in local instruction theory (LIT) on research in mathematics education (one set of procedures and conditions that lead to the achievement of learning targets) as shown in Fig. 1. Further, the characteristics of LIT on research in mathematics education provides a quirk in the design principles that produced.

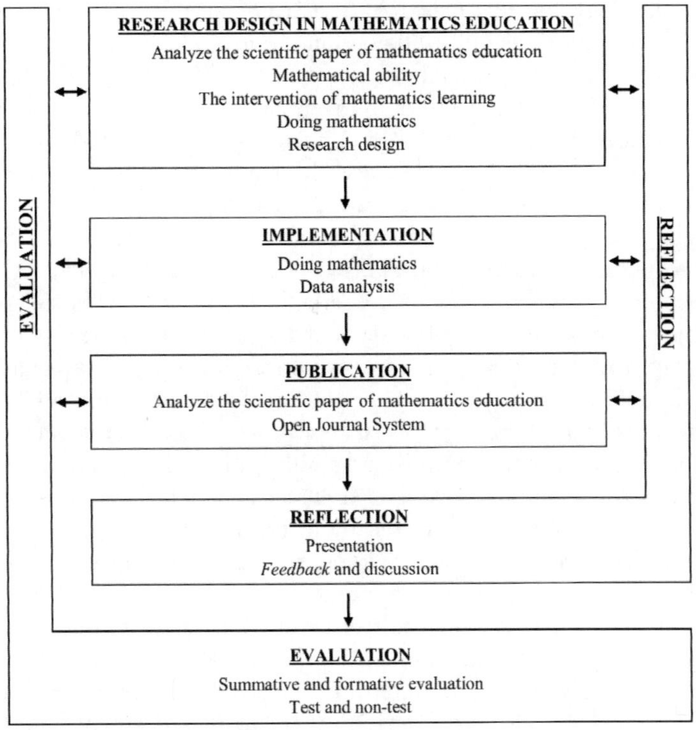

Figure 1. Didactic trajectory of research in mathematics education

The design principles of the learning trajectory of research in mathematics education namely didactic trajectory in mathematics education research is the research-based learning activities were carried out in 16 meetings in Research Methodology course. The activities were divided into two major parts of the learning activities in the classroom (Mathematics Education Research Design, Publication, Reflection, and Evaluation) and data collection in schools (Implementation). During the learning activities in the classroom, students are given the opportunity to learn about how to examine the scientific article mathematics education research, formulate research problems of mathematics education that consist of facts, procedures, principles, and mathematical concepts. Students study some mathematical ability and its indicators, such as the ability of mathematical understanding, mathematical problem solving, and the mathematical representation. Lastly, students conduct research data processing, making inferences, writing scientific articles based on research data that collected, and submitting scientific articles in national journals indexed Google Scholar using the Open Journal System. Activities reflection carried out at each learning activities in the classroom to provide feedback and understanding of the material that is being studied in the form of discussions and presentations, while the evaluation activities carried out on all activities to see achievement in each learning activity.

The didactic trajectory on research in mathematics education was developed based on the principles of research-based learning to cultivate the students' skills in conducting research and writing scientific papers. Those principles are: (1) teaching material enriched by the faculty research that can be used as a reference in conducting research for students; (2) a number of the findings of recent studies on mathematics education research trends, use and find the history of inventions; (3) learning activities enriched with contemporary research issues in the world of mathematics education; (4) the material taught mathematics education research methods in the early learning process; (5) learning activities enriched with mathematics education research activities on a small scale is done in groups; (6) enriched learning activities with the involvement of students in mathematics education research activities that carry the name of the institution (Student Creativity Program Research); (7) enriched learning activities by encouraging students to become part of the culture of research in Mathematics Education courses; and (8) the learning activities enriched with values that must be owned by a researcher [see e.g. [4, 5, 7, 11, 17, 18, 19, 20, 21, 22, 23, 24, 25, 26, 27]].

In addition to the principles above, there are some characteristics of the didactic trajectory that has been developed to cultivate the skills of researching and writing scientific papers. These characteristics are (1) the student as a center of learning; (2) the role of a facilitator of learning is held fully by the lecturer; (3) the student has a project on the study of mathematics education; (4) the results of the research are integrated with learning materials; (5) The student is required to play an active role in the implementation of the mathematics education research; (6) the research instruments used during the learning process; (7) the research context developed inclusively (the students learn the procedures and results of research to understand the intricacies of synthesis); and (8) the results disseminated in scientific forums or scientific journals of mathematics education by students, facilitated by lecturers [11].

The whole subject of research succeeded in doing mathematics education research and publish the results of their studies in the form of scientific articles. These results show that the didactic trajectory in mathematics education research that consists of 7 activity [23, 28] has provided knowledge about the various problems faced by the students about their ability to solve mathematically and measures formulated in the form of mathematics education research. Students also gain experience teaching activities (provide treatment) in the classroom to resolving the problems that hypothesized. Lastly, students gain experience in research data processing until the conclusion of the results of research and then write it in the form of scientific articles. Students feel their skills in doing research and writing scientific papers grew and developed after completion of all activities in the learning trajectory of research in mathematics education that has been in the design.

5. Conclusion

This study develops a learning trajectory of research in mathematics education using design research methodology to enhance research and academic writing skills for pre-service mathematics teachers. The product of development is didactic trajectory on research in mathematics education that gives real contribution to improving their skills. Initial data analysis of 35 pre-service teachers research and academic writing skills conducted in six stages and the learning trajectories on this topic identified. The 35 pre-service teachers were divided into ten groups and did research independently to produce ten scientific articles that published in ten different Nasional Journal. The retrospective analysis occurred in the teaching experiment showed that these pre-service teachers were on the level 3 (20%) and 4 (80%), based on research skill indicators developed by Willison & O'Regan [5]. The academic writing skill category of these pre-service teachers was excellent (30%) and good (70%), based on academic writing skill indicators developed by Supriyadi [28].

Acknowledgment

The authors would like to thank the Indonesian Directorate General of Higher Education that supported and funded this research.

References

[1] Peraturan Menteri Pendidikan dan Kebudayaan (PERMENDIKBUD) Republik Indonesia Nomor 49 Tahun 2014 Tentang Standar Nasional Pendidikan Tinggi.

[2] Tim Penyusun KKNI Dikti 2013*Kerangka Kualifikasi Nasional Indonesia dan Implikasinya Pada Dunia Kerja dan Pendidikan Tinggi* (Jakarta: DIKTI) p 7

[3] Webb F, Smith C and Worsfold K 2011 *Research Skills Toolkit* (Queensland: Griffith Institute for Higher Education) p 5

[4] Widayati D T, Luknanto D, Rahayuningsih E, Sutapa G, Harsono, Sancayaningsing R P and Sajarwa 2010 *Pedoman Umum Pembelajaran Berbasis Riset* (Yogyakarta: Universitas Gadjah Mada) pp 4–7

[5] Willison J and O′Regan K 2007 Commonly known, commonly not known, totally unknown: A framework for students becoming researchers *The Higher Education Research and Development* **26** p 393

[6] Suhaimi Z, Shahrill M, Abbas N A H, Tengah K A, Roslan R and Yusof N 2017 Exploring the use of journal writing in mathematics classroom *International Journal on Emerging Mathematics Education* **1** p 41

[7] Waris A 2009 Model pembelajaran berbasis riset (PBR) di Program Studi Fisika ITB *Berita Pembelajaran* **6** p 1

[8] Majelis Profesor Riset Lembaga Ilmu Pengetahuan Indonesia 2007 *Kode Etika Peneliti* (Jakarta: Lembaga Ilmu Pengetahuan Indonesia) p 1

[9] Young J R 2017 Technology integration in mathematics education: Examining the quality of meta-analytic research *International Journal on Emerging Mathematics Education* **1**p 71

[10] Dowse C and Howie S 2013 Promoting academic research writing with South African masters students in the field of education *Educational Design Research - Part B: Illustrative Cases* eds T Plomp and N Nieveen (Enschede: SLO) pp 851–879

[11] Prahmana R C I and Kusumah Y S 2016 The hypothetical learning trajectory on research in mathematics education using research-based learning *Pedagogika* **123** p 42

[12] Prahmana R C I 2013 *Proc. in the First South East Asia Design/Development Research (SEA-DR) Conf.* (Palembang: Universitas Sriwijaya) pp 391–398

[13] Akker J V D, Gravemeijer K, McKenney S and Nieveen 2006 *Educational Design Research* (London: Routledge Taylor and Francis Group) p 4

[14] Gravemeijer K 2004 Local instructional theories as means of support for teacher in reform mathematics education *Mathematical Thinking and Learning* **6** p 105

[15] Gravemeijer K and Eerde D V 2009 Design research as a means for building a knowledge base for teaching in mathematics education *The Elementary School Journal* **109** p 510

[16] Plomp T 2013 Educational design research: An introduction *Educational Design Research* eds T Plomp and N Nieveen (Enschede: SLO) pp 10–51

[17] Appel J 1995 *Roles of Teachers: A Case Study Based on Diary of a Language Teacher* (Autora: Universitat JaumeI) p 3

[18] Baldwin G 2005 *The Teaching-Research Nexus: How Research Informs and Enhances Learning and Teaching in the University of Melbourne* (Melbourne: Centre for the Study of Higher Education, the University of Melbourne) pp 3–4

[19] Cryer P 2006 *The Research Students Guide to Success, Third Edition* (London: McGraw-Hill Open University Press) p 26

[20] Farkhan M 2008 *Research-Based Learning* (Jakarta: UIN Syarif Hidayatullah) p 9

[21] Griffith Institute for Higher Education (GIHE) 2008 *Research-Based Learning: Strategies for Successfully Linking Teaching and Research* (Queensland: University of Griffith) p 5

[22] Healey M 2005 Linking Research and Teaching: Disciplinary Spaces *Reshaping the University: New Relationships between Research, Scholarship, and Teaching* ed R Barnett (Maidenhead: McGraw-Hill/Open University Press) pp 30–42

[23] Prahmana R C I 2015 Hubungan antara keterampilan meneliti dan pembuatan skripsi mahasiswa pendidikan matematika *Jurnal Numeracy* **2** 70

[24] Prahmana R C I 2015 *Penelitian Pendidikan Matematika* (Yogyakarta: Matematika) p 23

[25] Prahmana R C I, Kusumah Y S and Darhim 2016 Keterampilan mahasiswa dalam melakukan penelitian pendidikan matematika melalui pembelajaran berbasis riset *Beta* **9** 1

[26] Umar M K, Yusuf M, Supartin, Uloli R, Abjul T and Ntobuo N E 2011 *Laporan Hasil Penelitian Pengembangan Pembelajaran Berbasis Riset di Program Studi Pendidikan Fisika FMIPA Universitas Negeri Gorontalo* (Gorontalo: Universitas Negeri Gorontalo) pp 22–40

[27] Prahmana R C I 2017 *Design Research (Teori dan Implementasinya: Suatu Pengantar)* (Jakarta: Rajawali Pers) p 87

[28] Supriyadi 2013 *Menulis Karya Ilmiah dengan Pendekatan Konstruktivisme: Pembelajaran Menulis Karya Ilmiah yang Inovatif dan Konstruktif* (Gorontalo: UNG Press) pp 166–174

Reflective thinking in solving an algebra problem: a case study of field independent-prospective teacher

Agustan S[1,2], Dwi Juniati[3], Tatag Yuli Eko Siswono[3]

[1] Faculty of Teacher Training and Education, Universitas Muhammadiyah Makassar, Indonesia
[2] Doctoral Program Student of Mathematics Education, Indonesia
[3] Department of Mathematics, Universitas Negeri Surabaya, Indonesia

E-mail: agustan@unismuh.ac.id, juniati@yahoo.com, tatagsiswono@unesa.ac.id

Abstract. Nowadays, reflective thinking is one of the important things which become a concern in learning mathematics, especially in solving a mathematical problem. The purpose of this paper is to describe how the student used reflective thinking when solved an algebra problem. The subject of this research is one female student who has field independent cognitive style. This research is a descriptive exploratory study with data analysis using qualitative approach to describe in depth reflective thinking of prospective teacher in solving an algebra problem. Four main categories are used to analyse the reflective thinking in solving an algebra problem: (1) formulation and synthesis of experience, (2) orderliness of experience, (3) evaluating the experience and (4) testing the selected solution based on the experience. The results showed that the subject described the problem by using another word and the subject also found the difficulties in making mathematical modelling. The subject analysed two concepts used in solving problem. For instance, geometry related to point and line while algebra is related to algebra arithmetic operation. The subject stated that solution must have four aspect to get effective solution, specifically the ability to (a) understand the meaning of every words; (b) make mathematical modelling; (c) calculate mathematically; (d) interpret solution obtained logically. To test the internal consistency or error in solution, the subject checked and looked back related procedures and operations used. Moreover, the subject tried to resolve the problem in a different way to compare the answers which had been obtained before. The findings supported the assertion that reflective thinking provides an opportunity for the students in improving their weakness in mathematical problem solving. It can make a grow accuracy and concentration in solving a mathematical problem. Consequently, the students will get the right and logic answer by reflective thinking.

1. Introduction

The competence in solving mathematical problems is one of the competencies which must be owned by a student in learning mathematics [1]. To solve a mathematical problem, it is required a high level thinking skills which it can encourage students's curiosity in order that there is a desire to solve mathematical problems encountered. One of the thinking skills which can be applied in solving mathematical problem is reflective thinking [2] because reflective thinking is a type of high-level thinking which can encourage students's curiosity [3]. This is in line with demands of curriculum 2013 in Indonesia which is found key words on main and basic competencies as science behavioral includes curiosity, objective, fair, thorough, meticulous, diligent, careful, responsible, open, critical, creative, innovative and caring environment. Moreover the skills required to face the 21st century challenges, namely cognitive skills, intrapersonal skills, and interpersonal skills. It confirms that reflective thinking is a future competence that should be taught to students to adapt changes and respond demands of the 21st century. In addition, Goodell [4] and Ville [5] stated that one of goals of the educational personnel and institutions creates a prospective teacher who is responsible and capable reflective thinking. King and Kitchener stated that, by reflective thinking, student can solve more complex problems because student's sense will be directed and solution of the problem being solved tend to be correct and appropriate [6]. In addition, Ambrose [7], Gelter [8] and Koszalka [6] stated that reflective thinking gives opportunities for students to justify misconceptions by helping students to think about what they did and why they did it. This indicates that the reflective thinking occurs after mathematical problem solving conducted with the aim checking the error of the concepts used and try to justify these misconceptions, so that it can develop students' skills in using mathematical concepts [9]. Every student has a unique character so as to learn, master, process information, solve problems, and to teach the material well, in this case mathematics, someone will do it in also different way [10]. Every student has cognitive style. Differences in student's cognitive styles affect students' skills in reflective thinking and solving problems. Cognitive styles correlate with intellectual and perceptual behaviour. Intellectual refers to a person's ability to think, while perceptually refers to a person's ability to view or interpret anything. Another concept of cognitive style developed by Witkin, et.al [19] based on continuous global analytical. Based on this concept, cognitive styles are divided into field-independent (FI) and field-dependent (FD). Individuals who are field-independent (FI) prefer to separate the parts of a number of patterns and analysis patterns based on their components. Another hand, individuals who are field-dependent (FD) tend to see a pattern as a whole, not separate it into parts.

Based on the explanation above, reflective thinking is one of the important tools that help students to improve their curiosity. It gives also opportunities for students in justifying their mistakes or misconceptions when they have solved mathematical problem and try to improve their weakness. As a consequence, in this research paper, the researchers will describe how the subject uses her reflective thinking after solving algebra problem who has field independent cognitive style. It is interesting to investigate subject's reflective thinking to give us description and insight how reflective thinking can be applied in solving mathematical problem to get the best answer logically. As a result, we can choose the best way to teach the students in providing a chance to look for their mistakes in solving mathematical problem and to justify them so that their answers tends to be true.

2. Literature Review

Currently reflective thinking has became the most prominent issues in the literature, in particular on professional education of teachers [11], [12]. Reflective thinking is active, persistent, and careful consideration of any belief or suppose from of knowledge in the light of the grounds that support it and the conclusion to which it tends [13]. Furthermore, Dewey [14] suggested that there are six phases in reflective thinking, namely: (a) an experience; (b) spontaneous interpretation of the experience; (c) naming the problem or question that arise out of the experience; (d) generating possible explanations for the problem or question posed; (e) ramifying the explanation into full-blown hypotheses. Meanwhile, according to Lee [15], there are five phases reflective thinking, namely: (a) problem context; (b) Problem definition; (c) Seeking possible solution; (d) experimentation; (e) evaluation; (f) acceptance/rejection. Rodgers [16] argues that there are four phases in the process of reflective thinking as follows: (a) presence to experience; (b) description of experience; (c) analysis of experience; (d) intelligent action/experimentation. Zehavi and Mann [17] stated that the reflective thinking for the level of meta-cognitive referred to four categories: electoral techniques, monitoring process of solutions, insight or ingenuity, and conceptualization which connects the concept and meaning. To gain an overview of the process of reflective thinking in solving mathematical problems, previously it is need to be synchronized several components of reflective thinking processes that exist, namely the components of the process of reflective thinking by John Dewey [13] abbreviated **PBRD**; by Lee [15] abbreviated **PBRL**; by Rodgers [16] abbreviated **PBRR**, by Zehavi and Mann [17] abbreviated **PBRZM**. From the components of reflective thinking processes served by each figures who examined reflective thinking process, there are similarities indicators. It is obtained results of the construction of thinking reflective process contained in the following table I which consists of four stages. These four stages are: (1) *Formulation and synthesis of the experience*; (2) *Orderliness of experience*; (3) *Evaluation of experience*; and (4) *Testing the selected solution based on the experience.*

TABLE I
CONSTRUCTION OF REFLECTIVE THINKING PROCESS

Dewey (1998) (PBRD)	Lee (2000) (PBRL)	Rodgers (2002) (PBRR)	Zehavi & Mann (2006) (PBRZM)	Reflective Thinking
An experience *Spontaneous interpretation of the experience* *Naming the problem*	*Problem context* *Problem definition/ Reframing*	*Presence to experience* *Description of experience*	*Selection of techniques* *Monitoring of the solution process*	*Formulation and synthesis of experience*
Generating possible explanations for the problem	*Seeking possible solution*	*Analysis of experience*	*Conceptualization*	*Orderliness of experience*
Ramifying the explanations into full-blown Hypotheses			*Insight or ingenuity*	*Evaluating the experience*
Experimenting or testing the selected hypotheses	*Experimentation* *Evaluation* *Acceptance/ rejection*	*Intelligent action/ Experimentation*		*Testing selected solution based on the experience*

Everyone has a different character (unique) so as to learn, master, process information, solve problems, and to teach the material well, someone will do it in a different way also [10]. Every student has a characteristic which consistent when organizing and processing the information obtained. This characteristic is known as cognitive style. Tennant [18] and Witkin et.al [19] stated that cognitive style is a characteristic of thinking process which is consistent and reflected on students. Furthermore, Witkin, et.al [19] had developed a concept of cognitive style which is based on continuous global analyst. Based on this concept, cognitive styles are divided into field-independent and field-dependent. Field independent are individuals who are more analytical, they can choose the stimulus based on the situation, so that their perception is only a small part affected when there is a change in the situation. They prefer to separate the parts of a number of patterns and analyze patterns based on their components. While, Field dependent are individuals who tend to get difficulties in differentiating stimulus through situations that owned so that their perception is easily influenced by the manipulation of the surroundings. They tend to see a pattern as a whole, it is not separate it into parts. It can be concluded that the individuals who are Field Independent (FI) have a tendency to respond to stimulus by using their own perceptions, more analytical, and they analyze patterns based its components. Meanwhile, individuals who are Field Dependent (FD) have a tendency to respond a stimulus by using the environment as a basic requirement in perception, and they tend to see a pattern as a whole (global), not separate parts.

3. Method

This paper aims to explore and investigate student's reflective thinking process in solving algebra problem. As a consequence, the researchers adopted exploration approach to analyze the data by using qualitative approach. [20]. To get this aims, the participant was given a task of solving algebra problem and then the participant was interviewed. The participant is a student in Department of Mathematics Education, Universitas Muhammadiyah Makassar, Indonesia. To determine the participant, it will be used Group Embedded Figure Test (GEFT) which consists of 25 items. The test is divided into three sessions. The first session consists of 7 items as session of practice, the second session consists of nine items and the third session consists of nine items. The second and the third session serve as session assessed. The criteria used in the selection of the particpant using the criteria according to Kepner and Neimark [21], participant who can give right answer 0-9 classed as Field Dependent (FD) and 10-18 classed as Field Independent (FI). Furthermore, it was determined 1 participant who is female and has Field Independent (FI) cognitive styles. In addition, the participant was also considered able to verbally communicate well.

4. Result

Based on the construction of reflective thinking process of the subject in solving an algebra problem that is divided into four stages, the researchers investigated and found out what happened on every stage related to reflective thinking in solving algebra problem which had been done by the participant. The participant had been interviewed deeply on matters concerning the four stages and linkages that may affect it. The results of the interview toward the participant on data collection illustrating the process of reflective thinking in solving algebra problem by a prospective teacher who has field independent cognitive style.

The first stage, **formulation and synthesis of the experience**, the participant described problem by using his own words and explained the problem clearly related to the main point of the problem. The participant also found the concepts which were related to the problem given. For more clearly, the following is a transcript of the interview excerpt based on this stage.

Researcher: *What did you understand the problem?*

Participant: *I understood that Sukri is a trader. The distance his home and his shop is 20 km. There are 60 mangoes must bring to the shop. ... Taufan always eats one mango per kilometre. We are demanded to find out a strategy in bringing mangoes as much as possible. This problem related to the concept of algebra arithmetic operation (addition, subtraction, division and multiplication), linear equations, linear inequalities and geometry. When I wanted to solve this, I got difficulties to make a mathematical modelling ....*

From the transcript, it can be explained that the participant described the problem by using another word however participant's explanation still has the same idea from the problem. Furthermore, when the subject identified the concept or subject matter which is related to the problem, the participant mentioned not only the concept that ever the participant used in solving a problem before, but also the participant explained that the concept which will be used in solving algebra problem. Moreover, the participant found the difficulties to change from the problem to mathematical modelling.

The second stage, **orderliness of experience**, the participant described a strategy in solving algebra problem given. The participant gave an explanation related to steps in choosing an operation well and the difficulties when applying strategies. The participant also investigated two concepts used in solving problem and explained impact development of his strategy to solve other problems. For more clearly, the following is the excerpt of interview from participant.

Researcher: *Based on your experience, please tell me about your strategies to solve this problem!*

Participant: *In my opinion, this problem related to a concept of algebra and geometry. For algebra, ... For geometry, ... I tried to solve this by drawing a sketch. I found difficulties to apply my strategies (drawing). I was less conscientious and less concentration. It made my answer almost wrong. If I met other problem, I will apply this strategies and develop it. As a consequence, it can help me to solve it easier. However, I used much time.*

From the transcript, it can be explained that participant analyzes specifically two concepts used in solving problem. For instance, geometry is related to point and line and algebra is related to algebra arithmetic operation. The participant made drawing as a way to find answer or solution from the problem presented. The difficulties which were earned by the participant applying her strategies are a lack of concentration and not careful making her answer almost wrong. Moreover, the participant explained that she tended to use the same way to solve other problems. The participant also adapted her strategies in order to develop her solution to get the right answer. Furthermore, development of strategies was based on problem context where it gave effect not only positive impact but also a negative impact. the positive impact facilitated to take the right answer and a negative impact took much time in solving problem.

The third stage, **evaluation of experience**, the participant explains how to determine criteria of a good and proper solution. The participant find also strengths and weaknesses from subject's solutions and stated how to improve the weaknesses which are owned or carried when solving problems. For more clearly, the following is an excerpt of the interview from the participant.

Researcher: *Please, define a criteria of a good settlement* Participant: *Firstly, the criteria of a good settlement must be obvious and done step by step. The second, accuracy in understanding problem. There are four aspects to get an effective solution namely the ability to understand meaning of words, making mathematical modelling, performing mathematical calculations, interpreting the results obtained. I used many ways to solve this so that I believed my answer tended to be true. I sometimes made mistakes when I didn't have enough time. It can become that I were not careful and I didn't concentrate. However, I will exercise myself to solve other problem. In order that, it made me habitual to solve problem.*

From the transcript, it can be concluded that participant can give the explanation about the criteria of solution, namely (a) coherent; (b) accuracy in solving problem. Furthermore, participant makes opinion that solution must have four aspect to get effective answer, specifically (a) skill to understand meaning every words; (b) skill to make mathematical modelling; (c) skill to calculate mathematically; (d) skill to interpret solution obtained logically. The participant stated that less conscientious and less concentration become weaknesses in solving problem and to revise and improve the weakness then it is need a habit to solve other problem. As a consequence, someone will not be rigid when facing problems.

The last stage, **testing the selected solution based on the experience**, the participant explained how to make a decision in using operation and how to test internal consistency or mistake in operation or procedure. Clearly, the following is an excerpt of interview from the participant.

Researcher: *How would you define that procedure which you used had been appropriate with the principle of mathematical operation?*

Participant: *In my opinion, my procedure is correct. I noted the information known, asked, then how is completion and to make me sure with my answer. I tried to solve it in several times and in many ways. ... I also checked the operation which I used and interpreted my answer logically by linking with the context of the problem.*

From the transcript, it can be concluded that the participant illustrated that it was needed decision to choose a way or procedure which can help to find out the right answer. The procedure used was suitable with principle and the rule of mathematical operation by taking down all informations. Furthermore, to test the internal consistency or error in operation or in solution, the participant checked and looked back related procedures and operations which had been used. Moreover, the participant tried to resolve the problem in a different way to compare the answers which had been obtained before.

Based on explanation above, it can be described that the participant states that the first way, Taufan eats one mango per kilometre so that there is no mango arrives in the shop. The second way, it is still the same case, however, when Taufan arrives in the shop, the last mango is not ate. The last way, the participant solved by making stopover in the street which is passed by Taufan. If participant divides 2 parts the street, she will get 1 stopover and 10 mangoes can be in the shop. If participant divided 4 parts, she will get 3 stopovers and 15 mangoes. If participant divided 5 parts, she will get 4 stopovers and 16 mangoes. The last, if participant divides 10 parts the street, participant will get 9 stopovers and 16 mangoes.

Therefore, the participant made conclusion that 16 mangoes is the right answer. The first reason is that there are two the same answers which describe consistency of participant's answer. The second one is the participant interpreted her answers and connected to the context of problem.

5. Conclusion

Based on the result, it can be concluded that reflective thinking can be investigated after someone did a mathematical problem solving. Reflective thinking provides an opportunities for the students to improve their weakness. When they get some difficulties and find some mistakes or some misconceptions related to their answer, they will correct it and try to interpret it to obtain logical answer. Reflective thinking also can make a grow accuracy and concentration when the students solve a mathematical problem. Consequently, the students will get the right and logic answer by reflective thinking. Above all, this research paper provides insight into how promote reflective thinking that could be trained to prospective teacher that impact when prospective teacher become a real teacher in society.

References

[1] NCTM 2000 *Principle and Standards for School Mathematics* (Reston: The National Council of Teacher Mathematics, Inc.)
[2] Odafe V U 2008 *Teaching and Learning Mathematics: Student Reflective Adds a New Dimension* (USA: Bowling Green State University, Huron)
[3] Lipman M 2003 *Thinking in Education* (Cambridge: Cambridge University Press)
[4] Goodell J 2000 Learning to teach Mathematics for understanding: the role of reflection *J. Math. Teacher Education and Development* **2** pp 48–60
[5] Ville P A 2010 Mentoring reflective thinking practice in pre-service teachers: a reconstructions through the voices of australian science teachers *J. College Teaching and Learning* **7**(9)
[6] Koszalka T 2001 KaAMS: a PBL environment facilitating reflective thinking *Learning and Instruction Section* NY
[7] Ambrose R 2004 Initiating change in prospective elementary school teachers' orientations to mathematics teaching by building on beliefs *J. Math. Teacher Education* **7** pp 91–119
[8] Gelter H 2003 Why is reflective thinking uncommon? *J. Reflective Practice* **4**(3)
[9] Betne P 2009 Reflection as a learning tool in mathematics *J. Math., Eng. and Comp. Sci.* pp 93–101
[10] Stiff L V and Curcio F R 1999 *Developing Mathematical Reasoning in Grades K-12* (United Stated of America: General Year Book)
[11] Lim L Y 2011 A comparison of students' reflective thinking across different years in a problem-based learning environment *J. Instructional Science* **39** pp 171–188
[12] Amidu AR 2012 *Exploring real estate students' learning approaches reflective thinking and academic performance Proc. 48th ASC Anuual Int. Conf* (UK: The Associated of Construction)
[13] Dewey J 1998 *How We Think: A Restatement of The Relation of Reflective Thinking to The Educative Process* (Boston: Houghton-Mifflin)
[14] Roh K H and Lee Y H 2010 Promoting students' reflective thinking of multiple quantifications via the mayan activity *J. Educational Studies in Math*
[15] Lee H J 2005 Understanding and assessing preservice teachers' reflective thinking *J. Teaching and Teacher Education* **21**(6) pp 699–715
[16] Rodgers C 2002 Defining reflection: another look at john dewey and reflective thinking *J. Teachers College Record* **104**(4) pp 842–866
[17] Zehavy N and Mann G 2006 Instrumented techniques and reflective thinking in analytic geometry *J. The Montana Mathematics Enthusiast* **2**(2) pp 1551–3440
[18] Tennant M 1998 *Psychology and Adult Learning* (London: Routledge)
[19] Witkin H A et al. 1971 *A Manual for The Embedded Figures Tests* (Palo Alto, Calif: Consulting Psycologists Press)
[20] Creswell J W 2007 *Qualitative Inquiry and Research Design: Choosing Among Five Approach* 2nd Edition (London: Sage Publication)
[21] Kepner M D and Neimark E D 1984 Test-retest reliability and differential patterns of score change on the group embedded figure test *J. Personality and Social Psychology* **46**(6) pp 1405–1413

IOP Conf. Series: Journal of Physics: Conf. Series **893** (2017) 012003 doi:10.1088/1742-6596/893/1/012003

Intangible heritage for sustainable future: mathematics in the paddy field

Stanley P. Dewanto[1], Dianne A. Kusuma[1], Budi Nurani Ruchjana[1], and Atje Setiawan Abdullah[2]

[1] Department of Mathematics, Faculty of Mathematics and Natural Sciences
Universitas Padjadjaran Bandung, Indonesia
[2] Department of Computer Science Faculty of Mathematics and Natural Sciences
Universitas Padjadjaran Bandung, Indonesia

E-mail: stanleydewanto.mpd@unpad.ac.id

Abstract. Mathematics, as the only general language, can describe all phenomena on earth. Mathematics not only helps us to understand these phenomena, but it also can sustain human activities, consequently ensure that the future development is sustainable. Indonesia, with high cultural diversity, should aware to have its understanding, skills, and philosophies developed by certain societies, with long histories of interaction with their natural surroundings, which will provide a foundation for locally appropriate sustainable development. This paper discussed the condition and situation on certain area in Cigugur, Indonesia, and what skills, knowledge, and concept can be transmitted, regarding simple mathematics (arithmetic). Some examples are provided.

1. Introduction

Sustainable development (SD) becomes a major role today, as historically it is credited to Brundtland Commission: "Sustainable development is development that meets the needs of the present without compromising the ability of future generations to meet their own needs" [4]. SD is the term given to the process of developing the society to move from the existing state now to a state of sustainability. Ambali [1] stated without sustainability, development itself is ultimately counterfeit and counter productive. Undertaking SD and achieving this sustainability requires the development of knowledge, skills and attitudes. Knowledge that can be transmitted are culture-specific mathematical concepts and models; and culture-specific systems and tools for calculating, measuring and recording [4]. Skills that can be transmitted are in counting, sequencing, estimating, quantifying, matching, sorting, discerning patterns and rhythms, in process thinking, lateral thinking and problem solving [4].

This paper discussed local wisdom of indigenous people in a village in West java, namely Cigugur, particularly in determining cultivation and harvest time of rice, with simple arithmetic.

2. Main Result

Indonesia, considered as the largest archipelagic island country, with more than 13,000 islands, has an estimated population of over 258 million people, consists of hundreds of distinct native ethnic and linguistic groups. Java, as the most populous island is Java, contains more than half of the countrys population.

IOP Conf. Series: Journal of Physics: Conf. Series **893** (2017) 012003 doi:10.1088/1742-6596/893/1/012003

West Java, as part of the Pacific Ring of Fire, has mountains and volcanoes more than any other provinces in Indonesia. Because of these volcanoes mountain, West Java has rich and fertile volcanic soil, and in effect, the main way of life of the traditional Sundanese people (one of the society in West Java) is famous with the traditional rice cultivation, wet (sawah) or dry (ladang or huma). The landscape of West java is one of the volcanic mountains, steep terrain, forest, mountains, rivers, fertile agricultural land, and natural sea harbors (fisheries).

One of the provinces in Indonesia, West Java province (see picture below) has been subdivided into 21 regencies (*kabupaten*), and 620 districts (*kecamatan*), comprises of 1576 villages (*kelurahan*) and 4301 rural villages (*desa*).

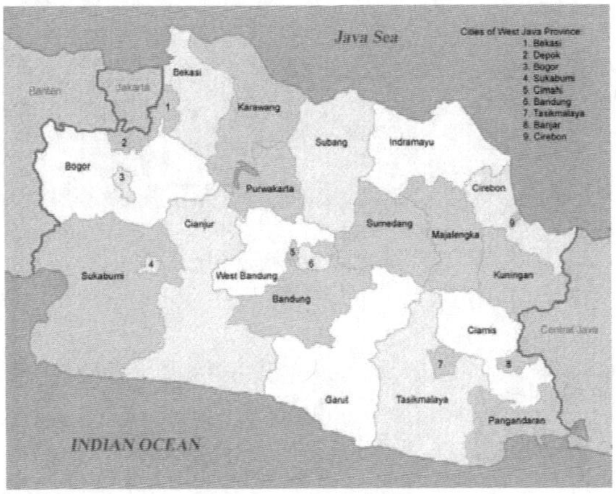

Figure 1. Map of Regencies at West Java Province

2.1. Indigenous knowledge in Cibubur

One of the villages is Cigugur, located in Kuningan regency, has a population of 7804 inhabitants, and has mostly wet paddy field (or rice field) as the people job on it. Wet paddy field will happen in the rainy time come, so the field has a lot of water.

Indigenous people believe that their traditional-cultural calculation in managing their paddy fields, called local wisdom, is the best for cultivating and harvesting rice. This local wisdom is believed a discourse that is wise, has a life value, belong and follow by the local society. Philosophically, a local wisdom means an indigenous knowledge system, which is empiric and pragmatic. Empirically, because its land product from facts in the life of the society, pragmatically because the whole concept is build from their believe in the indigenous knowledge system, to solve daily life problem solving.

It can be understood, that local wisdom is knowledge, developed by their ancestors, to survive in life surrounds them, and becomes a part of its culture from generation to generation. This knowledge appears in legends, stories, songs, rituals, and local regularities. Every part in their society daily life will be conducted by this knowledge, and well-internalized to the people. Inevitably, this local wisdom put in effect to young people (students) in learning, as Sukadi [3] stated that traditional and cultural values are the base for people to think, learn, and interact.

2.2. Cultivation time

The indigenous knowledge above is owned for generations and is used in agriculture, particularly in terms of planting rice in paddy fields. These people know which of the circulatory system of stars in the sky, namely the Orion star, which is used to determine the beginning of cultivating rice in paddy fields. At roughly the beginning of November, in the dawn of time the Orion star is visible in the east, and it is considered by farmers as a guide to start of the cultivation of their field. Then approximately in April, the Orion star at dusk on the western horizon seemed upside down. It is considered by them as a sign to save the plow, and harvesting rice or harvesting season is over [3].

Figure 2. Cultivating rice

Figure 3. Ready to Harvest Rice

The calculation in this paddy field management is based on *palintangan* (calculation), that is *etangan* weton (calculation based on the birth date) of the landlord (owner) of the paddy field, traditionally in the AKUR (Adat Karuhun Urang) society. Prior to the rice cultivation, there are 4 philosophies to be considered, i.e. *Sungut* (mouth); *Irung* (nose); *Mata* (eyes), *Ceuli* (ear), which means:

- *Sungut* (mouth), one cannot speak or talk too much, since the result will be unsatisfactory, thus people must do their job persistently.
- *Irung* (nose), sniffed by mice or several types of leafhoppers (wereng), a warning to always be alert for these animals.
- *Mata* (eyes), the rice that is planted will be seen by the mice and leafhoppers, so people must always be careful and be preventive.
- *Ceuli* (ears), means mung kapireng wungkul (only heard), so that the paddy field will be only heard, and these animals have no idea where the paddy field is.

The best calculation in cultivating rice is *numbuk* di *Ceuli*, which means at time when those animals will only heard, and it is based on calculating the days (each day has a certain value): Sunday = 5, Monday = 4, Tuesday = 3, Wednesday = 7, Thursday = 8, Friday = 6, and Saturday = 9.

Then, this value will be added with mancakalima (the cultural days in old tradition): Manis = 5, Pahing = 9, Puhun = 7, Wage = 4, and Kaliwon = 8.

For example: suppose today is Wednesday Wage, so its value is 7 + 4 = 11, means that one can cultivate rice 11 more days from today, which is believed will be the *numbuk* di *Ceuli* (the time where animals can only heard).

2.3. Harvest time

In agriculture, beside cultivation time, there is absolutely harvest time. According to the culture, there also two stages of calculation: the palintangan, that is a calculation based on *numbuk*, and etangan weton (calculation based on the birth date) of the landlord of the paddy field. Slightly different than cultivation time, in harvest time the calculation of *numbuk* is based upon:

- *Sri* (seed), that harvesting in *Sri* cannot be optimal, since it will produce more seeds instead of rice.
- *Pang* (stalk), similarly as *Sri*, harvesting will produce more stalks, rather than rice.
- *Godong* (leaf), similarly as *Sri*, harvesting will produce more leaves instead of rice.
- *Woh* (fruit), from its name, clearly this the best time to harvest.

As in cultivation time, the time *numbuk* will be combined with *naptu* (days), i.e. Sunday = 5, Monday = 4, Tuesday = 3, Wednesday = 7, Thursday = 8, Friday = 6, and Saturday = 9, and add with these number accordingly: Manis = 5, Pahing = 9, Puhun = 7, Wage = 4, and Kaliwon = 8. For example, when the owner thinks that this is the season to harvest, he/she looks at the date, if today is Monday Puhun, then its corresponding value is 4 + 7 = 11, thus 11 more days after today will be *Woh* time, the best time to harvest.

For example, when the owner thinks that this is the season to harvest, he/she looks at the date, if today is Monday Puhun, then its corresponding value is 4 + 7 = 11, thus 11 more days after today will be *Woh* time, the best time to harvest.

In a certain time, before harvesting, people gather and pray to God, a an elders (senior) do a pray (by singing):

Sampurasun ka Rumuhung
Ka anu ngersakeun
Nu linggih di Buwana Nyungcung
Nu Lenggah di Buwana Mandala Hyang
Pun ... sapun sampurasun !
Ka anu lenggah di Bumi Suci Alam Padang
Pwah Aci Sanghyang Asri
Di kiwari jisim kuring, seja amit mipit
Ngajayak lenggah Nyi Pwah Aci Sanghyang Asri
Ti Weweg Sampeg mandala Pageuh
Anu direpokkeun di bumi pertiwi
Kiwari jisim kuring, eudeuk ngajayak
Lantaran hirup turun ti anu rahayu,
Hurip ngancik na Pwah Aci
Anu kahalimpukeun di Mandala Tapa
Bari naggoan wayah
Balik ka Bumi Suci
Pulang ka Mandala Hyang
Panceg hanca di Buwana
Dina waruga manusa
Pun, muga Rahayu !

This pray tells the meaning of graciousness and asks God the Almighty, and asks that harvest can be launched and kept away from undesirable things, so everything goes smoothly, and begged harvest done can be launched and kept away from things that are not desirable and grateful for harvests that can be refined into a resource for human. Rice is an important thing for the AKUR society, because it is the main food resources, as well as agricultural society as Indonesia.

3. Conclusion

Within indigenous society in Indonesia, such as in Cigugur village, West Java, the understandings, skills and philosophies developed by the society with long histories of interaction with their natural surroundings, the indigenous knowledge systems provides decision making about fundamental aspects of day-to-day life, ritually and spiritually. This unique knowledge provides a foundation for locally appropriate sustainable development.

Acknowledgment

The work reported in this paper is supported by the Academic Leadership Grant Unpad 2016, so the authors thank to Rector Universitas Padjadjaran and Directorate Research and Community Services Unpad for supporting funding of research and paper dissemination.

References

[1] Ambali A G 2014 Real deal: mathematics education for sustainable development Text of the keynote Address delivered at the opening Ceremony of the 51st Annual Conference of the Mathematical Association of Nigeria at the University of Ilorin, Nigeria
[2] Subagja R 2014 Suku Sunda di kabupaten Kuningan.*http : //rikisubagja*47*.blogspot.co.id/*2014*/*06*/suku − sunda − di − kabupaten − kuningan*$_4$984*.html*
[3] Sukadi 2007 Guru Powerful Guru Masa Depan Bandung: Kolbu.
[4] United Nations Educational, Scientific and Cultural Organization 2015 Learning with Intangeible Heritage for a Sustainable Future. Unesco Bangkok Office.
[5] World Commission on Environment and Development 1987 Our Common Future Oxford: Oxford University Press

IOP Conf. Series: Journal of Physics: Conf. Series **893** (2017) 012004 doi:10.1088/1742-6596/893/1/012004

Minimizing student's faults in determining the design of experiment through inquiry-based learning

D P E Nilakusmawati, M Susilawati

Department of Mathematics, Udayana University, Bali, Indonesia

E-mail: {nilakusmawati, madesusilawati}@unud.ac.id

Abstract. The purpose of this study were to describe the used of inquiry method in an effort to minimize student's fault in designing an experiment and to determine the effectiveness of the implementation of the inquiry method in minimizing student's faults in designing experiments on subjects experimental design. This type of research is action research participants, with a model of action research design. The data source were students of the fifth semester who took a subject of experimental design at Mathematics Department, Faculty of Mathematics and Natural Sciences, Udayana University. Data was collected through tests, interviews, and observations. The hypothesis was tested by t-test. The result showed that the implementation of inquiry methods to minimize of students fault in designing experiments, analyzing experimental data, and interpret them in cycle 1 students can reduce fault by an average of 10.5%. While implementation in Cycle 2, students managed to reduce fault by an average of 8.78%. Based on t-test results can be concluded that the inquiry method effectively used to minimize of student's fault in designing experiments, analyzing experimental data, and interpreting them. The nature of the teaching materials on subject of Experimental Design that demand the ability of students to think in a systematic, logical, and critical in analyzing the data and interpret the test cases makes the implementation of this inquiry become the proper method. In addition, utilization learning tool, in this case the teaching materials and the students worksheet is one of the factors that makes this inquiry method effectively minimizes of student's fault when designing experiments.

1. Introduction

The demands for the quality of graduates and competition between institutions in improving graduate GPA sometimes into competition, however, as an educational institution should not be stuck in an increase GPA with a high value without going through a learning process that focuses on the gradual periodic competency.

The real conditions in the Mathematics Department of Udayana University, shows that the achievement of student learning outcomes for the course of Experimental Design in the last three years shows that the student grades is not increased. In addition, based on the results of the evaluation of the learning process is done by Quality Assurance Team of the Department and lecturer of the courses, found that the low of student grades on the course of Experimental Design, which are caused due to: (1) mistake in determining the appropriate design for the given case; (2) difficulty distinguishing between repetition and groups; (3) difficulty distinguishing between the treatment and groups; (4) difficulties in manual counting; (5) the limited laboratory facilities to support the practicum.

In spite of conditions mentioned above, the determination of appropriate learning methods is very important to be done in an effort to improve student results on this subject in particular, and to improve the quality and GPA of graduates in general quality of the learning process. One method that can be applied to minimize student's faults in the designing of the experiment is a method of inquiry-based learning.

The purpose of this study were to describe the used of inquiry method in an effort to minimize student's faults in designing an experiment and to determine the effectiveness of the implementation of the inquiry method in minimizing student's faults in designing experiments on subjects experimental design.

Inquiry learning is rooted in constructivist view. [1] explains that the inquiry-based classroom as:

> recognizes the diverse needs of students and employs the research-based strategies that help to keep all students engaged in learning. It is a community of inquiry where students and teachers share responsibility for learning, and where they collaborate on constructing new knowledge.

The statement above encourage teachers to develop learning inquiry on subjects where students are actively involved with mathematical concepts. [2], applying Inquiry-based learning to explore how researchers in this case are a teacher, can use Inquiry based learning encourages higher order thinking among students.

Five steps to be taken in carrying out the inquiry or discovery approach by [3, p.155], namely: (a) the formulation of the problem to be solved by students; (b) set a temporary answer, or better known as hypothesis; (c) students to find information, data, facts needed to answer the problems or hypothetical; (d) deduce the answer; and (e) apply the conclusions/ generalizations in new situations.

2. Research method

This type of research is action research participants, with a model of actionresearch design. The data source were students of the fifth semester who took acourse of Experimental Design at Mathematics Department, Faculty of Mathematicsand Natural Sciences, Udayana University.

Implementation of the overall study carried out for six months, while the implementation of the provision of treatment carried out over four months. Data was collected through tests, interviews, and observations. The data obtained from the test results were analyzed using statistical techniques t test, to test scores of students before and after the action learning groups. Hypothesis proposed in this study are: Inquiry method effective used in an effort to minimize the student's faults in designing experiments on courses of Experimental Design at the Mathematics Department, Faculty of Mathematics and Natural Sciences, Udayana University.

Action procedures designed in two cycles, in each cycle was conducted for two months. The subject matter is given on the implementation of the action cycles 1 and 2 for this course of Experimental Design are: Complete Random Design in cycle 1 and Randomized Block Design in cycle 2. At the end of the cycle of actions carried out subsequent monitoring and evaluation of the results of the evaluation reflected for planning their next action. The observations were made using the observation, interviews, and learning achievement test.

3. Results and discussion

The result showed that the implementation of inquiry methods to minimize the student's faults in designing experiments, analyzing experimental data, and interpret them in cycle 1 students can reduce faults by an average of 10.5%. While implementation in Cycle 2, students managed to reduce faults by an average of 8.78%.

The faults made by the students in designing experiments verified by observation, interview and assessment test results of students work in resolving cases of existing experiments in the Student Worksheet. There are some types of fault made by students in designing experiments, include: (1) mistake in determining the right environmental design in a case of experiment; (2) mistake in writing and explain precise linear model of an experiment; and (3) mistake in interpreting the results of data analysis in a case of experiment.

The effectiveness of methods of Inquiry in minimizing student's faults in experimental design, are tested using the t test. Results of testing the normality of the data as shown in Figure 1.

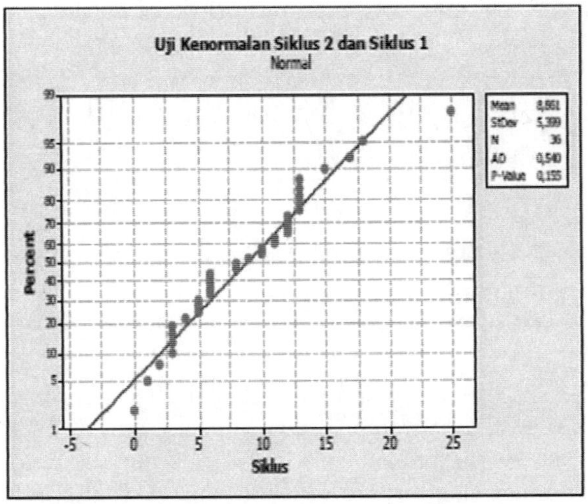

Figure 1. Plot of normality test data difference cycle 1 to cycle 2.

Anderson Darling test results get the value of 0.540 with P value = 0.155, if the value of P is compared with the value of $\alpha = 0.05$, P value is greater than the value of $\alpha = 0.05$, this means that H_0 is accepted, indicate that data normal spread.

Results of testing the assumption of homogeneity of variance seen in the MINITAB output in Table 1, showing that the value of P is equal to 0.571 greater than the value of $\alpha = 0.05$, which means that H_0 is accepted. This indicates that the data are homogeneous.

Table 1. Results of test assumptions of homogeneity variance.

Method	DF1	DF2	Test Statistic	P-Value
Levene's Test	1	70	0.32	0.571

Since both assumptions are met, the test can be performed with the t test, the results are presented in Table 2. The hypothesis tested is: $H_0 : \mu_1 = \mu_2$ (The mean of student learning outcomes with the implementation of inquiry method in cycle 1 and cycle 2 are the same) $H_1 : \mu_1 < \mu_2$ (The mean of student learning outcomes with the implementation of inquiry the method cycle 1 is smaller than the cycle 2).

The t test results are presented in Table 2. The results of the t test obtained t value = 9.85 with P value of 0.000 which is smaller than the value $\alpha = 0.05$, this means that H_0

refused and accept H_1. The decision to reject H_0 implies that the action of implementation of inquiry method managed to increase the average score of the students, this means that the Inquiry method effectively improve the student learning outcomes in working on experiment cases given, also means reducing the students' mistakes in analyzing the data and interpretation of experiment one factor in Experimental Design courses. Based on explanations above can be concluded that the inquiry method effectively used to minimize of student's faults in designing experiments, analyzing experimental data, and interpreting them.

Table 2. Results of t-test difference data cycle 1 and cycle 2.

Cycle	N	Mean	StDev	SE Mean
Cycle 1	36	65.36	7.27	1.21
Cycle 2	36	74.22	6.66	1.11
95% CI for mean difference: (7,034; 10,688)				
T-Test of mean difference = 0 (vs not = 0):				
T-Value = 9,85 P-Value = 0,000				

The nature of the teaching materials on subject of Experimental Design that demand the ability of students to think in a systematic, logical, and critical in analyzing the data and interpret the test cases makes the implementation of this inquiry become the proper method. This is supported by [4] about the purpose of the inquiry learning is to develop the ability to think in a systematic, logical, critical, or develop intellectual abilities as part of the mental process. Further explained that in the inquiry learning students are not only required to master the subject matter, but how they can use their potential. Man who just learned the lesson is not necessarily able to develop the ability to think optimally. Instead, students will be able to develop the capacity to think when he can master the subject matter.

The ability of students to think in a systematic, logical, and critical as required in the course of this experiment design, making the inquiry method proper applied to minimize student's faults in designing experiments. This is supported by [2], applying the inquiry based learning to explore how researchers in this case are a teacher, can use inquiry-based learning encourages higher order thinking among students.

In addition, utilization learning tool, in this case the teaching materials and the students worksheet is one of the factors Based on t test results can be concluded that that makes this inquiry method effectively minimizes of student's faults when designing experiments.

4. Conclusion

The implementation of inquiry method to minimize student's faults in designing experiments, analyzing experimental data, and interpret it in Cycle 1 students can reduce faults by an average of 10.5%. While implementation in Cycle 2 students managed to reduce faults by an average of 8.78%. The faults can be minimized are (1) mistake in determining the right environmental design in a case of experiment; (2) mistake in writing and explain precise linear model of an experiment; and (3) mistake in interpreting the results of data analysis in a case of experiment.

The inquiry method effectively used to minimize of student's faults in designing experiments, analyzing experimental data, and interpreting them on subject Complete Random Design and Randomized Block Design of Experimental Design course in the Department of Mathematics, University of Udayana.

Acknowledgments
High appreciation to the Udayana University, which has funded this Hibah Unggulan Program Studi (HUPS), through funding DIPA BLU Udayana University, the Grant Contract No. 1319/UN14.1.28.1/PP/2015.

References
[1] Jarret D 1997 *Inquiry Strategies for Science and Mathematics Learning: Its Just Good Teaching* (Northwest Regional Laboratory)
[2] Rooney C 2012 How am I using inquiry-based learning to improve my practice and to encourage higher order thinking among my students of mathematics? *Educational Journal of Living Theories* **5** pp 99–127
[3] Sudjana N 2000 *Dasar-dasar Proses Belajar Mengajar* (Bandung: Sinar Baru Algensindo)
[4] Sudrajat A Pembelajaran inkuiri: Pengertian, ciri-ciri, prinsip-prinsip dan langkah-langkah. artikel berita dan opini pendidikan. Accessed on 11 January 2015. [Online]. Available: https://akhmadsudrajat.wordpress.com/2011/09/12/pembelajaran-inkuiri

IOP Conf. Series: Journal of Physics: Conf. Series **893** (2017) 012005 doi:10.1088/1742-6596/893/1/012005

An intriguing application of telescoping sums

Julius Fergy T. Rabago

Department of Mathematics and Computer Science, College of Science
University of the Philippines Baguio
Governor Pack Road, Baguio City 2600, Benguet, PHILIPPINES

E-mail: jfrabago@gmail.com

Abstract. This note offers a simple, yet very intriguing application of telescoping sums. Particularly, the cancellation technique which is known as the method of differences is employed to establish analytically the closed-form solutions of some systems of nonlinear difference equations. The results delivered here, in addition, generalizes several results found in recent literatures.

1. Introduction

There has been an enormous amount of published work dealing with solution forms of some systems of nonlinear difference equations that are solvable in closed-forms. This type of equations appear very simple in form, however, in some situations, the structure and behavior of their solutions are quite difficult to formulate and completely understand. Nevertheless, several studies have dealt with the solution forms of several solvable systems of nonlinear difference equations, see, for instance, a recent work of Elsayed [3], Elsayed and El-Metwally [4], Elsayed and Ibrahim [5], Tollu et al. [12, 13], Touafek [14], Yazlik [16] and the papers cited therein. In these aforementioned works, we noticed that the results exhibited, particularly the solution forms of the systems being studied, were established through a mere application of induction principle. (in fact, at some point, the proofs of some of the statements presented are insufficient and are even completely omitted). We stress that this method can only validate analytically the solution form of a particular system of difference equation. However, it does not give much information on how these formulas are obtained. In an effort to explain theoretically some intriguing results in this research line, the author have revisited and re-examined in [8, 9, 10] and also in [11], with Bacani, several systems of nonlinear difference equations which have been studied previously in [1, 6, 14, 15, 16].

In this work, we turn our attention to a two-dimensional system of nonlinear difference equation which could be seen as a generalization of recent investigations. In particular, we shall show how the solution form of the system

$$x_{n+1} = \frac{x_n y_{n-k}}{y_{n-k+1}(\pm 1 - x_n y_{n-k})}, \qquad y_{n+1} = \frac{y_n x_{n-k}}{x_{n-k+1}(\pm 1 \pm y_n x_{n-k})}, \tag{1.1}$$

with real initial conditions $\{x_n\}_{n=-k}^{0} := \{x_n\}_{-k}^{0}$ and $\{y_n\}_{-k}^{0}$, can be established analytically. Our technique uses appropriate transformation on the phase variables x_n and y_n, reducing the system into a linear type and employs the "method of differences" or, in particular, the idea of

IOP Conf. Series: Journal of Physics: Conf. Series **893** (2017) 012005 doi:10.1088/1742-6596/893/1/012005

"telescoping sums". This method has been used effectively by the author, along with Bacani [11], in dealing with the solution forms of several solvable systems of nonlinear difference equations, covering also the equations considered in [14]. The solution form for the case when $k = 1$ has been established by Elsayed in [5] through the principle of induction. For this reason, we have inteded to re-examine and approach the problem in a different perspective which will perhaps provide interested readers some new ideas on how to attack similar types of problems.

The paper is organized as follows. In the next section (Section 2), we discuss shortly the idea behind the method of differences and prove a lemma which will be essential for our main result. Our main contribution is presented in Section 3 where we will exhibit the solution forms of system (1.1) through the use of telescoping. We end our paper with a short summary in Section 4.

2. Preliminaries

Let $\{a_n\}_1^\infty$ be a number sequence. A telescoping sum is a sum in which subsequent terms cancel each other, leaving only its initial and final terms. For instance,

$$\sum_{n=1}^{N}\{a_n - a_{n-1}\} = a_N - a_0, \tag{P.1}$$

is a telescoping sum. This idea, obviously, can be extended into products. For example, we have the following relation

$$\prod_{n=1}^{N}\left\{\frac{b_n}{b_{n-1}}\right\} = \frac{b_N}{b_0}, \tag{P.2}$$

where $\{b_n\}_1^\infty$ is some number sequence with nonzero terms. As alluded in the Introduction, we are interested in the application of these identities especially in obtaining closed-form solutions of some solvable systems on nonlinear difference equations. With this set objective, the following results are essential.

Lemma 2.1 ([7]). *Let $n \in \mathbb{N}_0$. Then, we have the following results:*

(i) *the equation $z_{n+1} - z_n = \pm 1$ has the solution $z_n = z_0 \pm n$,*

(ii) *the equation $z_{n+1} + z_n = \pm 1$ has the solution $z_n = (-1)^n[z_0 \mp \xi(n)]$, where $\xi(n) = 0$ when n is even and $\xi(n) = 1$ when n is odd.*

Proof. The proof follows directly from telescoping. Indeed, for (i), we have $z_n - z_0 = \sum_{j=0}^{n-1}\{z_{j+1} - z_j\} = \pm\sum_{j=0}^{n-1} 1 = \pm n$. Rearranging the equation, we get the desired equation. Meanwhile, for (ii), we have $(-1)^{n-1}z_n + z_0 = \sum_{j=0}^{n-1}(-1)^j\{z_{j+1}+z_j\} = \pm\sum_{j=0}^{n-1}(-1)^j$. The right hand side (RHS) equates to 0 (resp. 1) when n is even (resp. odd). Thus, $z_n = (-1)^n[z_0 \mp \xi(n)]$, where $\xi(n) = 0$ when n is even and $\xi(n) = 1$ when n is odd. □

Proposition 2.2. *Let n be a non-negative integer and*

$$\xi(n) = \begin{cases} 0, & \textit{if } n \textit{ is even,} \\ 1, & \textit{if } n \textit{ is odd.} \end{cases}$$

Then, $\xi(a + b) = \xi(\xi(a) + \xi(b))$ for all non-negative integers a and b. Moreover, if $a > 0$ is an odd integer, then we have $\xi(an) = \xi(n)$.

Proof. To see the first result, we consider two possibilities: (i) a and b are of the same parity and (ii) a and b are of different parity. For the first possibility, we note that $\xi(a) = \xi(b)$ which implies that $\xi(a) + \xi(b) = 2\xi(b)$. Hence, $\xi(\xi(a) + \xi(b)) = \xi(2\xi(b)) = 0$. However, since a and b are of the same parity, then $a+b$ is always even. Thus, $\xi(a+b)$ and so $\xi(a+b) = \xi(\xi(a)+\xi(b))$. Meanwhile, if a and b are of different parity, then we can assume without-loss-of-generality that a is even while b is odd. Hence, $\xi(a) = 0$ and $\xi(b) = 1$. It follows that $\xi(a) + 1 = \xi(b)$ and $\xi(b) - 1 = 0$. Adding $\xi(b) - 1$ on both sides of the former equation, we get $\xi(a) + \xi(b) = 2\xi(b) - 1$. Therefore, $\xi(\xi(a) + \xi(b)) = \xi(2\xi(b) - 1) = 1$. But a and b are of different parity, then $a + b$ is always odd. Thus, we have $\xi(a + b) = 1$ which in turn implies that $\xi(a + b) = \xi(\xi(a) + \xi(b))$.

Now, for the second result, we note that the parity of the product $a \cdot n$ for any odd integer $a > 0$ depends on the parity of n. Thus, $\xi(an) = \xi(n)$ for all non-negative integers a and n where a is odd. This proves the proposition. □

3. Main Results

In the following, we derive the solution form of the system (1.1), taking into account the following substitutions on the phase variables:

$$w_n = \frac{1}{x_n y_{n-k}} \quad \text{and} \quad z_n = \frac{1}{y_n x_{n-k}}, \quad \text{for all } n \in \mathbb{N}_0. \tag{T}$$

Our discussion proceeds according to the following particular cases of system (1.1):

$$x_{n+1} = \frac{x_n y_{n-k}}{y_{n-k+1}(1 - x_n y_{n-k})}, \qquad y_{n+1} = \frac{y_n x_{n-k}}{x_{n-k+1}(1 \pm y_n x_{n-k})}, \quad n \subset \mathbb{N}_0; \tag{S.1}$$

$$x_{n+1} = \frac{x_n y_{n-k}}{y_{n-k+1}(-1 - x_n y_{n-k})}, \qquad y_{n+1} = \frac{y_n x_{n-k}}{x_{n-k+1}(1 \pm y_n x_{n-k})}, \quad n \in \mathbb{N}_0; \tag{S.2}$$

$$x_{n+1} = \frac{x_n y_{n-k}}{y_{n-k+1}(-1 - x_n y_{n-k})}, \qquad y_{n+1} = \frac{y_n x_{n-k}}{x_{n-k+1}(-1 \pm y_n x_{n-k})}, \quad n \in \mathbb{N}_0. \tag{S.3}$$

3.1. *Solution form of system* (S.1)

Using the substitution (T), system (S.1) can be transformed into the following equations:

$$w_{n+1} = w_n - 1, \qquad z_{n+1} = z_n \pm 1, \qquad n \in \mathbb{N}_0.$$

In view of (i) of Lemma 2.1, we have

$$\forall n \in \mathbb{N}_0 : \quad w_n = w_0 - n \quad \text{and} \quad z_n = z_0 \pm n,$$

or equivalently, in reference to equation (T),

$$\forall n \in \mathbb{N}_0 : \begin{cases} x_n y_{n-k} &= \dfrac{x_0 y_{-k}}{1 - n x_0 y_{-k}} \\[2ex] y_n x_{n-k} &= \dfrac{y_0 x_{-k}}{1 \pm n y_0 x_{-k}}. \end{cases} \tag{3.1}$$

With these equations at hand, we could easily compute for $\{x_{2kn+i}\}_{i=-(2k-1)}^{0}$ and $\{y_{2kn+i}\}_{i=-(2k-1)}^{0}$. For instance, to find for $x_{2kn-(2k-1)}$ we can proceed as follows. First, replace n by $2kn + (2k-1)$ (resp. by $2kn + k - 1$) in the first (resp. second) equation in (3.1) to obtain

$$x_{2kn+2k-1} y_{2kn+k-1} = \frac{x_0 y_{-k}}{1 - (2kn + 2k - 1) x_0 y_{-k}}$$

and

$$y_{2kn+k-1}x_{2kn-1} = \frac{y_0x_{-k}}{1 \pm (2kn+k-1)y_0x_{-k}}.$$

Then, we can eliminate the second phase variable y_n by taking the ratio of these two equations; that is, we have

$$\frac{x_{2kn+2k-1}}{x_{2kn-1}} = \frac{x_{2kn+2k-1}y_{2kn+k-1}}{y_{2kn+k-1}x_{2kn-1}} = \frac{x_0y_{-k}(1 \pm (2kn+k-1)y_0x_{-k})}{y_0x_{-k}(1-(2kn+2k-1)x_0y_{-k})}.$$

After that, we replace n by j and then take the product of the resulting equation from $j = 0$ to $j = n-1$, to obtain

$$\frac{x_{2kn+2k-1}}{x_{-1}} = \prod_{j=0}^{n-1}\left\{\frac{x_{2kj+(2k-1)}}{x_{2kj-1}}\right\} = \prod_{j=0}^{n-1}\left\{\frac{x_0y_{-k}(1 \pm (2kj+k-1)y_0x_{-k})}{y_0x_{-k}(1-(2kj+2k-1)x_0y_{-k})}\right\}$$

Finally, by rearranging this equation, with an adjustment on the index on the RHS of the equation, we get

$$\forall n \in \mathbb{N}_0: \quad x_{2kn-1} = x_{-1}\left(\frac{x_0y_{-k}}{y_0x_{-k}}\right)^n \prod_{j=0}^{n-1}\left\{\frac{1 \pm (2kj+k-1)y_0x_{-k}}{1-(2kj+2k-1)x_0y_{-k}}\right\}.$$

Following the same inductive lines as above, the solution forms for the other phase variables can easily be established. Thus, we have the following theorem.

Theorem 3.1. *Every solution $\{(x_n, y_n)\}_1^\infty$ of system* (S.1) *takes, for all $n \in \mathbb{N}_0$, the form*

$$x_{2kn+i} = x_i\left(\frac{x_0y_{-k}}{y_0x_{-k}}\right)^n \prod_{j=0}^{n-1}\left\{\frac{1 \pm (2kj+k+i)y_0x_{-k}}{1-(2kj+2k+i)x_0y_{-k}}\right\},$$

for all $i = \overline{-2(k-1), 0}$ and

$$x_{2kn+1} = \frac{x_0y_{-k}}{y_{-k+1}(1-x_0y_{n-k})}\left(\frac{x_0y_{-k}}{y_0x_{-k}}\right)^n \times \prod_{j=0}^{n-1}\left\{\frac{1 \pm (2kj+k+1)y_0x_{-k}}{1-(2kj+2k+1)x_0y_{-k}}\right\},$$

and

$$y_{2kn+i} = y_i\left(\frac{y_0x_{-k}}{x_0y_{-k}}\right)^n \prod_{j=0}^{n-1}\left\{\frac{1 \pm (2kj+k+i)x_0y_{-k}}{1-(2kj+2k+i)y_0x_{-k}}\right\},$$

for all $i = \overline{-2(k-1), 0}$ and

$$y_{2kn+1} = \frac{y_0x_{-k}}{x_{-k+1}(1-y_0x_{n-k})}\left(\frac{y_0x_{-k}}{x_0y_{-k}}\right)^n \times \prod_{j=0}^{n-1}\left\{\frac{1 \pm (2kj+k+1)x_0y_{-k}}{1-(2kj+2k+1)y_0x_{-k}}\right\}.$$

3.2. Solution form of system (S.2)

In deriving the solution form for system (S.2), we apply the same argument used in the previous section. Through the substitution (T), system (S.2) can be transformed into the following equations:

$$w_{n+1} = -w_n - 1, \qquad z_{n+1} = z_n \pm 1, \qquad n \in \mathbb{N}_0.$$

In reference to Lemma 2.1, the solution to these equations can readily be found and are given by

$$\forall n \in \mathbb{N}_0: \quad w_n = (-1)^n[w_0 + \xi(n)] \quad \text{and} \quad z_n = z_0 \pm n,$$

respectively. Using the substitution defined in (T), we then have

$$\forall n \in \mathbb{N}_0: \begin{cases} x_n y_{n-k} &= \dfrac{(-1)^n x_0 y_{-k}}{1+\xi(n) x_0 y_{-k}}, \\ y_n x_{n-k} &= \dfrac{y_0 x_{-k}}{1 \pm n y_0 x_{-k}}. \end{cases} \tag{3.2}$$

Again, with these relations at hand, we can easily derive the solution form for $\{x_{2kn+i}\}_{i=-(2k-1)}^{0}$ and $\{y_{2kn+i}\}_{i=-(2k-1)}^{0}$. This time we only show how to compute for y_{2kn-k}. The formulas for the other phase variables, however, can be established in a similar fashion. Now, to proceed, we replace n by $2kn$ (resp. by $2kn+k$) in the first (resp. second) equation in (3.2) to obtain

$$\forall n \in \mathbb{N}_0: \begin{cases} x_{2kn} y_{2kn-k} &= \dfrac{x_0 y_{-k}}{1+\xi(2kn) x_0 y_{-k}}, \\ y_{2kn+k} x_{2kn} &= \dfrac{y_0 x_{-k}}{1 \pm (2kn+k) y_0 x_{-k}}, \end{cases}$$

and so we have, using the definition of $\xi(n)$ in Lemma 2.1,

$$\frac{y_{2kn+k} x_{2kn}}{x_{2kn} y_{2kn-k}} = \frac{y_0 x_{-k}}{x_0 y_{-k}(1 \pm (2kn+k) y_0 x_{-k})}.$$

Replacing n by j and then taking the product of the resulting expressions from $j=0$ to $j=n-1$ yields

$$\frac{y_{2kn-k}}{y_{-k}} = \prod_{j=0}^{n-1} \left\{ \frac{y_0 x_{-k}}{x_0 y_{-k}(1 \pm (2kj+k) y_0 x_{-k})} \right\}.$$

Rearranging this equation and after some algebra, we get

$$y_{2kn-k} = \frac{y_0^n x_{-k}^n}{x_0^n y_{-k}^{n-k} \prod_{j=0}^{n-1} \{1 \pm (2kj+k) y_0 x_{-k}\}}.$$

In concluding, we have proved analytically the following theorem.

Theorem 3.2. *Every solution $\{(x_n, y_n)\}_1^\infty$ of system (S.2) takes, for all $n \in \mathbb{N}_0$, the form*

$$x_{2kn+i} = \begin{cases} \dfrac{x_i}{(\pm 1 \pm x_0 y_{-k})^n} \left(\dfrac{x_0 y_{-k}}{y_0 x_{-k}} \right)^n \times \displaystyle\prod_{j=0}^{n-1} \{\mp 1 - (2kj+k+i) y_0 x_{-k}\}, & \text{for odd } i, \\ x_i \left(\dfrac{x_0 y_{-k}}{y_0 x_{-k}} \right)^n \displaystyle\prod_{j=0}^{n-1} \{1 \pm (2kj+k+i) y_0 x_{-k}\}, & \text{for even } i, \end{cases}$$

for all $i = \overline{-2(k-1), 0}$ and

$$x_{2kn+1} = \frac{x_0 y_{-k}(-1 - x_0 y_{-k})^{-1}}{y_{-k+1}(\pm 1 \pm x_0 y_{-k})^n} \left(\frac{x_0 y_{-k}}{y_0 x_{-k}} \right)^n \times \prod_{j=0}^{n-1} \{\mp 1 - (2kj+k+1) y_0 x_{-k},\}$$

and

$$y_{2kn+i} = \begin{cases} y_i \left(\dfrac{y_0 x_{-k}}{x_0 y_{-k}}\right)^n \dfrac{(\pm 1 \pm x_0 y_{-k})^n}{\prod_{j=0}^{n-1} \{\mp 1 - (2kj + 2k + i) y_0 x_{-k}\}}, & \text{for odd } i, \\ y_i \left(\dfrac{y_0 x_{-k}}{x_0 y_{-k}}\right)^n \prod_{j=0}^{n-1} \{1 \pm (2kj + 2k + i) y_0 x_{-k}\}^{-1}, & \text{for even } i, \end{cases}$$

for all $i = \overline{-2(k-1), 0}$ *and*

$$y_{2kn+1} = \frac{y_0 x_{-k}}{x_{-k+1}(1 \pm y_0 x_{-k})} \left(\frac{y_0 x_{-k}}{x_0 y_{-k}}\right)^n \times \frac{(\pm 1 \pm x_0 y_{-k})^n}{\prod_{j=0}^{n-1} \{\mp 1 - (2kj + 2k + 1) y_0 x_{-k}\}}.$$

3.3. Solution form of system (S.3)

Now we turn our attention to system (S.3) with initial conditions $x_{-k} y_0 \neq 1$ and $x_0 y_{-k} \neq 1$. Again, our approach follows the same argument as in the previous two cases. Using the substitution (T), system (S.3) can be express into the following equivalent forms:

$$w_{n+1} = -w_n - 1, \qquad z_{n+1} = -z_n \pm 1, \qquad n \in \mathbb{N}_0.$$

In view of Lemma 2.1, the solution to these equations are given by

$$\forall n \in \mathbb{N}_0 : \; w_n = (-1)^n [w_0 + \xi(n)] \text{ and } z_n = (-1)^n [z_0 \mp \xi(n)],$$

respectively. Once again, using the substitution defined in (T), we then have

$$\forall n \in \mathbb{N}_0 : \begin{cases} x_n y_{n-k} = \dfrac{(-1)^n x_0 y_{-k}}{1 + \xi(n) x_0 y_{-k}}, \\ y_n x_{n-k} = \dfrac{(-1)^n y_0 x_{-k}}{1 \mp \xi(n) y_0 x_{-k}}. \end{cases} \tag{3.3}$$

Through these equations, we can easily find for the form of the phase variables $\{x_{2kn+i}\}_{i=-(2k-1)}^0$ and $\{y_{2kn+i}\}_{i=-(2k-1)}^0$ in terms of n and the initial conditions $\{x_n\}_{-k}^0$ and $\{y_n\}_{n=-k}^0$. In the sequel, we shall give complete details in deriving the solution form for x_{2kn} and x_{2kn-1}. The other formulas, however, can be achieved using similar arguments. Now, for x_{2kn}, we replace n by $2kn$ (resp. by $2kn - k$) in the first (resp. second) equation in (3.3) to obtain

$$\forall n \in \mathbb{N}_0 : \begin{cases} x_{2kn} y_{2kn-k} = \dfrac{x_0 y_{-k}}{1 + \xi(2kn) x_0 y_{-k}}, \\ y_{2kn-k} x_{2kn-2k} = \dfrac{(-1)^k y_0 x_{-k}}{1 \mp \xi(k(2n-1)) y_0 x_{-k}}. \end{cases}$$

Hence, using the definition of $\xi(n)$ from Lemma 2.1, we have

$$\frac{x_{2kn}}{x_{2k(n-1)}} = \frac{x_{2kn} y_{2kn-2k}}{y_{2kn-2k} x_{2kn-2k}} = \frac{x_0 y_{-k} (-1)^k (1 \mp \xi(k) y_0 x_{-k})}{y_0 x_{-k}}$$

Replacing n by j and then taking the product of the resulting expressions from $j=1$ to $j=n$ yields the relation

$$\frac{x_{2kn}}{x_0}=\prod_{j=1}^{n}\left\{\frac{x_0y_{-k}(-1)^k(1\mp\xi(k)y_0x_{-k})}{y_0x_{-k}}\right\}=\left\{\frac{x_0y_{-k}(-1)^k(1\mp\xi(k)y_0x_{-k})}{y_0x_{-k}}\right\}^n$$

or equivalently,

$$x_{2kn}=x_0(-1)^{kn}(1\mp\xi(k)y_0x_{-k})^n\left(\frac{x_0y_{-k}}{y_0x_{-k}}\right)^n.$$

On the other hand, to get the form of x_{2kn-1}, we replace n by $2kn+2k-1$ (resp. by $2kn+k-1$) in the first (resp. second) equation in (3.3) to obtain, $\forall n\in\mathbb{N}_0$:

$$\begin{cases} x_{2kn+2k-1}y_{2kn+k-1} &= -\dfrac{x_0y_{-k}}{1+\xi(2kn+2k-1)x_0y_{-k}},\\[2ex] y_{2kn+k-1}x_{2kn-1} &= -\dfrac{(-1)^ky_0x_{-k}}{1\mp\xi(2kn+k-1)y_0x_{-k}}.\end{cases}$$

Using the definition of $\xi(n)$ from Lemma 2.1 and utilizing Proposition 2.2, we get

$$\frac{x_{2kn+2k-1}}{x_{2kn-1}}=\frac{x_{2kn+2k-1}y_{2kn+k-1}}{y_{2kn+k-1}x_{2kn-1}}=\frac{(-1)^kx_0y_{-k}(1\mp\xi(k-1)y_0x_{-k})}{y_0x_{-k}(1+x_0y_{-k})}.$$

Replacing n by j and then taking the product of the resulting expressions from $j=0$ to $j=n-1$ yields the relation

$$\frac{x_{2kn-1}}{x_{-1}}=\prod_{j=1}^{n}\left\{\frac{(-1)^kx_0y_{-k}(1\mp\xi(k-1)y_0x_{-k})}{y_0x_{-k}(1+x_0y_{-k})}\right\}=\left\{\frac{(-1)^kx_0y_{-k}(1\mp\xi(k-1)y_0x_{-k})}{y_0x_{-k}(1+x_0y_{-k})}\right\}^n$$

or equivalently,

$$x_{2kn-1}=x_{-1}(-1)^{kn}\left(\frac{1\mp\xi(k-1)y_0x_{-k}}{1+x_0y_{-k}}\right)^n\left(\frac{x_0y_{-k}}{y_0x_{-k}}\right)^n.$$

After computing for the other formulas through the same inductive lines used above, we'll finally arrive at the following theorem.

Theorem 3.3. *If $x_0y_{-k}\neq-2$ and $y_0x_{-k}\neq\pm2$, then every solution $\{(x_n,y_n)\}_1^\infty$ of system* (S.3) *takes the form*

$$x_{2kn+i}=\begin{cases}(-1)^{kn}x_i\left(\dfrac{1\mp\xi(k-1)y_0x_{-k}}{1+x_0y_{-k}}\right)^n\left(\dfrac{x_0y_{-k}}{y_0x_{-k}}\right)^n, & \text{for odd } i,\\[2ex] (-1)^{kn}x_i(1\mp\xi(k)y_0x_{-k})^n\left(\dfrac{x_0y_{-k}}{y_0x_{-k}}\right)^n, & \text{for even } i,\end{cases}$$

for all $i=\overline{-2(k-1),0}$ and

$$x_{2kn+1}=\frac{(-1)^{kn}}{y_{-k+1}}\left(\frac{x_0y_{-k}}{-1-x_0y_{-k}}\right)^{n+1}\times\left(\frac{-1\pm\xi(k-1)y_0x_{-k}}{y_0x_{-k}}\right)^n$$

and

$$y_{2kn+i} = \begin{cases} (-1)^{kn} y_i \left(\dfrac{1+\xi(k-1)x_0y_{-k}}{1 \mp y_0x_{-k}} \right)^n \left(\dfrac{y_0x_{-k}}{x_0y_{-k}} \right)^n, & \text{for odd } i, \\ \dfrac{(-1)^{kn} y_i}{(1 \mp \xi(k)x_0y_{-k})^n} \left(\dfrac{y_0x_{-k}}{x_0y_{-k}} \right)^n, & \text{for even } i, \end{cases}$$

for all $i = \overline{-2(k-1), 0}$ *and*

$$y_{2kn+1} = \frac{(-1)^{kn}}{x_{-k+1}} \left(\frac{y_0x_{-k}}{-1 \pm y_0x_{-k}} \right)^{n+1} \times \left(\frac{-1-\xi(k-1)x_0y_{-k}}{x_0y_{-k}} \right)^n.$$

It can be observed easily that the solutions of system (S.3) are all unbounded, except for the case when $x_0y_{-k} = -2$ and $y_0x_{-k} = \pm 2$. If, instead, we have the case $x_0y_{-k} = -2$ and $y_0x_{-k} = \pm 2$, then the above formulas suggest that system (S.3) has a periodic solution. More precisely, we have the following results.

Corollary 3.4. *Every solution* $\{(x_n, y_n)\}_1^\infty$ *of the system*

$$x_{n+1} = \frac{x_ny_{n-k}}{y_{n-k+1}(-1 - x_ny_{n-k})}, \qquad y_{n+1} = \frac{y_nx_{n-k}}{x_{n-k+1}(-1 + y_nx_{n-k})},$$

for all $n \in \mathbb{N}_0$, *is periodic with period* $2k$ *if and only if* $x_0y_{-k} = -2$ *and* $y_0x_{-k} = 2$.

Corollary 3.5. *Every solution* $\{(x_n, y_n)\}_1^\infty$ *of the system*

$$x_{n+1} = \frac{x_ny_{n-k}}{y_{n-k+1}(-1 - x_ny_{n-k})}, \qquad y_{n+1} = \frac{y_nx_{n-k}}{x_{n-k+1}(-1 - y_nx_{n-k})},$$

for all $n \in \mathbb{N}_0$, *is periodic with period* $4k$ *if and only if* $x_0y_{-k} = y_0x_{-k} = -2$.

The above results can be verified using the solution forms of the systems given in Theorem 3.3 from which their structure can also be obtained.

Remark 3.6. Some immediate consequences of our main results can be found in earlier papers such as in [4] and [5]. In these previous works, however, the results were established merely through induction principle and no further explanations at all were given by the authors to explain how the solution forms exhibited appear in such structures.

4. Summary

It was shown that the method of differences provides an elegant approach in dealing with the solution form of some particular systems of nonlinear difference equations. This work, in turn, has provided interested readers a new way to attack similar type of problems. It is possible that the technique used here can be applied to other classes of difference equations whose closed-form solutions are, in structure, similar to the ones obtained here – this provides considerable interest to continue investigating these type of equations.

Acknowledgment

The author would like to thank the anonymous referees for carefully handling and examining his manuscript, and for their valuable comments and suggestions that help improve the quality of the paper. He is also grateful to CIMPA for the financial support awarded to him for his attendance to The Asian Mathematical Conference 2016 (AMC2016). Finally, the author wishes to thank the organizers of AMC2016 for the opportunity to present the paper during the conference and to publish this paper in this proceeding.

References

[1] Bayram M, Daş S E 2010 On the positive solutions of the difference equation system $x_{n+1} = 1/y_{n-k}, y_{n+1} = x_{n-k}/y_{n-k}$ *Appl. Math. Sci. (Hikari)* **4**(7) pp 817–821
[2] Elsayed E M 2015 The expressions of solutions and periodicity for some nonlinear systems of rational difference equations *Adv. Stud. in Contemp. Math.* **25**(3) pp 341–367
[3] Elsayed E M 2014 Solution for systems of difference equations of rational form of order two *Comp. Appl. Math.* **33**(3) pp 751–765
[4] Elsayed E M and El-Metwally H 2013 On the solutions of some nonlinear systems of difference equations *Adv. Diff. Equ.* **161** 14 pages
[5] Elsayed E M and Ibrahim T F 2015 Periodicity and solutions for some systems of nonlinear rational difference equations *Hacet. J. Math. Stat.* **44**(6) pp 1361–1390
[6] Elsayed E M 2010 On the solutions of a rational system of difference equations *Fasc. Math.* **45** pp 25–36
[7] Rabago J F T 2016 Note on a paper of E. M. Elsayed (*submitted for publication*)
[8] Rabago J F T 2017 *Effective methods on determining the periodicity and form of solutions of some systems of nonlinear difference equations* *Int. J. Dyn. Syst. Differ. Equ* **7**(2) pp 112–135
[9] Rabago J F T 2016 On an open question concerning a product type difference equation (*submitted for publication*)
[10] Rabago J F T 2017 On the closed-form solution of a nonlinear difference equation and another proof to Sroysang's conjecture *Iran. J. Math. Sci. Inform.* to appear
[11] Rabago J F T and Bacani J B 2016 New techniques on solving some solvable systems of nonlinear difference equations (*preprint*)
[12] Tollu D T, Yazlik Y and Taskara N 2014 On fourteen solvable systems of difference equations *Appl. Math. Comp.* **233** pp.310–319
[13] Tollu D T , Yazlik Y and Taskara N 2013 On the solutions of two special types of Riccati difference equation via Fibonacci numbers *Adv. Differ. Equ.* **174** 7 pages
[14] Touafek N 2014 On some fractional systems of difference equations *Iran. J. Math. Sci. Inform.* **9**(2) pp 303–305
[15] Yang Y, Chen L and Shi Y G 2011 On solutions of a system of rational difference equations *Acta Math. Univ. Comenian. (N.S.)* **80**(1) pp 63–70
[16] Yazlik Y, Tollu D T and Taskara N 2013 On the solutions of difference equation systems with Padovan numbers *Applied Mathematics*, **4**(12A) pp 15–20

IOP Conf. Series: Journal of Physics: Conf. Series **893** (2017) 012006 doi:10.1088/1742-6596/893/1/012006

The nonabelian tensor square of a crystallographic group with quaternion point group of order eight

Siti Afiqah Mohammad, Nor Haniza Sarmin, and Hazzirah Izzati Mat Hassim

Department of Mathematical Sciences, Faculty of Science, Universiti Teknologi Malaysia 81310 UTM Johor Bahru, Johor, Malaysia

E-mail: afiqahmohammad91@gmail.com, nhs@utm.my, hazzirah@utm.my

Abstract. A crystallographic group is a discrete subgroup of the set of isometries of Euclidean space where the quotient space is compact. A torsion free crystallographic group, or also known as a Bieberbach group has the symmetry structure that will reveal its algebraic properties. One of the algebraic properties is its nonabelian tensor square. The nonabelian tensor square is a special case of the nonabelian tensor product where the product is defined if the two groups act on each other in a compatible way and their action is taken to be conjugation. Meanwhile, Bieberbach group with quaternion point group of order eight is a polycyclic group. In this paper, by using the polycyclic method, the computation of the nonabelian tensor square of this group will be shown.

1. Introduction

A crystallographic group is a description on the symmetrical pattern of a crystal. It is a symmetry group which has configuration in space. When a crystallographic group is torsion free where all elements are of infinite order except its identity then it is called a Bieberbach group. It is an extension of a free abelian group of finite rank by a finite point group. Research on homological invariants has been increasing in number since it is related to the study of the properties of the crystal using mathematical approach. One of homological invariants is the nonabelian tensor square of the group. The nonabelian tensor square is requisite in determining the other properties of the group. The study of the nonabelian tensor square was first introduced by Brown and Loday [1]. He introduced the nonabelian tensor product, $G \otimes H$, for two groups G and H. The nonabelian tensor square is a special case of the nonabelian tensor product in which $G = H$ and the action is conjugation. In 1987, the nonabelian tensor squares of all nonabelian groups of order up to 30 are computed [2]. Kappe *et al.* [3] determined the nonabelian tensor squares of 2-generator 2-groups of class 2 in 1999. Rashid *et al.* [4] computed the nonabelian tensor square of groups of order $8q$ while Zainal *et al.* [5] determined the nonabelian tensor squares of some groups of p-power order. In this research, the crystallographic groups being considered are taken from Crystallographic, Algorithms and Table (CARAT) package [6]. By using the technique developed by Blyth and Morse [7], these groups are transformed from matrix representation to polycyclic presentation before their nonabelian tensor squares are computed. It is crucial to perform the consistency check for those polycyclic presentations so that we can proceed to find the homological invariants of the groups. Recently, Tan *et al.* [8] used this

IOP Conf. Series: Journal of Physics: Conf. Series **893** (2017) 012006 doi:10.1088/1742-6596/893/1/012006

method to find the nonabelian tensor square of crystallographic groups with symmetric point group of order six. Therefore, in this research, the nonabelian tensor square of a crystallographic group with quaternion point group of order eight will be explicated.

1.1. Some Preliminaries

In this section, some basic definitions and preparatory results that are used in the computations of the nonabelian tensor square are presented.

Definition 1 *[9] The nonabelian tensor square of a group G, denoted as $G \otimes G$, is generated by the symbols $g \otimes h$ for all $g, h \in G$ subject to the relations*

$$gg' \otimes h = ({}^{g}g' \otimes {}^{g}h)(g \otimes h) \text{ and } g \otimes hh' = (g \otimes h)({}^{h}g \otimes {}^{h}h')$$

for all $g, g', h, h' \in G$, where ${}^{h}g = hgh^{-1}$.

Definition 2 *[10] Let G be a group with presentation $\langle G|R\rangle$ and let G^{φ} be an isomorphic copy of G via the mapping $\varphi : g \to g^{\varphi}$ for all $g \in G$. The group $\nu(G)$ is defined to be*

$$\nu(G) = \langle G, G^{\varphi}|R, R^{\varphi}, {}^{x}[g, h^{\varphi}] = [{}^{x}g, ({}^{x}h)^{\varphi}] = {}^{x^{\varphi}}[g, h^{\varphi}], \forall x, g, h \in G\rangle.$$

Definition 3 *[11] Let F_n be a free group on generators $g_i, \dots, g_n$ and R be a set of relations of group F_n. The relations of a polycyclic presentation ${}^{F_n}/_{R}$ have the form:*

$$\begin{aligned} g_i^{e_i} &= g_{i+1}^{x_{i,i+1}} \cdots g_n^{x_{i,n}} && \textit{for } i \in I, \\ g_j^{-1} g_i g_j &= g_{j+1}^{y_{i,j,j+1}} \cdots g_n^{y_{i,j,n}} && \textit{for } j < i, \\ g_j g_i g_j^{-1} &= g_{j+1}^{z_{i,j,j+1}} \cdots g_n^{z_{i,j,n}} && \textit{for } j < i \textit{ and } j \notin I \end{aligned}$$

for some $I \subseteq \{1, \dots n\}$, certain exponents $e_i \in \mathbb{N}$ for $i \in I$ and $x_{i,j}, y_{i,j,k}, z_{i,j,k} \in \mathbb{Z}$ for all i, j and k.

Definition 4 *[11] Let G be a group generated by $g_1, \dots, g_n$ and the consistency relations in G can be evaluated in the polycyclic presentation of G using the collection from the left as in the following:*

$$\begin{aligned} g_k(g_j g_i) &= (g_k g_j) g_i && \textit{for } k > j > i, \\ (g_j^{e_j}) g_i &= g_j^{e_j - 1}(g_j g_i) && \textit{for } j > i, j \in I, \\ g_j(g_i^{e_i}) &= (g_j g_i) g_i^{e_i - 1} && \textit{for } j > i, i \in I, \\ (g_i^{e_i}) g_i &= g_i (g_i^{e_i}) && \textit{for } i \in I, \\ g_j &= (g_j g_i^{-1}) g_i && \textit{for } j > i, i \notin I \end{aligned}$$

for some $I \subseteq \{1, \dots, n\}$, $e^i \in \mathbb{N}$. Then, G is said to be given by a consistent polycyclic presentation.

Next, some properties of the group $\nu(G)$ are presented. Besides, the relations of the homological functors of a group G with the subgroups of $\nu(G)$ such as $[G, G^{\varphi}]$ is also shown.

Theorem 1 *[7] If G is polycyclic then $\nu(G)$ is polycyclic.*

Theorem 2 *[10] Let G be a group. The map $\sigma : G \otimes G \to [G, G^{\varphi}] \lhd \nu(G)$ defined by $\sigma(g \otimes h) = [g, h^{\varphi}]$ for all g, h in G is an isomorphism.*

Theorem 3 *[7] Let G be a polycyclic group with a polycyclic generating sequence $g_1, \ldots, g_k$. Then $[G, G^{\varphi}]$, a subgroup of $\nu(G)$, is given by*

$$[G, G^{\varphi}] = \langle [g_i, g_i^{\varphi}], [g_i^{\delta}, (g_j^{\varphi})^{\varepsilon}], [g_i, g_j^{\varphi}][g_j, g_i^{\varphi}] \rangle$$

for $1 \leq i < j \leq k$, where

$$\varepsilon = \begin{cases} 1 & \text{if } |\mathfrak{g}_i| < \infty; \\ \pm 1 & \text{if } |\mathfrak{g}_i| = \infty \end{cases}$$

and

$$\delta = \begin{cases} 1 & \text{if } |\mathfrak{g}_j| < \infty; \\ \pm 1 & \text{if } |\mathfrak{g}_j| = \infty. \end{cases}$$

Theorem 4 *[12] Let $Q(6)$ be a Bieberbach group of dimension 6 with quaternion point group of order eight, then its polycyclic presentations is established as:*

$$\begin{aligned} Q(6) = \langle a, b, c, l_1, l_2, l_3, l_4, l_5, l_6 | a^2 &= cl_6, b^2 = cl_5l_6^{-1}, b^a = bcl_5^{-2}l_6^2, c^2 = l_5l_6^{-1}, \\ c^a &= cl_5l_6^{-1}, c^b = c, l_1^a = l_4^{-1}, l_1^b = l_3^{-1}, l_1^c = l_1^{-1}, l_2^a = l_3, \\ l_2^b &= l_4^{-1}, l_2^c = l_2^{-1}, l_3^a = l_2^{-1}, l_3^b = l_1, l_3^c = l_3^{-1}, l_4^a = l_1, \\ l_4^b &= l_2, l_4^c = l_4^{-1}, l_5^a = l_6, l_5^b = l_5, l_5^c = l_5, l_6^a = l_5, l_6^b = l_6, l_6^c = l_6, \\ l_j^{l_i} &= l_j, l_j^{l_i^{-1}} = l_j \text{ for } j > i, \ 1 \leqslant i, \ j \leqslant 6 \rangle. \end{aligned}$$

This presentation is proven to be consistent.

2. Main Results

In this section, the main result of this research which is the nonabelian tensor square of a crystallographic group with quaternion point group of order eight is presented as follows:

Theorem 5 *Let $Q(6)$ be a Bieberbach group of dimension 6 with quaternion point group of order eight, then its nonabelian tensor square is established as:*

$$\begin{aligned} Q(6) \otimes Q(6) = \langle & a \otimes a, b \otimes b, l_1 \otimes l_1, a \otimes b, a \otimes c, a \otimes l_1, a \otimes l_2, b \otimes l_1, b \otimes l_2, \\ & (a \otimes b)(b \otimes a), (a \otimes c)(c \otimes a), (a \otimes l_1)(l_1 \otimes a), (a \otimes l_2)(l_2 \otimes a), \\ & (a \otimes l_6)(l_6 \otimes a), (b \otimes l_1)(l_1 \otimes b), (b \otimes l_2)(l_2 \otimes b) \rangle. \end{aligned}$$

Proof : By Theorem 2, $G \otimes G \cong [G, G^{\varphi}]$, therefore we can find $G \otimes G$ by finding $[G, G^{\varphi}]$. By Theorem 3, $[G, G^{\varphi}] = \langle [g_i, g_i^{\varphi}], [g_i^{\delta}, (g_j^{\varphi})^{\varepsilon}], [g_i, g_j^{\varphi}][g_j, g_i^{\varphi}] \rangle$.
First, we have,

$$\langle [g_i, g_i^{\varphi}] \rangle = \{[a, a^{\varphi}], [b, b^{\varphi}], [c, c^{\varphi}], [l_1, l_1^{\varphi}], [l_2, l_2^{\varphi}], [l_3, l_3^{\varphi}], [l_4, l_4^{\varphi}], [l_5, l_5^{\varphi}], [l_6, l_6^{\varphi}]\}.$$

Next, we have,

$$\begin{aligned} \langle [g_i^{\varepsilon}, (g_j^{\varphi})^{\partial}] \rangle = \{ & [[a^{\pm 1}, (b^{\varphi})^{\pm 1}], [[a^{\pm 1}, (c^{\varphi})^{\pm 1}], [[a^{\pm 1}, (l_1^{\varphi})^{\pm 1}], [[a^{\pm 1}, (l_2^{\varphi})^{\pm 1}], [[a^{\pm 1}, (l_3^{\varphi})^{\pm 1}], \\ & [[a^{\pm 1}, (l_4^{\varphi})^{\pm 1}], [[a^{\pm 1}, (l_5^{\varphi})^{\pm 1}], [[a^{\pm 1}, (l_6^{\varphi})^{\pm 1}], [[b^{\pm 1}, (c^{\varphi})^{\pm 1}], [[b^{\pm 1}, (l_1^{\varphi})^{\pm 1}], \\ & [[b^{\pm 1}, (l_2^{\varphi})^{\pm 1}], [[b^{\pm 1}, (l_3^{\varphi})^{\pm 1}], [[b^{\pm 1}, (l_4^{\varphi})^{\pm 1}], [[b^{\pm 1}, (l_5^{\varphi})^{\pm 1}], [[b^{\pm 1}, (l_6^{\varphi})^{\pm 1}], \\ & [[c^{\pm 1}, (l_1^{\varphi})^{\pm 1}], [[c^{\pm 1}, (l_2^{\varphi})^{\pm 1}], [[c^{\pm 1}, (l_3^{\varphi})^{\pm 1}], [[c^{\pm 1}, (l_4^{\varphi})^{\pm 1}], [[c^{\pm 1}, (l_5^{\varphi})^{\pm 1}], \\ & [[c^{\pm 1}, (l_6^{\varphi})^{\pm 1}], [[l_1^{\pm 1}, (l_2^{\varphi})^{\pm 1}], [[l_1^{\pm 1}, (l_3^{\varphi})^{\pm 1}], [[l_1^{\pm 1}, (l_4^{\varphi})^{\pm 1}], [[l_1^{\pm 1}, (l_5^{\varphi})^{\pm 1}], \\ & [[l_1^{\pm 1}, (l_6^{\varphi})^{\pm 1}], [[l_2^{\pm 1}, (l_3^{\varphi})^{\pm 1}], [[l_2^{\pm 1}, (l_4^{\varphi})^{\pm 1}], [[l_2^{\pm 1}, (l_5^{\varphi})^{\pm 1}], [[l_2^{\pm 1}, (l_6^{\varphi})^{\pm 1}], \\ & [[l_3^{\pm 1}, (l_4^{\varphi})^{\pm 1}], [[l_3^{\pm 1}, (l_5^{\varphi})^{\pm 1}], [[l_3^{\pm 1}, (l_6^{\varphi})^{\pm 1}], [[l_4^{\pm 1}, (l_5^{\varphi})^{\pm 1}], [[l_4^{\pm 1}, (l_6^{\varphi})^{\pm 1}], \\ & [[l_5^{\pm 1}, (l_6^{\varphi})^{\pm 1}]\}. \end{aligned}$$

IOP Conf. Series: Journal of Physics: Conf. Series **893** (2017) 012006 doi:10.1088/1742-6596/893/1/012006

Lastly, we have,

$$\begin{aligned}\langle [g_i, g_j^{\varphi}][g_j, g_i^{\varphi}]\rangle = \{&[a,b^{\varphi}][b,a^{\varphi}], [a,c^{\varphi}][c,a^{\varphi}], [a,l_1^{\varphi}][l_1,a^{\varphi}], [a,l_2^{\varphi}][l_2,a^{\varphi}], [a,l_3^{\varphi}][l_3,a^{\varphi}],\\ &[a,l_5^{\varphi}][l_5,a^{\varphi}], [a,l_6^{\varphi}][l_6,a^{\varphi}], [b,c^{\varphi}][c,b^{\varphi}], [b,l_1^{\varphi}][l_1,b^{\varphi}], [b,l_2^{\varphi}][l_2,b^{\varphi}],\\ &[b,l_3^{\varphi}][l_3,b^{\varphi}], [b,l_4^{\varphi}][l_4,b^{\varphi}], [b,l_5^{\varphi}][l_5,b^{\varphi}], [b,l_6^{\varphi}][l_6,b^{\varphi}], [c,l_1^{\varphi}][l_1,c^{\varphi}],\\ &[c,l_2^{\varphi}][l_2,c^{\varphi}], [c,l_3^{\varphi}][l_3,c^{\varphi}], [c,l_4^{\varphi}][l_4,c^{\varphi}], [c,l_5^{\varphi}][l_5,c^{\varphi}], [c,l_6^{\varphi}][l_6,c^{\varphi}],\\ &[l_1,l_2^{\varphi}][l_2,l_1^{\varphi}], [l_1,l_3^{\varphi}][l_3,l_1^{\varphi}], [l_1,l_4^{\varphi}][l_4,l_1^{\varphi}], [l_1,l_5^{\varphi}][l_5,l_1^{\varphi}],\\ &[l_1,l_6^{\varphi}][l_6,l_1^{\varphi}], [l_2,l_3^{\varphi}][l_3,l_2^{\varphi}], [l_2,l_4^{\varphi}][l_4,l_2^{\varphi}], [l_2,l_5^{\varphi}][l_5,l_2^{\varphi}],\\ &[l_2,l_6^{\varphi}][l_6,l_2^{\varphi}], [l_3,l_4^{\varphi}][l_4,l_3^{\varphi}], [l_3,l_5^{\varphi}][l_5,l_3^{\varphi}], [l_3,l_6^{\varphi}][l_6,l_3^{\varphi}],\\ &[l_4,l_5^{\varphi}][l_5,l_4^{\varphi}], [l_4,l_6^{\varphi}][l_6,l_4^{\varphi}], [l_5,l_6^{\varphi}][l_6,l_5^{\varphi}]\}.\end{aligned}$$

Thus, there are 189 elements of $[G, G^{\varphi}]$ based on Theorem 3, but they are not independent. Some of these elements are identities and some of those are products of power of the other elements. Hence, some of them can be eliminated. Some calculations of the elimination are shown in the following:

i. $[l_3, l_3^{\varphi}] = [l_3, l_3^{\varphi}]^b$
$= [l_1, l_1^{\varphi}]$ by relations of $Q(6)$.

ii. $[l_4, l_1^{\varphi}] = [l_4, l_1^{\varphi}]^a$
$= [l_1, l_4^{-1\varphi}]$ by relations of $Q(6)$
$= [l_1, l_4^{\varphi}]^{-1}$.
Therefore, $[l_1, l_4^{\varphi}][l_4, l_1^{\varphi}] = 1$.

iii. $[b, l_4^{\varphi}][l_4, b^{\varphi}] = ([b, l_4^{\varphi}][l_4, b^{\varphi}])^b$
$= [b, l_4^{\varphi}]^b [l_4, b^{\varphi}]^b$
$= [b^b, (l_4^b)^{\varphi}][l_4^b, (b^b)^{\varphi}]$
$= [b, l_2^{\varphi}][l_2, b^{\varphi}]$ by relations of $Q(6)$.

iv. $[a, c^{-\varphi}] = [c^{-1}, [a, c]][a, c^{\varphi}]^{-1}$
$= [c^{-1}, (l_6^{-1} l_5)^{\varphi}]$
$= [c^{-1}, l_5^{\varphi}][c^{-1}, l_6^{-\varphi}][[c^{-1}, l_6^{-1}], l_5^{\varphi}][a, c^{\varphi}]^{-1}$
$= [a, c^{\varphi}]^{-1}$ since c commute with l_5 and l_6.

v. $[a, l_6^{\varphi}] = [a, c^{-1}a^2]$ by relations of $Q(6)$
$= [a, a^{\varphi}][a, (c^{-1}a)^{\varphi}][[a, c^{-1}a], a^{\varphi}]$
$= [a, a^{\varphi}][a, a^{\varphi}][a, (c^{-1})^{\varphi}][[a, (c^{-1})], a^{\varphi}], [l_6^{-1} l_5, a^{\varphi}]$
$= [a, a^{\varphi}]^2 [a, (c^{-1})^{\varphi}][l_5^{-1} l_6, a^{\varphi}][l_6^{-1} l_5, a^{\varphi}]$
$= [a, a^{\varphi}]^2 [a, (c^{-1})^{\varphi}][l_5^{-1}, a^{\varphi}][[l_5^{-1}, a], l_6^{\varphi}][l_6, a^{\varphi}]$
$[l_6^{-1} l_5, a^{\varphi}]$
$= [a, a^{\varphi}]^2 [a, (c^{-1})^{\varphi}][l_5^{-1}, a^{\varphi}][[l_5 l_6^{-1}, l_6^{\varphi}][l_6, a^{\varphi}]$
$[l_6^{-1} l_5, a^{\varphi}]$
$= [a, a^{\varphi}]^2 [a, (c^{-1})^{\varphi}][l_5^{-1}, a^{\varphi}][l_5, l_6^{\varphi}][[l_5, l_6], l_6^{-\varphi}]$
$[l_6^{-1}, l_6^{\varphi}][l_6, a^{\varphi}][l_6^{-1} l_5, a^{\varphi}]$ $= [a, a^{\varphi}]^2 [a, (c^{-1})^{\varphi}][l_5^{-1}, l_6^{\varphi}][l_5^{-1}, l_5^{\varphi}][l_5, a^{\varphi}]^{-1}$
$[l_5, l_6^{\varphi}][l_6^{-1}, l_6^{\varphi}][l_6, a^{\varphi}][l_6^{-1} l_5, a^{\varphi}]$
$= [a, a^{\varphi}]^2 [a, (c^{-1})^{\varphi}][l_5^{-1}, l_5^{\varphi}]$
$[l_6^{-1}, l_6^{\varphi}][l_6^{-1} l_5, a^{\varphi}]$ by relations of $Q(6)$
$= \ldots\ldots$
$= [a, a^{\varphi}]^2 [a, c^{\varphi}]^{-1}$

Until at the end the elements can be reduced to 16. Thus,

$$\begin{aligned}[Q(6), Q(6)^{\varphi}] =& \langle [a, a^{\varphi}], [b, b^{\varphi}], [l_1, l_1{}^{\varphi}], [a, b^{\varphi}], [a, c^{\varphi}], [a, l_1{}^{\varphi}], [a, l_2{}^{\varphi}], [b, l_1{}^{\varphi}], \\ & [b, l_2{}^{\varphi}], [a, b^{\varphi}][b, a^{\varphi}], [a, c^{\varphi}][c, a^{\varphi}], [a, l_1{}^{\varphi}][l_1, a^{\varphi}], \\ & [a, l_2{}^{\varphi}][l_2, a^{\varphi}], [a, l_6{}^{\varphi}][l_6, a^{\varphi}], [b, l_1{}^{\varphi}][l_1, b^{\varphi}], [b, l_2{}^{\varphi}][l_2, b^{\varphi}]\rangle.\end{aligned}$$

Therefore, by Theorem 2,

$$Q\,(6) \otimes Q\,(6) \cong [Q(6), (Q(6)^{\varphi}].$$

Hence,

$$\begin{aligned}Q(6) \otimes Q(6) = & \langle a \otimes a, b \otimes b, l_1 \otimes l_1, a \otimes b, a \otimes c, a \otimes l_1, a \otimes l_2, b \otimes l_1, b \otimes l_2, \\ & (a \otimes b)(b \otimes a), (a \otimes c)(c \otimes a), (a \otimes l_1)(l_1 \otimes a), (a \otimes l_2)(l_2 \otimes a), \\ & (a \otimes l_6)(l_6 \otimes a), (b \otimes l_1)(l_1 \otimes b), (b \otimes l_2)(l_2 \otimes b)\rangle. \qquad \square\end{aligned}$$

3. Conclusion

In this research, the nonabelian tensor square of a crystallographic group with quaternion point group of order eight, $Q(6) \otimes Q(6)$ is computed.

Acknowledgment

The authors would like to express their appreciation for the support of the sponsor; Ministry of Higher Education (MOHE) Malaysia for the financial funding for this research through Research University Grant (GUP), Vote no: 13H79 from Research Management Centre (RMC) Universiti Teknologi Malaysia (UTM) Johor Bahru. The first author is also indebted to UTM for her Zamalah Scholarship. The third author is also grateful to UTM and MOHE for her postdoctoral scholarship in University of Leeds, United Kingdom.

References

[1] Brown R and Loday J L 1987 *Topology* **26** pp 311–35
[2] Brown R, Johnson D L and Robertson E F 1987 *J. Algebra* **111** pp 177–202
[3] Kappe L C, Sarmin N H and Visscher M P 1999 *Glasg. Math. J.* **41** pp 417–30
[4] Rashid S, Sarmin N H, Erfanian A, Mohd Ali N M and Zainal R 2013 *Indag. Math.* **24** pp 581–588
[5] Zainal R, Mohd Ali N M, Sarmin N H and Rashid S 2013 *AIP Conf. Proc.* **1522** pp 1039–1044
[6] The CARAT homepage. (http://wwwb.math.rwth-aachen.de/carat/).
[7] Blyth R D and Morse R S 2009 *J. Algebra* **321** pp 2139–2148
[8] Tan Y T, Mohd Idrus N, Masri R, Wan Mohd Fauzi W N F, Sarmin N H and Mat Hassim H I 2016 *Jurnal Teknologi* **78**(1) pp 189–193
[9] Bacon M R and Kappe L C 2003 *Illinois J. Math.* **47** pp 49–62
[10] Rocco N R 1991 *Bol. Soc. Brasil. Mat. (N. S.)* **22**(1) pp 63–79
[11] Eick B and Nickel W 2008 *J. Algebra* **320**(1) pp 927–44
[12] Mohammad S A, Sarmin N H and Mat Hassim H I 2015 *Jurnal Teknologi* **77**(3) pp 151–156

IOP Conf. Series: Journal of Physics: Conf. Series **893** (2017) 012007 doi:10.1088/1742-6596/893/1/012007

Equivalent weavings

Aliw-iw Zambrano,Eden Delight Miro and Agnes Garciano

Mathematics Department, Ateneo de Manila University
Loyola Heights, Diliman, Quezon City, Philippines

E-mail: aliw-iw.zambrano@obf.ateneo.edu

Abstract. A weaving is defined as the interlacement of the overlapping simple connected graphs. In particular, weavings may be constructed from colorings of the plane. A definition of equivalent weavings constructed from colorings of the plane as well as some examples are presented in this paper.

1. Introduction

In this paper, we present a definition for equivalent weavings. We begin with a summary of the method for constructing weavings of overlapping nets from 2-colorings of the plane and a definition of weavings as presented by Miro, Garciano and Zambrano in [2]. We then present a definition of combinatorial and weaving equivalence and give some examples of equivalent weavings from colorings of the plane.

In [1], Batten and Robson used weavings to study molecular entanglements. Meanwhile, Ramsden, Robins, and Hyde, in [4], showed that three-dimensional crystalline euclidean nets may be constructed from weavings on the hyperbolic plane.

The ultimate goal is to present a methodology for constructing weavings that can be used to construct three-periodic patterns using the methodology given in [4].

2. Methodology for constructing weavings

Let $\triangle$ be any triangle on the plane $\mathbb{P}$ with interior angles $\pi/p, \pi/q, \pi/r$ where p, q, r are integers greater than or equal to 2. Repeatedly reflecting the triangle on its sides results in a triangle tiling $\mathcal{T} := \mathcal{T}(p, q, r)$ of the plane by copies of the triangle $\triangle$. Let P, Q, R denote the reflections on the side of $\triangle$ opposite $\pi/p, \pi/q$ and π/r, respectively. The group $G := G(p, q, r)$ of isometries generated by P, Q, R is called a *triangle group* with group presentation

$$\langle P, Q, R \mid P^2 = Q^2 = R^2 = (QR)^p = (RP)^q = (PQ)^r \rangle.$$

By construction , each $t \in \mathcal{T}$ is the image of $\triangle$ under some unique element $g \in G$. The group G acts transitively on the tiling. Hence, the mapping $g \to g\triangle$ is a bijection. Note that,

(i) $\bigcup_{g \in G} g\triangle = \mathbb{P}$ and

(ii) $\triangle^\circ \cap g\triangle^\circ = \varnothing$, for every $g \in G \backslash \{I\}$.

Thus, $\triangle$ is a *fundamental region of the triangle group* $G(p, q, r)$.

A coloring of $\mathcal{T}$ is a labeling of the tiles in $\mathcal{T}$ by colors. We use the method presented by Provido, De las Peñas and Felix in [3] to color triangle tilings.

Consider an index 2 subgroup H of G and let $\{g_1 = I, g_2\}$ be a complete set of left coset representatives of H in G. Then $G = H \cup g_2 H$ and $H \cap g_2 H = \varnothing$. Consequently, $\mathcal{T} = H\triangle \cup g_2 H\triangle$. Also, since $H \cap g_2 H = \varnothing$ and the mapping $g \mapsto g\triangle$ is a bijection, then $H\triangle \cap g_2 H\triangle = \varnothing$. Suppose c_1, c_2 is a set of distinct colors, where c_1 represents the color white and c_2 represents the color black. A 2-coloring of $\mathcal{T}$ can be defined by assigning color c_1 to $H\triangle$ and c_2 to $g_2 H\triangle$. This results in a two-coloring of the triangle tiling $\mathcal{T}$ by H which we denote by $\mathcal{T}_H(p, q, r)$.

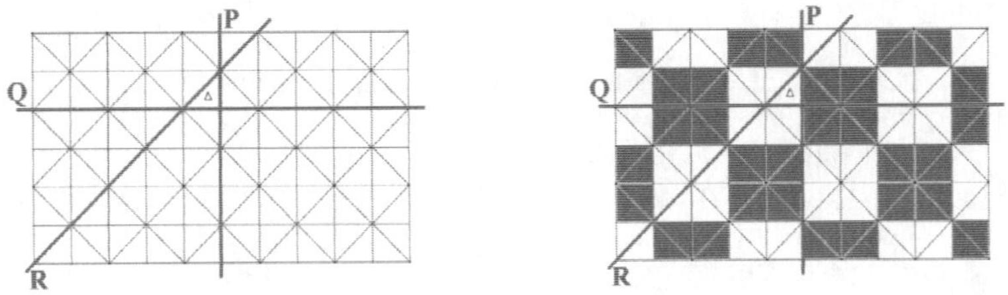

(a) (b)

Figure 1. (a) The tiling $\mathcal{T}(4, 4, 2)$, and (b) its 2-coloring by $H_2 = \langle R, Q, PQP, PRP \rangle$

In Fig 1, we illustrate a coloring of $\mathcal{T}(4, 4, 2)$ by a subgroup $H_2 = \langle R, Q, PQP, PRP \rangle$ of $G(4, 4, 2)$.

In [2], we introduced a concept of black and white patches of a colored tiling $\mathcal{T}_H$. Intuitively, a (black) B-patch of a colored tiling is composed of edge-adjacent black tiles of the same color all having a unique point in common. A (white) W-patch is similarly defined. For example, the B-patches of the colored tiling $\mathcal{T}_{H_2}$ where $H_2 = \langle R, Q, PQP, PRP \rangle$ is illustrated in Figure 2(b). A B-patch is a union of edge-adjacent black tiles about the common vertex of 1- and 2-edges; W-patches are formed similarly.

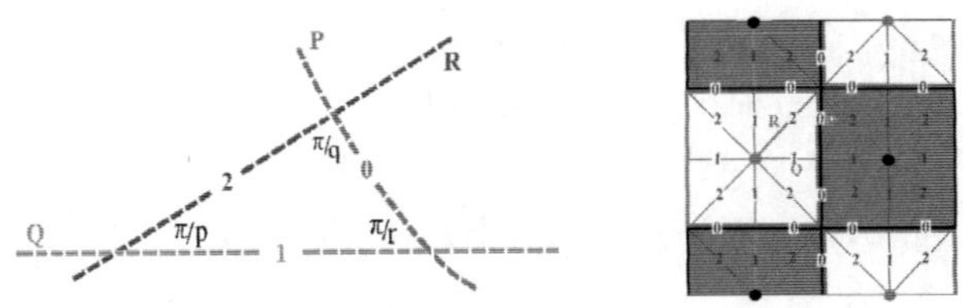

(a) (b)

Figure 2. (a) Labeling the edges opposite the angles $\pi/p, \pi/q$, and π/r by $0, 1$ and 2, respectively, and (b) The B- and W-patches of $\mathcal{T}_{H_2}$

We now construct two disjoint simple connected graphs, $\mathcal{N}_B = (\mathcal{V}, \mathcal{E})$ and $\mathcal{N}_W = (\mathcal{V}', \mathcal{E}')$ which we refer to as B- and W-nets, respectively. The vertex sets $\mathcal{V}_B$ and the $\mathcal{V}_W$ are the set of centroids of the B-patches and W-patches of $\mathcal{T}_H$ respectively. In Figure 2(b), the black and green points are B- and W-vertices of $\mathcal{T}_{H_2}$.

Given an index 2 subgroup $H = \langle h_1, h_2, ..., h_l \rangle$ of G, we can determine a *constructible fundamental region* $\triangle_H = \triangle \cup g\triangle$ with respect to the given the generating set of H. This is elaborated in [2]. The vertices in the constructible fundamental region are called the initial B-vertex $b := b_H$ and the initial W-vertex $w := w_H$ of $\triangle_H$.

Now for b and each generator h_i, construct edges $\{b, h_i b\}$ and $\{b, h_i^{-1} b\}$ using black line segments. The B-motif $\triangle_{bH}$ consists of the B-vertex b together with the segments of the edges $\{b, h_i b\}$, $\{b, h_i^{-1} b\}$ within $\triangle_H$. The W-motif $\triangle_{mH}$ is similarly constructed.

The elements of the set $\mathcal{E}_B$ are of the form $h\{b, h_i b\} := \{hb, hh_i b\}$, such that $h \in H, h_i \in \{h_1^j, h_2^j ..., h_l^j\}$ and $j \in \{1, -1\}$. The set $\mathcal{E}_W$ is similarly defined. The above construction yields two disjoint nets $\mathcal{N}_B = (\mathcal{V}, \mathcal{E})$ and $\mathcal{N}_W = (\mathcal{V}', \mathcal{E}')$. An *overlapping net* $\mathcal{O}_H = \mathcal{N}_B \cup \mathcal{N}_W$ is the union of $\mathcal{N}_B$ and $\mathcal{N}_W$.

We call the points where the edges of the nets $\mathcal{N}_B$ and $\mathcal{N}_W$ intersect, the *weaving points* of the overlapping net and denote the set of all weaving points in $\mathcal{O}_H$ by $\mathscr{I}_{\mathcal{O}_H}$. We define a weaving map $\omega_H : \mathscr{I}_{\mathcal{O}_H} \to \{\oplus, \ominus\}$ by first assigning the $\oplus$ or $\ominus$ values to the weaving points p within the fundamental region $\triangle_H$ and since every element q of $\mathscr{I}_{\mathcal{O}_H}$ is of the form $q = hp$ for some $h \in H$, let $\omega(q) = \omega(p)$. Thus, we obtain a weaving $\mathcal{W}_H = (\mathcal{O}_H, \omega_H)$. Also, for a weaving point $(e_b, e_w) \in \mathscr{I}_{\mathcal{O}_H}$, if $\omega(e_b, e_w) = \oplus$, then edge e_b is above edge e_w while edge e_b is below edge e_w if $\omega(e_b, e_w) = \ominus$. If $\omega_H(\mathscr{I}_{\mathcal{O}_H}) = \{\oplus, \ominus\}$, then $\mathcal{W}_H$ is a *proper weaving*. This method is outlined in Figure 3 and discussed in detail in [2].

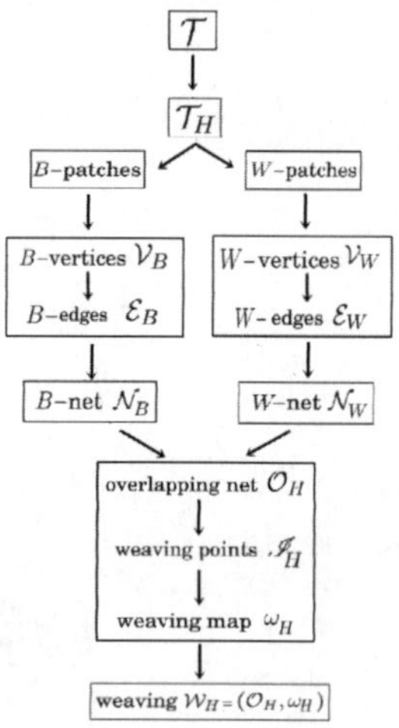

Figure 3. Weaving construction outline

Consider the weaving $\mathcal{W}_{H_2} = (\mathcal{O}_{H_2}, \omega)$ constructed from the index 2 subgroup $H_2 = \langle R, Q, PQP, PRP \rangle$ of $G(4, 4, 2)$.

Since $\omega_{H_2}(\mathscr{I}) = \{\oplus\}$, $\mathcal{W}_{H_2}$ is not a proper weaving. Consider index 2 subgroups of H_2, whose fundamental regions consist of two copies of $\triangle_{H_2}$ resulting in additional weaving points. By inspection, only the subgroup $N = \langle R, PRP, QRQ, PQRQP \rangle$ of H_2 results in a proper weaving as shown in Figure 4.

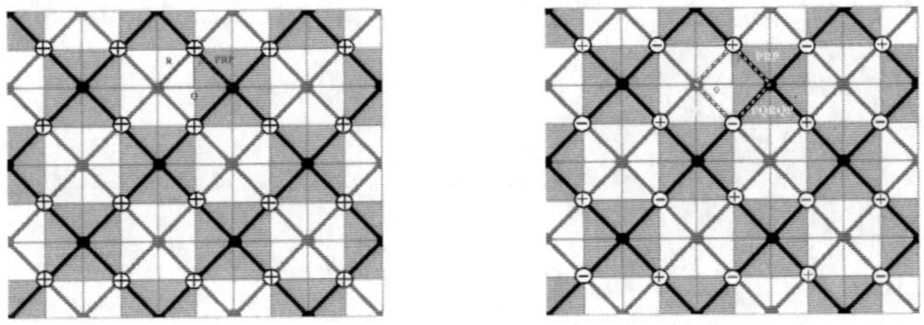

(a) (b)

Figure 4. (a) Nonproper weaving constructed by subgroup H_2 and (b) the weaving constructed by subgroup N of H_2

3. The Combinatorial Equivalence of Overlapping Nets

We first introduce the concept of the combinatorial equivalence between two overlapping nets.

Definition 1. *Let $\mathcal{O} = \mathcal{N}_1 \cup \mathcal{N}_2$ and $\mathcal{O}' = \mathcal{M}_1 \cup \mathcal{M}_2$ be two overlapping nets, where $\mathcal{N}_i = (\mathcal{U}_i, \mathcal{E}_i)$ and $\mathcal{M}_i = (\mathcal{V}_i, \mathcal{F}_i)$ for $i = 1, 2$. We say that the overlapping nets $\mathcal{O}$ and $\mathcal{O}'$ are* ***combinatorially equivalent*** *if and only if there exists a bijective mapping $\phi : \mathcal{U}_1 \cup \mathcal{U}_2 \to \mathcal{V}_1 \cup \mathcal{V}_2$ such that any vertices u, v are adjacent in $\mathcal{O}$ if and only if $\phi(u)$ and $\phi(v)$ are adjacent in $\mathcal{O}'$, and any two edges $e_1 = \{u_1, u_2\}$ and $e_2 = \{u_1', u_2'\}$ intersect in $\mathcal{O}$ if and only if $\phi(e_1) = \{\phi(u_1), \phi(u_2)\}$ and $\phi(e_2) = \{\phi(v_1), \phi(v_2)\}$ intersect in $\mathcal{O}'$.*

Proposition 1. *The relation of combinatorial equivalence is an equivalence relation on the set of all overlapping nets on the plane.*

Proof. Let $\mathcal{O} = \mathcal{N}_1 \cup \mathcal{N}_2$ be an overlapping net where $\mathcal{N}_i = (\mathcal{U}_i, \mathcal{E}_i)$. Clearly, the trivial isometry $I : \mathcal{U}_1 \cup \mathcal{U}_2 \to \mathcal{U}_1 \cup \mathcal{U}_2$ fixing each vertex is a bijection that preserves the adjacency of vertices and the intersection of edges. Hence $\mathcal{O}$ is combinatorially equivalent to itself.

Now, suppose $\mathcal{O} = \mathcal{N}_1 \cup \mathcal{N}_2$ is combinatorially equivalent to $\mathcal{O}' = \mathcal{M}_1 \cup \mathcal{M}_2$. Then there exists a bijection $\phi : \mathcal{U}_1 \cup \mathcal{U}_2 \to \mathcal{V}_1 \cup \mathcal{V}_2$ which preserves adjacency of the vertices and intersection of edges. Using its inverse function ϕ^{-1}, it follows that $\mathcal{O}'$ is combinatorially equivalent to $\mathcal{O}$.

Lastly, let $\mathcal{O}'' = \mathcal{L}_1 \cup \mathcal{L}_2$ be overlapping nets such that $L_i = (\mathcal{Y}_i, \mathcal{D}_i)$ for $i = 1, 2$. Furthermore, suppose that there exist a bijective mapping $\phi : \mathcal{U}_1 \cup \mathcal{U}_2 \to \mathcal{V}_1 \cup \mathcal{V}_2$ such that $\mathcal{O}$ and $\mathcal{O}'$ are combinatorially equivalent and a bijective mapping $\alpha : \mathcal{V}_1 \cup \mathcal{V}_2 \to \mathcal{Y}_1 \cup \mathcal{Y}_2$ such that $\mathcal{O}'$ and $\mathcal{O}''$ are combinatorially equivalent. Since the product of two bijective functions is also bijective, then $\alpha\phi : \mathcal{U}_1 \cup \mathcal{U}_2 \to \mathcal{Y}_1 \cup \mathcal{Y}_2$ is a bijection. Furthermore, since $\mathcal{O}$ and $\mathcal{O}'$ are combinatorially equivalent, two vertices u, v are adjacent vertices in $\mathcal{U}_1 \cup \mathcal{U}_2$ if and only if $\phi(u)$ and $\phi(v)$ are adjacent in $\mathcal{V}_1 \cup \mathcal{V}_2$. Since $\mathcal{O}'$ and $\mathcal{O}''$ are combinatorially equivalent, then $\phi(u)$ is adjacent to $\phi(v)$ if and only if $\alpha\phi(u)$ and $\alpha\phi(v)$ are adjacent in $\mathcal{Y}_1 \cup \mathcal{Y}_2$. Thus, adjacency of vertices in $\mathcal{U}_1 \cup \mathcal{U}_2$ is preserved by $\alpha\phi$. A similar argument can be used to show that two edges e_1 and e_2 intersect in $\mathcal{O}$ if and only if $\alpha\phi(e_1)$ and $\alpha\phi(e_2)$ intersect in $\mathcal{O}''$. Thus, combinatorial equivalence of overlapping nets is transitive and therefore an equivalence relation. □

As an example, consider the overlapping nets constructed from $H_2 = \langle R, Q, PQP, PRP \rangle$ and $H_4 = \langle RQR, RPR, P, Q \rangle$ as shown in Figure 5. If T is a product of a 135° about the origin indicated by the red dot and a translation with a scaling, then T will be the bijective mapping such that $\mathcal{O}_{H_2}$ and $\mathcal{O}_{H_4}$ are combinatorially equivalent.

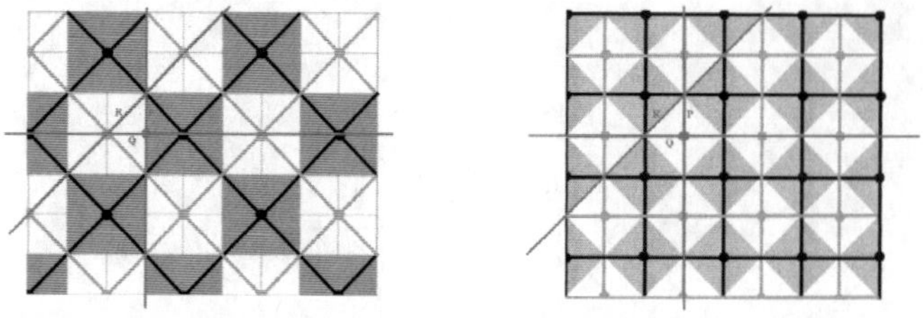

(a) Overlapping net $\mathcal{O}_{H_2}$ (b) Overlapping net $\mathcal{O}_{H_4}$

Figure 5. Overlapping nets by $H_2 = \langle R, Q, PQP, PRP\rangle$ and $H_4 = \langle RQR, RPR, P, Q\rangle$ of $G(4, 4, 2)$

4. The Weaving Equivalence

We now present a definition of equivalent weavings..

Definition 2. *Two weavings $\mathcal{W} = (\mathcal{O}, \omega)$ and $\mathcal{W}' = (\mathcal{O}', \omega')$ are* ***equivalent weavings*** *if the following conditions are satisfied:*

(i) there exists a bijection $\phi : \mathcal{O} \to \mathcal{O}'$ under which $\mathcal{O}$ and $\mathcal{O}'$ are combinatorially equivalent, and

(ii) $\omega(e_b, e_w) = \omega'(\phi(e_b, e_w))$
or $\omega(e_b, e_w) = \overline{\omega'(\phi(e_b, e_w))}$ for all $(e_b, e_w) \in \mathscr{I}$ where $\overline{\oplus} = \ominus$ and $\overline{\ominus} = \oplus$.

Proposition 2. *The relation of weaving equivalence is an equivalence relation.*

Proof. Suppose $\mathcal{W} = (\mathcal{O}, \omega)$ is a weaving where $\mathcal{O} = \mathcal{N}_1 \cup \mathcal{N}_2$ such that $\mathcal{N}_i = (\mathcal{U}_i, \mathcal{E}_i)$ for $i = 1, 2$ and let (e_b, e_w) be an element of the set of weaving points $\mathscr{I}$ of $\mathcal{O}$. Consider the mapping $I : \mathcal{U}_1 \cup \mathcal{U}_2 \to \mathcal{U}_1 \cup \mathcal{U}_2$ where I is the trivial isometry that sends $\mathcal{O}$ to itself. Since I is a bijection, and preserves the adjacency of the vertices and the intersection of edges, then the overlapping net $\mathcal{O}$ is combinatorially equivalent to itself. Also, since $I(\mathscr{I}) = \mathscr{I}$, then $\omega(e_b, e_w) = \omega(I(e_b, e_w))$ for all $(e_b, e_w) \in \mathscr{I}$.

Suppose a weaving $\mathcal{W} = (\mathcal{O}, \omega)$ is equivalent to a weaving $\mathcal{W}' = (\mathcal{O}', \omega')$ where $\mathcal{O}' = \mathcal{M}_1 \cup \mathcal{M}_2$ and $\mathcal{M}_i = (\mathcal{V}_i, \mathcal{F}_i)$ for $i = 1, 2$. Then there exists a bijective mapping $\phi : \mathcal{U}_1 \cup \mathcal{U}_2 \to \mathcal{V}_1 \cup \mathcal{V}_2$ such that $\mathcal{O}$ and $\mathcal{O}'$ are combinatorially equivalent. Since edges intersect in $\mathcal{O}$ if and only if their images under ϕ intersect in $\mathcal{O}'$, then it follows that $\phi(\mathscr{I}) = \mathscr{I}'$ where $\mathscr{I}'$ is the set of weaving points in $\mathcal{O}'$. From the second condition of the definition one may immediately deduce that $\omega'(\phi(e_b, e_w)) = \omega(e_b, e_w)$ or $\omega'(\phi(e_b, e_w)) = \overline{\omega(e_b, e_w)}$ $\forall (e_b, e_w) \in \mathscr{I}$. Thus, $\mathcal{W}' = (\mathcal{O}', \omega')$ is equivalent to $\mathcal{W} = (\mathcal{O}, \omega)$. Therefore, weaving equivalence is symmetric.

Let $\mathcal{W}'' = (\mathcal{O}'', \omega'')$ where $\mathcal{O}'' = \mathcal{L}_1 \cup \mathcal{L}_2$ and $\mathcal{L}_i = (\mathcal{Y}_i, \mathcal{D}_i)$. Suppose that weaving $\mathcal{W}' = (\mathcal{O}', \omega')$ is equivalent to weaving $\mathcal{W}''$ and $\mathcal{W}$ is equivalent to $\mathcal{W}'$. This implies that $\mathcal{O}$ and $\mathcal{O}'$ are combinatorially equivalent under a bijection ϕ and $\mathcal{O}'$ and $\mathcal{O}''$ are combinatorially equivalent under a bijection α. By the transitive property of combinatorial equivalence, $\mathcal{O}$ and $\mathcal{O}''$ are combinatorially equivalent under the bijective mapping $\alpha\phi$ from $\mathcal{U}_1 \cup \mathcal{U}_2$ to $\mathcal{Y}_1 \cup \mathcal{Y}_2$.

Since $\phi(\mathscr{I}) = \mathscr{I}'$ where $\omega(e_b, e_w) = \omega'(\phi(e_b, e_w))$ or $\omega(e_b, e_w) = \overline{\omega'(\phi(e_b, e_w))}$, and $\alpha(\mathscr{I}') = \mathscr{I}''$ where $\omega'(\phi(e_b, e_w)) = \omega''(\alpha\phi(e_b, e_w))$ or $\omega'(\phi(e_b, e_w)) = \overline{\omega''(\alpha\phi(e_b, e_w))}$ for all $(e_b, e_w) \in \mathscr{I}$. Then $\alpha\phi(\mathscr{I}) = \mathscr{I}''$, and $\omega(e_b, e_w) = \omega''(\alpha\phi(e_b, e_w))$ or $\omega(\phi(e_b, e_w)) = \overline{\omega''(\alpha\phi(e_b, e_w))}$ for all $(e_b, e_w) \in \mathscr{I}$.This implies that $\mathcal{W} = (\mathcal{O}, \omega)$ is equivalent to $\mathcal{W}'' = (\mathcal{O}'', \omega'')$. This shows that weaving equivalence is an equivalence relation. □

The equivalence classes of weavings with respect to this equivalence relation are called weaving patterns.

5. Examples

In Section IV, we noted that $\mathcal{O}_{H_2}$ and $\mathcal{O}_{H_4}$ are combinatorially equivalent. To show that $\mathcal{W}_{H_4} = \{\mathcal{O}_{H_4}, \omega\}$ and $\mathcal{W}_{N_4} = \{\mathcal{O}_{H_2}, \omega'\}$ satisfy condition 2 of equivalent weavings, consider Figure 6.

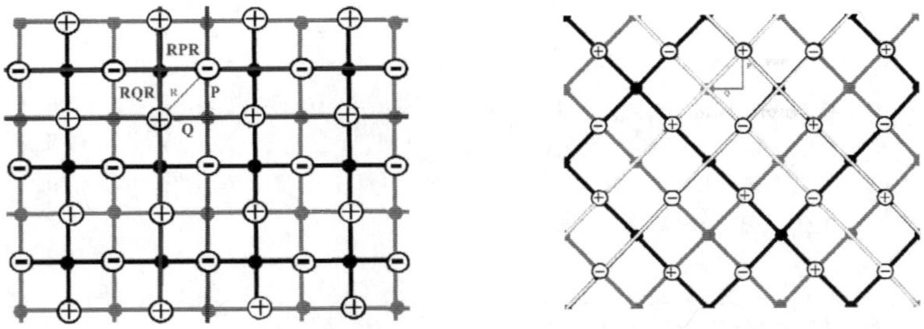

(a) Weaving by H_4 (b) Weaving by N_4

Figure 6. Weavings constructed from subgroup $H_4 = \langle RQR, RPR, P, Q\rangle$ of $G(4,4,2)$ and an index 2 subgroup $N_4 = \langle R, PRP, QRQ, PQRQP\rangle$ of H_2, respectively.

For this example, first consider the effect of the transformation T to the motif $\triangle_{N_4}$ of $\mathcal{W}_{N_4}$ as shown in Figure 7.

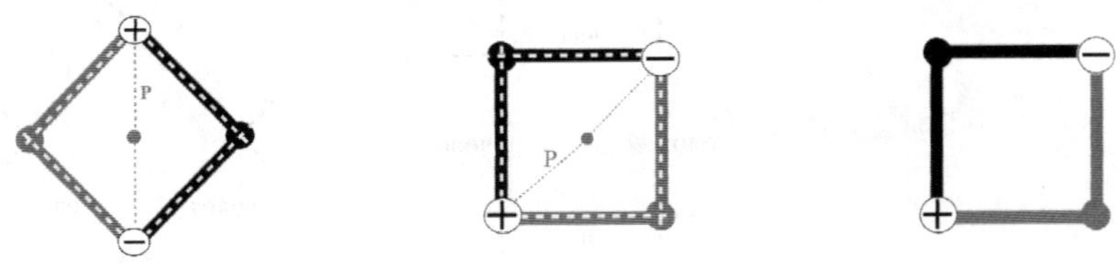

(a) (b) (c)

Figure 7. (a) Motif $\triangle_{N_4}$, (b) motif $\triangle_{N_4}$ rotated 135° about the origin and (c) motif $\triangle_{N_4}$ translated and scaled

Notice that both motifs $\triangle_{H_4}$ and $\triangle_{N_4}$ of overlapping nets $\mathcal{O}_{H_4}$ and $\mathcal{O}_{N_4}$ have a b-vertex and a w-vertex. Figure 8 shows that the motifs have the same structure and the weaving points of one can be made to correspond with the weaving points of the other.

(a) motif $\triangle_{H_4}$ (b) motif $\triangle_{N_4}$

Figure 8. Equivalent weavings by $H_4 = \langle RQR, RPR, P, Q\rangle$ and its index 2 subgroup $N_4 = \langle R, PRP, QRQ, PQRQP\rangle$, respectively.

Since $\mathcal{O}_{H_4}$ and $\mathcal{O}_{N_4}$ are combinatorially equivalent and both weavings $\mathcal{W}_{H_4}$ and $\mathcal{W}_{N_4}$ are constructed by 4 reflection lines along the sides of the motif, then subgroup H_4 and N_4 result in weavings such that $\omega(e_b, e_w)_{H_4} = \omega'(\phi(e_b, e_w))\ \forall(e_b, e_w)_{H_4} \in \mathscr{I}_{H_4}$. This implies that $\mathcal{W}_{H_4}$ and $\mathcal{W}_{N_4}$ are equivalent weavings.

The previous example illustrates that if one can find an isometry of the plane that sends a weaving to another weaving, then the two weavings are equivalent.

In our study, we are interested in listing down weaving patterns or non-equivalent weavings that we can derive from the subgroups of a triangle group. In fact, this is the motivation why we came up with the concept of equivalent weavings. In the next example, we give some weaving patterns derived from a subgroup of the triangle tiling $G(6,4,2)$, which corresponds to a triangle tiling of the hyperbolic plane, since $\frac{\pi}{6} + \frac{\pi}{4} + \frac{\pi}{2} < \pi$.

Now, consider the overlapping net constructed from the colored tiling $\mathcal{T}(6,4,2)$ by subgroup $H_2 = \langle R, Q, PQP, PRP \rangle$ of $G(6,4,2)$ as shown in Figure 9.

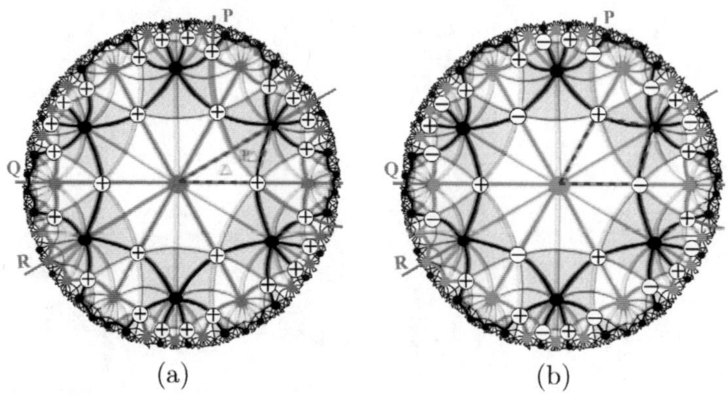

Figure 9. (a) The nonproper weaving constructed by H_2 from the colored tiling $\mathcal{T}(6,4,2)$ and (b) the only proper weaving constructed from an index 2 subgroup of H_2

Since the motif $\triangle_{mH_2}$ has only one weaving point, then clearly subgroup H_2 does not yield a proper weaving. This is what we call a *No Weaving Scenario*. When this scenario occurs, consider an index 2 subgroup N_i of H, identify the motif $\triangle_{N_i}$ of N_i in $\mathcal{T}(6,4,2)$ and assign values to the weaving points in $\triangle_{N_i}$. This method of overcoming a No Weaving Scenario is discussed in detail in [2].

For the seven index 2 subgroups N_i of $H_2 = \langle R, Q, PQP, PRP \rangle$, only subgroup $N_2 = \langle Q, PQP, RQR, PRQRP \rangle$ results in a proper weaving as shown in Figure 9(b).

Consider the index 4 subgroups of $H_2 = \langle R, Q, PQP, PRP \rangle$. For weavings constructed from the same overlapping net, we need only satisfy condition 2 of the definition on weaving equivalence. By inspection we have determined two nonequivalent weaving patterns from the index 4 subgroups K_i of H_2: $K_1 = \langle Q, R, PQRQRQP, PQPRQRQRPQP \rangle$ and $K_2 = \langle Q, RQR, PRQRQP \rangle$ as shown in Figure 10.

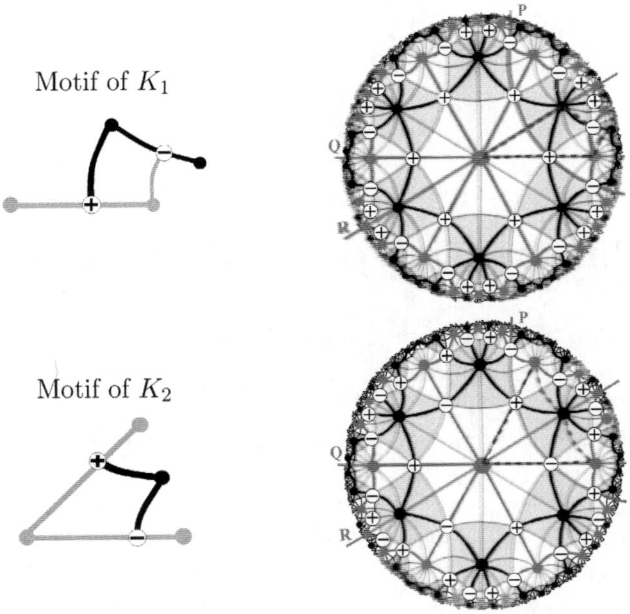

Figure 10. Weavings patterns from subgroups of $H_2 = \langle R, Q, PQP, PRP \rangle$

6. Recommendation and Conclusion

In this paper, we developed a concept of weaving equivalence. We used this concept to get a collection of weavings constructed using our methodology which are not combinatorially identical.

As a continuation of the study, we are looking at extending the definition of B- and W-vertices to include more points other than the centroids of the B- and W-patches. We also consider constructing weavings from n-colorings of triangle tilings for $n > 2$. Ultimately, our goal is to use these weavings to construct three-periodic patterns using the methodology given in [4].

Acknowledgments

Aliw-iw Zambrano would like to thank the Mathematics Department of Ateneo de Manila University and the Asian Mathematical Conference Committee for the travel fund given to her to present the paper in the Asian Mathematical Conference 2016. Eden Delight Miro and Agnes Garciano would also like to thank the Ateneo de Manila University, the Loyola Schools and the Department of Mathematics, as well as the School of Science and Engineering for the travel fund and support.

References

[1] Batten S and Robson R 1998 *Angewandte Chemie International Edition* **37** pp 1490–1494

[2] Miro E D, Garciano A and Zambrano A 2014 From colorings to weavings *ACTA Phys Pol A* **126** pp 560–563

[3] Provido E D, De Las Peñas M L A N and Felix R 2013 On subgroups of crystallographic coxeter groups *ACTA Cryst A* **69** pp 445–451

[4] Ramsden S J, Robins V and Hyde S T 2009 3D euclidean nets from 2D hyperbolic tilings: Kaleidoscopic examples *ACTA Cryst A* **65** pp 81–108

On X-sub-linearly independent modules

Fitriani[1,2], B Surodjo[1], and I E Wijayanti[1]

[1] Department of Mathematics, Universitas Gadjah Mada
[2] Department of Mathematics, Universitas Lampung

E-mail: `fitriani27@mail.ugm.ac.id, {surodjo_b, ind_wijayanti}@ugm.ac.id`

Abstract. The notion of X-sub-exact sequence of modules is a generalization of exact sequences. Let K, L, M be R-modules and X a submodule of L. The triple (K, L, M) is said to be X-sub-exact at L if $K \to X \to M$ is exact at X. The exact sequence is a special case of X-sub-exact by taking $X = L$. We introduce an X-sub-linearly independent module which is a generalization of linearly independent relative to an R-module M by using the concept of X-sub-exact sequence.

1. Introduction

Let R be a ring and let M be an R-module, A subset $S \subseteq M$ is R-linearly dependent if there exist distinct $x_1, x_2, .., x_n$ in S and elements $a_1, a_2, .., a_n$ of R, not all of which are 0, such that $a_1x_1 + a_2x_2 + ... + a_nx_n = 0$. A set that is not R-linearly dependent is said to be R-linearly independent [1]. Let N be a left R-module, then N is said linearly independent to R (or N is $R-$linearly independent) if there exists a monomorphism $\varphi : R^{(\Lambda)} \to N$ [5].

Suprapto [6] introduced a generalization of linearly independency relative to an R-module M as follows: Let M be an R-module. The family of R-modules $\mathcal{N} = \{N_\lambda\}_\Lambda$ is said to be linearly independent to M if there exist a monomorphism $f : \coprod_\Lambda N_\lambda \to M$. If $\{N_\lambda = N\}_\Lambda$, then $f : N^{(\Lambda)} \to M$. We can say that $\mathcal{N} = \{N_\lambda\}_\Lambda$ is linearly independent to M if the sequence $0 \to \coprod_\Lambda N_\lambda \xrightarrow{f} M$ is exact at $\coprod_\Lambda N_\lambda$.

Let R be a ring and let $A \xrightarrow{f} B \xrightarrow{g} C$ be an exact sequence of R-modules, i.e. $Im\ f = Ker\ g (= g^{-1}(\{0\}))$. We can generalize the submodule $\{0\}$ to any submodule $U \subseteq C$ as we refer to [2] in which Davvaz and Parnian-Garamaleky introduced the concept of quasi-exact sequences. A sequence of R-modules and R-homomorphisms $A \xrightarrow{f} B \xrightarrow{g} C$ is quasi-exact in B or U-exact in B if there exists a submodule U in C such that $Im\ f = g^{-1}(U)$.

Then, Anvariyeh dan Davvaz [7] proved further results about quasi-exact sequences and introduced generalization of Schanuel Lemma. Moreover, they obtained some relationships between quasi-exact sequences and superfluous (or essential) submodules.

Furthermore, Davvaz and Shabani-Solt introduced a generalization of some notions in homological algebra [3]. They gave a generalization of the Lambek Lemma, Snake Lemma, connecting homomorphism and exact triangle and they established new basic properties of the U-homological algebra. In [8], Anvariyeh and Davvaz studied U-split sequences and established several connections between U-split sequences and projective modules.

Let K, L, M be R-modules and X a submodule of L. The triple (K, L, M) is said to be an X-sub-exact at L if

$$K \to X \to M$$

is exact, i.e. $Im\ f = Ker\ g$. The exact sequence is a special case of X-sub-exact by taking $X = L$ [4].

In this paper, we introduce an X-sub-linearly independent module which is a generalization of linearly independent relative to an R-module M by using the concept of X-sub-exact sequence.

Let M be an R-module. The family of R-modules $\mathcal{N} = \{N_\lambda\}_\Lambda$ is said to be X-sub-linearly independent to M if the triple $(0, \coprod_\Lambda N_\lambda, M)$ is X-sub-exact (where X is a submodule of $\coprod_\Lambda N_\lambda$). Then, we collect all submodules X of $\coprod_\Lambda N_\lambda$ such that $\mathcal{N}$ is X-sub-linearly independent to M, we denote it by $\sigma(0, \coprod_\Lambda N_\lambda, M)$. In this paper, we give some basic properties of X-sub-linearly independent modules and $\sigma(0, \coprod_\Lambda N_\lambda, M)$. We will show that $\sigma(0, \coprod_\Lambda N_\lambda, M)$ is closed under submodules and intersections. Furthermore, $\sigma(0, \coprod_\Lambda N_\lambda, M)$ always has a maximal element, for every family of R-modules $\mathcal{N}$ and R-module M. In other words, for every family of R-modules $\mathcal{N} = \{N_\lambda\}_\Lambda$ and R-module M, there exist a submodule X maximal such that $\mathcal{N}$ is an X-sublinearly independent.

2. Main Results

As a generalization of linearly independent relative to an R-module M, we define X-sub-linearly independent by using the concept of X-sub-exact sequence as follows:

Definition 2.1 *Let M be an R-module. The family of R-modules $\mathcal{N} = \{N_\lambda\}_\Lambda$ is said to be X-sublinearly independent to M if the triple $(0, \coprod_\Lambda N_\lambda, M)$ is X-sub-exact (where X is a submodule of $\coprod_\Lambda N_\lambda$), i.e. the sequence*

$$0 \to X \to M$$

is exact.

Example 2.1 Let $\mathcal{N} = \{\mathbb{Z}_2, \mathbb{Z}_5\}$ the family of $\mathbb{Z}$-modules and let $\mathbb{Z}_6$ be $\mathbb{Z}$-module. We define $f : \mathbb{Z}_2 \to \mathbb{Z}_6$, where $f(0) = 0$ and $f(1) = 3$. So, f is a monomorphism. Hence, the sequence

$$0 \to \mathbb{Z}_2 \xrightarrow{f} \mathbb{Z}_6$$

is exact. Therefore, the triple $(0, \mathbb{Z}_2 \oplus \mathbb{Z}_5, \mathbb{Z}_6)$ is $\mathbb{Z}_2$-sub-exact. So, $\mathcal{N}$ is $\mathbb{Z}_2$-sub-linearly independent to $\mathbb{Z}_6$.

Assume f is a monomorphism from $\mathbb{Z}_2 \oplus \mathbb{Z}_5$ to $\mathbb{Z}_6$. Then,

$$0 = f(0,0) = f(5(0,1)) = 5f(0,1).$$

We get $f(0,1) = f(0,0) = 0$, a contradiction. So, we can not define a monomorphism from $\mathbb{Z}_2 \oplus \mathbb{Z}_5$ to $\mathbb{Z}_6$. Hence $\mathcal{N}$ is not linearly independent to $\mathbb{Z}_6$.

Example 2.1 shows that if the family of R-modules $\mathcal{N}$ is an X-sub-linearly independent to an R-module M, for some submodule X of $\coprod_\Lambda N_\lambda$, $N_\lambda \in \mathcal{N}$, for all $\lambda \in \Lambda$, then $\mathcal{N}$ is not necessary linearly independent to M.

We already know that any set that containing 0 is linearly dependent since $1.0 = 0$. In the following Proposition, we want to show that the family of R-modules $\mathcal{N}$ is 0-sub-linearly independent to M, for any R-module M.

IOP Conf. Series: Journal of Physics: Conf. Series **893** (2017) 012008 doi:10.1088/1742-6596/893/1/012008

Proposition 2.1 *Let $\mathcal{N} = \{N_\lambda\}_\Lambda$ be a family of R-modules. Then $\mathcal{N}$ is 0-sub-linearly independent to M, for any R-module M.*

Proof. Since the sequence $0 \to 0 \to M$ is exact, the triple $(0, \coprod_\Lambda N_\lambda, M)$ is 0-sub-exact at $\coprod_\Lambda N_\lambda$. Hence, $\mathcal{N}$ is 0-sub-linearly independent to M. □

In fact, we can define a monomorphism from R-module M to itself. So, Any R-module M is M-sub-linearly independent relative to M. We already know that any subset of a linearly independent set is linearly independent. In the following proposition, we will prove that M is X-sub-linearly independent to M, for every submodule X of M.

Proposition 2.2 *For any R-module M, M is X-sub-linearly independent to M, for every submodule X of M.*

Proof. Let M be an R-module and let X be a submodule of M. We have the inclusion $i : X \to M$ such that the sequence $0 \to X \xrightarrow{i} M$ is exact. Hence, the triple $(0, M, M)$ is X-sub-exact. Therefore M is X-sub-linearly independent to M. □

Then, we will give some properties of X-sub-linearly independent relative to an R-module M.

Clearly, we can define a monomorphism from N_λ to $\coprod_\Lambda N_\lambda$. So, we have the following proposition:

Proposition 2.3 *Let $\mathcal{N} = \{N_\lambda\}_\Lambda$ be a family of R-modules. Then $\mathcal{N}$ is N_λ-sub-linearly independent to $\coprod_\Lambda N_\lambda$, for every $\lambda \in \Lambda$.*

Proof. For every $\lambda \in \Lambda$, we have the inclusion $i : N_\lambda \to \coprod_\Lambda N_\lambda$ such that the sequence $0 \to N_\lambda \xrightarrow{i} \coprod_\Lambda N_\lambda$ is exact. Therefore, the triple $(0, \coprod_\Lambda N_\lambda, \coprod_\Lambda N_\lambda)$ is N_λ-sub-exact at $\coprod_\Lambda N_\lambda$. So, $\mathcal{N}$ is N_λ-sub-linearly independent to $\coprod_\Lambda N_\lambda$, for every $\lambda \in \Lambda$. □

Since for any submodule X of N_λ, we can define a monomorphism from X to $\coprod_\Lambda N_\lambda$, we have the following proposition.

Proposition 2.4 *Let $\mathcal{N} = \{N_\lambda\}_\Lambda$ be a family of R-modules. Then, for every $\lambda \in \Lambda$, $\mathcal{N}$ is X-sub-linearly independent to N_λ for any submodule X of N_λ.*

Proof. Let X be a submodule of $N_\lambda \subset \coprod_\Lambda N_\lambda$. We have the inclusion $i : X \to N_\lambda$ such that the sequence $0 \to X \to N_\lambda$ is exact. This implies the triple $(0, \coprod_\Lambda N_\lambda, N_\lambda)$ is X-sub-exact sequence at $\coprod_\Lambda N_\lambda$. Hence $\mathcal{N}$ is X-sublinearly independent to N_λ. □

Let K, L, M be R-modules. We define

$$\sigma(K, L, M) = \{X \leq L | (K, L, M) \ X\text{-sub-exact at } L\}.$$

Then $\sigma(K, L, M) \neq \emptyset$ since $0 \in \sigma(K, L, M)$.

Let $\mathcal{N}$ be a family of R-modules. If we take $K = 0$, $L = \coprod_\Lambda N_\lambda$ and $K = M$, then

$$\begin{aligned}\sigma(0, L, M) &= \{X \leq L | (0, L, M) \text{ is } X\text{-sub-exact at } L\} \\ &= \{X \leq L | \mathcal{N} \text{ is } X\text{-sublinearly independent to } M\}.\end{aligned}$$

We recall the properties of $\sigma(0, L, M)$ as follows:

Proposition 2.5 *[4] Let L, M be two R-modules and X_λ be a submodule of L, for every $\lambda \in \Lambda$. If $X_\lambda \in \sigma(0, L, M)$, for every $\lambda \in \lambda$, then $\cap_\Lambda X_\lambda \in \sigma(0, L, M)$.*

In the following Proposition, we will prove that $\sigma(0, \coprod_\Lambda N_\lambda, M)$ is closed under intersections, i.e. if $X_i \in \sigma(0, \coprod_\Lambda N_\lambda, M)$, for every $i \in I$, then $\cap_{i\in I} X_i \in \sigma(0, \coprod_\Lambda N_\lambda, M)$.

Proposition 2.6 *Let $\mathcal{N} = \{N_\lambda\}_\Lambda$ be a family of R-modules and let M be an R-module. If $\mathcal{N}$ is X_i-sub-linearly independent to M, for every $i \in I$, then $\mathcal{N}$ is $\cap_{i\in I} X_i$-sub-linearly independent to M. In other words, $\sigma(0, \coprod_\Lambda N_\lambda, M)$ is closed under intersections.*

Proof. Since $\mathcal{N}$ is X_i-sub-linearly independent to M, for every $i \in I$, then the triple $(0, \coprod_\Lambda N_\lambda, M)$ is X_i-sub-exact. Therefore, $X_i \in \sigma(0, \coprod_\Lambda N_\lambda, M)$, for every $i \in I$. By Proposition 2.5, we get $\cap_{i\in I} X_i \in \sigma(0, \coprod_\Lambda N_\lambda, M)$. So, $\mathcal{N}$ is $\cap_{i\in I} X_i$-sub-linearly independent to M. □

Furthermore, in the following proposition, we want to show that $\sigma(0, \coprod_\Lambda N_\lambda, M)$ is closed under submodules.

Proposition 2.7 *Let $\mathcal{N} = \{N_\lambda\}_\Lambda$ be a family of R-modules and M be an R-module. If $\mathcal{N}$ is X-sub-linearly independent to M, then $\mathcal{N}$ is X'-sub-linearly independent to M, for every submodule X' of X. In other words, $\sigma(0, \coprod_\Lambda N_\lambda, M)$ is closed under submodules.*

Proof. Since $\mathcal{N}$ is X-sub-linearly independent to M, then there is a monomorphism $f : X \to M$. Let X' be a submodule of X. Then, we can define the inclusion $i : X' \to X$. So, $f \circ i : X' \to M$ is a monomorphism. Hence, $\mathcal{N}$ is X'-sub-linearly independent to M. Therefore, if $X \in \sigma(0, \coprod_\Lambda N_\lambda, M)$, then $X' \in \sigma(0, \coprod_\Lambda N_\lambda, M)$, for every submodule X' of X. □

We already know that a basis for a free R-module F is a maximal linearly independent set in R-module F. So, we will investigate the maximal element of $\sigma(0, \coprod_\Lambda N_\lambda, M)$, i.e the maximal subset X of $\coprod_\Lambda N_\lambda$ such that $\mathcal{N}$ is X-sub-linearly independent to an R-module M. If there is a monomorphism $f : \coprod_\Lambda N_\lambda \to M$, then $\coprod_\Lambda N_\lambda$ is the maximal element in $\sigma(0, \coprod_\Lambda N_\lambda, M)$, i.e. for every $X \in \sigma(0, \coprod_\Lambda N_\lambda, M)$, if $\coprod_\Lambda N_\lambda \subseteq X$, then $\coprod_\Lambda N_\lambda = X$. But, $\coprod_\Lambda N_\lambda$ is not necessary belong to $\sigma(0, \coprod_\Lambda N_\lambda, M)$.

The most important criterion for the existence of maximal elements in a partially ordered set is Zorn's lemma. We recall Zorn's lemma as follows:

Proposition 2.8 *(**Zorn's Lemma**)[1] Let X be a partially ordered set and assume that every chain in X has an upper bound. Then X has a maximal element.*

By using Zorn's lemma, we want to show that there exist a submodule X in $\coprod_\Lambda N_\lambda$ maximal such that $\mathcal{N}$ is an X-sublinearly independent to M, for every family of R-modules $\mathcal{N}$ and R-module M.

Theorem 2.1 *Let $\mathcal{N} = \{N_\lambda\}_\Lambda$ be a family of R-modules and M be an R-module. Then there exist a submodule X in $\coprod_\Lambda N_\lambda$ maximal such that $\mathcal{N}$ is an X-sublinearly independent to M. In other words, $\sigma(0, \coprod_\Lambda N_\lambda, M)$ always has a maximal element.*

Proof. Let $\mathcal{X} = \{X | X \subseteq \coprod_\Lambda N_\lambda \text{ and } X \subseteq M\}$. The set $\mathcal{X}$ is not empty since $0 \in \mathcal{X}$. Let $\{X_i\}_{i \in I}$ be a chain (totally ordered set) in $\mathcal{X}$. Let $Y = \bigcup_{i \in I} X_i$, where $X_i \in \mathcal{X}$, for all $i \in I$. As a set, Y certainly contains all the X_i's. Since a union of submodules is not usually a submodule, we will show that Y is a submodule of $\coprod_\Lambda N_\lambda$.

If x and y are in Y, then $x \in X_i$ and $y \in X_j$, for two of the submodules X_i and X_j of $\coprod_\Lambda N_\lambda$. Since the set of submodules $\{X_i\}_i \in I$ is totally ordered,

$$X_i \subset X_j \text{ or } X_j \subset X_i.$$

Without loss of generality, $X_i \subset X_j$. Therefore x and y are in X_j, so $x + y \in X_j \subset Y$ and $rx \in X_j \subset Y$, for every $r \in R$. We can conclude that $Y = \cup_{i \in I} X_i$ is a submodule of $\coprod_\Lambda N_\lambda$. Similarly, we obtain Y is a submodule of M.

Since Y contains every X_i, for all $i \in I$, Y is an upper bound on the totally ordered set $\{X_i\}_{i \in I}$. By Zorn's lemma, $\mathcal{X}$ contains a maximal element. This maximal element is a submodule of $\coprod_\Lambda N_\lambda$ and M that is maximal for inclusion among all submodule of $\coprod_\Lambda N_\lambda$ and M. We can conclude that there exist a submodule X in $\coprod_\Lambda N_\lambda$ maximal such that $\mathcal{N}$ is an X-sublinearly independent to M or $\sigma(0, \coprod_\Lambda N_\lambda, M)$ always has a maximal element. □

3. Conclusion

The family of R-modules $\mathcal{N} = \{N_\lambda\}_\Lambda$ is an X-sub-linearly independent to M if the triple $(0, \coprod_\Lambda N_\lambda, M)$ is X-sub-exact (where X is a submodule of $\coprod_\Lambda N_\lambda$). If we take $X = \coprod_\Lambda N_\lambda$, then $\mathcal{N} = \{N_\lambda\}_\Lambda$ is linearly independent. Hence, sub-linearly independent module is a generalization of linearly independent module.

Then, we collect all submodules X of $\coprod_\Lambda N_\lambda$ such that $\mathcal{N}$ is X-sub-linearly independent to M, we denote it by $\sigma(0, \coprod_\Lambda N_\lambda, M)$. We have proved that $\sigma(0, \coprod_\Lambda N_\lambda, M)$ is closed under submodules and intersections. Furthermore, for every family of R-modules $\mathcal{N} = \{N_\lambda\}_\Lambda$ and R-module M, there exist X maximal such that $\mathcal{N}$ is an X-sublinearly independent. In other words, $\sigma(0, \coprod_\Lambda N_\lambda, M)$ always has a maximal element, for every family of R-modules $\mathcal{N}$ and R-module M.

Acknowledgment

We would like to express deep gratitude to Professor Alexander Zimmermann for his contribution to this research work. We would also like to thank the Indonesian Mathematical Society (IndoMS) which support the first author to join the Asian Mathematical Conference (AMC) 2016.

References

[1] Adkins W A and Weintraub S H 1992 *Algebra : An Approach via Module Theory* (New York : Springer-Verlag)
[2] Davvaz B and Parnian-Garamaleky Y A 1999 A note on exact sequences *Bull.Malays. Math. Sci. Soc.* **22**(1) p 53
[3] Davvaz B and Shabani-Solt H 2001 A generalization of homological algebra*J. Korean Math. Soc.* **39**(6) p 881
[4] Fitriani, Surodjo B and Wijayanti I E 2016 on X-sub-exact sequences, *Far East J. Math. Sci* **100**(7) p 1055
[5] Maclane S and Birkhoff G 1979 *Algebra* (New York : Mac Millan Publishing Co)

[6] Suprapto 2015 *Structure of Family of Linearly Independent Modules and Cohereditary Modules on a Category* $\tau[M]$ Dissertation Universitas Gadjah Mada
[7] Anvariyeh S M and Davvaz B 2005 on quasi-exact sequences *Bull. Korean Math. Soc.* **42**(1) p 149
[8] Anvariyeh S M and Davvaz B 2002 U-split-exact sequences *Far East J. Math. Sci.* **4**(2) p 209
[9] Wisbauer R 1991 *Foundation of Module and Ring Theory* (USA : Gordon and Breach)

An algebraic study of the matrix meta-population model

Sisilia Sylviani, Ema Carnia, and A.K. Supriatna

Department of Mathematics, Faculty of Mathematics and Natural Sciences
Universitas Padjadjaran

E-mail: sisilia.sylviani@unpad.ac.id

Abstract. In studying the population living in some patches, a single patch model usually generalizes into a meta-population model. Roughly speaking a meta-population is population consisting of some sub-populations, such as sub-population reflecting different life stages of the population or different habitat of patches. In a meta-population framework the dynamics of population within and between patches can be easily studied. This paper will discuss the matrix meta-population model that describes the growth and reproduction within the patches denoted by F. The matrix F is obtained by multiplying three matrices; the transpose of a permutation matrix, a block diagonal matrix in which its diagonal blocks is a matrix A_j, and the permutation matrix. The matrix A_j is the real non-negative matrix that represents the dynamics of population growth in the patch j. The permutation matrix has a special form that has properties that are more specific than the general permutation matrix. This paper will also gives a brief application for the matrix F. This will provide a space for exploring the characterization of the matrix F from algebraic point of view.

1. Introduction

A meta-population model described a population consisting of some sub-populations, such as sub-population reflecting different life stages of the population or different habitat of patches. One example of the population living in more than one spatial location is the salmon population. The dynamics of population within and between patches in a meta-population framework can be easily studied in algebraic point of view. This paper will discuss the matrix meta-population model that describes the growth and reproduction within the patches denoted by F. Before proceed into a further discussion, first we will give a brief explanation about block diagonal matrix, permutation matrix, and Kronecker Product.

Let A_j be a square matrix with nonnegative entries. The matrix $A_1 \oplus A_2 \oplus \cdots \oplus A_n$ defined as a block diagonal matrix with diagonal blocks $A_1, A_2, \cdots, A_n$ or can be written as follows [4]

$$A_1 \oplus A_2 \oplus \cdots \oplus A_n = \begin{pmatrix} A_1 & 0 & 0 & \cdots & 0 \\ 0 & A_2 & 0 & \cdots & 0 \\ \vdots & \vdots & \vdots & \ddots & \vdots \\ 0 & 0 & 0 & \cdots & A_n \end{pmatrix}. \tag{1}$$

An $n \times n$ matrix S with non-negative entries is column sub stochastic if all its columns sum are less than or equal to +1, or can be written as follows [4]

$$\begin{pmatrix} s_{11} & s_{12} & \cdots & s_{1n} \\ s_{21} & s_{22} & \cdots & s_{2n} \\ \vdots & \vdots & \ddots & \vdots \\ s_{n1} & s_{n2} & \cdots & s_{nn} \end{pmatrix}. \tag{2}$$

where $\sum_{i=1}^{n} |s_{ij}| \le 1, \forall j = 1, ..., n$

The matrix $\mathbf{S}$ describes the movement of a population between patches and can be written as follow

$$\mathbf{S} = S_1 \oplus S_2 \oplus \cdots \oplus S_n, \tag{3}$$

where $S_1, S_2, ..., S_n$ is substochastic matrices [4].

The matrix F is a matrix that represents the growth and reproduction within the patches and can be written as follows

$$F = P^t(A_1 \oplus A_2 \oplus \cdots \oplus A_n)P \tag{4}$$

where P is a permutation matrix. Horn & Johnson defines a permutation matrix as a matrix in which every row and column have exactly a single 1 and 0 everywhere else [3].

Let $A = (a_{ij})$ is a $m \times n$ matrix and $B = (b_{ij})$ is a $p \times q$ matrix, then Kronecker Product $A \otimes B$ is an $mp \times nq$ where [4]

$$A \otimes B = \begin{pmatrix} a_{11}B & a_{12}B & \cdots & a_{1n}B \\ a_{21}B & a_{22}B & \cdots & a_{2n}B \\ \vdots & \vdots & \ddots & \vdots \\ a_{n1}B & a_{n2}B & \cdots & a_{nn}B \end{pmatrix}. \tag{5}$$

Matrix E_{mn} is defined as a square matrix that its entries are zero except in m-th row and n-th column, or it can be written as $E_{mn} = (e_{ij})$ where [4]

$$e_{ij} = \begin{cases} 1, & \text{if } i = m, n = j \text{ ;} \\ 0, & \text{otherwise.} \end{cases}$$

Another terms that is used in this paper is the Column Sum Norm and the Spectral Radius. The Column Sum Norm of a square matrix M is defined as follow [4]

$$\|M\|_c = max\{\sum_{i=1}^{n} |a_{ij}| : 1 \le j \le n\}. \tag{6}$$

Let M be a square matrix, the spectral radius of M is the non-negative number $\rho(M)$ which is defined as follow [3]

$$\rho(M) = max_{1 \le i \le n} |\lambda_i|,$$

where λ_i is eigenvalues for M.

2. Main Result

In the previous section it has been mentioned that the matrix F is a matrix that represents the growth and reproduction within the patches and can be written as follows

$$F = P^t(A_1 \oplus A_2 \oplus \cdots \oplus A_n)P \tag{7}$$

where P is a permutation matrix. It also has been shown in [6] that the permutation matrix has a specific form as follows

$$P = \begin{pmatrix} E_{11} & E_{21} & \cdots & E_{(n-1)1} & E_{n1} \\ E_{12} & E_{22} & \cdots & E_{(n-1)2} & E_{n2} \\ \vdots & \vdots & \ddots & \vdots & \vdots \\ E_{1n} & E_{2n} & \cdots & E_{(n-1)n} & E_{nn} \end{pmatrix}. \tag{8}$$

If we expand the matrix F, then we have

$$\begin{aligned} F &= P^t(A_1 \oplus A_2 \oplus \cdots \oplus A_n)P \\ &= \begin{pmatrix} E_{11} & E_{21} & \cdots & E_{n1} \\ E_{12} & E_{22} & \cdots & E_{n2} \\ \vdots & \vdots & \ddots & \vdots \\ E_{1n} & E_{2n} & \cdots & E_{nn} \end{pmatrix}^t \begin{pmatrix} A_1 & & & \\ & A_2 & & \\ & & \ddots & \\ & & & A_n \end{pmatrix} \\ & \quad \begin{pmatrix} E_{11} & E_{21} & \cdots & E_{n1} \\ E_{12} & E_{22} & \cdots & E_{n2} \\ \vdots & \vdots & \ddots & \vdots \\ E_{1n} & E_{2n} & \cdots & E_{nn} \end{pmatrix} \\ &= \begin{pmatrix} x^1_{11} & \cdots & x^1_{1n} & 0 & 0 & \cdots & \cdots & \cdots & 0 \\ 0 & \cdots & 0 & x^2_{11} & \cdots & x^2_{1n} & 0 & \cdots & 0 \\ \vdots & \vdots & \vdots & \vdots & \vdots & \ddots & & & \vdots \\ 0 & \cdots & \cdots & \cdots & \cdots & \cdots & x^n_{11} & \cdots & x^n_{1n} \\ x^1_{21} & \cdots & x^1_{2n} & 0 & 0 & \cdots & \cdots & \cdots & 0 \\ 0 & \cdots & 0 & x^2_{21} & \cdots & x^2_{2n} & 0 & \cdots & 0 \\ \vdots & \vdots & \vdots & \vdots & \vdots & \ddots & \vdots & \vdots & \vdots \\ 0 & \cdots & \cdots & \cdots & \cdots & \cdots & x^n_{21} & \cdots & x^n_{2n} \\ \vdots & \vdots & \vdots & \vdots & \vdots & \vdots & \vdots & \vdots & \vdots \\ x^1_{n1} & \cdots & x^1_{nn} & 0 & 0 & \cdots & \cdots & \cdots & 0 \\ 0 & \cdots & 0 & x^2_{n1} & \cdots & x^2_{nn} & 0 & \cdots & 0 \\ \vdots & \vdots & \vdots & \vdots & \vdots & \ddots & \vdots & \vdots & \vdots \\ 0 & \cdots & \cdots & \cdots & \cdots & \cdots & x^n_{n1} & \cdots & x^n_{nn} \end{pmatrix} \\ & \quad \begin{pmatrix} E_{11} & E_{21} & \cdots & \cdots & \cdots & \cdots & \cdots & \cdots & E_{n1} \\ E_{12} & E_{22} & \cdots & \cdots & \cdots & \cdots & \cdots & \cdots & E_{n2} \\ \vdots & \vdots & \vdots & \vdots & \vdots & \vdots & \vdots & \cdots & \vdots \\ \vdots & \vdots & \vdots & \vdots & \vdots & \vdots & \vdots & \cdots & \vdots \\ \vdots & \vdots & \vdots & \vdots & \vdots & \vdots & \vdots & \cdots & \vdots \\ \vdots & \vdots & \vdots & \vdots & \vdots & \vdots & \vdots & \cdots & \vdots \\ E_{1n} & E_{2n} & \cdots & \cdots & \cdots & \cdots & \cdots & \cdots & E_{nn} \end{pmatrix} \end{aligned}$$

IOP Conf. Series: Journal of Physics: Conf. Series **893** (2017) 012009 doi:10.1088/1742-6596/893/1/012009

$$
= \begin{pmatrix}
x_{11}^1 & 0 & \cdots & \cdots & x_{1n}^1 & \cdots & \cdots & 0 \\
0 & x_{11}^2 & 0 & \cdots & \cdots & x_{1n}^2 & \cdots & 0 \\
\vdots & & \ddots & & & & \ddots & \vdots \\
0 & \cdots & \cdots & x_{11}^n & 0 & \cdots & 0 & x_{1n}^n \\
x_{21}^1 & 0 & \cdots & \cdots & x_{22}^1 & 0 & \cdots & 0 \\
\vdots & \ddots & & & & & \ddots & \vdots \\
0 & \cdots & x_{21}^n & \cdots & 0 & \cdots & \cdots & x_{2n}^n \\
\vdots & & & & & & & \vdots \\
x_{n1}^1 & 0 & \cdots & \cdots & x_{nn}^1 & 0 & \cdots & 0 \\
\vdots & \ddots & & & & & & \vdots \\
0 & \cdots & x_{n1}^n & 0 & \cdots & \cdots & 0 & x_{nn}^n
\end{pmatrix}
$$

$$
= \begin{pmatrix}
x_{11}^1 & 0 & \dots & x_{12}^1 & 0 & \dots & x_{1n}^1 & 0 & \dots & 0 \\
0 & 0 & \dots & 0 & 0 & \dots & 0 & 0 & \dots & 0 \\
0 & 0 & \dots & 0 & 0 & \dots & 0 & 0 & \dots & 0 \\
\vdots & \vdots & \vdots & \vdots & \vdots & \vdots & \vdots & \vdots & \vdots & \vdots \\
x_{21}^1 & 0 & \dots & x_{22}^1 & 0 & \dots & x_{2n}^1 & 0 & \dots & 0 \\
0 & 0 & \dots & 0 & 0 & \dots & 0 & 0 & \dots & 0 \\
0 & 0 & \dots & 0 & 0 & \dots & 0 & 0 & \dots & 0 \\
\vdots & \vdots & \vdots & \vdots & \vdots & \vdots & \vdots & \vdots & \vdots & \vdots \\
x_{n1}^1 & 0 & \dots & x_{n2}^1 & 0 & \dots & x_{nn}^1 & 0 & \dots & 0 \\
0 & 0 & \dots & 0 & 0 & \dots & 0 & 0 & \dots & 0 \\
0 & 0 & \dots & 0 & 0 & \dots & 0 & 0 & \dots & 0 \\
\vdots & \vdots & \vdots & \vdots & \vdots & \vdots & \vdots & \vdots & \vdots & \vdots \\
0 & 0 & \dots & 0 & 0 & \dots & 0 & 0 & \dots & 0
\end{pmatrix} + \cdots +
\begin{pmatrix}
0 & 0 & \dots & 0 & 0 & \dots & 0 & 0 & \dots & 0 \\
0 & 0 & \dots & 0 & 0 & \dots & 0 & 0 & \dots & 0 \\
\vdots & \vdots & \vdots & \vdots & \vdots & \vdots & \vdots & \vdots & \vdots & \vdots \\
x_{11}^n & 0 & \dots & x_{12}^n & 0 & \dots & x_{1n}^n & 0 & \dots & 0 \\
0 & 0 & \dots & 0 & 0 & \dots & 0 & 0 & \dots & 0 \\
0 & 0 & \dots & 0 & 0 & \dots & 0 & 0 & \dots & 0 \\
\vdots & \vdots & \vdots & \vdots & \vdots & \vdots & \vdots & \vdots & \vdots & \vdots \\
x_{21}^1 & 0 & \dots & x_{22}^1 & 0 & \dots & x_{2n}^n & 0 & \dots & 0 \\
0 & 0 & \dots & 0 & 0 & \dots & 0 & 0 & \dots & 0 \\
0 & 0 & \dots & 0 & 0 & \dots & 0 & 0 & \dots & 0 \\
\vdots & \vdots & \vdots & \vdots & \vdots & \vdots & \vdots & \vdots & \vdots & \vdots \\
x_{n1}^1 & 0 & \dots & x_{n2}^1 & 0 & \dots & x_{nn}^n & 0 & \dots & 0
\end{pmatrix}
$$

$$
= \begin{pmatrix}
x_{11}^1 & \dots & x_{1n}^1 \\
\vdots & \ddots & \vdots \\
x_{n1}^1 & \dots & x_{nn}^1
\end{pmatrix} \otimes \begin{pmatrix}
1 & \dots & 0 \\
\vdots & \ddots & \vdots \\
0 & \dots & 0
\end{pmatrix} +
\begin{pmatrix}
x_{11}^2 & \dots & x_{1n}^2 \\
\vdots & \ddots & \vdots \\
x_{n1}^2 & \dots & x_{nn}^2
\end{pmatrix} \otimes \begin{pmatrix}
0 & 0 & \dots & 0 \\
\vdots & 1 & \ddots & \vdots \\
0 & 0 & 0\dots & 0
\end{pmatrix} + \cdots +
\begin{pmatrix}
x_{11}^n & \dots & x_{1n}^n \\
\vdots & \ddots & \vdots \\
x_{n1}^n & \dots & x_{nn}^n
\end{pmatrix} \otimes \begin{pmatrix}
0 & 0 & \dots & 0 \\
\vdots & \dots & \ddots & \vdots \\
0 & 0 & 0\dots & 1
\end{pmatrix}
$$

$$
\begin{aligned}
&= A_1 \otimes E_{11} + A_1 \otimes E_{22} + A_1 \otimes E_{11} + \\
&\quad + \cdots + A_n \otimes E_{nn} \\
&= \sum_{j=1}^{n} A_j \otimes E_{jj}
\end{aligned}
$$

Hence the matrix F can also be written as a Kronecker Product between A_j and E_{jj}, or

$$F = \sum_{j=1}^{n} A_j \otimes E_{jj}. \tag{9}$$

Furthermore, from [5] we have the following theorem.

Theorem 1 [5] Let M be an $n \times n$ block matrix where

$$M = \begin{pmatrix} X & Y \\ Z & W \end{pmatrix}$$

then the transpose of M is

$$M^T = \begin{pmatrix} X^T & Z^T \\ Y^T & W^T \end{pmatrix}.$$

The theorem above can be generalized into $n \times n$ matrix.

Consider the permutation matrix described in equation (8) as follow

$$P = \begin{pmatrix} E_{11} & E_{21} & \cdots & E_{n1} \\ E_{12} & E_{22} & \cdots & E_{n2} \\ \vdots & \vdots & \ddots & \vdots \\ E_{1n} & E_{2n} & \cdots & E_{nn} \end{pmatrix}$$

Using a generalized form of theorem 1 we obtain

$$P^T = \begin{pmatrix} E_{11}^T & E_{12}^T & \cdots & E_{1n}^T \\ E_{21}^T & E_{22}^T & \cdots & E_{2n}^T \\ \vdots & \vdots & \ddots & \vdots \\ E_{n1}^T & E_{n2}^T & \cdots & E_{nn}^T \end{pmatrix}$$

Note that the matrix E_{mn} has a property, that is,

$$E_{mn}^T = \begin{cases} E_{mn}, & \text{if } m = n \ ; \\ E_{nm}, & \text{otherwise.} \end{cases} \tag{10}$$

since we have (10) then we obtain

$$\begin{aligned} P^T &= \begin{pmatrix} E_{11}^T & E_{12}^T & \cdots & E_{1n}^T \\ E_{21}^T & E_{22}^T & \cdots & E_{2n}^T \\ \vdots & \vdots & \ddots & \vdots \\ E_{n1}^T & E_{n2}^T & \cdots & E_{nn}^T \end{pmatrix} \\ &= \begin{pmatrix} E_{11} & E_{21} & \cdots & E_{n1} \\ E_{12} & E_{22} & \cdots & E_{n2} \\ \vdots & \vdots & \ddots & \vdots \\ E_{1n} & E_{2n} & \cdots & E_{nn} \end{pmatrix} \\ &= P. \end{aligned}$$

Generally, a permutation matrix is not always a symmetric matrix. However, from the result above, it can be concluded that, in this case, the permutation matrix P is a symmetric matrix.

In ecological models, the population may consist of individuals in different age classes (e.g. juveniles, sub-adults, adults) living in different spatial locations [1]. Similarly, epidemiological models often account for different classes of individuals (e.g. susceptible, exposed, infected, removed) as well as spatial structure [2]. The matrix F also can be used to study a population having some stages of life and occupying some different patches. The matrix, which is the product of two matrices $\mathbf{S}$ and F,or can be called an $\mathbf{S}F$ matrix multiplication model. The matrix $\mathbf{S}$ is a diagonal block matrix in which its block is a sub-stochastic column matrix. The matrix S represents the movement of a population between locations (patches). The multiplication of $\mathbf{S}$ and F is also a matrix model that represent a biological model that account for spatial structure.

The discussion about the $\mathbf{S}F$ matrix multiplication model in this paper, will focus on a certain spectral radius properties of the $\mathbf{S}F$ matrix multiplication model. The spectral radius of a matrix is used to determine the rate of growth of a population that described in a matrix model. It will be shown that the spectral radius of $\mathbf{S}F$ matrix model is less than or equal to the spectral radius of F. However, before we proceed to that, it will be explained first some theorem that will be used to proof that properties.

Theorem 2 [3] The largest eigenvalue of a substochastic matrix is 1 or in the other words is that the radius spectral of a substochastic matrix is equal to 1.

It also has been known that, for a square matrix, let say A, then we have property $\rho(A) \leq \|A\|$. The proposition bellow describes the spectral radius of a multiplication matrix.

Proposition 3 [3] Let D be a diagonal nonnegative matrix and S a column substochastic matrix. Then

$$\rho(SD) \leq \rho(D)$$

Proof.
Based on the lemma above we have

$$\rho(SD) \leq \|SD\| \leq \|S\|\|D\|.$$

Since the S is column sub stocasthic matrix, then all of the columns sum are less than or equal to +1. Furthermore, the spectral radius of D is equal to its column sum norm. Thus, we have

$$\rho(SD) \leq \|D\| = \rho(D).$$

It proves the lemma.

Proposition 3 has some important meanings, especially for biological models that account for spatial structure [4]. One of them is, from that fact one can know whether population growth occurs before dispersal or population growth occurs after dispersal.

However, the conditions can not always apply for the $\mathbf{S}F$ matrix model. Suppose $\mathbf{S} = S_1 \oplus ... \oplus S_n$ where $S_1, ..., S_n$ is substochastic matrices then for every block matrix F we can not always have condition of $\rho(\mathbf{S}F) \leq \rho(\mathbf{S}F)$.

For example let $A_1 = \begin{pmatrix} 0 & 2 \\ 0 & 0 \end{pmatrix}$, and $A_2 = \begin{pmatrix} 0 & 0 \\ 1 & 0 \end{pmatrix}$, we obtain

$$\begin{aligned} F &= A_1 \otimes E_{11} + A_2 \otimes E_{22} \\ &= \begin{pmatrix} 0 & 2 \\ 0 & 0 \end{pmatrix} \otimes \begin{pmatrix} 1 & 0 \\ 0 & 0 \end{pmatrix} + \begin{pmatrix} 0 & 0 \\ 1 & 0 \end{pmatrix} \otimes \begin{pmatrix} 0 & 0 \\ 0 & 1 \end{pmatrix} \\ &= \begin{pmatrix} 0 & 0 & 2 & 0 \\ 0 & 0 & 0 & 0 \\ 0 & 0 & 0 & 0 \\ 0 & 1 & 0 & 0 \end{pmatrix} \end{aligned}$$

and we also have $\rho(F) = 0$.

Meanwhile, Let $S_1 = \begin{pmatrix} 0 & 0 \\ 1 & 1 \end{pmatrix}$ and $S_2 = \begin{pmatrix} 1 & 1 \\ 0 & 0 \end{pmatrix}$, then

$$\begin{aligned}\mathbf{S} &= S_1 \oplus S_2 \\ &= \begin{pmatrix} 0 & 0 & 0 & 0 \\ 1 & 1 & 0 & 0 \\ 0 & 0 & 1 & 1 \\ 0 & 0 & 0 & 0 \end{pmatrix}.\end{aligned}$$

Multiplying $\mathbf{S}$ and F we have

$$\begin{aligned}\mathbf{S}F &= \begin{pmatrix} 0 & 0 & 0 & 0 \\ 1 & 1 & 0 & 0 \\ 0 & 0 & 1 & 1 \\ 0 & 0 & 0 & 0 \end{pmatrix}\begin{pmatrix} 0 & 0 & 2 & 0 \\ 0 & 0 & 0 & 0 \\ 0 & 0 & 0 & 0 \\ 0 & 1 & 0 & 0 \end{pmatrix} \\ &= \begin{pmatrix} 0 & 0 & 0 & 0 \\ 0 & 0 & 2 & 0 \\ 0 & 1 & 0 & 0 \\ 0 & 0 & 0 & 0 \end{pmatrix}\end{aligned}$$

and we have $\rho(\mathbf{S}F) = \sqrt{2} > 0$. In other words

$$\rho(\mathbf{S}F) > \rho(F).$$

Hence, in general we can't always have

$$\rho(\mathbf{S}F) \leq \rho(F).$$

There are some contrains in order the property can be applied to $\mathbf{S}F$ matrix model.

3. Conclusion

It has been shown that the matrix F can be written in the form of the sum of the product of Kronecker of matrix A_j and matrix E_{jj} and also it has been shown that the permutation in the model have is also a symmetric matrix.

The spectral radius from SD matrix model is less than or equal to the spectral radius of the matrix D, where D is a diagonal nonnegative matrix. That relation has some important meanings, one of them is that one can know whether population growth occurs before dispersal or after dispersal. However for $\mathbf{S}F$ matrix model, the condition does not always apply.

Acknowledgments

This research is funded by the Academic Leadership Grant (2015-2019) of Universitas Padjadjaran and Direktorat Riset dan Pengabdian pada Masyarakat (DRPM) of Universitas Padjadjaran.

References

[1] Caswell H 2001 *Matrix Population Models: Construction, Analysis and Interpretation*, Second Edition. Sunderland:Sinauer Associates, Inc.
[2] Diekmann O J A P, Heesterbeek 2000 *Mathematical Epidemiology of Infectious Disease*, Wiley Series in
[3] Horn R A, Johnson C R 1985 *Matrix Analysis*, Cambridge: University Press Cambridge
[4] Li, Chi-Kwong, Schreiber S J 2006 *On Dispersal and Population Growth for Multistate Matrix Models*, Linear Algebra and Its Applications, **418** pp 900–912
[5] Strang G 2006 *Linear Algebra and Its Applications*, Thomson Brooks/cole
[6] Sylviani S, Carnia E, Supriatna A K 2016 *An Algebraic Problem Arising in Biomathematics*, Jurnal Teknologi pp 75–79

IOP Conf. Series: Journal of Physics: Conf. Series **893** (2017) 012010 doi:10.1088/1742-6596/893/1/012010

On representation of a ring on a free module over a commutative ring with identity

N Hijriati[1,2], S Wahyuni[3], and I E Wijayanti[3]

[1] Doctoral Student of Mathematics Universitas Gadjah Mada, Yogyakarta Indonesia
[2] Dept. of Mathematics, Univ. Lambung Mangkurat Banjarmasin, Indonesia
[3] Dept. of Mathematics Universitas Gadjah Mada, Yogyakarta, Indonesia

E-mail: naimah.hijriati@mail.ugm.ac.id, {swahyuni, ind.wijayanti}@ugm.ac.id

Abstract. Let R be a commutative ring with identity and M be a free R-module then we always have a representation of R, that is homomorphism ring $\mu\colon R \to End_R(M)$, with $\mu(r) := \mu_r\colon M \to M$ and $\mu_r(m) = rm$ for all $r \in R$ and for all $m \in M$. In this paper, we will present some properties of representations of ring R on R-module, based on some notions in representation of R on vector space, such as admissible submodule, equivalence of two representations, decomposable representation and completely reducible representation. It will be shown that if M, N are two free R-modules then two representations $\mu\colon R \to End_R(M)$ and $\varphi\colon R \to End_R(N)$ are equivalent if and only if there is a module isomorphism $T\colon M \to N$. If R is a principle ideal domain(PID), then it will be shown that every submodule of M is an admissible submodule of M, every representation of ring R on a free R-module is decomposable, and a representation of R on M is completely reducible if and only if M is semisimple.

1. Introduction

Let R be a commutative ring with identity and V be a vector space over F with finite dimension. Representation of R is a ring homomorphism from R to a ring of endomorphism on V ($End_F(V)$). Furthermore, V is called as a representation space. There are some notion on representation of ring such as equivalence of two representations, intertwining mapping, admissible submodule, top constituent, bottom constituent, decomposable representation and completely reducible representation.

Let $\rho\colon R \to End_F(V)$ be a representation of ring R, with $\rho(r) := \rho_r\colon V \to V$ for all $r \in R$. Subspace U is said to be an admissible submodule of V, if $\rho_r(U) \subseteq U$ for all $r \in R$. Two representations $\rho\colon R \to End_F(V)$ and $\varphi\colon R \to End(W)$ are equivalent if there is exist an isomorphism $T\colon V \to W$, such that $T\rho_r = \varphi_r T$ for any $r \in R$. Every Linear transformation $T\colon V \to W$ such that $T\rho_r = \varphi_r T$ for any $r \in R$ is called an intertwining mapping. Furthermore representation ρ of R is called a completely reducible if V is a finite direct sum of admissible submodules of V and they are irreducible [2].

Let R be a commutative ring with identity and M be an R-module, then we can construct a module homomorphism $\mu_r\colon M \to M$, denoted by $\mu_r(m) = rm$ for all $r \in R$ and for all $m \in M$. Since for any $r_1, r_2 \in R$, we have $\mu_{r_1+r_2}(m) = (r_1 r_2)m = r_1 m + r_2 m = \mu_{r_1}(m) + \mu_{r_2}(m) = (\mu_{r_1} + \mu r_2)(m)$ and $\mu_{r_1 r_2}(m) = (r_1 r_2)m = r_1(r_2 m) = \mu_{r_1}(\mu_{r_2}(m)) = (\mu_{r_1}\mu_{r_2})(m)$ for any $m \in M$, then $\mu\colon R \to End_R(M)$ defined by $\mu(r) := \mu_r$ for all $r \in R$ is a ring homomorphism. Hence from definition of ring representation on vector space, we have that $\mu\colon R \to End_R(M)$ is generalization of ring representation on vector space. Since every vector space and every subspace of vector space have a base, then R-module M is assumed to be a free module. Furthermore, M is called as a representation module, and degree of μ equal to the rank of M ($rank(M) < \infty$).

In this paper, we will present some properties of representations of commutative ring R on free R-module M with $rank(M) < \infty$, based on the nations above. The properties of representation of ring on free module over a commutative ring with identity in this paper are different from properties that stated by Kosan, Lee&Zhou [5]. In this paper, we present M as free R-module with R is a commutative ring. But in [5], Kosan, Lee & Zhou present properties R-module M that connected to finite group G. It is called module MG over group ring RG.

2. Main Result

Let M and N be two free R-modules. If $T\colon M \to N$ is an R-module homomorphism, then for any $m \in M$ and for any $r \in R$, $T(rm) = rT(m)$ [1]. Hence by definition of the equivalent of two representations, we have this following

Proposition 2.1 *Let M, N be two free R-modules, and let $\mu\colon R \to End_R(M)$ and $\varphi\colon R \to End_R(N)$ be two representations of R defined by $\mu_r(m) = rm$ and $\varphi_r(n) = rn$ for any $r \in R$, for any $n \in N$ and for any $m \in M$. μ and φ are equivalent if and only if there is an R-module isomorphism $T\colon M \to N$.*

Proof. If μ and φ are equivalent, then there is an R-module isomorphism $T\colon M \to N$ such that $T\mu_r = \varphi_r T$ for any $r \in R$. Conversely If there is an R-module isomorphism T, then $(T\mu_r)(m) = T(\mu_r(m)) = T(rm) = rT(m) = \varphi_r(T(m)) = (\varphi_r T)(m)$ for any $r \in R$, for any $m \in M$. In another word μ and φ are equivalent. ■

From definition of intertwining mapping and from Proposition 2.1, we have that intertwining mapping from μ to φ is all a module homomorphisms from M to N.

Example 2.2 *Let $M = \{\begin{pmatrix} a & b \\ c & d \end{pmatrix} \mid a, b, c, d \in Z\}$ and $N = \mathbb{Z}^4$ be a free $\mathbb{Z}$-modules. Then $\mu\colon \mathbb{Z} \to End_R(M)$ and $\varphi\colon \mathbb{Z} \to End_R(N)$ are representations of $\mathbb{Z}$. Because there is an isomorphism $T\colon M \to N$ defined by $T(\begin{pmatrix} a & b \\ c & d \end{pmatrix}) = (a, b, c, d)$ and $rank(M) = rank(N) = 4$, then by Proposition 2.1 μ and φ are equivalent.*

Properties of equivalence in representations of ring R on free R-modules is still work on commutative ring with identity. Different situation when we speak about admissibility, decomposability, and complete reducibility. The general situation will not always work because the submodule of representation module is not always free. This situation will work when we assume R is a principle ideal domain(PID). Relating with the notion of admissible submodules, we have this following proposition :

Proposition 2.3 *Let R be a PID, M be a free R-module, and $\mu\colon R \to End_R(M)$ be a representation of R defined by $\mu_r(m) = rm$ for any $r \in R$ and for any $m \in M$. Then every submodule of M is an admissible submodule.*

Proof. Suppose that P is any submodule of M. Then $rp \in P$ for any $r \in R$ and for any $p \in P$. Hence by definition of μ, we have for any $r \in R$ and for any $p \in P$, $\mu_r(p) = rp \in P$. In another word P is an admissible submodule. ■

Example 2.4 *Let $M = \mathbb{Z}^2$ be a free $\mathbb{Z}$-module and $\mu\colon \mathbb{Z} \to End_{\mathbb{Z}}(M)$ be a representation of $\mathbb{Z}$ with degree two defined by $\mu_r(a, b) = r(a, b) = (ra, rb)$ for all $r \in \mathbb{Z}$ and $(a, b) \in M$. Let $P = \{(a, a) | a \in \mathbb{Z}^2\} \subseteq M$. Because of $r_1(a, a) + r_2(b, b) = (r_1a + r_2b, r_1a + r_2b) \in P$ for any $(a, a), (b, b) \in P$ and $r_1, r_2 \in \mathbb{Z}$, then P is a submodule of M. Hence by Proposition 2.3 P is an admissible submodule.*

Example 2.5 *Let* $M = \{\begin{pmatrix} a & b \\ c & d \end{pmatrix} \mid a,b,c,d \in \mathbb{Z}\}$ *be a* $\mathbb{Z}$*-module.* M *be a free module over* $\mathbb{Z}$ *with* $\mu(M) = 4$*. Then* $\mu\colon \mathbb{Z} \to End_{\mathbb{Z}}(M)$ *is a representation of* $\mathbb{Z}$ *defined by* $\mu_r(\begin{pmatrix} a & b \\ c & d \end{pmatrix}) = r\begin{pmatrix} a & b \\ c & d \end{pmatrix} = \begin{pmatrix} ra & rb \\ rc & rd \end{pmatrix}$*, for all* $r \in \mathbb{Z}$ *and for all* $\begin{pmatrix} a & b \\ c & d \end{pmatrix} \in M$*. Let* $N = \{\begin{pmatrix} a & 0 \\ 0 & b \end{pmatrix} \mid a,b \in \mathbb{Z}\} \subseteq M$*.*
Because of

$$r_1\begin{pmatrix} a & 0 \\ 0 & b \end{pmatrix} + r_2\begin{pmatrix} c & 0 \\ 0 & d \end{pmatrix} = \begin{pmatrix} r_1a + r_2c & 0 \\ 0 & r_1b + r_2d \end{pmatrix} \in N$$

for any $\begin{pmatrix} a & 0 \\ 0 & b \end{pmatrix}, \begin{pmatrix} c & 0 \\ 0 & d \end{pmatrix} \in M$ *and* $r_1, r_2 \in R$*, then* N *is a submodule of* M*. Hence by Proposition 2.3,* N *is an admissible submodule.*

From Proposition 2.3, we have every submodule of an R-module M is an admissible submodule. Of course M always has an admissible submodule, i.e $\{0\}$ and M. If $\{0\}$ and M are the only admissible submodule of M, then M is called irreducible and the ring representation on M is called an irreducible representation. Let P be an admissible submodule of M, then we can construct another representation of R, by restricting representation on a submodule of M, that is a ring homomorphism $\mu'\colon R \to End_R(P)$ defined by $\mu'(r) = \mu(r)$ for all $r \in R$ and it is said to be a top constituent. Another representation of R that we can construct by using an admissible submodule P is a ring homomorphism $\bar{\mu}\colon R \to End_R(M/P)$ defined by $\bar{\mu}_r(m+P) = \mu_r(m)+P$ for all $r \in R$ and for all $m + P \in M/P$, and it is called as a bottom constituent. Furthermore, by using Properties 2.3 we will be shown the decomposability and complete reducibility of ring representation on free modules.

Proposition 2.6 *Every representation of* R *(*R *is PID) on a free* R*-module is decomposable.*

Proof. Let M be a free R-module and let $\mu\colon R \to End_R(M)$ be any representation of R with a finite degree. Let P be any admissible submodule of M, then P is a submodule. Since R is a PID, then P is complemented (Proposition 3.8.2 in [1]), so there exists a submodule $Q \subseteq M$, such that $M = P \oplus Q$. From Proposition 2.3 Q is an admissible submodule. Hence μ is decomposable. ∎

Let R be a ring with identity and M be an R-module. A module M is said to be simple, if $\{0\}$ and M the only submodule. If M is a finite direct sum of simple submodules of M then M is said to be semisimple [1]. From this, we have this following:

Proposition 2.7 *Every representation of* R *on free* R*-module* M *is completely reducible if and only if* M *is a semisimple* R*-module.*

Proof. Let $\mu\colon R \to End_R(M)$ be any representation of R with finite degree. By the notion of completely reducible representations, we then have μ is completely reducible if and only if $M = M_1 \oplus M_2 \oplus ... \oplus M_k$, with M_i is an irreducible admissible submodule for $1 \leq i \leq k$. Then by the notion of admissible submodules and Proposition 2.3, we have that M_i is a simple submodule of M for $1 \leq i \leq k$. Hence M is a semisimple R-module. Conversely, if M is a semisimple R-module then $M = M_1 \oplus M_2 \oplus ... \oplus M_k$, with M_i is a simple submodule for $1 \leq i \leq k$. From Proposition 2.3, we have that M_i is an admissible submodule of M. Since M_i is a simple submodule then M_i is an irreducible admissible submodule for $1 \leq i \leq k$. Hence μ is completely reducible. ∎

IOP Conf. Series: Journal of Physics: Conf. Series **893** (2017) 012010 doi:10.1088/1742-6596/893/1/012010

Example 2.8 *In Example 2.5, $\mu\colon \mathbb{Z} \to End_R(M)$ is a representation of $\mathbb{Z}$. Because* $M = \mathbb{Z}\begin{pmatrix}1&0\\0&0\end{pmatrix} \oplus \mathbb{Z}\begin{pmatrix}0&1\\0&0\end{pmatrix} \oplus \mathbb{Z}\begin{pmatrix}0&0\\1&0\end{pmatrix} \oplus \mathbb{Z}\begin{pmatrix}0&0\\0&1\end{pmatrix}$ *and* $\langle\begin{pmatrix}1&0\\0&0\end{pmatrix}\rangle, \langle\begin{pmatrix}0&1\\0&0\end{pmatrix}\rangle, \langle\begin{pmatrix}0&0\\1&0\end{pmatrix}\rangle, \langle\begin{pmatrix}0&0\\0&1\end{pmatrix}\rangle$ *are simple submodules of M over $\mathbb{Z}$, then M is a semisimple submodule over $\mathbb{Z}$. Hence by Proposition 2.7 μ is completely reducible.*

Let R be a commutative ring with identity, then R is an additive group and R is a left and a right free R-module ([1], [3]). Then we can construct a representation of R on representation module R, i.e ring homomorphism $\theta\colon R \to End_R(R)$ such that for any $x \in R$, $\theta_r(x) = rx$. This representation is called as a regular representation. Furthermore every submodule of R is an ideal of R and it is a finitely generated R-modul. If R is a PID, then ideal of R is a free R-module [1]. Hence every submodule of R (R is a PID) is a free R-module. From this we will be shown the connection between a regular representation of R (R is a PID) and an irreducible representation of R on a free R-module by the following proposition:

Proposition 2.9 *Let R be a PID. Every irreducible representation of R on a free R-module is equivalent to a bottom constituent of regular representation of R.*

Proof. Let $\mu\colon R \to End_R(M)$ defined by $\mu_r(m) = rm$ be any irreducible representation of R. Since μ is not the zero representation then $RM \neq \{0\}$ and there is $m \in M$ such that $Rm \neq \{0\}$. We know that Rm is a submodule of M and μ is irreducible then $Rm = M$. Let $\phi\colon R \to M$ be a mapping defined by $\phi(r) = rm$ for any $r \in R$ and for any $m \in M$. Because of

$$\phi(r + r') = (r + r')m = rm + r'm = \phi(r) + \phi(r')$$

and

$$\phi(rr') = (rr')m = r(r'm) = r\phi(r')$$

then ϕ is an R-module epimorphism. Let K be a kernel of ϕ. We know that K is an R-submodule of R. Since R is a PID then K is an admissible submodule of R and we can construct bottom constituent from regular representation of R, i.e representation of R on a free R-module R/K. From the first isomorphism theorem of module theory in [1] and [3], we have that $R/K \cong M$. Therefore the representation of R on R-module R/K is equivalent to μ. Because the representation of R on an R-module R/K is a bottom constituent of regular representation of R then the lemma is proved. ∎

Example 2.10 *From Example 2.8, we have* $M_1 = \langle\begin{pmatrix}1&0\\0&0\end{pmatrix}\rangle$ *is a simple submodule of M, so by Proposition 2.3, we have M_1 is admissible submodule. Because M_1 is a simple submodule, then submodules of M_1 are $\{0\}$ and M_1 or in another word $\mu^1\colon \mathbb{Z} \to End_R(M_1)$ is irreducible. Since $\mathbb{Z}$ is a free $\mathbb{Z}$-module then $\psi\colon \mathbb{Z} \to End(\mathbb{Z})$ is a regular representation of $\mathbb{Z}$ and $\psi^1\colon \mathbb{Z} \to End(\mathbb{Z}/\{0\})$ is a bottom constituent of ψ. We defined $T\colon \mathbb{Z} \to M_1$ by $T(x) = xm$ for any $x \in \mathbb{Z}$ and for any $m \in M$. Since T is a isomorphism and $Ker(T) = \{0\}$ then μ^1 and ψ^1 are equivalent.*

3. Conclusion

Let R be a commutative ring with identity and let two free R-modules M and N be representation modules of R. From the main result we have to conclude that : (1) Every two representations of R on free R-modules M and N are equivalent if and only if M and N are isomorphic; (2) if R is a principle ideal domain then every submodule of M is an admissible submodule; (3) Every representation of a principle ideal domain R on free R-module M is decomposable; (4) A representation of a principle ideal domain R on free R-module M is completely reducible if and only if M is a semisimple R-module; (5) Every irreducible representation of a principle ideal domain R on a free R-module is equivalent to a bottom constituent of regular representation of R.

IOP Conf. Series: Journal of Physics: Conf. Series **893** (2017) 012010 doi:10.1088/1742-6596/893/1/012010

Acknowledgment
The authors thanks to Korean Mathematical Society (KMS) and the Faculty of Mathematics and Natural Sciences Universitas Lambung Mangkurat which support the first author to join the AMC 2016.

References

[1] Adkins W A and Weintraub S H 1992 *Algebra : an Approach via Module theory* (New York : Springer-Verlag)
[2] Borrow M 1967 *Representation Theory of Finite Groups* (New York : Academic Press)
[3] Dummit DS and Foote R 1999, *Abstract Algebra* (New York : John Wiley & Sons Inc)
[4] Jacobsons N 1989 *Basic Algebra II* (New York : W.H. Freemen and Company)
[5] Kosan M T, Lee T K and Zhou Y 2014 On modules over group rings *Algebra Represent Theor* **17** p 87

IOP Conf. Series: Journal of Physics: Conf. Series **893** (2017) 012011 doi:10.1088/1742-6596/893/1/012011

On representation of polynomial ring on a vector space via a linear transformation

S Wahyuni[1], **I E Wijayanti**[1], **and N Hijriati**[2]

[1] Dept. of Mathematics, Universitas Gadjah Mada, Yogyakarta, Indonesia
[2] Doctoral Student of Mathematics, Universitas Gadjah Mada, Yogyakarta Indonesia
[3] Dept. of Mathematics, Univ. Lambung Mangkurat, Banjarmasin, Indonesia

E-mail: {swahyuni, ind.wijayanti}@ugm.ac.id, naimah.hijriati@mail.ugm.ac.id

Abstract. A representation of a ring R is a ring homomorphism from R to the ring of all linear transformations from V to V ($End_F(V)$). From the field F, we can form a polynomial ring $F[X]$. A representation of $F[X]$ is a ring homomorphism $\varphi\colon F[X] \to End_F(V)$ via linear transformation $T\colon V \to V$ with $\varphi(f(X)) = f(T)$ for all $f(X) \in F[X]$. For general ring representation $\rho\colon R \to End_F(V)$, we have notions of admissibility submodule, and completely reducible and simple admissible submodule. In this paper, we will show that admissible submodules of V are invariant under T, and a representation of $F[X]$ is completely reducible.

1. Introduction

A representation of a ring R with an identity on a vector space V over a field F is a ring homomorphism $\rho\colon R \to End_F(V)$. Let $F[X] = \{a_0 + a_1X + ... + a_nX^n | a_i \in F, i = 1, 2, ..., n\}$ be the set of polynomials with indeterminate X over a field F. It has been known that $F[X]$ is a ring with addition and multiplication operations defined by $\sum_{i=0}^{n} a_iX^i + \sum_{i=0}^{n} b_iX^i = \sum_{i=0}^{n}(a_i+b_i)X^i$ and $(\sum_{i=0}^{n} a_iX^i)(\sum_{i=0}^{n} b_iX^i) = \sum_{i=0}^{n}(\sum_{j=0}^{i}(a_jb_{i-j}))X^i$, for all $\sum_{i=0}^{n} a_iX^i, \sum_{i=0}^{n} b_iX^i \in F[X]$. The ring $F[X]$ is called a polynomial ring over a field F.

If $T \in End_F(V)$, we can define a function $\varphi\colon F[X] \to End_F(V)$ by $\varphi(f(X)) := \varphi(a_0 + a_1X + ... + a_nX^n) = a_0I + a_1T + ... + a_nT^n = f(T)$, and it has well known that the function φ is a ring homomorphism. Hence by the definition of a ring representation in [2], φ is a representation of $F[X]$ on vector space V (via linear transformation $T\colon V \to V$).

There are some important notions related to general representation of ring such as admissible submodule, equivalence of two representations, intertwining mapping, and completely reducible representation. Let $\rho\colon R \to End_F(V)$ be a general representation of ring R, and U be a subspace of V. Subspace U is said to be an admissible submodule of V, if $\rho_r(U) \subseteq U$ for all $r \in R$. Here ρ_r is a notion of $\rho(r) \in End_F(V)$. A Representation $\rho\colon R \to End_F(V)$ and $\rho'\colon R \to End_F(W)$ are called equivalent if there exist isomorphism $S\colon V \to W$, such that $S\rho_r = \rho'_rS$ for any $r \in R$, and S is called an intertwining mapping (S not necessarily an isomorphism). Furthermore representation ρ of R is called an completely reducible if V is a finite direct sum of admissible submodules of V and they are irreducible [2].

In this paper we will show the properties representation of the polynomial ring on a vector space via a linear transformation, associated with the notions.

IOP Conf. Series: Journal of Physics: Conf. Series **893** (2017) 012011 doi:10.1088/1742-6596/893/1/012011

2. Main Result

Let $T \in End_F(V)$. Any subspace U of V such that $T(U) \subseteq U$, is called T-invariant ([3]). Relating with the notion of invariant submodule we then have :

Proposition 2.1 *Let $\varphi\colon F[X] \to End_F(V)$ be a representation of $F[X]$ via a linear transformation T. A subspace U of V, is an admissible submodule of V if U is T-invariant.*

Proof : Let $f(X)$ be an any element of $F[X]$. For any $u \in U$, we have

$$\begin{aligned}\varphi_{f(X)}(u) &= (a_0 1_d + a_1 T + \dots + a_n T^n)(u)\\ &= a_0 u + a_1 T(u) + \dots + a_n T^n(u)\end{aligned}$$

by definition admissible submodule, a subspace U is an admissible submodule, if $\varphi_{f(X)}(u) \in U$ for all $u \in U$. If U is a T-invariant subspace then $T^i(u) \in U$ for any $u \in U$. Hence $\varphi_{f(X)}(u) \in U$ for any $u \in U$. □

From Proposition 2.1, if V has a T-invariant subspace, then we can construct another representation of $F[X]$, by restricting representation on T-invariant, and it is said to be top constituent (Definition 4.5 [2]).

Example 2.2 *Let $\varphi\colon F[X] \to End_F(V)$ representation of $F[X]$ via linear transformation T. Because $Ker(T)$ and $Im(T)$ is T-invariant ([4]) then $\varphi_1\colon F[X] \to End(Ker(T))$ and $\varphi_2\colon F[X] \to End(Im(T))$ are top constituents of $F[X]$*

Example 2.3 *Let $\varphi\colon F[X] \to End_{\mathbb{F}}(V)$ representation of $F[X]$ via linear transformation $T\colon V \to V$. If T is an identity mapping or $T = 0$, then any subspace of V is an admissible submodule of V.*

Example 2.4 *Let $V = \mathbb{C}^2$ be a vector space over field of complex number $\mathbb{C}$ with standard base B. The representation ring $\varphi\colon \mathbb{C}[X] \to End_{\mathbb{C}}(V)$ of $\mathbb{C}[X]$ via a linear transformation T with $T(a,b) = (b,a)$. Subspaces $U = \{(a,a) \in V | a \in \mathbb{C}\}$ and $W = \{(-a,a) \in V | a \in \mathbb{C}\}$ are admissible submodules of V.*

Let $T\colon V \to V$ and $S\colon V \to W$ be two linear transformations, then $S \circ T\colon V \to W$ is a linear transformation. If S is invertible, then we can construct another linear transformation $T'\colon W \to W$, by $T' = S \circ T \circ S^{-1}$, such that $T'S = ST$. Using these properties, we will show the equivalence of two representations of polynomial $F[X]$.

Proposition 2.5 *Let $\varphi\colon F[X] \to End_F(V)$ and $\varphi'\colon F[X] \to End_F(W)$ be two representations of $F[X]$ with the same degree, where $\varphi_{f(X)} = f(T)$ and $\varphi'_{f(X)} = f(T')$ for some $T \in End_F(V)$ and some $T' \in End_F(W)$. Representations φ and φ' are equivalent if there exists isomorphism $S\colon V \to W$ such that $ST = T'S$.*

Proof. Assume there exists a isomorphism $S\colon V \to W$ such that $ST = T'S$, then we have

$$\begin{aligned}S\varphi_{f(X)} = Sf(T) &= S(a_0 1_d + a_1 T + \dots + a_n T^n)\\ &= a_0 S 1_d + a_1 ST + \dots + a_n ST^n\\ &= a_0 1_d S + a_1 T'S + \dots + a_n (T')^n S\\ &= (a_0 1_d + a_1 T' + \dots + a_n (T')^n)S\\ &= f(T')S = \varphi'_{f(X)} S\end{aligned}$$

It is proved that φ and φ' are equivalent. □

Example 2.6 *Let* $V = \mathbb{C}^2$ *and* $W = \{\begin{pmatrix} a & 0 \\ 0 & b \end{pmatrix} \mid a, b \in \mathbb{C}\}$ *be vector space over* $\mathbb{C}$. *Suppose that* $\varphi\colon \mathbb{C}[X] \to End_{\mathbb{C}}(V)$ *is a representation of* $\mathbb{C}[X]$ *via a linear transformation* $T\colon V \to V$ *denoted by* $T(a,b) = (b,a)$ *for all* $(a,b) \in \mathbb{C}^2$ *and* $\varphi'\colon \mathbb{C}[X] \to End_{\mathbb{C}}(W)$ *is a representation of* $\mathbb{C}[X]$ *via a linear transformation* $T'\colon W \to W$ *defined by* $T(\begin{pmatrix} a & 0 \\ 0 & b \end{pmatrix}) = \begin{pmatrix} b & 0 \\ 0 & a \end{pmatrix}$. *Because there is a linear transformation* $S\colon V \to W$ *namely by* $S(a,b) = \begin{pmatrix} a & 0 \\ 0 & b \end{pmatrix}$ *for all*$(a,b) \in \mathbb{C}^2$. *It can be shown* $ST = T'S$ *and* S *is an isomorphism then* φ *and* φ' *are equivalent.*

If S in Proposition 2.5, is not invertible and the dimension of V and W are not the same then S is said to be an intertwining mapping from ρ to ρ'. The collection of all intertwining mapping from ρ to ρ' is denoted by $Hom(\rho,\rho')$ with $Hom(\rho,\rho') = \{S \in Hom(V,W) | ST = T'S\}$.

A Representation space of a representation of R is a right module over R (Lemma 3.1 in [2]). If $R = F[X]$, then vector space V over field F is a $F[X]$-module (left module over $F[X]$) (Example 3.5.12 in [1]). Hence we have this following :

Remarks 2.7 *Let* $\varphi\colon F[X] \to End_F(V)$ *and* $\varphi'\colon F[X] \to End_F(W)$ *be two representations of* $F[X]$ *with the same degree, where* $\varphi_{f(X)} = f(T)$ *and* $\varphi'_{f(X)} = f(T')$ *for some* $T \in End_F(V)$ *and for some* $T' \in End_F(W)$, *then we have*

(i) Every admissible submodule of V *as a vector space over* F *is an admissible submodule over* $F[X]$.
(ii) Every $S \in Hom(\varphi,\varphi')$ *is a module homomorphism over* $F[X]$.
(iii) $Hom(\varphi,\varphi')$ *is a subspace* $Hom(V,W)$.

By Proposition 2.1 and Remarks 2.7, if U is a submodule over $F[X]$ under scalar multiplication, then U is an admissible submodule.

If a vector space V over a field F is a right $F[X]$-module, then there is a unique representation of $F[X]$ (Lemma 3.1 in [2]). Using properties of V as a right $F[X]$-module, we will show that every representation of $F[X]$ is decomposable and completely reducible.

A representation of $F[X]$ is said to be completely reducible if representation space is a semisimple module over $F[X]$. Before we give the prove of these properties. First we will prove that every representation of $F[X]$ are decomposable.

Proposition 2.8 *Every representation of* $F[X]$ *via linear transformation with finite degree is decomposable*

Proof. Let $\varphi\colon F[X] \to End_F(V)$ be any representation of $F[X]$ via linear transformation T. Because V is a right module over $F[X]$ then by Proposition 4.4.3[1], V is a finitely generated torsion over $F[X]$. So by Theorem 3.7.1 in [1], V can be written as direct sum of $k(= \mu(V))$ cyclic submodule over $F[X]$.

$$V = v_1F[X] \oplus v_2F[X] \oplus ... \oplus v_kF[X]$$

such that $Ann(v_i) = \langle f_i(X)\rangle$ for $1 \le i \le k$ and

$$\langle f_1(X)\rangle \supseteq \langle f_2(X)\rangle \supseteq ... \supseteq \langle f_k(X)\rangle$$

Furthermore by Remarks 2.7, $\langle v_i\rangle$ for $i \le i \le k$ is an admissible submodule of V. So φ decomposable. □

IOP Conf. Series: Journal of Physics: Conf. Series **893** (2017) 012011 doi:10.1088/1742-6596/893/1/012011

The monic polynomial $f_1(X), f_2(X), ..., f_k(X)$ in Equation (3) are called the invariant factors of linear transformation T. The invariant factor $f_k(X)$ of T is called the minimal polynomial $m_T(X)$ of T, and the characteristic polynomial $c_T(X)$ of T is the product of all the invariant factors of T.

Lemma 2.9 *Let $\varphi\colon F[X] \to End_F(V)$ be a representation of $F[X]$ via linear transformation. Then $Ker(\varphi) = \langle m_T(X)\rangle$*

Proof. We know that V is a right module over $F[X]$. Then by Lemma 4.8 in [1] $Ann(V) = \langle m_T(X)\rangle$. For any $f(X) \in Ker(\varphi)$, $\varphi(f(X))(v) = f(T)(v) = 0$ if and only if $v \cdot f(X) = 0$, then $f(X) \in Ann(V)$ and $Ann(V) = \langle m_T(X)\rangle$. So we have $Ker(\varphi) \subseteq \langle m_T(X)\rangle$. Conversely for any $f(X) \in \langle m_T(V)\rangle = Ann(V)$, then $v \cdot f(X) = f(T)(v) = \varphi_{f(X)}(v) = 0$, so we have $f(X) \in Ker(\varphi)$ in another word $\langle m_T(V)\rangle \subseteq Ker(\varphi)$. Hance $Ker(\varphi) = \langle m_T(X)\rangle$. □

Theorem 2.10 *Let $\varphi\colon F[X] \to End_F(V)$ be a representation of $F[X]$ with n degree via linear transformation $T \in End_F(V)$. Representation φ is a completely reducible if and only if the minimal polynomial $m(T)$ of T is a product of distinct irreducible factors. If F algebraically closed, then φ is a completely reducible if T diagonalizable.*

Proof. Assume a representation space V is a right module over $F[X]$. Because $F[X]$ is a principle ideal domain, then by Theorem 7.1.39 in [1], V is semisimple if only if minimal polynomial $m_T(X)$ of T is a product of distinct irreducible factors, and by Lemma 4.4.11 in [1], V is a simple module over $F[X]$ if and only if a minimal polynomial $m_T(X)$ is equal to its characteristic polynomial $c_T(X)$ of T. Furthermore, by Theorem 4.4.30 in [1], if F ia algebraically closed, then V is semisimple if and only if T is diagonalizable and, by Corollary 4.4.32 in [1], V is simple if and only if V is one dimension. □

From Theorem 2.10, if F is algebraically closed then a representation of a polynomial ring $F[X]$ on a vector space V via a linear transformation T is said to be completely reducible if the representation matrix of T is similar to a diagonal matrix, i.e there is a basis of V consisting of eigenvector of the representation of T by Corollary 4.3.4 in [1]

Next we give the example of a completely reducible representation of polynomial ring

Example 2.11 *Based on example 2.4, we have $m_T(X) = X^2 - 1 = c_T(X)$ of T and eigenvectors of representation of T are U and W. Because U and W is admissible submodule and $V = U \oplus W$, then ρ completely reducible.*

By Propositions 2.1, Proposition 2.5, Proposition 2.8, and Theorem 2.10, it can be concluded that the properties of the representation of $F[X]$ is closely related to the properties of modules over $F[X]$

3. Conclusion

Let $\varphi\colon F[X] \to End_F(V)$ be e representation of $F[X]$ via a linear transformation $T\colon V \to V$. The properties of φ are (1) Every subspace of V is an admissible submodule of V; (2) If there is another representation $\varphi'\colon F[X] \to End_F(W)$ of $F[X]$ via linear transformation $T'\colon W \to W$, then φ and φ' are equivalent if and only if there is a bijective linear transformation $S\colon V \to W$ such that $ST = T'S$; (3) The representation φ is a completely reducible in V if and only if the minimal polynomial $m(T)$ of T is a product of distinct irreducible factors. If F algebraically closed, then φ is a completely reducible if T diagonalizable.

Acknowledgment

The authors would like to thank to the Dep. of Mathematics, Universitas Gadjah Mada for the support of this research.

IOP Conf. Series: Journal of Physics: Conf. Series **893** (2017) 012011 doi:10.1088/1742-6596/893/1/012011

References

[1] Adkins W A and Weintraub S H 1992 *Algebra : an Approach via Module Theory* (New York : Springer-Verlag)
[2] Borrow M 1967 *Representation Theory of Finite Groups* (New York : Academic Press)
[3] Dummit DS and Foote R 1999, *Abstract Algebra* (New York : John Wiley & Sons Inc)
[4] Clark P L 2013, Linear algebra : invariant subspace (http://math.uga.edu/ pete/invariant_subspaces.pdf)

Invertibility of some circulant matrices

Bustomi and A Barra
Algebra Research Group, Faculty of Mathematics and Natural Sciences
Institut Teknologi Bandung, Bandung 40132, Indonesia
E-mail: bustomi@students.itb.ac.id, barra@math.itb.ac.id

Abstract. We consider several cases of circulant matrices and determine the invertibility of these matrices. We give a criterion of the invertibility of matrices of the form $circ(a, b, c, ..., c)$ and $circ(a, b, c, ..., c, a)$ where a, b and c are real numbers. By using a different approach, we are able to generalize the result of Carmona.

1. Introduction

Circulant matrices have many connections to many branches in mathematics including but not limited to coding theory, image processing, numerical analysis, number theory and differential equation. One of the important problem is to determine the invertibility of circulant matrices. Carmona(2015) in her paper gave the necessary and sufficient condition for some special circulant matrices, that are matrices of the form $circ(a, b, c, ..., c)$ and $circ(a, b, c, ..., c, a)$ where a, b and c are real numbers. In this paper, we will give a new proof of some results in the paper of Carmona and generalize her result to include circulant matrices of the similar type where a and b are in other position.

2. Basic properties of circulant matrices

We define a circular matrix by using linear operators on row vector space. Let T be the shift operator on $\mathbb{C}^n$ (written as row vectors) defined by

$$T : \mathbf{x} = (x_1, x_2, x_3 \dots, x_n) \mapsto (x_n, x_1, x_2, \dots, x_{n-1}).$$

Definition 1. Let A be an $n \times n$ complex matrix and $\mathbf{a_i}$ be the ith row of A. The matrix A is called *circulant* if $T(\mathbf{a_i}) = \mathbf{a_{i+1}}$ where the indices are taken in modulo n. Then we denote the circulant matrix A by $circ(\mathbf{a_1})$.

Definition 2. Let $A = circ(a_1, \dots, a_n)$. The polynomial $p_A(X) = \sum_{i=1}^{n} a_i X^{i-1}$ is called the *representer polynomial* of the circulant matrix A.

In what follows we list some basic properties of circulant matrix that we will use throughout this paper. We denote by $s(A)$ the sum of all entries in the first row of A.

Proposition 1. *Let* $A = circ(\boldsymbol{a}) = circ(a_1, a_2, \dots, a_n)$.

(i) $(1, 1, \dots, 1)A = s(A)(1, 1, \dots, 1)$.

(ii) If $s(A) \neq 0$ *then* $(1, 1, \dots, 1) = \frac{1}{s(A)} \sum_{i=1}^{n} T^i(\boldsymbol{a})$ *is in the row space of* A.

(iii) The circulant matrix A is invertible if and only if $(1,0,\ldots,0)$ is in the row space of A. Moreover if $\mathbf{e}_1 = (1,0,\ldots,0) = \sum_{i=1}^{n} \alpha_i T^i(\boldsymbol{a})$ for some real numbers α_i then

$$A^{-1} = circ(\alpha_1, \alpha_2, \ldots, \alpha_n).$$

(iv) Let $p_A(X)$ be the representer matrix of A and $\epsilon_j = e^{\frac{2\pi i(j-1)}{n}}$. Then A is diagonalizable by the matrix

$$F = \begin{pmatrix} 1 & 1 & \ldots & 1 \\ 1 & \epsilon_2 & \ldots & \epsilon_n \\ \vdots & \vdots & \ddots & \vdots \\ 1 & \epsilon_2^{n-1} & \ldots & \epsilon_n^{n-1} \end{pmatrix}$$

and

$$A = F\, Diag\,(p_A(\epsilon_1), p_A(\epsilon_2), \ldots, p_A(\epsilon_n))\ F^*.$$

(v) A is invertible if and only if $\gcd(p_A(X), X^n - 1) = 1$.

From the first two results we have the following corollary which are very useful to determine if a circulant matrix is not invertible.

Corollary 1. *If $s(A) = 0$ then A is singular.*

Proof. By Proposition 1.(1), $xA = 0$ has a non zero solution. □

Corollary 2. *If $s(A) \neq 0$ and $(1,\ldots,1) = \sum_{i=1}^{n} \alpha_i T^i(\mathbf{a})$ where $\alpha_{i_0} \neq \frac{1}{s(A)}$ for some i_0, then A is singular.*

Proof. By Proposition 1.(2) we have two different ways to write $(1,1,\ldots,1)$ as a linear combination of rows of A. □

Corollary 3. $\det(A) = \prod_{i=1}^{n} p_A(\epsilon_i)$ *and if A then*

$$A^{-1} = F\ diag\left(p_A^{-1}(\epsilon_1), p_A^{-1}(\epsilon_2), \ldots, p_A^{-1}(\epsilon_n)\right)\ F^*.$$

Notice that even though we can compute the determinant of a circulant matrix by the above corollary and hence we can determine its invertibility but as n getting bigger it is not easy to compute these determinant. Because of that in this paper we concentrate on some classes of circulant matrices whose invertibility are easy to determine.

3. Special Classes of Circulant Matrices

First we consider circulant matrix which its first row is of the form $(1,1,\ldots,1,0,0,\ldots,0)$, that is the first k components are all 1 and the rest are zero.

Theorem 1. *The matrix $A = circ(\underbrace{1,1,\ldots,1}_{k},0,0,\ldots,0)$ is invertible if only if $(k,n) = 1$.*

Proof. Clearly $p_A(1) = k \neq 0$. Suppose $(n,k) \neq 1$. Let $\nu \neq 1$ be an (n,k)th roof of unity. Then ν is and nth root of unity and also a kth root of unity. Since $(1-X)p_A(X) = X^k - 1$, then $p_A(\nu) = 0$. Therefore A is not invertible.

IOP Conf. Series: Journal of Physics: Conf. Series **893** (2017) 012012 doi:10.1088/1742-6596/893/1/012012

If $(k,n)=1$ then there exist $a,b\in\mathbb{Z}$ such that $ak-bn=1$. Let $\mathbf{a}=(1,1,\ldots,1,0,\ldots,0)$ be the first row of A. Notice that

$$\mathbf{a}+T^k(\mathbf{a})+\cdots+T^{(a-1)k}(\mathbf{a})=(b+1,b,b,\ldots,b).$$

But then $(1,0,\ldots,0)=(b+1,b,b,\ldots,b)-b(1,1,\ldots)$ is in the row space of A, hence A is invertible. □

In the case of $(n,k)=1$ above we can write

$$(1,0,\ldots,0)=\sum_{j=0}^{a-1}T^{jk}(\mathbf{a})-\frac{b}{k}\sum_{i=1}^{n}T^{i-1}(\mathbf{a}),$$

hence by Proposition 1.(4) we can find the explicit invers of A.

Theorem 2. *Let $\boldsymbol{a}=(1,0,\ldots,0,-q,0,\ldots,0)$ be the first row of a circulant matrix A. Let $-q$ be on the $k+1$th position. Then $A=circ(\boldsymbol{a})$ is invertible if only if q is not $\frac{n}{(k,n)}$-root of unity. In the case that A is invertible we have*

$$A^{-1}=\frac{1}{1-q^{\frac{n}{d}}}circ\left(\left(1,q,q^2,\ldots,q^{\frac{n}{d}-1},0,\ldots,0\right)circ_k(\boldsymbol{e}_1)\right)$$

with $circ_k(\boldsymbol{e}_1)=\left(\boldsymbol{e}_1,T^k(\boldsymbol{e}_1),T^{2k}(\boldsymbol{e}_1),\ldots,T^{(n-1)k}(\boldsymbol{e}_1)\right)^t$.

Proof. Let $(k,n)=d$. Define

$$\mathbf{b}=\mathbf{a}+qT^k(\mathbf{a})+q^2T^{2k}(\mathbf{a})+\cdots+q^{\frac{n}{d}-1}T^{(\frac{n}{d}-1)k}(\mathbf{a})$$

nonzero entry of $\mathbf{b}$ are 1 in the first coordinate dan $-q^{\frac{n}{d}}$ in the $(k+1)+(\frac{n}{d}-1)k=\frac{n}{d}k+1=1$th coordinate. Hence $\mathbf{b}=(1,0,\ldots,0,-q^{\frac{n}{d}},0,0,\ldots,0)$. Therefore A invertible if and only if q is not an $\frac{n}{(k,n)}$th root of unity. If A is invertible then by using $\mathbf{b}$ we can write $\mathbf{e}_1$ as linear combination of rows of A. Hence we get the formula for A^{-1} as written above. □

Corollary 4. *Let $a,b\in\mathbb{R}$, $k,l\in\mathbb{Z}$ with $k<l$. Let $A=circ(\underbrace{a,\ldots,a}_{k},\underbrace{0,\ldots,0}_{l-k},\underbrace{b,\ldots,b}_{k},0,\ldots,0)$*

If $(k,n)=1$ and $|\frac{b}{a}|\neq 1$ then A invertible.

Proof. Note that

$$\begin{aligned}p_A(X)&=a\sum_{i=1}^{k}X^{i-1}+bX^{l+1}\sum_{j=1}^{k}X^{j-1}\\&=a\left(1+\frac{b}{a}X^{l+1}\right)\left(1+X+\cdots+X^{k-1}\right).\end{aligned}$$

The polynomial $1+\frac{b}{a}X^{l+1}$ is the representer polynomial of $circ(1,0,\ldots,0,\frac{b}{a},0,0,\ldots,0)$ and the polynomial $1+X+\cdots+X^{k-1}$ is representer polynomial of the $circ(1,\ldots,1,0,\ldots,0)$. Now the result follows from Theorem 1 and Theorem 2. □

Theorem 3. *Let $\boldsymbol{a}=(\underbrace{1,0,\ldots,0}_{k},\underbrace{-2q,0,\ldots,0}_{k},1,0\ldots,0)$ Then $A=circ(\boldsymbol{a})$ is invertible if only if $q\neq\cos(\frac{2\pi jk}{n})$ for $j=0,1,\ldots,\frac{n}{(k,n)}-1$.*

Proof. We have

$$\begin{aligned} p_A(X) &= 1 - 2qX^k + X^{2k} \\ &= \left(1 - (q + i\sqrt{1-q^2})X^k\right)\left(1 - (q - i\sqrt{1-q^2})X^k\right). \end{aligned}$$

The result follows. □

4. Main Results

In what follows J will be the circulant matrix $J = circ(1, 1, \ldots, 1)$. We will consider several types of circulant matrices that can be written as $A + cJ$ where A is an invertible circulant matrix and c is scalar. First we have some preliminary result.

Proposition 2. *The matrix $A = circ(0, 1, 2, \ldots, n-1)$ is invertible and*

$$A^{-1} = \frac{2}{n^2(n-1)}J - \frac{1}{n}circ(1, -1, 0, \ldots, 0).$$

Proof. Let $\mathbf{a} = (0, 1, 2, \ldots, n-1)$. Then

$$-(\mathbf{a} - T(\mathbf{a})) = (n-1, -1, -1, \ldots, -1)$$

$$\frac{2}{n(n-1)}\sum_{i=0}^{n-1} T^i(\mathbf{a}) - (\mathbf{a} - T(\mathbf{a})) = (n, 0, 0, \ldots, 0).$$

Therefore A invertible and

$$A^{-1} = \frac{2}{n^2(n-1)}J - \frac{1}{n}circ(1, -1, 0, \ldots, 0).$$

□

Theorem 4. *Let A be a circulant matrix of size n and c is scalar. If A is invertible and $s(A) + cn \neq 0$ then $A + cJ$ invertible and*

$$(A + cJ)^{-1} = A^{-1} - \frac{c}{s(A)(s(A) + cn)}J.$$

Proof. Since A is invertible then all eigen values of A are nonzero (including $s(A)$. Then we can write $A = F\ diag(s(A), \lambda_2, \ldots, \lambda_n)F^*$ where $\lambda_i \neq 0$. Now

$$\begin{aligned} A + cJ &= F\ diag(s(A), \lambda_2, \ldots, \lambda_n)F^* + F\ diag(cn, 0, \ldots, 0)F^* \\ &= F\ diag((s(A) + cn), \lambda_2, \ldots, \lambda_n)F^*. \end{aligned}$$

Therefore $A + cJ$ invertible.

Write $(A + cJ)^{-1} = A^{-1} + X$. Then

$$\begin{aligned} (A + cJ)(A^{-1} + X) &= I \\ I + cJA^{-1} + (A + cJ)X &= I \\ (A + cJ)X &= -cJA^{-1} \\ X &= (A + cJ)^{-1}(-\frac{c}{s(A)}J) \\ X &= -\frac{c}{s(A)}\frac{1}{s(A) + cn}J. \end{aligned}$$

It follows that $(A + cJ)^{-1} = A^{-1} - \frac{c}{s(A)(s(A)+cn)}J$. □

By using the above theorems we will generalize the results of Carmona et.al ([1])

Theorem 5. *Let* $\mathbf{a} = (a, c, \ldots, c, b, c, \ldots, c)$ *where* a *is in the first position and* b *is in the* $k+1th$ *position while the rest are all* c *where* $a, b, c \in \mathbb{R}$. *If* $a - c \neq 0, s(A) \neq 0$ *and* q *is not a* $\frac{n}{(k,n)}th$ *root of unity then* $A = circ(\mathbf{a})$ *invertible and*

$$A^{-1} = \frac{circ\left(\left(1, q, q^2, \ldots, q^{\frac{n}{(k,n)}-1}, 0, \ldots, 0\right) circ_k(\mathbf{e}_1)\right)}{(a-c)(1-q^{\frac{n}{(k,n)}})} - \frac{c}{(a+b-2c)s(A)}J$$

for $q = -\frac{b-c}{a-c}$.

Proof. Write A as

$$\begin{aligned} A &= circ(a, c, \ldots, c, b, c, \ldots, c) \\ &= (a-c)circ\left(1, 0, \ldots, 0, -\left(-\frac{b-c}{a-c}\right), 0, \ldots, 0\right) + cJ. \end{aligned}$$

Now the result follows from Theorem 3 and Theorem 4. □

In similar manner we can proof the following result.

Theorem 6. *Let* $a, b, c \in \mathbb{R}$ *and* $\mathbf{a} = (a, \underbrace{c, \ldots, c}_{k-1}, b, \underbrace{c \ldots, c}_{k-1}, a, c, \ldots, c)$. *If* $a - c \neq 0, s(A) \neq 0$ *and* q *is not a* $\frac{n}{(k,n)}th$ *root of unity then* $A = circ(\mathbf{a})$ *is invertible and*

$$A^{-1} = \frac{circ\,(1, 0\ldots, 0, -2q, 0\ldots, 0, 1, 0\ldots, 0)^{-1}}{(a-c)} - \frac{c}{(2a+b-3c)(2a+b-(n-3)c)}J$$

for $q = -\frac{b-c}{2(a-c)}$.

Corollary 5. *Let* $A = circ(\underbrace{a, \ldots, a}_{k}, \underbrace{b, \ldots, b}_{k}, c, \ldots, c)$ *where* $a, b, c \in \mathbb{R}$. *If* $(k, n) = 1$ *dan* $s(A) \neq 0$ *dan* $-\frac{a-c}{b-c}$ *is not an nth root of unity, then* A *invertible.*

Proof. Write A as $A = (a-c)BC + cJ$ where

$$\begin{aligned} B &= circ(1, \ldots, 1, 0, \ldots, 0) \text{ and} \\ C &= circ\left(1, 0, \ldots, 0, -\left(-\frac{b-c}{a-c}\right), 0, \ldots, 0\right). \end{aligned}$$

□

Acknowledgment
This research is supported by Riset KK ITB 2016.

References

[1] Carmona A, Encinas A M, Gago S, Jiménez M J and Mitjana M 2015 The inverses of some circulant matrices *Appl. Math. Comput* **270** pp 785–793
[2] Kra I and Simanca S R 2012 On circulant matrices *Notices of the AMS* **59**(3) pp 368–377

IOP Conf. Series: Journal of Physics: Conf. Series **893** (2017) 012013 doi:10.1088/1742-6596/893/1/012013

The β-dual of the Cesàro sequence spaces defined on a generalized Orlicz space

Haryadi,[1] **Supama**[2], **and Atok Zulijanto**[2]

[1] Department of Mathematics, Universitas Muhammadiyah, Palangkaraya, ID
[2] Department of Mathematics, Universitas Gadjah Mada, Yogyakarta, ID

E-mail: haryadi_ump@yahoo.co.id

Abstract. In this paper we characterized the β-dual of the Cesàro sequence space with terms in a generalized Orlicz space. Further,we find that the dual is a generalization of the dual of the Cesàro space in the classical Banach space L_p for$1 < p < \infty$.

1. Introduction and Preliminaries

The Kothe-Toeplitz dual is a useful tool in the characterization of class of the matrix transformations between sequence spaces. The recent study of the class of the matrix transformations between sequence spaces can be found in Tripathy and Sen [14]. Khan and Khan [3], and Rahman and Karim [11], examined the characterization of the class of matrix transformations on the scalar-valued Cesàro sequence spaces. Maddox [6, 7], characterized a generalized Kothe-Toeplitz of the vector-valued sequence spaces. In [15], Yilmaz and Ozdemir extend the definition of Maddox [6] and examined Kothe-Toeplitz duals of some vector-valued Orlicz sequence spaces. The properties of some new Cesàro sequence spaces were studied by Bala [1], Kubiak [5], Petrot and Suantai [10], and Sever and Altay [13]. Bhardwaj and Gupta [2] studied Kothe-Toeplitz dual and matrix map on the Cesàro summable difference sequence spaces.

In this paper, we characterize the β-dual of the Cesàro sequence spaces constructed by the modular ρ_ϕ with terms in a generalized Orlicz space.

We begin by recalling the definition of the Orlicz function. Let $\phi : \mathbb{R} \to [0, \infty)$ be an Orlicz function, that is, ϕ is even, $\phi(x) = 0$ if and only if $x = 0$, $\lim_{x\to\infty} \phi(x) = \infty$, continuous, and convex. If in addition, ϕ satisfies the conditions $\lim_{x\to 0} \phi(x)/x = 0$ and $\lim_{x\to\infty} \phi(x)/x = \infty$, then it is called N-function. The complementary to the Orlicz function ϕ is a function ψ such that $|xy| \leq \phi(x) + \psi(y)$, for every $x, y \in \mathbb{R}$. For any Orlicz function ϕ, the function ψ defined by $\psi(y) = \sup\{|y|x - \phi(x) : x \geq 0\}$ is an Orlicz function complementary to ϕ. An Orlicz function ϕ is said to satisfy the Δ_2-condition if there is a $K > 0$ such that $\phi(2x) \leq K\phi(x)$ for each $x \geq 0$ (see e.g., [4]). We denote by ϕ^{-1}, the inverse function of the Orlicz function ϕ in the non-negative values argument.

Let $\mathcal{S}$ denote the collection of all Lebesgue measurable real valued functions on $[a, b]$. Let (ϕ, ψ) be a pair of complementary Orlicz functions. The generalized Orlicz space $\mathcal{L}_\phi$ is defined as the space of all $u \in \mathcal{S}$ such that $|\int_a^b u(x)v(x)dx| < \infty$ for every v with $\int_a^b \psi(v(x))dx < \infty$. It is easy to show that

$$\rho_\phi(u) = \int_a^b \phi(u(x))dx$$

is a modular on $\mathcal{L}_\phi$, i.e. ρ_ϕ satisfies : (1) $\rho_\phi(u) = 0 \Leftrightarrow u = \theta$, where θ is the zero function in $\mathcal{L}_\phi$, (2) $\rho_\phi(-u) = \rho_\phi(u)$ for each $u \in \mathcal{L}_\phi$, and (3) $\rho_\phi(\alpha u + \beta v) \leq \rho_\phi(u) + \rho_\phi(v)$ for each $u, v \in \mathcal{L}_\phi$ and $\alpha, \beta \geq 0$, $\alpha + \beta = 1$. In addition, the convexity of ϕ implies the convexity of ρ_ϕ.

For any Orlicz function ϕ, $\mathcal{L}_\phi$ is a Banach space with respect to the Orlicz norm

$$\|f\|_\phi = \sup\left\{\left|\int_a^b f(x)g(x)dx\right| : \rho_\psi(g) \leq 1\right\},$$

where ψ is complementary to ϕ. In addition, it is true that (i) $\|f\|_\phi \leq \rho_\phi(f)+1$ for every $f \in \mathcal{L}_\phi$ and (ii) $\rho_\phi(f) \leq 1$ if and only if $\|f\|_\phi \leq 1$ (see, e.g., [4, 12]).

A sequence (f_k) in $\mathcal{L}_\phi$ is said to be ρ_ϕ-convergent if $\rho_\phi(f_k - f_0) \to 0$ as $k \to \infty$; it is called ρ_ϕ-Cauchy if $\rho_\phi(f_k - f_l) \to 0$ as $k, l \to \infty$. It can be shown (see e.g. [4, 9]), if the Orlicz function ϕ satisfies Δ_2-condition, then the norm convergence is equivalent to ρ_ϕ-convergence in $\mathcal{L}_\phi$.

Let $\omega(\mathcal{L}_\phi)$ denotes the space of all sequences with terms in $\mathcal{L}_\phi$. The member of $\mathcal{L}_\phi$ is written as $f = (f_k)$. We modify the definition of Maddox [6], the generalized Cesàro sequence space, as follows:

$$\mathcal{W}_{0,\phi} = \{(f_k) : (\exists\lambda > 0) \lim_{N\to\infty} \frac{1}{N}\sum_{k=1}^{N}\int_a^b \phi(\lambda f_k(x))dx = 0\}.$$

In relation to the sequence space above, we also defined the following spaces:

$$\mathcal{W}_\phi = \{(f_k) : (f_k - f_0) \in \mathcal{W}_{0,\phi} \text{ for some } f_0 \in \mathcal{L}_\phi\}$$

$$\mathcal{W}_{0,\|\cdot\|} = \{(f_k) \subset \mathcal{L}_\phi : \lim_{N\to\infty} \frac{1}{N}\sum_{k=1}^{N} \|f_k\|_\phi = 0\}.$$

We denote by ω, the set of all sequence in $\mathbb{R}$. Further, if the Orlicz function ϕ satisfies the Δ_2-condition, we define

$$w_{0,\phi} = \left\{(\lambda_k) \in \omega : \lim_{N\to\infty} \frac{1}{N}\sum_{k=1}^{N} \phi(\lambda_k) = 0\right\}.$$

It is clear that $\mathcal{W}_{0,\phi} \subset \mathcal{W}_\phi$. Note that any ρ_ϕ-convergent sequence in $\mathcal{L}_\phi$ belongs to $\mathcal{W}_\phi$. Especially, for any p, $1 \leq p < \infty$, every mean convergent sequence in L_p, belongs to $\mathcal{W}_\phi$ for the Orlicz function $\phi(x) = |x|^p$.

Using the fact that ρ_ϕ is a modular, it is easy to establish that the function $\rho : \omega(\mathcal{L}_\phi) \to [0, \infty]$ defined by

$$\rho(f) = \sup_N \left\{\frac{1}{N}\sum_{k=1}^{N}\int_a^b \phi(f_k(x))dx\right\}$$

is a modular. Furthermore, $\rho(f) < \infty$ for every $f \in \mathcal{W}_\phi$. Note that the convexity of ϕ implies the convexity of ρ. Further, each of the space $\mathcal{W}_{0,\phi}$ and $\mathcal{W}_\phi$ is a Banach space with respect to the Luxemburg norm

$$\|f\|_\rho = \inf\left\{t > 0 : \rho\left(\frac{f}{t}\right) \leq 1\right\}.$$

Let $\mathcal{L}^*_\phi$ be the space of bounded linear functionals on $\mathcal{L}_\phi$. We write u^*u, the value of linear functional u^* at u. The β-dual of $X \subset \omega(\mathcal{L}_\phi)$ is defined as

$$[X]^\beta = \{(f^*_k) \subset \mathcal{L}^*_\phi : \sum_{k=1}^{\infty} f^*_k f_k \text{ convergent for each } (f_k) \in X\}.$$

2. The Results

Our results mainly concern on the Cesàro sequence space which generated by the Orlicz function satisfying Δ_2-condition.

Theorem 1. *If the Orlicz function ϕ satisfies the Δ_2-condition, then*

$$\mathcal{W}_{0,\phi} = \left\{(f_k) : \lim_{N\to\infty} \frac{1}{N}\sum_{k=1}^{N}\int_a^b \phi(f_k(x))dx = 0\right\} \tag{1}$$

$$\mathcal{W}_{\phi} = \left\{(f_k) : \lim_{N\to\infty} \frac{1}{N}\sum_{k=1}^{N}\int_a^b \phi(f_k(x) - f_0(x))dx = 0 \text{ for some } f_0 \in \mathcal{L}_\phi\right\}. \tag{2}$$

Proof. To prove part (1), let we write the right hand side of (1) by $\mathcal{W}'_{0,\phi}$. It is enough to prove that $\mathcal{W}_{0,\phi} \subset \mathcal{W}'_{0,\phi}$. Let $(f_k) \in \mathcal{W}_{0,\phi}$, then there exists $\lambda > 0$ such that $\frac{1}{N}\sum_{k=1}^{\infty} \rho_\phi(\lambda f_k) \to 0$ as $N \to \infty$. For $\lambda \geq 1$, the monotonicity of ϕ implies $f \in \mathcal{W}'_{0,\phi}$. For $\lambda < 1$, since ϕ satisfies the Δ_2-condition, then there exist $K > 0$ and $l_0 \in \mathbb{N}$ such that $\phi(\frac{u}{\lambda}) \leq K^{l_0}\phi(u)$. Hence,

$$\begin{aligned}\frac{1}{N}\sum_{k=1}^{N}\int_a^b \phi(f_k(x))dx &= \frac{1}{N}\sum_{k=1}^{N}\int_a^b \phi(\frac{\lambda}{\lambda} f_k(x))dx \\ &\leq \frac{K^{l_0}}{N}\sum_{k=1}^{N}\int_a^b \phi(\lambda f_k(x))dx \to 0\end{aligned}$$

as $N \to \infty$, i.e. $(f_k) \in \mathcal{W}'_{0,\phi}$.

The proof of part (2) is similar. □

The Cesàro sequence space constructed by the modular ρ_ϕ is contained in the space defined by the Orlicz norm, as stated by the following lemma.

Lemma 1. *If the Orlicz function ϕ satisfying the Δ_2-condition then $\mathcal{W}_{0,\phi} \subset \mathcal{W}_{0,\|\cdot\|}$.*

Proof. Let $\varepsilon > 0$ be arbitrary and $(f_k) \in \mathcal{W}_{0,\phi}$. Take $n_0 \in \mathbb{N}$ such that $2^{-n_0+1} < \varepsilon$, and let $K > 0$ such that $\phi(2^{n_0}(f_k(x)) \leq K^{n_0}\phi(f_k(x))$. Take $N_0 \in \mathbb{N}$ such that

$$\frac{1}{N}\sum_{k=1}^{N}\int_a^b \phi(f_k(x))dx < \frac{1}{K^{n_0}}, \quad \text{for each } N \geq N_0,$$

Then for each $N \geq N_0$,

$$\frac{1}{N}\sum_{k=1}^{N}\int_a^b \phi(2^{n_0}(f_k(x))dx \leq \frac{K^{n_0}}{N}\sum_{k=1}^{N}\int_a^b \phi(f_k(x))dx \leq 1.$$

Hence, for each $N \geq N_0$ we have

$$\frac{1}{N}\sum_{k=1}^{N}\|2^{n_0}f_k\|_\phi \leq \frac{1}{N}\sum_{k=1}^{N}\int_a^b \phi(2^{n_0}(f_k(x))dx + 1 \leq 2$$

which implies $\frac{1}{N}\sum_{k=1}^{N}\|f_k\|_\phi \leq 2^{-n_0+1} < \varepsilon$. □

Note that this lemma implies $\frac{1}{2^v}\sum_{k=2^v}^{2^{v+1}-1}\|f_k\|_\phi \to 0$ as $v \to \infty$ for every $(f_k) \in \mathcal{W}_{0,\phi}$. The converse of Lemma 1 is not necessarily true in general. This is clear from the following example.

Example. Let $\phi(x) = |x|^2$ and (f_k) be a sequence of real function on $[0,1]$ with $f_k(x) = \frac{1}{k}$ for $k \neq j^3$ and $f_k(x) = \sqrt[3]{k}$ for $k = j^3$. Then $\frac{1}{N}\sum_{k=1}^{N}\|f_k\|_\phi \to 0$, but $\frac{1}{N}\sum_{k=1}^{N}\rho_\phi(f_k) \not\to 0$.

Lemma 2. *If the Orlicz function ϕ satisfying the Δ_2-condition and $(f_k) \in \mathcal{W}_{0,\phi}$, then*

$$\sup\left\{\frac{\|f_k\|_\phi}{\phi^{-1}(\rho_\phi(f_k))} : \|f_k\|_\phi > 1\right\} < \infty. \tag{3}$$

Proof. Let $\varepsilon > 0$ and $(f_k) \in \mathcal{W}_{0,\phi}$. There exists $N_0 \in \mathbb{N}$ such that $\frac{1}{N}\sum_{k=1}^{N}\rho_\phi(f_k) < \varepsilon$ for each $N \geq N_0$, which implies

$$\rho_\phi(f_k) < \varepsilon \cdot k, \quad \forall k \geq N_0. \tag{4}$$

Suppose (3) does not hold. Then for any natural number $n \geq N_0$ there exists $\|f_k\|_\phi > 1$ such that

$$n\phi^{-1}(\rho_\phi(f_k) < \|f_k\|_\phi.$$

Since $1 < \|f_k\|_\phi \leq \rho_\phi(f_k)$, then

$$\left(\frac{n}{k}\right)\phi^{-1}(1)\cdot k < \rho_\phi(f_k),$$

contradic to (4). □

Corollary 1. *Let the Orlicz function ϕ satisfying the Δ_2-condition and $(f_k) \in \mathcal{W}_{0,\phi}$. There exists $c > 0$ such that*

$$\phi(\|f_k\|_\phi) \leq c\rho_\phi(f_k), \quad \forall\ \|f_k\|_\phi > 1.$$

From Lemma 1 and Corollary 1, it can be shown the following corollary.

Corollary 2. *Let the Orlicz function ϕ satisfying the Δ_2-condition. If $(f_k) \in \mathcal{W}_{0,\phi}$, then $(\|f_k\|_\phi) \in w_{0,\phi}$.*

A linear space $X \subset \omega(\mathcal{L}_\phi)$ is said to be normal if for every f with $|f| \leq |g|$ for some $g \in X$, implies $f \in X$. The space $\mathcal{W}_{0,\phi}$ is normal. For, let $g \in \mathcal{W}_{0,\phi}$ and $|f| \leq |g|$ in the sense $|f_k| \leq |g_k|$ a.e. on $[a,b]$ for each $k \in \mathbb{N}$. Obviously $f_k \in \mathcal{S}$ and the monotonicity of ϕ implies $\phi \circ f_k \leq \phi \circ g_k$. Hence $f \in \mathcal{W}_{0,\phi}$.

In addition, $\mathcal{W}_\phi$ is not a normal space, let $\phi(u) = |u|$ and define $f_1(x) = 1,\ f_2(x) = 0$ and

$$f_k(x) = \begin{cases} 1, & 2^{r-1} < k \leq 3 \cdot 2^{r-2},\ r = 2, 3, \cdots \\ 0, & otherwise \end{cases}$$

for each $x \in [a, b]$. Then $\rho_\phi(f_k) \leq 1$ but $(f_k) \notin \mathcal{W}_\phi$.

Theorem 2. *Let the Orlicz function ϕ satisfies the Δ_2-condition. Then $(f_k^*) \in [\mathcal{W}_{0,\phi}]^\beta$ if and only if*

$$(\|f_k^*\|) \in [w_{0,\phi}]^\alpha. \tag{5}$$

Proof. Suppose that $\sum_{k=1}^\infty f_k^* f_k$ convergent for every $(f_k) \in \mathcal{W}_{0,\phi}$. Since $f_k^* \in \mathcal{L}_\phi^*$ then there exists $f_k \in \mathcal{L}_\phi$ with $\|f_k\|_\phi \leq 1$ such that

$$\|f_k^*\| \leq 2|f_k^* f_k| \tag{6}$$

for each $k \in \mathbb{N}$. It is easy to see that there exists $c > 0$ such that $\rho_\phi(\lambda f_k) \leq c\phi(\lambda)$ for each $\lambda \in \mathbb{R}$ if $\|f_k\|_\phi \leq 1$. Let $(\lambda_k) \in w_{0,\phi}$, and define $g_k = |\lambda_k| sgn(f_k^* f_k) f_k$. Then

$$\frac{1}{N} \sum_{k=1}^N \rho_\phi(g_k) = \frac{1}{N} \sum_{k=1}^N \rho_\phi(\lambda_k f_k) \leq \frac{c}{N} \sum_{k=1}^N \phi(\lambda_k) \to 0$$

as $N \to \infty$, i.e. $(g_k) \in \mathcal{W}_{0,\phi}$. Consequently

$$\sum_{k=1}^\infty |\lambda_k| \cdot |f_k^* f_k| = \sum_{k=1}^\infty f_k^* g_k$$

is convergent, which implies that $\sum_{k=1}^\infty \|f_k^*\| \cdot |\lambda_k| \leq 2 \sum_{k=1}^\infty |\lambda_k| \cdot |f_k^* f_k| < \infty$, i.e. $(\|f_k^*\|) \in w_{0,\phi}^\alpha$.

For the sufficiently, let $(\|f_k^*\|) \in [w_{0,\phi}]^\alpha$ and $f \in \mathcal{W}_{0,\phi}$. Since $(\|f_k\|_\phi) \in w_{0,\phi}$, then $\sum_{k=1}^\infty \|f_k^*\| \|f_k\|_\phi < \infty$. Hence, for each $\varepsilon > 0$, there exist k_0 such that

$$|\sum_{k=m}^n f_k^* f_k| \leq \sum_{k=m}^n \|f_k^*\| \|f_k\|_\phi < \varepsilon, \quad \forall m, n \geq k_0,$$

i.e. $(\sum_{k=1}^n f_k^* f_k)$ is a Cauchy sequence in $\mathbb{R}$. Hence $\sum_{k=1}^n f_k^* f_k$ convergent. □

Corollary 3. *Let the Orlicz function ϕ satisfies the Δ_2-condition. Then*

$$[\mathcal{W}_\phi]^\beta = [\mathcal{W}_{0,\phi}]^\beta.$$

Proof. Since $\mathcal{W}_{0,\phi} \subset \mathcal{W}_\phi$, it is enough to show that $[\mathcal{W}_{0,\phi}]^\beta \subset [\mathcal{W}_\phi]^\beta$.

Let $(f_k^*) \in [\mathcal{W}_{0,\phi}]^\beta$ and $(f_k) \in \mathcal{W}_\phi$. Let $f_0 \in \mathcal{L}_\phi$ such that $(f_k - f_0) \in \mathcal{W}_{0,\phi}$. By Theorem 2, $(\|f_k^*\|) \in [w_{0,\phi}]^\alpha$. Since $(\|f_k - f_0\|_\phi) \in w_{0,\phi}$, then

$$\sum_{k=1}^\infty |f_k^*(f_k - f_0)| \leq \sum_{k=1}^\infty \|f_k^*\| \|f_k - f_0\|_\phi < \infty.$$

Since ϕ satisfies Δ_2-condition, then there exists $1 < p < \infty$ such that $\phi(x) \le c|x|^p$ (see [4]), and hence, $w_{0,p} \subset w_{0,\phi}$. Since $w^{\beta}_{0,\phi} \subset w^{\beta}_{0,p}$, then $\sum_{r=0}^{\infty} 2^{r/p}(\sum_{k=2^r}^{2^{r+1}-1} \|f_k^*\|^q)^{1/q} < \infty$, $1/p + 1/q = 1$ (see [8]). Hence,

$$\sum_{k=1}^{\infty} |f_k^* f_0| \le \sum_{k=1}^{\infty} \|f_k^*\| \|f_0\|_\phi \le \sum_{r=0}^{\infty} 2^{r/p} \left(\sum_{k=2^r}^{2^{r+1}-1} \|f_k^*\|^q \right)^{1/q} \left(\frac{1}{2^r} \sum_{k=2^r}^{2^{r+1}} \|f_0\|_\phi^p \right)^{1/p} < \infty.$$

Finally, we have

$$\sum_{k=1}^{\infty} |f_k^* f_k| \le \sum_{k=1}^{\infty} |f_k^*(f_k - f_0)| + \sum_{k=1}^{\infty} |f_k^* f_0| < \infty,$$

and hence, $\sum_{k=1}^{\infty} f_k^* f_k$ converges.

□

Acknowledgments
The author would like to thank to the reviewer for making some useful corrections which improve the paper.

References

[1] Bala I 2012 *Communications in Mathematics and Applications* **3**(2) pp 197–204
[2] Bhardwaj V K and Gupta S 2013 *J. Inequalities and App.* **315** pp 1–9
[3] Khan F M and KhanM A 1994 *Indian J. Pure Appl. Math.* **25**(6) pp 641–645
[4] Krasnosel'skii MA and RutickiiY B 1961 *Convex Functions and Orlicz Spaces* (Netherlands: P. Noordhoff Ltd)
[5] Kubiak D 2009 *J. Math. Anal. and App.* **349** pp 291–296
[6] Maddox I J 1980 *J. Math. Math. Sci.* **3** pp 423–432
[7] Maddox I J 1974 *Comp. Math.* **29** pp 35–42
[8] Malkowsky E and Rakocevic V 2000 *An Introdukction to the theory of sequence spaces and measures of noncompactness* (Zbornik radova, Matematicki institut SANU, Beograd)
[9] Musielak J 1983 *Orlicz spaces and modular space* (Berlin Heidelberg New York Tokyo: Springer Verlag)
[10] Petrot N and Suantai S 2005 *Sci. Asia* **31** pp 173–177
[11] Rahman MF and Karim A B R M 2015 *Pure and App. Math. J.* **4**(6) pp 237–241
[12] Rao M M and Ren Z D 1991 *Theory of Orlicz Spaces* (New York: Marcel Dekker Inc.)
[13] Sever Y and Altay B 2014 *Filomat* **28**(7) pp 1417–1424
[14] Tripathy B C and Sen M 2006 *Tamkang J. Math.* **37**(2) pp 155–162
[15] Yilmaz Y and Ozdemir M K 2005 *Soochow J. Math.* **31**(3) pp 389–402.

The boundedness of generalized Bessel-Riesz operators on generalized Morrey spaces

Mochammad Idris and Hendra Gunawan
Department of Mathematics, Bandung Institute of Technology, Bandung 40132, Indonesia
E-mail: idemath@gmail.com, hgunawan@math.itb.ac.id

Abstract. The purpose of this paper is to prove the boundedness of generalized Bessel-Riesz operators on generalized Morrey spaces. The kernel of the operators contain some parameters, one of which is related to Bessel decay. As usual, we use the usual dyadic decomposition, Hölder's inequality, a Hedberg-type inequality for the operators, and the boundedness of Hardy-Littlewood maximal operator in the proofs. In addition, we also exploit the relationship between the parameters of the kernel and of the space. We obtain that the norm of the operators is dominated by the norm of the kernels

Keywords: generalized Bessel-Riesz operators, Hardy-Littlewood maximal operator, generalized Morrey spaces.

MSC 2010: Primary 42B20; Secondary 26A33, 42B25, 26D10, 47G10.

1. Introduction

In this paper, we shall discuss about an integral operator. The following definition formulates the operator.

Definition 1.1. Let $\gamma \in (0,\infty)$ and $\rho : \mathbb{R}^+ \to \mathbb{R}^+$ be an increasing function satisfying the doubling condition ($\frac{1}{2} \leq \frac{r_1}{r_2} \leq 2$ implies $C_1 \leq \frac{H(r_1)}{H(r_2)} \leq C_2$, where $C_1, C_2 > 0$). *The generalized Bessel-Riesz operator* $I_{\rho,\gamma}$ is defined by

$$I_{\rho,\gamma} f(x) := \int_{\mathbb{R}^n} K_{\rho,\gamma}(x-y)\, f(y)\, dy$$

for every $f \in L^p_{loc}(\mathbb{R}^n)$, $1 \leq p < \infty$, where $K_{\rho,\gamma}(x) := \frac{\rho(|x|)}{|x|^n (1+|x|)^\gamma}$, for every $x \in \mathbb{R}^n$.

Here, $K_{\rho,\gamma}$ is called *generalized Bessel-Riesz kernel.* If $\rho(r) := r^\alpha$, for every $r > 0$, where $0 < \alpha < n$, then we have $I_{\rho,\gamma} = I_{\alpha,\gamma}$ (Bessel-Riesz operator [6]).

Definition 1.2. Let $\phi : \mathbb{R}^+ \to \mathbb{R}^+$ be a decreasing function satisfying the doubling condition. For $1 \leq p < \infty$, *The generalized Morrey spaces* $L^{p,\phi}(\mathbb{R}^n)$ is defined to be the set of all functions $f \in L^p_{loc}(\mathbb{R}^n)$ such that $\|f\|_{L^{p,\phi}} := \sup_{B=B(a,R)} \frac{\left(\int_B |f(x)|^p dx\right)^{1/p}}{\phi(R)|B|^{1/p}} < \infty$, where $|B|$ denotes *Lebesgue measure* of ball $B(a,R)$, where $a \in \mathbb{R}^n, R > 0$.

If $\phi(r) = r^{-n/q}$, $p \le q < \infty$, for every $r > 0$, then we have the Morrey space $L^{p,q}(\mathbb{R}^n)$. In particular, we know that $L^{p,p}(\mathbb{R}^n) = L^p(\mathbb{R}^n)$ is the Lebesgue space. Moreover, one may check that $r \mapsto r^{-n/q}$ satisfy the doubling condition.

On Lebesgue spaces, the boundedness of $I_{\alpha,\gamma}$ could be shown via Young's inequality [4]. For $0 < \gamma$ and $0 < \alpha < n$, we have $\|I_{\alpha,\gamma} f\|_{L^q} \le \|K_{\alpha,\gamma}\|_{L^t} \|f\|_{L^p}$ for every $f \in L^p(\mathbb{R}^n)$, $1 \le p < t'$, where $\frac{1}{q} = \frac{1}{p} - \frac{1}{t'}$, $\frac{n}{n+\gamma-\alpha} < t < \frac{n}{n-\alpha}$. The membership of $K_{\alpha,\gamma}$ on Lebesgue spaces was also proved [6]. On Morrey spaces, the boundedness of $I_{\alpha,\gamma}$ was proved by Hardy-Littlewood maximal operator. The proof is explained in [6]. Furthermore, $I_{\alpha,\gamma}$ is also bounded on generalized Morrey spaces [7]. To show the boundedness of $I_{\alpha,\gamma}$ on (generalized) Morrey spaces, $K_{\alpha,\gamma}$ can be viewed as a member of Lebesgue spaces or Morrey spaces.

Meanwhile for $\gamma = 0$, we have $I_{\rho,0} = I_\rho$, *the generalized fractional integral operator [5]*. In 2002, Eridani and Gunawan [2] proved the boundedness of I_ρ on generalized Morrey spaces. Using different assumptions, Eridani, Gunawan, and Nakai [3] also proved it.

The Hardy-Littlewood maximal operator M, which is used to prove the boundedness of $I_{\alpha,\gamma}$ and I_ρ on Morrey spaces and generalized Morrey spaces, is defined by

$$Mf(x) := \sup_{x\in B=B(a,R)} \frac{1}{|B|}\int_B |f(y)|\,dy,$$

for every $f \in L^p_{loc}(\mathbb{R}^n)$. We know that M is bounded on Morrey spaces [1]. In [8], Nakai proved the boundedness of M on generalized Morrey spaces. We shall use this fact to prove our results.

2. Main Results

In this section, the boundedness of $I_{\rho,\gamma}$ on generalized Morrey spaces will be proved. We must show that $K_{\rho,\gamma}$ is contained in Lebesgue spaces.

Lemma 2.1. *Let $\gamma > 0$ and ρ satisfy $\int_{0<r\le R} \frac{\rho^t(r)}{r^{(nt+1)-n}}dr + \int_{R<r<\infty} \frac{\rho^t(r)}{r^{(nt+\gamma t+1)-n}}dr < \infty$, for some $t \in [1,\infty)$, for every $R > 0$. Then $K_{\rho,\gamma}$ is a member of $L^t(\mathbb{R}^n)$.*

Proof. Suppose that $\gamma > 0$ and ρ satisfy $\int_{0<r\le R} \frac{\rho^t(r)}{r^{(nt+1)-n}}dr + \int_{R<r<\infty} \frac{\rho^t(r)}{r^{(nt+\gamma t+1)-n}}dr < \infty$, for some $t \in [1,\infty)$, for every $R > 0$. Observe that

$$\int_{\mathbb{R}^n} K^t_{\rho,\gamma}(x)dx = \int_{|x|\le R} \frac{\rho^t(|x|)}{|x|^{nt}(1+|x|)^{\gamma t}}dx + \int_{R<|x|} \frac{\rho^t(|x|)}{|x|^{nt}(1+|x|)^{\gamma t}}dx$$

and we estimate $\int_{\mathbb{R}^n} K^t_{\rho,\gamma}(x)dx \le C\int_{0<r\le R} \frac{\rho^t(r)}{r^{(nt+1)-n}}dr + C\int_{R<r<\infty} \frac{\rho^t(r)}{r^{(nt+\gamma t+1)-n}}dr < \infty$. So we have $\int_{\mathbb{R}^n} K^t_{\rho,\gamma}(x)dx < \infty$. Hence $K_{\rho,\gamma} \in L^t(\mathbb{R}^n)$. □

By Lemma 2.1 and the inclusion property of Morrey spaces, $K_{\rho,\gamma} \in L^t(\mathbb{R}^n) \subseteq L^{s,t}(\mathbb{R}^n)$, where $1 \le s \le t$. Consequently, $\frac{\left(\int_{2^kR<|x|\le 2^{k+1}R} K^s_{\rho,\gamma}(x)dx\right)^{1/s}}{(2^kR)^{n/s-n/t}} \le C\|K_{\rho,\gamma}\|_{L^{s,t}}$ for every integer k and $R > 0$. Because ρ satisfies the doubling condition, $\frac{\left(\sum_{k=-1}^{-\infty} K^s_{\rho,\gamma}(2^kR)(2^kR)^n\right)^{1/s}}{R^{n/s-n/t}} \le C\|K_{\rho,\gamma}\|_{L^{s,t}}$ holds, for every $R > 0$. Here, one may check that $1 - \frac{\ln\rho(R_1)}{n\ln R_1} < \frac{1}{t} < 1 + \frac{\gamma}{n} - \frac{\ln\rho(R_1)}{n\ln R_1}$, for every $R_1 > 1$. If $\rho(R) = R^\alpha$ for every $R > 0$ where $0 < \alpha < n$, then $1 - \frac{\alpha}{n} < \frac{1}{t} < 1 + \frac{\gamma-\alpha}{n}$.

Theorem 2.2. *Let $\gamma > 0$ and ρ satisfy $\frac{1}{R^{\gamma t}}\int_{0<r\le R} \frac{\rho^t(r)r^n}{r^{(nt+1)}}dr + \int_{R<r<\infty} \frac{\rho^t(r)r^{n-1}}{r^{(n+\gamma)t}}dr \le \frac{\rho^t(R)R^n}{R^{(n+\gamma)t}}$, for some $t \in [1,\infty)$, for every $R > 0$. Suppose that ϕ is surjective and*

$$\int_{0<r\le R} r^{\frac{n}{t'}-1}dr + \frac{1}{\phi(R)}\int_{R<r<\infty} \phi(r)r^{\frac{n}{t'}-1}dr \le \phi^{\frac{p_1}{p_2}-1}(R),$$

where $1 < p_1 < p_2 < \infty$. *Then* $\|I_{\rho,\gamma}f\|_{L^{p_2,\phi^{p_1/p_2}}} \leq C_{p_1,\phi} \|K_{\rho,\gamma}\|_{L^{s,t}} \|f\|_{L^{p_1,\phi}}$, *for every* $f \in L^{p_1,\phi}(\mathbb{R}^n)$, *where* $1 \leq s \leq t$.

Proof. Consider $1 < p_1 < p_2 < \infty$. Suppose that all assumptions hold. Next, take $f \in L^{p_1,\phi}(\mathbb{R}^n)$ where ϕ is surjective and $\int_{0<r\leq R} r^{\frac{n}{t'}-1}dr + \frac{\int_{R<r<\infty}\phi(r)r^{\frac{n}{t'}-1}dr}{\phi(R)} \leq \phi^{\frac{p_1}{p_2}-1}(R)$. Split the following formula into two parts, $I_{\rho,\gamma}f(x) := I_1(x) + I_2(x)$, where $I_1(x) := \int_{|x-y|<R} K_{\rho,\gamma}(x-y)f(y)dy$ and $I_2(x) := \int_{|x-y|\geq R} K_{\rho,\gamma}(x-y)f(y)dy$. Each part will be estimated by dyadic decomposition. The positive constants are denoted by the letter C which may vary from line to line. To estimate I_1, we have $|I_1(x)| \leq C\sum_{k=-\infty}^{-1} \frac{\rho(2^kR)\int_{2^kR\leq|x-y|<2^{k+1}R}|f(y)|dy}{(2^kR)^n(1+2^kR)^\gamma}$. Use maximal operator to obtain $|I_1(x)| \leq Mf(x)\sum_{k=-\infty}^{-1} \frac{\rho(2^kR)(2^kR)^{n/s}}{(2^kR)^n(1+2^kR)^\gamma}(2^kR)^{n/s'}$. By Hölder's inequality, we obtain $|I_1(x)| \leq CMf(x)\left(\sum_{k=-\infty}^{-1}(2^kR)^n\right)^{1/s'}\left(\sum_{k=-\infty}^{-1}K_{\rho,\gamma}^s(2^kR)(2^kR)^n\right)^{1/s}$. Now, we state $|I_1(x)| \leq CMf(x)\frac{\left(\sum_{k=-\infty}^{-1}(2^kR)^n\right)^{1/s'}}{R^{n/t-n/s}}\frac{\left(\sum_{k=-\infty}^{-1}K_{\rho,\gamma}^s(2^kR)(2^kR)^n\right)^{1/s}}{R^{n/s-n/t}}$. By the above lemma that $K_{\rho,\gamma} \in L^{s,t}(\mathbb{R}^n)$ and $R^{n/t'} \leq \phi^{\frac{p_1-p_2}{p_2}}(R)$, so we get $|I_1(x)| \leq C\|K_{\rho,\gamma}\|_{L^{s,t}} Mf(x)\phi^{\frac{p_1-p_2}{p_2}}(R)$.

Meanwhile for I_2, we have $|I_2(x)| \leq C\sum_{k=0}^{\infty}\frac{\rho(2^kR)\int_{2^kR\leq|x-y|<2^{k+1}R}|f(y)|dy}{(2^kR)^n(1+2^kR)^\gamma}$. Use again Hölder inequality, so we get $|I_2(x)| \leq C\sum_{k=0}^{\infty}\frac{\rho(2^kR)(2^kR)^{n/p_1'}}{(2^kR)^n(1+2^kR)^\gamma}\left(\int_{2^kR\leq|x-y|<2^{k+1}R}|f(y)|^{p_1}dy\right)^{1/p_1}$ and the norm of f dominates I_2, that is $|I_2(x)| \leq C\|f\|_{L^{p_1,\phi}}\sum_{k=0}^{\infty}\frac{\rho(2^kR)\phi(2^kR)(2^kR)^n}{(2^kR)^n(1+2^kR)^\gamma}$. Because $\frac{\rho(2^kR)(2^kR)^{n/t}}{(2^kR)^n(1+2^kR)^\gamma} \leq \|K_{\rho,\gamma}\|_{L^{s,t}}$, for every $k \in \mathbb{Z}$, so

$$\sum_{k=0}^{\infty}\frac{\rho(2^kR)\phi(2^kR)(2^kR)^n}{(2^kR)^n(1+2^kR)^\gamma} \leq C\|K_{\rho,\gamma}\|_{L^{s,t}}\sum_{k=0}^{\infty}\phi\left(2^kR\right)\left(2^kR\right)^{n/t'}.$$

We also have $\sum_{k=0}^{\infty}\phi(2^kR)(2^kR)^{n/t'} \leq \phi^{\frac{p_1}{p_2}}(R)$, consequently

$$|I_2(x)| \leq C\|K_{\rho,\gamma}\|_{L^{s,t}}\|f\|_{L^{p_1,\phi}}\phi^{\frac{p_1}{p_2}}(R).$$

By summing of the estimates of I_1 and I_2, $|I_{\rho,\gamma}f(x)| \leq C\|K_{\rho,\gamma}\|_{L^{s,t}}\phi^{\frac{p_1}{p_2}}(R)\left(\frac{Mf(x)}{\phi(R)} + \|f\|_{L^{p_1,\phi}}\right)$, holds for each $x \in \mathbb{R}^n$. We know that ϕ is surjective. So, choose $R > 0$, such that $\phi(R) = \frac{Mf(x)}{\|f\|_{L^{p_1,\phi}}}$. Now, we get $|I_{\rho,\gamma}f(x)| \leq C\|K_{\rho,\gamma}\|_{L^{s,t}}\|f\|_{L^{p_1,\phi}}^{1-p_1/p_2}Mf(x)^{p_1/p_2}$. Take arbitrary $r > 0$ and $a \in \mathbb{R}^n$ to obtain $\left(\int_{|x-a|<r}|I_{\rho,\gamma}f(x)|^{p_2}dx\right)^{1/p_2} \leq C\|K_{\rho,\gamma}\|_{L^{s,t}}\|f\|_{L^{p_1,\phi}}^{1-p_1/p_2}\left(\int_{|x-a|<r}|Mf(x)|^{p_1}dx\right)^{(1/p_2)}$. The two sides are divided by $\phi(r)^{p_1/p_2}r^{n/p_2}$ and take supremum over $r > 0$ and $a \in \mathbb{R}^n$ to get $\|I_{\rho,\gamma}f\|_{L^{p_2,\phi^{p_1/p_2}}} \leq C\|K_{\rho,\gamma}\|_{L^{s,t}}\|f\|_{L^{p_1,\phi}}^{1-p_1/p_2}\|Mf\|_{L^{p_1,\phi}}^{p_1/p_2}$. By the boundedness of maximal operator on generalized Morrey spaces, $\|I_{\rho,\gamma}f\|_{L^{p_2,\phi^{p_1/p_2}}} \leq C_{p_1,\phi}\|K_{\rho,\gamma}\|_{L^{s,t}}\|f\|_{L^{p_1,\phi}}$ holds. □

By Theorem 2.2 and the inclusion property of Morrey spaces,

$$\|I_{\rho,\gamma}f\|_{L^{p_2,\phi^{p_1/p_2}}} \leq C_{p_1,\phi}\|K_{\rho,\gamma}\|_{L^{s,t}}\|f\|_{L^{p_1,\phi}} \leq C_{p_1,\phi}\|K_{\rho,\gamma}\|_{L^t}\|f\|_{L^{p_1,\phi}}.$$

The above inequality shows that norm of $I_{\rho,\gamma}$ is dominated by $\|K_{\rho,\gamma}\|_{L^{s,t}}$ and $\|K_{\rho,\gamma}\|_{L^t}$. Furthermore, we want to obtain the best upper bound. We shall reinvestigate $K_{\rho,\gamma}$. The following Lemma explains that $K_{\rho,\gamma}$ is a member of generalized Morrey spaces.

Lemma 2.3. *If $\gamma > 0$ and ρ, σ satisfy $\int_{0<r\leq R} \rho^s(r) r^{-(ns+1)+n} dr \leq C\sigma^s(R)R^n$, for some $s \in [1,\infty)$, for every $R > 0$, then $K_{\rho,\gamma}$ is a member of $L^{s,\sigma}(\mathbb{R}^n)$.*

Proof. Suppose that $0 < \gamma$. Assume that ρ, σ satisfies $\int_{0<r\leq R} \rho^s(r) r^{-(ns+1)+n} dr \leq C\sigma^s(R)R^n$, for some $s \in [1,\infty)$, for every $R > 0$. Now, observe that

$$\int_{|x-0|\leq R} K^s_{\rho,\gamma}(x)dx \leq C \int_{0<r\leq R} \frac{\rho^s(r)}{r^{(ns+1)-n}} dr \leq C\sigma^s(R)R^n.$$

We get $\sup_{R>0} \frac{\left(\int_{|x-0|\leq R} K^s_{\rho,\gamma}(x)dx\right)^{1/s}}{\sigma(R)R^{n/s}} < \infty$. Hence $K_{\rho,\gamma} \in L^{s,\sigma}(\mathbb{R}^n)$. □

By Lemma 2.3 we also have $\frac{\left(\int_{2^kR<|x|\leq 2^{k+1}R} K^s_{\rho,\gamma}(x)dx\right)^{1/s}}{\sigma(2^kR)\left(2^kR\right)^{n/s}} \leq C\left\|K_{\rho,\gamma}\right\|_{L^{s,\sigma}}$, for every integer k and $R > 0$. Moreover, we obtain $\frac{\left(\sum_{k=-1}^{-\infty} K^s_{\rho,\gamma}(2^kR)\left(2^kR\right)^n\right)^{1/s}}{\sigma(R)(R)^{n/s}} \leq C\left\|K_{\rho,\gamma}\right\|_{L^{s,\sigma}}$. Observe that $1 \leq s \leq \frac{n \ln R_1}{-\ln\sigma(R_1)}$ for every $R_1 > 1$. In particular, for $\sigma(R) = R^{-n/t}$, we have $K_{\rho,\gamma} \in L^{s,t}(\mathbb{R}^n)$, where $1 \leq s \leq t$.

Theorem 2.4. *Let $\gamma > 0$ and ρ, σ satisfies $\int_{0<r\leq R} \frac{\rho^s(r)}{r^{(ns+1)-n}} dr \leq C\sigma^s(R)R^n$, for some $s \in [1,\infty)$, for every $R > 0$. If ϕ is surjective and $\int_{0<r\leq R} \frac{\sigma(r)}{r^{-n+1}} dr + \frac{\int_{R<r<\infty} \frac{\phi(r)\sigma(r)}{r^{-n+1}} dr}{\phi(R)} \leq \phi^{\frac{p_1}{p_2}-1}(R)$, where $1 < p_1 < p_2$. Then for every $f \in L^{p_1,\phi}(\mathbb{R}^n)$, we have $\left\|I_{\rho,\gamma} f\right\|_{L^{p_2,\phi^{p_1/p_2}}} \leq C_{p_1,\phi}\left\|K_{\rho,\gamma}\right\|_{L^{s,\sigma}}\left\|f\right\|_{L^{p_1,\phi}}$.*

Proof. To process the proof, we do similar steps as in the proof of Theorem 2.2. □

The norm of $I_{\rho,\gamma}$ have some upper bounds. Choosing the membership of the kernels on the function spaces influences the upper bound of the norm. The upper bound in Theorem 2.4 is smaller than in Theorem 2.2, while $R^{-n/t} < \sigma(R)$ for every $R > 0$. So we can choose the best of σ to obtain the smallest upper bound.

3. Concluding Remarks

In this paper, we have shown that the boundedness of $I_{\rho,\gamma}$ from $L^{p_1,\phi}(\mathbb{R}^n)$ to $L^{p_2,\phi^{p_1/p_2}}(\mathbb{R}^n)$ where $1 < p_1 < p_2 < \infty$. In the next study, the boundedness of $I_{\rho,\gamma}$ will be investigated from $L^{1,\phi}(\mathbb{R}^n)$ to $L^{p_2,\phi^{p_1/p_2}}(\mathbb{R}^n)$ using different methods.

Acknowledgments

The both authors are supported by ITB Research & Innovation Program 2016.

References

[1] Chiarenza F and Frasca M 1987 *Morrey spaces and Hardy-Littlewood maximal function*, Rend. Mat. **7** pp 273–279
[2] Eridani and GunawanH 2002 *On generalized fractional integrals*, J. Indones. Math. Soc. (MIHMI) **8(3)** pp 25–28
[3] Eridani, Gunawan H and Nakai E 2004 *On generalized fractional integral operators*, Sci. Math. Jpn. **60**
[4] Grafakos L 2008 *Classical Fourier Analysis*, Graduate Texts in Matemathics, Vol. 249, *Springer*, New York
[5] Gunawan H 2003 *A note on the generalized fractional integral operators*, J. Indones. Math. Soc. **9**
[6] Idris M, Gunawan H, Lindiarni J and Eridani 2016 *The boundedness of Bessel-Riesz operators on Morrey spaces*, AIP Conf. Proc., **1729, 02000** , doi: 10.1063/1.4946909
[7] Idris M, Gunawan H and Eridani 2016 *The boundedness of Bessel-Riesz operators on generalized Morrey spaces*, Aust. J. Math. Anal. Appl., **13**
[8] Nakai E 1994 *Hardy-Littlewood maximal operator, singular integral operators and Riesz potentials on generalized Morrey spaces*, Math. Nachr. **166** pp 95–103

IOP Conf. Series: Journal of Physics: Conf. Series **893** (2017) 012015 doi:10.1088/1742-6596/893/1/012015

An inclusion property of Orlicz-Morrey spaces

Al Azhary Masta, Hendra Gunawan, Wono Setya-Budhi

Analysis and Geometry Group, Faculty of Mathematics and Natural Sciences
Bandung Institute of Technology, Jl. Ganesha 10, Bandung 40132, Indonesia

E-mail: alazhari.masta@upi.edu, {hgunawan, wono}@math.itb.ac.id

Abstract. The Orlicz-Morrey spaces $L_{\phi,\Phi}$ (where Φ is a Young function and ϕ is a parameter for the Morrey spaces) are generalizations of Orlicz spaces and Morrey spaces. Inclusion properties between Orlicz spaces L_Φ and between Morrey spaces $\mathcal{M}^p_\psi$ are well known. In this study we will investigate the inclusion relation between Orlicz-Morrey spaces L_{ϕ_1,Φ_1} and L_{ϕ_2,Φ_2} with respect to Young functions Φ_1, Φ_2 and parameters ϕ_1, ϕ_2. Also, we give sufficient and necessary conditions for the inclusion property of these spaces, which are obtained through norm estimates for the characteristic functions of balls in $\mathbb{R}^n$. In addition, we shall give a sufficient and necessary condition for generalized Hölder's inequality.

1. Introduction

Orlicz-Morrey spaces are generalizations of Orlicz spaces and Morrey spaces. There are two versions of Orlicz-Morrey spaces. One is defined by Nakai [2, 9] and another by Sawano, Sugano, and Tanaka [2, 12]. Here we are interested in studying the inclusion property of Orlicz-Morrey spaces which were introduced by Nakai.

First, we recall the definiton of Young functions (see [11, 9]). A function $\Phi : [0,\infty) \to [0,\infty)$ is called a Young function if Φ is convex, left-continuous, $\lim\limits_{t\to 0} = \Phi(t) = \Phi(0) = 0$, and $\lim\limits_{t\to\infty} \Phi(t) = \infty$. Let G_1 be the set of all functions $\phi : (0,\infty) \to (0,\infty)$ such that $\phi(r)$ is nondecreasing but $\frac{\phi(r)}{r}$ is nonincreasing. For $\phi_1, \phi_2 \in G_1$, we write $\phi_1 \sim \phi_2$ if there exists a constant $C > 1$ such that

$$C^{-1}\phi_1(t) \leq \phi_2(t) \leq C\phi_1(t)$$

for all $t > 0$.

Let Φ be a Young function and $\phi \in G_1$. The Orlicz-Morrey space $L_{\phi,\Phi}(\mathbb{R}^n)$ is the set of measurable functions $f \in L^1_{loc}(\mathbb{R}^n)$ such that for every open ball B in $\mathbb{R}^n$, the following

$$\| f \|_{(\phi,\Phi,B)} := \inf\{b > 0 : \frac{\phi(|B|)}{|B|} \int_B \Phi(\frac{|f(x)|}{b})dx \leq 1\}$$

is finite. We use the notation $\mathfrak{B}$ to denote the family of all open balls B in $\mathbb{R}^n$, and $|B|$ for its Lebesgue measure. $L_{\phi,\Phi}(\mathbb{R}^n)$ is a Banach space with respect to the norm

$$\| f \|_{L_{\phi,\Phi}(\mathbb{R}^n)} := \sup_{B\in\mathfrak{B}} \| f \|_{(\phi,\Phi,B)} .$$

For $\phi(t) = t$, $L_{\phi,\Phi}(\mathbb{R}^n) = L_\Phi(\mathbb{R}^n)$ is the Orlicz space. Meanwhile, for $\phi(t) = \frac{1}{\psi(t^{1/n})^p}$ (where $\psi : (0,\infty) \to (0,\infty)$ is almost decreasing and $t^{\frac{n}{p}}\psi(t)$ is almost increasing) and $\Phi(t) = t^p$ we have $L_{\phi,\Phi}(\mathbb{R}^n) = \mathcal{M}^p_\psi(\mathbb{R}^n)$, the generalized Morrey space introduced by Nakai in 1994.

Gunawan *et al.* [3] have proved an inclusion property of generalized Morrey spaces: If $1 \le p_1 \le p_2 < \infty$, then $\mathcal{M}^{p_2}_{\psi_2} \subset \mathcal{M}^{p_1}_{\psi_1}$ if and only if $\psi_2(t) \le C\psi_1(t)$ for all $t > 0$ and some $C > 0$. On the other hand, inclusion properties between Orlicz spaces $L_\Phi(\mathbb{R}^n)$ and between weak Orlicz spaces $wL_\Phi(\mathbb{R}^n)$ are well known (see [5, 6]).

Motivated by these results, the purpose of this study is to get the inclusion property of Orlicz-Morrey spaces $L_{\phi,\Phi}(\mathbb{R}^n)$.

The main results are presented in Section 2. In particular, Theorem 2.2 contains a necessary and sufficient condition for the inclusion relation between Orlicz-Morrey spaces. In Section 3, we have also given sufficent and necessary conditions for generalized Hölder's inequality.

To prove the results, we will use the same method as in [3] and [6] which pay attention to the characteristic functions of balls in $\mathbb{R}^n$ and use the inverse function of Φ, namely $\Phi^{-1}(s) := \inf\{r \ge 0 : \Phi(r) > s\}$.

In the following, we recall several lemmas which will be used in the next section.

Lemma 1.1. *[11, 6] Suppose that Φ is a Young function and $\Phi^{-1}(s) = \inf\{r \ge 0 : \Phi(r) > s\}$. We have*
(1) $\Phi^{-1}(0) = 0$.
(2) $\Phi^{-1}(s_1) \le \Phi^{-1}(s_2)$ for $s_1 \le s_2$.
(3) $\Phi(\Phi^{-1}(s)) \le s \le \Phi^{-1}(\Phi(s))$ for $0 \le s < \infty$.
(4) If, for some constants $C_1, C_2 > 0$, we have $\Phi_2^{-1}(s) \le C_1\Phi_1^{-1}(C_2 s)$, then $\Phi_1(\frac{t}{C_1}) \le C_2\Phi_2(t)$ for $t = \Phi_2^{-1}(s)$.

Lemma 1.2. *[2] Let Φ be a Young function, $a \in \mathbb{R}^n$ and $r > 0$. Then $\|\chi_{B(a,r)}\|_{L_{\phi,\Phi}(\mathbb{R}^n)} = \frac{1}{\Phi^{-1}(\frac{1}{\phi(|B(a,r)|)})}$, where $|B(a,r)|$ denotes the volume of $B(a,r)$.*

Lemma 1.3. *If $f \in L_{\phi,\Phi}(\mathbb{R}^n)$, then*

$$\frac{\phi(|B|)}{|B|}\int_B \Phi\left(\frac{|f(x)|}{\| f \|_{(\phi,\Phi,B)}}\right) dx \le 1$$

for any open ball $B \in \mathfrak{B}$. Furthermore, $\| f \|_{(\phi,\Phi,B)} \le 1$ if and only if $\frac{\phi(|B|)}{|B|}\int_B \Phi(|f(x)|)dx \le 1$ for any open ball $B \in \mathfrak{B}$.
Proof.

Let f be an element of $L_{\phi,\Phi}(\mathbb{R}^n)$ and take an arbitrary $\epsilon > 0$, then there exists $b_\epsilon > 0$ such that $b_\epsilon \le \| f \|_{(\phi,\Phi,B)} + \epsilon$ and $\frac{\phi(|B|)}{|B|}\int_B \Phi\left(\frac{|f(x)|}{b_\epsilon}\right) dx \le 1$ for any open ball $B \in \mathfrak{B}$. Because, Φ is increasing, we have

$\frac{\phi(|B|)}{|B|}\int_B \Phi\left(\frac{|f(x)|}{\|f\|_{(\phi,\Phi,B)}+\epsilon}\right) dx \le \int_B \Phi\left(\frac{|f(x)|}{b_\epsilon}\right) dx \le 1.$
Since $\epsilon > 0$ is arbitrary, we can conclude

$$\frac{\phi(|B|)}{|B|}\int_B \Phi\left(\frac{|f(x)|}{\| f \|_{(\phi,\Phi,B)}}\right) dx \le 1$$

for any open ball $B \in \mathfrak{B}$.

Next, if $\|f\|_{(\phi,\Phi,B)} \leq 1$ for any open ball $B \in \mathfrak{B}$, then

$$\frac{\phi(|B|)}{|B|}\int_B \Phi(|f(x)|)dx \leq \frac{\phi(|B|)}{|B|}\int_B \Phi\left(\frac{|f(x)|}{\| f \|_{(\phi,\Phi,B)}}\right)dx \leq 1.$$

Now, assume that $\frac{\phi(|B|)}{|B|}\int_B \Phi(|f(x)|)dx \leq 1$ holds for any open ball $B \in \mathfrak{B}$. Clearly, we have $\| f \|_{(\phi,\Phi,B)} \leq 1$. □

Corollary 1.4. *If $f \in L_{\phi,\Phi}(\mathbb{R}^n)$, then*

$$\frac{\phi(|B|)}{|B|}\int_B \Phi\left(\frac{|f(x)|}{\| f \|_{L_{\phi,\Phi}(\mathbb{R}^n)}}\right)dx \leq 1$$

for any open ball $B \in \mathfrak{B}$. Furthermore, $\| f \|_{L_{\phi,\Phi}(\mathbb{R}^n)} \leq 1$ if and only if $\frac{\phi(|B|)}{|B|}\int_B \Phi(|f(x)|)dx \leq 1$ for any open ball $B \in \mathfrak{B}$.

2. Main Results

Now, we come to the inclusion relation between Orlicz-Morrey space L_{ϕ_1,Φ_1} and L_{ϕ_2,Φ_2} with respect to Young functions Φ_1, Φ_2 and parameters ϕ_1, ϕ_2.

Theorem 2.1. *[7] Let Φ_1, Φ_2 be Young functions and $\phi_1, \phi_2 \in G_1$ such that $\phi_1 \sim \phi_2$. If $\Phi_1(x) \leq \Phi_2(Kx)$ for some $K > 0$ then $L_{\phi_2,\Phi_2}(\mathbb{R}^n) \subseteq L_{\phi_1,\Phi_1}(\mathbb{R}^n)$.*

We can prove the Lemma 2.1 by using similar arguments in the proof of [9, Proposition 3.2].

Theorem 2.2. *Let Φ_1, Φ_2 be Young functions and $\phi_1, \phi_2 \in G_1$ such that $\phi_1 \sim \phi_2$. Then the following statements are equivalent:*
(1) $\Phi_1(t) \leq \Phi_2(Ct)$, for every $t > 0$.
(2) $L_{\phi_2,\Phi_2}(\mathbb{R}^n) \subseteq L_{\phi_1,\Phi_1}(\mathbb{R}^n)$.
(3) There exists a constant $C > 0$ such that

$$\| f \|_{L_{\phi_1,\Phi_1}(\mathbb{R}^n)} \leq C \| f \|_{L_{\phi_2,\Phi_2}(\mathbb{R}^n)},$$

for every $f \in L_{\phi_2,\Phi_2}(\mathbb{R}^n)$.

Proof.

Assume that (1) holds. Let f be an element of $L_{\phi_2,\Phi_2}(\mathbb{R}^n)$. Since $\phi_1 \sim \phi_2$ and $\Phi_1(t) \leq \Phi_2(Ct)$, for every $t > 0$, by Theorem 2.1, we have $L_{\phi_2,\Phi_2}(\mathbb{R}^n) \subseteq L_{\phi_1,\Phi_1}(\mathbb{R}^n)$. Next, since $(L_{\phi_2,\Phi_2}(\mathbb{R}^n), L_{\phi_1,\Phi_1}(\mathbb{R}^n))$ is a Banach pair, it follows from [4, Lemma 3.3] that (2) and (3) are equivalent. It thus remains to show that (3) implies (1). Assume that (3) holds. By Lemma 1.2, for every $r_0 > 0$ we have

$$\begin{aligned}\frac{1}{\Phi_1^{-1}\left(\frac{1}{\phi_1(|B(a,r_0)|)}\right)} &= \| \chi_{B(a,r_0)} \|_{L_{\phi_1,\Phi_1}(\mathbb{R}^n)} \\ &\leq C \| \chi_{B(a,r_0)} \|_{L_{\phi_2,\Phi_2}(\mathbb{R}^n)} \\ &= \frac{C}{\Phi_2^{-1}\left(\frac{1}{\phi_2(|B(a,r_0)|)}\right)}\end{aligned}$$

whence $\Phi_2^{-1}(\frac{1}{\phi_1(|B(a,r_0)|)}) \leq C\Phi_1^{-1}(\frac{1}{\phi_2(|B(a,r_0)|)}) \leq C\Phi_1^{-1}(\frac{C_2}{\phi_1(|B(a,r_0)|)})$, for arbitrary $a \in \mathbb{R}^n$ and $r_0 > 0$. By Lemma 1.1 (4), $\phi_1 \sim \phi_2$, and Φ is convex, we have

$$\Phi_1(\frac{t_0}{C_2C}) \leq \frac{1}{C_2}\Phi_1(\frac{t_0}{C}) \leq \Phi_2(t_0)$$

where $t_0 = \Phi_2^{-1}(\frac{1}{\phi_2(|B(a,r_0)|)})$. Since r_0 is an arbitrary positive real number, we get $\Phi_1(t) \leq \Phi_2(Ct)$, for every $t > 0$. □

Corollary 2.3. *[6] Let Φ_1, Φ_2 be Young functions and $\phi_1(t) = \phi_2(t) = t$ for $t > 0$. Then the following statements are equivalent:*
(1) $\Phi_1(t) \leq \Phi_2(Ct)$, for every $t > 0$.
(2) $L_{\Phi_2}(\mathbb{R}^n) \subseteq L_{\Phi_1}(\mathbb{R}^n)$.
(3) There exists a constant $C > 0$ such that

$$\| f \|_{L_{\Phi_1}(\mathbb{R}^n)} \leq C \| f \|_{L_{\Phi_2}(\mathbb{R}^n)}$$

for every $f \in L_{\Phi_2}(\mathbb{R}^n)$.

Remark. Corollary 2.3 refines Corollary 2.11 in [10] which only states that (1) implies (2) for $w_1(x) = w_2(x) = 1$ and $X = \mathbb{R}^n$.

3. Generalized Hölder Inequality

In the following, we will give sufficient and necessary conditions for generalized Hölder inequality on Orlicz-Morrey spaces. To get the result we will give attention to estimate the norm of the characteristic function of ball in $\mathbb{R}^n$.

Theorem 3.1. *For $m \geq 2$. Let Φ_i be Young functions and $\phi_i \in G_1$, for $i = 1, 2, 3, ..., m$. If there exists a constant $C > 0$ such that $\prod_{i=1}^{m-1} \Phi_i^{-1}(\frac{s}{\phi_i(t)}) \leq C\Phi_m^{-1}(\frac{s}{\phi_m(t)})$ for $s, t > 0$, then for $f_i \in L_{\phi_i,\Phi_i}(\mathbb{R}^n)$, $i = 1, 2, 3, ..., m-1$ we have*

$$\prod_{i=1}^{m-1} f_i \in L_{\phi_m,\Phi_m}(\mathbb{R}^n)$$

with

$$\| \prod_{i=1}^{m-1} f_i \|_{L_{\phi_m,\Phi_m}(\mathbb{R}^n)} \leq (m-1)C \prod_{i=1}^{m-1} \| f_i \|_{L_{\phi_i,\Phi_i}(\mathbb{R}^n)} .$$

Proof.

Let f_i be an element of $L_{\phi_i,\Phi_i}(\mathbb{R}^n)$ for $i = 1, 2, 3, ..., m-1$. Then

$$\frac{\phi_i(|B|)}{|B|} \int_B \Phi_i(\frac{|f_i(x)|}{\| f_i \|_{(\phi_i,\Phi_i,B)}})dx \leq 1$$

for any ball $B \in \mathfrak{B}$ and $i = 1, 2, 3, ..., m-1$. Now fix B. For each $x \in B$, let

$$M(x) = \max\left\{\phi_i(|B|)\Phi_i(\frac{|f_i(x)|}{\| f_i \|_{(\phi_i,\Phi_i,B)}}) : 1 \leq i \leq m-1\right\}.$$

From $\Phi_i(\frac{|f_i(x)|}{\|f_i\|_{(\phi_i,\Phi_i,B)}}) \leq \frac{M(x)}{\phi_i(|B|)}$ and Lemma 1.1(3), we have

$$\frac{|f_i(x)|}{\| f_i \|_{(\phi_i,\Phi_i,B)}} \leq \Phi_i^{-1}(\Phi_i(\frac{|f_i(x)|}{\| f_i \|_{(\phi_i,\Phi_i,B)}})) \leq \Phi_i^{-1}(\frac{M(x)}{\phi_i(|B|)})$$

for $i = 1, 2, 3, ..., m-1$. Hence

$$\prod_{i=1}^{m-1} \frac{|f_i(x)|}{\| f_i \|_{(\phi_i,\Phi_i,B)}} \leq \prod_{i=1}^{m-1} \Phi_i^{-1}(\frac{M(x)}{\phi_i(|B|)}) \leq C\Phi_m^{-1}(\frac{M(x)}{\phi_m(|B|)})$$

and

$$\Phi_m(\frac{1}{(m-1)C} \prod_{i=1}^{m-1} \frac{|f_i(x)|}{\| f_i \|_{(\phi_i,\Phi_i,B)}}) \leq \frac{\Phi_m(\Phi_m^{-1}(\frac{M(x)}{\phi_m(|B|)}))}{(m-1)} \leq \frac{M(x)}{(m-1)\phi_m(|B|)}.$$

On the other hand, we have

$$\frac{M(x)}{(m-1)\phi_m(|B|)} \leq \frac{\sum_{i=1}^{m-1} \phi_i(|B|)\Phi_i(\frac{|f_i(x)|}{\|f_i\|_{(\phi_i,\Phi_i,B)}})}{(m-1)\phi_m(|B|)}$$

and

$$\sum_{i=1}^{m-1} \phi_i(|B|) \int_B \Phi_i(\frac{|f_i(x)|}{\| f_i \|_{(\phi_i,\Phi_i,B)}})dx \leq (m-1)|B|.$$

Therefore

$$\int_B \Phi_m(\frac{1}{(m-1)C} \prod_{i=1}^{m-1} \frac{|f_i(x)|}{\| f_i \|_{(\phi_i,\Phi_i,B)}})dx \leq \frac{(m-1)|B|}{(m-1)\phi_m(|B|)} = \frac{|B|}{\phi_m(|B|)}.$$

This shows that

$$\| \prod_{i=1}^{m-1} f_i \|_{(\phi_m,\Phi_m,B)} \leq (m-1)C \prod_{i=1}^{m-1} \| f_i \|_{(\phi_i,\Phi_i,B)}$$

for every open ball $B \in \mathfrak{B}$. Hence, we conclude that

$$\| \prod_{i=1}^{m-1} f_i \|_{L_{\phi_m,\Phi_m}(\mathbb{R}^n)} \leq (m-1)C \prod_{i=1}^{m-1} \| f_i \|_{L_{\phi_i,\Phi_i}(\mathbb{R}^n)} .$$

□

Remark 3.2. For $m = 3$, Theorem 3.1 reduces to Theorem 4.1 (p.200) in [7].

Corollary 3.3. *(Hölder's inequality on Orlicz spaces) Let Φ_i be Young functions, $i = 1, 2, 3$. If there exists a constant $C > 0$ such that $\Phi_1^{-1}(t)\Phi_2^{-1}(t) \leq C\Phi_3^{-1}(t)$ for $t > 0$, then for $f \in L_{\Phi_1}(\mathbb{R}^n)$ and $g \in L_{\Phi_2}(\mathbb{R}^n)$ we have $f \cdot g \in L_{\Phi_3}(\mathbb{R}^n)$ with*

$$\| f \cdot g \|_{L_{\Phi_3}(\mathbb{R}^n)} \leq 2C \| f \|_{L_{\Phi_1}(\mathbb{R}^n)} \| g \|_{L_{\Phi_2}(\mathbb{R}^n)} .$$

IOP Conf. Series: Journal of Physics: Conf. Series **893** (2017) 012015 doi:10.1088/1742-6596/893/1/012015

Theorem 3.4. *For $m \geq 2$. Let Φ_i be Young functions and $\phi_i \in G_1$, for $i = 1, 2, 3, ..., m$. If there exists a constant $K > 0$ such that $\| \prod_{i=1}^{m-1} f_i \|_{L_{\frac{\phi_m}{s},\Phi_m}(\mathbb{R}^n)} \leq K \prod_{i=1}^{m-1} \| f_i \|_{L_{\frac{\phi_i}{s},\Phi_i}(\mathbb{R}^n)}$ for $f_i \in L_{\frac{\phi_i}{s},\Phi_i}(\mathbb{R}^n)$, $i = 1, 2, 3, ..., m-1$, then we have $\prod_{i=1}^{m-1} \Phi_i^{-1}(\frac{s}{\phi_i(t)}) \leq K\Phi_m^{-1}(\frac{s}{\phi_m(t)})$ for every $s, t > 0$.*

Proof.

Observe that, for every open ball $B(a,r)$ we have

$$\begin{aligned}\| \chi_{B(a,r)} \|_{L_{\frac{\phi_m}{s},\Phi_m}(\mathbb{R}^n)} &= \| \prod_{i=1}^{m-1} \chi_{B(a,r)} \|_{L_{\frac{\phi_m}{s},\Phi_m}(\mathbb{R}^n)} \\ &\leq K \prod_{i=1}^{m-1} \| \chi_{B(a,r)} \|_{L_{\frac{\phi_i}{s},\Phi_i}(\mathbb{R}^n)} .\end{aligned}$$

By Lemma 1.2 we have

$$\frac{1}{\Phi_m^{-1}(\frac{s}{\phi_m(|B(a,r)|)})} \leq \frac{K}{\prod_{i=1}^{m-1} \Phi_i^{-1}(\frac{s}{\phi_i(|B(a,r)|)})}$$

or

$$\prod_{i=1}^{m-1} \Phi_i^{-1}(\frac{s}{\phi_i(|B(a,r)|)}) \leq K\Phi_m^{-1}(\frac{s}{\phi_m(|B(a,r)|)}).$$

Since $B(a,r)$ is arbitrary, we get

$$\prod_{i=1}^{m-1} \Phi_i^{-1}(\frac{s}{\phi_i(t)}) \leq K\Phi_m^{-1}(\frac{s}{\phi_m(t)})$$

for $s, t > 0$. □

Corollary 3.5. *Let $p_i \geq 1$ be real numbers, $i = 1, 2, 3, ..., m$. If $f_i \in L_{p_i}(\mathbb{R}^n)$, $i = 1, 2, 3, ..., m$ such that*

$$\| \prod_{i=1}^{m} f_i \|_{L_1(\mathbb{R}^n)} \leq \prod_{i=1}^{m} \| f_i \|_{L_{p_i}(\mathbb{R}^n)},$$

then $\frac{1}{p_1} + \frac{1}{p_2} + ... + \frac{1}{p_m} = 1$.

Proof.

By Theorem 3.4, we have $|x|^{\frac{1}{p_1}}|x|^{\frac{1}{p_2}}|x|^{\frac{1}{p_3}}|x|^{\frac{1}{p_m}} \leq |x|$ for every $x > 0$. Since x is an arbitrary positive real number, we conclude that $\frac{1}{p_1} + \frac{1}{p_2} + ... + \frac{1}{p_m} = 1$. □

Remark. Corollary 3.5 completes Theorem 2.11 in [1]. Theorem 2.1 states that $\frac{1}{p_1}+\frac{1}{p_2}+...+\frac{1}{p_m} = 1$ is a sufficient condition for generalized Hölder's inequality on Lebesgue spaces, while Corollary 3.5 states that it is also a necessary condition.

Corollary 3.6. *Let Φ_1, Φ_2, Φ_3 be Young functions. If there exists a constant $C > 0$ such that*

$$\| f \cdot g \|_{L_{\Phi_3}(\mathbb{R}^n)} \leq C \| f \|_{L_{\Phi_1}(\mathbb{R}^n)} \| g \|_{L_{\Phi_2}(\mathbb{R}^n)}$$

for $f \in L_{\Phi_1}(\mathbb{R}^n)$ *and* $g \in L_{\Phi_2}(\mathbb{R}^n)$, *then*

$$\Phi_1^{-1}(t)\Phi_2^{-1}(t) \le C\Phi_3^{-1}(t)$$

for every $t > 0$.

For each Young function Φ, we can define another convex function $\tilde{\Phi} : \mathbb{R} \to (0, \infty)$ having similiar properties, by $\tilde{\Phi}(y) = \sup_{x>0} \{x|y| - \Phi(x)\}$, $y \in \mathbb{R}$. Then $\tilde{\Phi}$ is called the complementary function to Φ.

Corollary 3.7. *If* Φ *is Young function and* $\tilde{\Phi}$ *is complementary function of* Φ, *then*

$$\Phi^{-1}(t)\tilde{\Phi}^{-1}(t) \le 2t$$

for every $t > 0$.

Proof.

It follows from [7, Theorem 2.3] that, for any $f \in L_\Phi(\mathbb{R}^n)$ and $g \in L_{\tilde{\Phi}}(\mathbb{R}^n)$ we have

$$\int_{\mathbb{R}^n} |f \cdot g| dx \le 2 \parallel f \parallel_{L_\Phi(\mathbb{R}^n)} \parallel g \parallel_{L_{\tilde{\Phi}}(\mathbb{R}^n)} .$$

By Corollary 3.5, we have $\Phi^{-1}(t)\tilde{\Phi}^{-1}(t) \le 2t$. □

Remark. We can prove the Corollary 3.7 directly by definition of Φ^{-1} and $\tilde{\Phi}^{-1}$. What we showed here is that we can obtain the result through the lens of Orlicz-Morrey spaces.

4. Conclusion

In this paper, we have discussed the inclusion relation between Orlicz-Morrey spaces for Nakai version. By estimating the norm of characteristic function of balls in $\mathbb{R}^n$, we obtain sufficient and necessary conditions for inclusion relation between Orlicz-Morrey spaces (Theorem 2.2), which were generalized for inclusion property of Orlicz spaces in [6]. Furthermore, by estimating the characteristic function of balls in $\mathbb{R}^n$ we obtain sufficient and necessary conditions for generalized Hölder's inequality on Orlicz-Morrey spaces (Theorem 3.1 and Theorem 3.4). Theorem 3.4 generalized Hölder's inequality in [7].

Acknowledgment

The first and second authors are supported by ITB Research and Innovation Program 2016.

References

[1] Cheung W S 2001 *Intl. J. Math. Math. Sci.* **26**(1) pp 7–10
[2] Gala S, Sawano Y, and Tanaka H 2015 *J. Approx. Theory* **198** pp 1–9
[3] Gunawan H, Hakim D I, Limanta K M, Masta A A 2017 *Math. Nachr.* **290** pp 332–340
[4] Kreĭn S G, Petunĭn Yu Ī, and Semënov E M 1982 *Interpolation of Linear Operators* (Translation of Mathematical Monograph vol. 54) (USA: American Mathematical Society)
[5] Maligranda L 1989 *Orlicz Spaces and Interpolation* (Brazil: Departamento de Matemática, Universidade Estadual de Campina)
[6] Masta A A, Gunawan H, Setya-Budhi W 2016 *J. Math. Fund. Sci.* **48**-3 pp 193–203
[7] Nakai E 2006 (research report: http://repository.kulib.kyoto-u.ac.jp/dspace/bitstream/2433/58769/1/1520-10.pdf, accessed on August 17, 2015.)
[8] Nakai E 2004 (research report: http://repository.kulib.kyoto-u.ac.jp/dspace/bitstream/2433/26035/1/ 1399-13.pdf, accessed on August 17, 2015.)
[9] Nakai E 2008 *Studia Math.* **188**-3 pp 193–221
[10] Osançliol A 2014 *J. Inequal. Appl.* **2014**-390 pp 1–8
[11] Rao M M and Ren Z D 1991 *Theory of Orlicz spaces* (Volume 146 of Monographs and Textbooks in Pure and Applied Mathematics) (New York: Marcel Dekker, Inc.)
[12] Sawano Y, Sugano S, and Tanaka H 2012 *Potential Anal.* **36**-4 pp 517–556

IOP Conf. Series: Journal of Physics: Conf. Series **893** (2017) 012016 doi:10.1088/1742-6596/893/1/012016

Bounded 2-linear functionals on the n-normed spaces

Harmanus Batkunde and Francis Y. Rumlawang

Department of Mathematics, Universitas Pattimura, Ambon, 97123, Indonesia

E-mail: {h.batkunde, rumlawang}@fmipa.unpatti.ac.id

Abstract. Using concepts of n-normed spaces and n-inner product spaces, we will discuss about bounded 2-linear functionals or bounded bilinear functionals on the n-normed spaces which can be defined in several ways (with respect to the type of boundedness). We shall be interested in evaluating the norm of these functionals as well as the dual spaces which are formed by the bounded 2-linear functionals defined earlier. Moreover, we will show an analogue of Riesz-Fréchet theorem of bounded 2-linear functionals on the n-normed spaces.

1. Introduction

The concept of n-normed spaces ($n \geq 2$) was initially introduced by S. Gähler, and that of n-inner product spaces ($n \geq 2$) was developed by Misiak.
As we know that the length of a vector is a subject we discuss on normed spaces, in n-normed spaces we speak about volume of parallelepiped which spanned by a set consist of n-vectors.

Definition 1.1. *Let n be a nonnegative integer and X is a real vector space with $dim(X) \geq n$. An n-norm is a function $\|\cdot, \dots, \cdot\| : X^n \to \mathbb{R}$ which satisfies these following conditions:*

i. $\|x_1, \cdots, x_n\| \geq 0$
$\|x_1, \cdots, x_n\| = 0$ *if and only if* $x_1, \cdots, x_n$ *linearly dependent,*

ii. $\|x_1, \cdots, x_n\|$ *is invariant under permutation,*

iii. $\|\alpha x_1, \dots, x_n\| = |\alpha| \ \|x_1, \dots, x_n\|$ *for any* $\alpha \in \mathbb{R}$,

iv. $\|x_1 + x_1', x_2, \dots, x_n\| \leq \|x_1, x_2, \dots, x_n\| + \|x_1', x_2, \dots, x_n\|$,

a pair of real vector X and its n-norm notated by $(X, \|\cdot, \dots, \cdot\|)$ is called an n-normed space.

The fact about n-normed spaces is that they are actually normed spaces. As an example, if $(X, \langle\cdot,\cdot\rangle)$ is an inner product space, we can define the standard n-norm on X by

$$\|x_1, \dots, x_n\|^S := \begin{vmatrix} \langle x_1, x_1\rangle & \cdots & \langle x_1, x_n\rangle \\ \vdots & \ddots & \vdots \\ \langle x_n, x_1\rangle & \cdots & \langle x_n, x_n\rangle \end{vmatrix}^{1/2}.$$

Geometrically, the value of $\|x_1, \dots, x_n\|^S$ represents the volume of the n-dimensional parallelepiped spanned by $x_1, \dots, x_n$ (see [3]).

Definition 1.2. *Let n be a nonnegative integer and X is a real vector space. Assume that $dim(X) = d \geq n+1$. A real-valued function $\langle\cdot,\cdot|\cdot,\dots,\cdot\rangle$ on X^{n+1} that satisfying these following conditions:*

i. $\langle x_1, x_1|x_2, \ldots, x_n\rangle \geq 0$ *and it is equal to 0 if and only if* $x_1, \ldots, x_n$ *are linearly dependent,*
ii. $\langle x_{i_1}, x_{i_1}|x_{i_2}, \ldots, x_{i_n}\rangle = \langle x_1, x_1|x_2, \ldots, x_n\rangle$ *for any permutation* $\{i_1, \ldots, i_n\}$ *of* $\{1, \ldots, n\}$,
iii. $\langle x, y|x_2, \ldots, x_n\rangle = \langle y, x|x_2, \ldots, x_n\rangle$,
iv. $\langle \alpha x, y|x_2, \ldots, x_n\rangle = \alpha\langle x, y|x_2, \ldots, x_n\rangle$ *for any* $\alpha \in \mathbb{R}$,
v. $\langle x + x', y|x_2, \ldots, x_n\rangle = \langle x, y|x_2, \ldots, x_n\rangle + \langle x', y|x_2, \ldots, x_n\rangle$,

is called an n-inner product on X, and the pair $(X, \langle\cdot,\cdot|\cdot,\ldots,\cdot\rangle)$ *is called an n-inner product space.*

If $(X, \langle\cdot,\cdot|\cdot,\ldots,\cdot\rangle)$ is an n-inner product space, then we can define an n-norm $\|\cdot,\ldots,\cdot\|$ on X by

$$\|x_1, x_2, \ldots, x_n\| := \langle x_1, x_1|x_2, \ldots, x_n\rangle^{1/2}.$$

We say this norm is induced by n-inner product. Here we also have the Cauchy-Schwarz inequality

$$|\langle x, y|x_2, \ldots, x_n\rangle| \leq \|x, x_2, \ldots, x_n\| \, \|y, x_2, \ldots, x_n\|.$$

In an n-normed space $(X, \|\cdot,\ldots,\cdot\|)$ we have [6]:

$$\|x_1, \ldots, x_n\| = \|x_1 + \alpha_2 x_2 + \alpha_n x_n, x_2, \ldots, x_n\| \tag{1}$$

for all $x_1, \ldots, x_n \in X$ and $\alpha_2, \ldots, \alpha_n \in \mathbb{R}$.

2. Main Result

In this section we will define bounded 2-linear functionals in several ways, evaluate its norm and also provide an anologue of Riesz-Fréchet theorem of bounded 2-linear funtionals.

2.1. Bounded 2-linear Functional (of 1st index)

Definition 2.1. *Let* $(X, \|\cdot,\ldots,\cdot\|_X)$ *and* $(Z, \|\cdot,\ldots,\cdot\|_Z)$ *be the n-normed spaces and fix linearly independent sets* $Y = \{y_1, \ldots, y_n\}$ *and* $W = \{w_1, \ldots, w_n\}$ *in X and Z, respectively. A 2-linear functional* $h : X \times Z \to \mathbb{R}$ *is said to be bounded of 1st index (with respect to a pair of sets Y and W) if there exist* $k \in \mathbb{R}$ *such that*

$$|h(x, z)| \leq k \left(\sum \|x, y_{i_2}, \ldots, y_{i_n}\|_X\right) \left(\sum \|z, w_{i_2}, \ldots, w_{i_n}\|_Z\right) \tag{2}$$

for all $x \in X$ *and* $z \in Z$*, where the sum is taken over* $\{i_2, \ldots, i_n\} \subset \{1, \ldots, n\}$ *with* $i_2 \leq \cdots \leq i_n$.

Let $(X \times Z)'_1$ be a dual space that contains all bounded 2-linear (of 1st index) in $X \times Z$, for every $h \in (X \times Z)'_1$ we define :

$$\|h\|_1 := \inf\{k > 0 \,:\, (2) \text{ holds}\}. \tag{3}$$

One can see that this formula is identical with

$$\|h\|_1 := \sup\left\{|h(x, z)| : \left(\sum \|x, y_{i_1}, \ldots, y_{i_n}\|_X\right) \leq 1, \left(\sum \|z, w_{i_1}, \ldots, w_{i_n}\|_Z\right) \leq 1\right\}. \tag{4}$$

Moreover, these identical formulas (3) and (4) define a norm on $(X \times Z)'_1$.
As an example of bounded 2-linear functional, let $(X, \langle\cdot,\cdot|\cdot,\ldots,\cdot\rangle_X)$ and $(Z, \langle\cdot,\cdot|\cdot,\ldots,\cdot\rangle_Z)$ be the n-inner product spaces and $h : X \times Z \to \mathbb{R}$ is defined by

$$h\,(x, z) = \left(\sum \langle x, y_{i_1}|y_{i_2}, \ldots, y_{i_n}\rangle_X\right) \left(\sum \langle z, w_{i_1}|w_{i_2}, \ldots, w_{i_n}\rangle_Z\right), \tag{5}$$

with $\{y_1,\ldots,y_n\}$ and $\{w_1,\ldots,w_n\}$ are linear independent sets on X and Z, respectively. Easily one can see that h is a 2-linear functional. For simplicity, all the sums in this paper are taken over $\{i_2,\ldots,i_n\}\subset\{1,\ldots,n\}$ with $i_2\leq\cdots\leq i_n$ and $i_1\in\{1,\ldots,n\}\setminus\{i_2,\ldots,i_n\}$.

Moreover, we'll show that (5) is bounded functional 2-linear functional of 1st index on this following fact.

Fact 2.2. *2-linear functional in (5) is bounded (of 1st index) with*

$$\|h\|_1=\|y_1,\ldots,y_n\|_X\ \|w_1,\ldots,w_n\|_Z. \tag{6}$$

Proof. We observe that for any $x\in X$ and $z\in Z$, by using triangle and Cauchy Schwars Inequality we'll have

$$|h(x,z)|\leq\|y_{i_1},\ldots,y_{i_n}\|_X\ \|w_{i_1},\ldots,w_{i_n}\|_Z\left(\sum\|x,y_{i_2},\ldots,y_{i_n}\|_X\right)\left(\sum\|z,w_{i_2},\ldots,w_{i_n}\|_Z\right),$$

Then h is a bounded 2-linear functional (of 1st index) with
$\|h\|_1\leq\|y_{i_1},y_{i_2},\ldots,y_{i_n}\|_X\ \|w_{i_1},w_{i_2},\ldots,w_{i_n}\|_Z$. To show the equality, just take

$$x=\|y_1,\ldots,y_n\|^{-1}y_1 \text{ and } z=\|w_1,\ldots,w_n\|^{-1}w_1,$$

then we have

$$\sum\|x,y_{i_2},\ldots,y_{i_2}\|=1 \text{ and} \sum\|z,w_{i_2},\ldots,w_{i_2}\|=1,$$

also

$$\begin{aligned}|h(x,z)|=\ &\|y_1,\ldots,y_n\|^{-1}\ \|w_1,\ldots,w_n\|^{-1}\\ &\left(\sum\langle y_1,y_{i_1}|y_{i_2},\ldots,y_{i_n}\rangle\right)\\ &\left(\sum\langle w_1,w_{i_1}|w_{i_2},\ldots,w_{i_n}\rangle\right).\end{aligned}$$

One can observe that when $i_1\neq1$ and $\{i_2,\ldots,i_n\}=\{1,\ldots,n\}\setminus\{i_1\}$, by using Cauchy-Schwarz inequality we have

$$|\langle y_1,y_{i_1}|y_{i_2},\ldots,y_{i_n}\rangle|\leq\|y_1,y_{i_2},\ldots,y_{i_n}\|\|y_{i_1},y_{i_2},\ldots,y_{i_n}\|=0,$$

it is analogue for $|\langle w_1,w_{i_1}|w_{i_2},\ldots,w_{i_n}\rangle|=0$, because one of $y_{i_2},\ldots,y_{i_n}$ must be equal to y_1 and one of $w_{i_2},\ldots,w_{i_n}$ must be equal to w_1. Then we have

$$\begin{aligned}|h(x,z)|=\ &\|y_1,\ldots,y_n\|^{-1}\ \|w_1,\ldots,w_n\|^{-1}\\ &\langle y_1,y_1|y_2,\ldots,y_n\rangle\ \langle w_1,w_1|w_2,\ldots,w_n\rangle\\ =\ &\|y_1,\ldots,y_n\|\ \|w_1,\ldots,w_n\|.\end{aligned}$$

Therefore, h is bounded (of 1st index) with
$\|h\|_1=\|y_1,\ldots,y_n\|\ \|w_1,\ldots,w_n\|$. □

2.2. Bounded 2-linear Functional (of p − th index)

Definition 2.3. *Let $(X,\|\cdot,\ldots,\cdot\|_X)$ and $(Z,\|\cdot,\ldots,\cdot\|_Z)$ be the n-normed spaces, $h:X\times Z\to\mathbb{R}$ and $1\leq p\leq\infty$, fix linearly independent sets $Y=\{y_1,\ldots,y_n\}$ and $W=\{w_1,w_2,\ldots,w_n\}$ in X and Z, respectively. A 2-linear functional $h:X\times Z\to\mathbb{R}$ is said to be bounded of p-th index (with respect to a pair of sets Y and W) if there exist $k\in\mathbb{R}$ such that*

$$|h(x,z)|\leq k\left(\sum\|x,y_{i_2},\ldots,y_{i_n}\|_X^p\right)^{\frac{1}{p}}\left(\sum\|z,w_{i_2},\ldots,w_{i_n}\|_Z^p\right)^{\frac{1}{p}} \tag{7}$$

for all $x\in X$ and $z\in Z$.

Let $(X \times Z)'_p$ be a dual space contains all bounded 2-linear (of p-th index) in $X \times Z$, for every $h \in (X \times Z)$ we define :

$$\|h\|_p := \inf\{k > 0 \,:\, (7) \text{ holds}\}. \tag{8}$$

One can see that (8) is identical with

$$\|h\|_p := \sup\{\sum \|x, y_{i_2}, \ldots, y_{i_n}\|_X \leq 1 \text{ and} \sum \|z, w_{i_2}, \ldots, w_{i_n}\|_Z \leq 1\} \tag{9}$$

Fact 2.4. *2-linear functional in (3) is bounded (of p-th index) with*

$$\|h\|_p := n^{\frac{2}{q}} \|y_1, \ldots, y_n\|_X \|w_1, \ldots, w_n\|_Z,$$

where $\frac{1}{p} + \frac{1}{q} = 1$.

Proof. The proof is similar with **Fact II.2.**

Theorem 2.5. *Let* $(X, \|\cdot, \ldots, \cdot\|_X)$ *and* $(Z, \|\cdot, \ldots, \cdot\|_Z)$ *be the n-normed spaces, if h is a 2-linear functional in* $X \times Z$ *then h is bounded of p-th index if and only if h is bounded* $1st$ *index.*

Proof. Let h be a bounded 2-linear functional of $p-th$ index (with respect to linear independent sets $Y = \{y_1, \ldots, y_n\}$ and $W = \{w_1, \ldots, w_n\}$ in X and Z, respectively) .
If $x \in X$ and $z \in Y$ satisfy

$$\sum \|x, y_{i_2}, \ldots, y_{i_n}\| \leq 1 \,\text{and} \sum \|z, w_{i_2}, \ldots, w_{i_n}\| \leq 1,$$

it means each term of the sum is less than 1, as a consequence

$$(\sum \|x, y_{i_2}, \ldots, y_{i_n}\|^p) (\sum \|z, w_{i_2}, \ldots, w_{i_n}\|^p) \leq 1,$$

then we have $|h(x, z)| \leq \|h\|_p$ or h is bounded of $1st$ index with

$$\|h\|_1 \leq \|h\|_p, \text{ for every } p \geq 1.$$

Conversely, let h be a bounded 2-linear functional of $1st$ index with respect to linear independent sets $Y = \{y_1, \ldots, y_n\}$ and $W = \{w_1, \ldots, w_n\}$.
If $x \in X$ and $z \in Y$ satisfy

$$\sum \|x, y_{i_2}, \ldots, y_{i_n}\|^p \leq 1,$$
$$\sum \|z, w_{i_2}, \ldots, w_{i_n}\|^p \leq 1,$$

then by using Hölder inequality we have

$$(\sum \|x, y_{i_2}, \ldots, y_{i_n}\|) (\sum \|z, w_{i_2}, \ldots, w_{i_n}\|) \leq n^{\frac{2}{q}}.$$

so it means

$$\left| h\left(\frac{x}{n^{\frac{1}{q}}}, \frac{z}{n^{\frac{1}{q}}}\right) \right| \leq \|h\|_1$$

or we can write

$$|h(x, z)| \leq n^{\frac{2}{q}} \|h\|_1.$$

Then h is bounded type-p with
$\|h\|_p \leq n^{\frac{2}{q}} \|h\|_1$. □

Moreover, this theorem states that $\|\cdot\|_p$ and $\|\cdot\|_1$ equivalent by $\|h\|_1 \leq \|h\|_p \leq n^{\frac{2}{q}} \|h\|_1$. Where $\frac{1}{p} + \frac{1}{q} = 1$ and $p \geq 1$.

Furthermore, dual spaces that formed by each type (or index) of bounded 2-linear functional are identical. We may simply use the word 'bounded' instead of 'bounded of p-th index' Unless we need to specify the index explicitly.

2.3. Duality Properties for $p = 2$.
Let $(X, \langle\cdot,\cdot|\cdot,\ldots,\cdot\rangle_X)$ and $(Y, \langle\cdot,\cdot|\cdot,\ldots,\cdot\rangle_Z)$ be the n-inner product spaces with $\|\cdot,\ldots,\cdot\|_X = \langle\cdot,\cdot|\cdot,\ldots,\cdot\rangle_X^{\frac{1}{2}}$ and $\|\cdot,\ldots,\cdot\|_Z = \langle\cdot,\cdot|\cdot,\ldots,\cdot\rangle_Z^{\frac{1}{2}}$ are induced norm on X and Z respectively (with respect to linear independent sets $Y = \{y_1,\ldots,y_n\}$, $W = \{w_1,\ldots,w_n\}$ in X and Z respectively). Define $h_\xi(x,z) : X \times Z \to \mathbb{R}$ by

$$h_\xi(x,z) = \left(\sum\langle x, y|y_{i_2},\ldots,y_{i_n}\rangle_X\right)\left(\sum\langle z, w|w_{i_2},\ldots,w_{i_n}\rangle_Z\right), \tag{10}$$

One can observe that h_ξ is a 2-linear functional and also see that (10) is a special case of (5) by taking $y = y_1 + y_2 + \cdots + y_n$ and $w = w_1 + w_2 + \cdots + w_n$. The norm of h_ξ will be stated in this following fact.

Fact 2.6. *Linear functional h_ξ which defined in* (10) *is bounded of* 2nd *index with* $\|h_\xi\| = \left(\sum\|y, y_{i_2},\ldots,y_{i_n}\|\right)_X^{\frac{1}{2}}\left(\sum\|w, w_{i_2},\ldots,w_{i_n}\|\right)_Z^{\frac{1}{2}}$

Proof. The proof is similar with **Fact II.2.**

We will show an analog of Riesz-Fréchet theorem for 2-linear functionals on the n-normed space which are bounded of 2nd index. To get that, we use this following theorem.

Theorem 2.7. *Let $(X \times Y, \langle\cdot,\cdot|\cdot,\ldots,\cdot\rangle)$ be an n-inner product space and $\|\cdot,\ldots,\cdot\| = \langle\cdot,\cdot|\cdot,\ldots,\cdot\rangle^{\frac{1}{2}}$ be the induced n-norm on $X \times Y$ with respect to linearly independent set $\{v_1,\ldots,v_n\}$ on $X \times Y$. The mapping $\langle\cdot,\cdot\rangle : (X \times Y)^2 \to \mathbb{R}$ is given by*

$$\langle v, u\rangle := \sum\langle v, u|v_{i_2},\ldots,v_{i_n}\rangle, \tag{11}$$

for all $v, u \in X \times Y$, defines an n-inner product on $X \times Y$ and its induced norm $\|\cdot\|_2 : X \times Y \to \mathbb{R}$, given by

$$\|v\|_2 := \left(\sum\|v, y_{i_2},\ldots,y_{i_n}\|^2\right)^{\frac{1}{2}}, \tag{12}$$

for all $v \in X \times Y$.

Here we have an analogue of Riesz-Fréchet theorem as a corollary of the theorem above.

Corollary 2.8. *If $(X \times Y, \langle\cdot,\cdot|\cdot,\ldots,\cdot\rangle)$ is complete with respect to the norm $\|\cdot\|_2$ in (12), then for every 2-linear functional h which is bounded of 2nd index on $X \times Y$ on there exist a unique $A \in B(X \times Y)$ such that*

$$h(x,y) = \langle Ax, y\rangle_{X\times Y}; x \in X, y \in Y.$$

Acknowledgments
We would like to thank all the reviewers for their helpful remarks. This research is supported by Department of Mathematics, Mathematics and Science Faculty, Universitas Pattimura.

References
[1] Batkunde H, Gunawan H and Pangalela Y E P 2013 *Acta univ. M. Belii Ser. Math.* **21**
[2] Gahler S *Math. Nachr.* **28** (1)
[3] Gantmacher F R 2000 *AMS Chelsea Publishing.* **1**
[4] Gozali S, Gunawan H and Neswan O 2010 *Annals of Functional Analysis* **1**
[5] Gunawan H 2001 *bull. Auth. Math. Soc.* **64** (137)
[6] Gunawan H and Mashadi H 2001 *Int. J. Math. Sci* **27** (631)
[7] Gunawan H and Mashadi H 2001 *Soochow J. Math* **27**
[8] Kreyszig E 1978 *Introductory Functional Analysis with Application*
[9] Misiak A 1989 *Math. Nachr.* **149**
[10] Pangalela Y E and Gunawan H 2013 *Math Bohem.* **138**
[11] Wibawa-Kusuma R A and Gunawan H 2013 *Math. Hungar.* **67**

IOP Conf. Series: Journal of Physics: Conf. Series **893** (2017) 012017 doi:10.1088/1742-6596/893/1/012017

Social and economic influences on human behavioural response in an emerging epidemic

P Phang[1,2], B Wiwatanapataphee[2], and Y H Wu[2]

[1] Faculty of Computer Science and Information Technology
Universiti Malaysia Sarawak, Sarawak, Malaysia
[2] Department of Mathematics and Statistics, Faculty of Science and Engineering
Curtin University, Perth, Australia

E-mail: pphang@unimas.my

Abstract. The human behavioural changes have been recognized as an important key in shaping the disease spreading and determining the success of control measures in the course of epidemic outbreaks. However, apart from cost-benefit considerations, in reality, people are heterogeneous in their preferences towards adopting certain protective actions to reduce their risk of infection, and social norms have a function in individuals' decision making. Here, we studied the interplay between the epidemic dynamics, imitation dynamics and the heterogeneity of individual protective behavioural response under the considerations of both economic and social factors, with a simple mathematical compartmental model and multi-population game dynamical replicator equations. We assume that susceptibles in different subpopulations have different preferences in adopting either normal or altered behaviour. By incorporating both intra- and inter-group social pressure, the outcome of the strategy distribution depends on the initial proportion of susceptible with normal and altered strategies in both subpopulations. The increase of additional cost to susceptible with altered behaviour will discourage people to take up protective actions and hence results in higher epidemic final size. For a specific cost of altered behaviour, the social group pressure could be a "double edge sword", though. We conclude that the interplays between individual protective behaviour adoption, imitation and epidemic dynamics are necessarily complex if both economic and social factors act on populations with existing preferences.

1. Introduction

The role of human behaviour in the spread and control of infectious diseases has received significant attention [1]. Various behavioural changes, including vaccination, taking antiviral drugs, health protective actions such as social distancing, wearing masks, reducing travels, practicing better hygiene and avoiding crowded places, have been incorporated into epidemic models. Without including behavioural changes, the disease spreading model will predict the worst possible scenario [2]. In the course of epidemic outbreak, individuals are not passive, they weigh up the costs and risks associated with certain protective actions with the benefit of reducing the infection risk, and then choose the best strategy that maximizes their own benefit. This individual-level decision making can be concisely described in the language of game theory.

In biology, strategies are inherited, but the underlying mechanism for the evolution of strategy adoption in epidemiology is imitation process. Imitation dynamics can be coupled into one-population susceptible-infective-recovered (SIR) compartmental model via the replicator dynamical equation. For instance, Bauch [3] used cost-benefit analysis to study the decision making by parents in vaccinating their children for childhood diseases. Bauch's model was extended in [4] by introducing social norm into it. In [5], the spontaneous behavioural changes in an emerging epidemic were modelled based on Bauch's framework for both natural selection and imitation process. With these simple well-mixed one-population models, rich epidemic dynamics, including oscillations and simulation results fitted with real disease incidence data, can be achieved.

In homogeneous population models, every individual has the same kind of contact pattern and strategy interactions. However, entities are heterogeneous and there also exists heterogeneity between groups. People that belong to certain group may be more inclined to certain behaviour or perceive the risk of disease in a specific way. These give rise to the development of multi-group epidemic and game-dynamical models. Among them, most of the studies are related to vaccination behaviour with specific investigations on different age groups [6], countries [7] and risk perceptions [8-9]. Also, multi-population model was used in [10] to study health protective behaviour during epidemic. However, all these studies assumed that individuals only make economic cost-benefit considerations without considering other social factors in their decision making. Hence, we focus on the health protective actions adopted by susceptibles in an emerging epidemic in which the imitation dynamics are constructed by taking into account of a combination of economic and social points of view in two subpopulations setting.

In two subpopulations (or groups), a player could be randomly paired with one of other players either from the same or different subpopulation. This leads to intra- and inter-subpopulation strategy interactions. Bi-matrix games are suitable for the situation whereby there are only inter-subpopulation interactions [11]. This specification rules out own subpopulation effects. However, since we are living in a modern society with people from different backgrounds, beliefs and cultural, both intra- and inter-subpopulation strategy interactions are important in shaping our decision making. Therefore, we chose the multi-population game-dynamical modelling framework proposed in [12-13] which allows strategy interactions within and across groups with different preferences in constructing our game-dynamical model.

2. The Model

Our proposed natural selection and imitation dynamics model with economic and social factors is given below.

$$\frac{dS}{dt} = -\beta XSI \tag{1a}$$

$$\frac{dI}{dt} = \beta XSI - \gamma I \tag{1b}$$

$$\frac{dR}{dt} = \gamma I \tag{1c}$$

$$\begin{aligned}\frac{dp}{dt} &= \beta p(X-1)I + p(1-p)\rho\{f\left[k(1-mI)+\Omega_n+\delta(2p-1)\right]+\\ &\quad (1-f)\left[k(1-mI)+\Omega_n+\delta(p(1-q)^2-(1-p)q^2)\right]\}\end{aligned} \tag{1d}$$

$$\begin{aligned}\frac{dq}{dt} &= \beta q(X-\alpha)I + q(1-q)\rho\{(1-f)\left[-k(1-mI)+\Omega_a+\delta(2q-1)\right]+\\ &\quad f\left[-k(1-mI)+\Omega_a+\delta(q(1-p)^2-(1-q)p^2)\right]\}\end{aligned} \tag{1e}$$

where $X = fp + \alpha f(1-p) + (1-f)(1-q) + \alpha(1-f)q$. S, I, R, p, q and hence X are time-dependent variables. Equations 1(a)-(c) are the rate equations for individuals with disease state susceptible (S), infective (I) and recovered (R), which govern the epidemic dynamics.

Susceptible individuals are subdivided into two subclasses, namely individuals adopting normal behaviour (or strategy) (S_n) and individuals adopting altered behaviour (S_a). β is the disease transmission rate, $1/\gamma$ is the average duration of infectivity and $\alpha \in (0,1)$ is the reduction of the force of infection performed by S_a.

Equations (1d) and (1e) govern the evolution of strategy frequencies in two subpopulations, respectively. Following [12-13], we assume that the preferred strategy in subpopulation 1 is normal behaviour and subpopulation 2 is altered behaviour. Hence, p (resp. q) is the proportion of susceptibles in subpopulation 1 (resp. 2) who adopt subpopulation 1's (resp. 2's) preferred behaviour, i.e. normal (resp. altered) behaviour. We have $S \equiv p\,S^{[1]} + (1-q)\,S^{[2]} + (1-p)\,S^{[1]} + q\,S^{[2]}$, where S with superscript [1] and [2] are the fraction of susceptible in two respective subpopulations at any time. We assume that the subpopulation of susceptibles does not change throughout the simulation, i.e. no migration between subpopulations.

The parameter ρ gives the speed of the imitation process with respect to the disease transmission process as well as a measure of how willing the players are to switch to new strategy based on the payoff difference. If $\rho = 0$, individuals do not include cost-benefit analysis in changing their strategy, and system (1) is reduced to the natural selection model. In two-subpopulation game dynamics, f (resp. $1 - f$) can be interpreted as the relative power of subpopulation 1 (resp. 2), which reflects how much influence a subpopulation has on the strategy adoption of susceptible individuals [13] and essentially the sizes of both subpopulations evolve in a constant ratio [11]. Hence, X in (1) reflects the population average reduction of force of infection which is determined by strategy frequencies and the relative power of both subpopulations.

The parameter k represents the additional cost of adopting altered behaviour as compared to doing nothing in the course of epidemic outbreak, and m weighs the perceived disease prevalence [5]. As the direct consequences of following [12-13], the $\pm$ sign in $+k(1 - mI)$ and $-k(1 - mI)$ reflects that the evolution of strategies frequency under cost-benefit analysis in one subpopulation is the opposite of the outcome on another subpopulation. In a way, this model the common belief that the increase (resp. decrease) of the cost of adopting altered behaviour will drive more susceptible in subpopulation 1 (resp. 2) adopting normal (resp. altered) behaviour. We assume that the susceptibles have full knowledge on disease prevalence which is certainly not realistic but unavoidable so as to use minimal number of equations in the model.

To reflect the incompatible preferences, we assume that one of the two strategies in a particular subpopulation is more attractive than the other. For each individual, if one strategy is more attractive, he/she would be better paid if choosing that strategy [14]. Unlike [12-13], the incompatible preferences in our model are reflected by adding an extra benefit Ω, as in [14], for susceptibles choosing the preferred strategy in their respective subpopulation. We incorporated social factor by assuming that individuals can apply two types of group pressure δ to support conformity and discourage dis-coordinated behaviour. First, similar to [4], $\delta(2p-1)$ and $\delta(2q-1)$ are the intra-subpopulation pressure imposed on subpopulation 1 and 2, respectively. Second, the appearance of $\delta(p(1-q)^2 - (1-p)q^2)$ and $\delta(q(1-p)^2 - (1-q)p^2)$ are the result of inter-subpopulation strategy interactions.

3. Results and Discussion

Figure 1 gives the vector fields which represent the direction and size of temporal change of strategy distribution. The initial conditions $(p(0), q(0))$ can be regarded as the previous history of proportion of normal and altered strategies in both subpopulations, which is indeed the existing preferences of particular subpopulation. By setting $\rho = 0$ (figure 1(a)), the strategy switching is only driven by natural selection and consequently the temporal change of strategy distribution in both subpopulations is minimal. If the imitation process is incorporated into the model by setting $\rho > 0$, then when the magnitude of social group pressure δ is small (figure

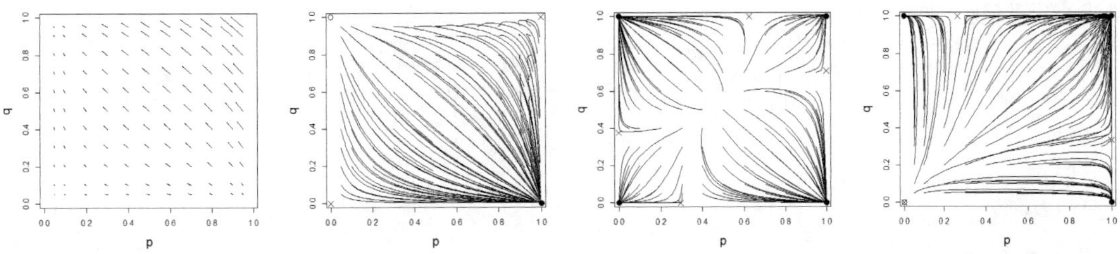

(a) (b) (c) (d)

Figure 1: The $(p(t), q(t))$ trajectories for natural selection model, imitation dynamics with different magnitudes of group pressure and extra benefit. (a) $\rho = 0$. (b) $\rho = 1.5$, $\delta = 0.0001$. (c) $\rho = 1.5$, $\delta = 0.4$. (d) $\Omega_n = 0.035$, $\Omega_a = 0.075$. Unless otherwise specified in the description for each of the subfigures, the following parameter values were used for all simulations and figures in this paper: $S(0) = 0.95$, $I(0) = 0.05$, $\beta = 0.9$, $\gamma = 0.5988$, $\alpha = 0.86$, $f = 0.5$, $\rho = 1.5$, $\delta = 0.1$, $k = 0.025$, $m = 0.1$, $\Omega_n = \Omega_a = 0$, and $0.05 \leq p(0), q(0) \leq 0.95$. The simulations end at $t = 80$ with $\triangle t = 1$. Legend: $\{\bullet, \circ, \times\}$ denote {stable, unstable, saddle point} respectively.

Table 1: The fixed points and their associated eigenvalues obtained from the local stability analysis.

l	Fixed point (p_l, q_l)	Associated eigenvalues, λ_l and v_l
1	$(0,0)$	$\lambda_1 : -\rho(f\delta - k)$
		$v_1 : -\rho(k + (1-f)\delta)$
2	$(1,0)$	$\lambda_2 : -\rho(k+\delta)$
		$v_2 : -\rho(k+\delta)$
3	$(0,1)$	$\lambda_3 : \rho(k-\delta)$
		$v_3 : \rho(k-\delta)$
4	$(1,1)$	$\lambda_4 : -\rho(f\delta + k)$
		$v_4 : -\rho(-k + (1-f)\delta)$
5	$(p_5 = \frac{f\delta - k}{\delta f + \delta}, 0)$	$\lambda_5 : p_5(1-p_5)\rho(f\delta + \delta)$
		$v_5 : \rho((1-f)(-k-\delta) + f(-k - \delta p_5^2))$
6	$(p_6 = \frac{\delta - k}{\delta f + \delta}, 1)$	$\lambda_6 : p_6(1-p_6)\rho(f\delta + \delta)$
		$v_6 : -\rho((1-f)(-k+\delta) + f(-k + \delta(1-p_6)^2))$
7	$(1, q_7 = \frac{k+\delta}{2\delta - \delta f})$	$\lambda_7 : -\rho(f(k+\delta) + (1-f)(k + \delta(1-q_7)^2))$
		$v_7 : q_7(1-q_7)\rho(2(1-f)\delta + \delta f)$
8	$(0, q_8 = \frac{k+\delta(1-f)}{2\delta - \delta f})$	$\lambda_8 : \rho(f(k-\delta) + (1-f)(k - \delta q_8^2))$
		$v_8 : q_8(1-q_8)\rho(2(1-f)\delta + \delta f)$

1(b)), all trajectories tend towards the fixed point $(1, 0)$ which means all susceptibles finally adopt normal behaviour as the epidemic is over. However, if δ is large (figure 1(c)), there exist four basins of attraction leading to four fixed points located at four respective corners of the pq-plane. The social group pressure could be balanced by extra benefit Ω given to individuals adopting preferred strategy in their respective subpopulation. In figure 1(d), both subpopulations end up with subpopulation-specific norms (fixed point $(1, 1)$ which is defined as coexisting "subcultures" [13]) whenever $p(0)$ and $q(0)$ are almost the same.

Following [12], we carried out local stability analysis for subsystem (1d) and (1e) without

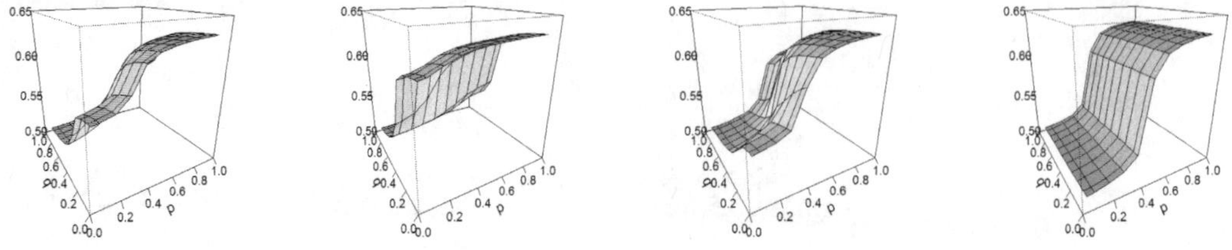

(a) (b) (c) (d)

Figure 2: The epidemic final size for all initial conditions $(p(0), q(0))$ with different relative powers (or strengths) of subpopulation, f. (a) $f = 0.5$, both subpopulations are equally strong. (b) $f = 0.2$, subpopulation 2 is more powerful than subpopulation 1. (c) $f = 0.7$, subpopulation 1 is more powerful than subpopulation 2. (d) $f = 0.99$, subpopulation 1 is extremely strong such that the preference of subpopulation 2 is negligible.

Table 2: The fixed points and their associated eigenvalues for figures 1(b) and 1(c).

		Figure 1(b)			Figure 1(c)		
l	Fixed point (p_l, q_l)	λ_l	v_l	Type of fixed point	λ_l	v_l	Type of fixed point
1	$(0, 0)$	0.0374	-0.0375	saddle point	-0.2625	-0.3375	stable
2	$(1, 0)$	-0.0376	-0.0376	stable	-0.6375	-0.6375	stable
3	$(0, 1)$	0.0373	0.0373	unstable	-0.5625	-0.5625	stable
4	$(1, 1)$	-0.0375	0.0374	saddle point	-0.3375	-0.2625	stable
5	$(p_5, 0) = (0.2916, 0)$	Note: All p_5, p_6, q_7 and q_8 are			0.1859	-0.3630	saddle point
6	$(p_6, 1) = (0.625, 1)$	out of range $[0, 1]$. Therefore, no			0.2109	-0.3047	saddle point
7	$(1, q_7) = (1, 0.7083)$	fixed points lie on the			-0.3630	0.1859	saddle point
8	$(0, q_8) = (0, 0.375)$	boundary of the pq-plane.			-0.3047	0.2109	saddle point

extra benefit ($\Omega_n = \Omega_a = 0$) at disease free equilibrium. We can determine the stability state of each of the fixed points based on the two eigenvalues of each fixed point given in table 1. With the combination parameter values used in figures 1(b) and 1(c), the nature of the fixed points were examined by evaluating their associated eigenvalues (table 2). The numerical simulation result is in good agreement with local stability analysis.

Figure 2 illustrates the relation between the previous history of strategy proportion and the epidemic final size for different relative strengths of subpopulation, f. When the two subpopulations are equally strong (figure 2(a)), the severity of epidemics is low (resp. high) if the initial proportion of S_a in both subpopulations are high (resp. low). When $f < 0.5$ (figure 2(b)), if the initial proportion of S_a is high, then the epidemic final size is low regardless of strategy proportion in another subpopulation. Figure 2(d) highlights the scenario of population with single preference whereby we purposely impose an extreme high relative strength on subpopulation 1 by setting $f = 0.99$. By doing so, the whole population can be regarded as comprising of only one single group with normal behaviour as their preference. We found that the epidemic final sizes evolve smoothly as compared to two subpopulations with different relative powers.

We explored the epidemic dynamics in parameter space δ-k for four different initial proportions of strategies in figure 3. All panels show the increase of additional cost to those

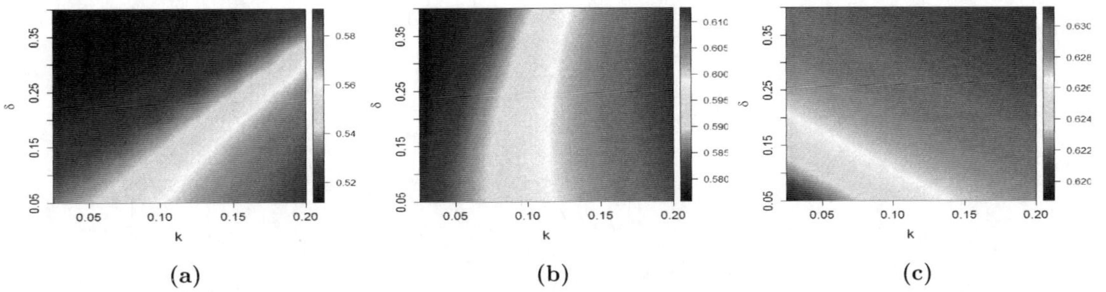

Figure 3: (Color online) The contour plots of epidemic final size in the parameter space δ-k, where $\delta \in (0.05, 0.4)$ and $k \in (0.025, 0.2)$, with different initial conditions $(p(0), q(0))$. (a) $p(0) = 0.2$, $q(0) = 0.8$, the initial proportion of S_a are high in both subpopulations. (b) $p(0) = q(0) = 0.8$ (or $p(0) = q(0) = 0.2$). (c) $p(0) = 0.8$, $q(0) = 0.2$, the initial proportion of S_n are high in both subpopulations.

adopting the altered behaviour k will discourage people take up protective actions and hence results in higher epidemic final size. For a specific k, the epidemic final size will be reduced if the intensity of group pressure δ is increased (figures 3(a)-(b)). However, interestingly, the group pressure could be a "double edged sword" if the initial proportions of S_n are already majority in both subpopulations. The increase of δ acting on S_n gives rise to higher epidemic final size (figure 3(c)). This finding aligns with that of [4] who claimed that social norms can either support or hinder immunization goals.

4. Conclusion

We conclude that the interplays between individuals protective behaviour, imitation and epidemic dynamics are necessarily complex if both economic and social factors act on population with existing preferences.

Acknowledgments

The first author would like to thank the financial support provided by the Faculty of Science and Engineering, Curtin University, Australia for attending the conference, and the Ministry of Education Malaysia for his postgraduate study.

References

[1] Funk S, Salathé M and Jansen V A A 2010 *J. Roy. Soc. Interface* **7** pp 1247–1256
[2] Valle S D, Hethcote H, Hyman J M and Castillo-Chavez C 2005 *Math. Biosci.* **195** pp 228–251
[3] Bauch C T 2005 *Proc. Roy. Soc.* B **272** pp 1669–1675
[4] Oraby T, Thampi V and Bauch C T 2014 *Proc. Roy. Soc.* B **281** 20133172
[5] Poletti P, Ajelli M and Merler S 2011 *PLOS ONE* **6** e16460
[6] Bhattacharyya S and Bauch C T 2010 *J. Theoret. Biol.* **267** pp 276–282
[7] Montopoli L, Bhattacharyya S and Bauch C T 2009 *Can. Appl. Math. Q.* **17** pp 317–338
[8] Manfredi P, Posta P D, d'Onofrio A, Salinelli E, Centrone F, Meo C and Poletti P 2009 *Vaccine* **28** pp 98–109
[9] Shim E, Grefenstette J J, Albert S M, Cakouros B E and Burke D S 2012 *J. Theoret. Biol.* **295** pp 194–203
[10] Reluga T C 2009 *J. Biol. Dyn.* **3** pp 515–531
[11] Cressman R 1995 *Games Econom. Behav.* **11** pp 237–253
[12] Helbing D and Johansson A 2010 *Phys. Rev.* E **81** 016112
[13] Helbing D and Johansson A 2010 *PLOS ONE* **5** e12530

IOP Conf. Series: Journal of Physics: Conf. Series **893** (2017) 012017 doi:10.1088/1742-6596/893/1/012017

[14] Ding F, Liu Y, Shen B and Si X M 2010 *Phys.* A **389** pp 1745-1752

IOP Conf. Series: Journal of Physics: Conf. Series **893** (2017) 012018 doi:10.1088/1742-6596/893/1/012018

Dynamical roguing model for controlling the spread of tungro virus via *Nephotettix Virescens* in a rice field

Nikki Blas and Guido David

Institute of Mathematics, College of Science, University of the Philippines - Diliman

E-mail: nikkitagabiblas@yahoo.com, gdavid@math.upd.edu.ph

Abstract. Rice tungro disease is described as a cancer due to its major impact on the livelihood of farmers and the difficulty of controlling it. Tungro is a semi-persistent virus transmitted by green leafhoppers called *Nephotettix Virescens*. In this paper, we presented a compartmental plant-vector model of the *Nephotettix Virescens* - rice plant interaction based on a system of ordinary differential equations to simulate the effects of roguing in controlling the spread of Tungro virus in a model rice field of susceptible rice variety (Taichung Native 1).

1. Introduction

Tungro is a complex disease of rice associated with dual infection by either deoxyribonucleic acid (DNA) or ribonucleic acid (RNA) rice tungro bacilliform virus (RTBV) and an RNA virus, rice tungro spherical virus (RTSV). The latter acts as a helper virus for the transmission of the dependent virus RTBV that is responsible for the severe tungro symptoms. The development of rice tungro disease (RTD) is a result of the interaction of RTSV and RTBV. Plants infected with both viruses (RTSV + RTBV) enhances the symptoms caused by single virus infection of RTBV [2]. Some of the symptoms and effects are: orange-yellow discoloration, panicles are small and not fully projected which results in partially filled grains, delayed flowering, and slow maturity that causes poor development of roots and severe stunting [1].

The roguing process to be considered in this paper is the identification of infected plants and removing a percentage of diseased plants. The aim of the study is to analyze the efficiency of roguing process in a field infected with rice tungro disease.

1.1. Virus-Vector Interaction

Nephotettix Virescens is the most efficient leafhopper that transmits tungro viruses because of its close biological relationship with irrigated rice. Rice tungro disease is transmitted in a semi-persistent manner, that is, the viruses stay in the mouth of the vector for a short period of time and does not enter the body of the vector, unlike rice dwarf disease [7].

2. The Model

The general model for transmission of rice tungro disease is described by the system of ordinary differential equations (1) and Figure 1. The system is composed of 8 non-overlapping compartments. P_0 and V_0 are the numbers of healthy, susceptible and virus-free rice plants and vectors, respectively. P_1 and V_1 are the numbers of plants and vectors infected with RTSV alone, while P_2 and V_2 are the numbers of plants and vectors infected with RTBV alone. Finally, the number of plants and vectors infected with both RTSV and RTBV are given by P_3 and V_3.

$$
\begin{aligned}
\frac{dP_0}{dt} &= r(K - N_P) - \frac{\alpha P_0 V_3}{N_P} - \frac{\gamma P_0 V_3}{N_P} - \frac{\tau P_0 V_3}{N_P} - \frac{\beta P_0 V_1}{N_P} - \frac{\sigma P_0 V_2}{N_P} - q_0 P_0, \\
\frac{dP_1}{dt} &= \frac{\beta P_0 V_1}{N_P} + \frac{\gamma P_0 V_3}{N_P} - \frac{\lambda P_1 V_3}{N_P} - q_1 P_1, \\
\frac{dP_2}{dt} &= \frac{\tau P_0 V_3}{N_P} + \frac{\sigma P_0 V_2}{N_P} - \frac{\delta P_2 V_3}{N_P} - q_2 P_2, \\
\frac{dP_3}{dt} &= \frac{\alpha P_0 V_3}{N_P} + \frac{\lambda P_1 V_3}{N_P} + \frac{\delta P_2 V_3}{N_P} - q_3 P_3, \\
\frac{dV_0}{dt} &= BN_V\left(1 - \frac{N_V}{V}\right) - \frac{aV_0 P_3}{N_P} - \frac{bV_0 P_1}{N_P} + fV_2 - \mu V_0, \\
\frac{dV_1}{dt} &= \frac{bV_0 P_1}{N_P} - \frac{gV_1 P_2}{N_P} - \mu V_1, \\
\frac{dV_2}{dt} &= cV_3 - fV_2 - \mu V_2, \\
\frac{dV_3}{dt} &= \frac{aV_0 P_3}{N_P} + \frac{gV_1 P_2}{N_P} - cV_3 - \mu V_3.
\end{aligned}
\tag{1}
$$

We assume constant total population of plants and vectors, N_P and N_V where $N_P = P_0 + P_1 + P_2 + P_3$ and $N_V = V_0 + V_1 + V_2 + V_3$, respectively.

We use the assumption of Zhang and Holt [6] in planting of new healthy plants, $r(K - N_P)$ where r is the replanting rate and K is the plant population capacity of the rice field. Birth rate of vectors, B, occurs in all four compartments. Population of vectors is contrained by a maximum value of V which is a constant multiple of K. The values of transmission rates $\alpha, \beta, \gamma, \sigma, \tau, \lambda$ and δ were determined from experimental data of Chowdhury et. al. [3]. The average lengths of time a vector acquires the virus from an infectious plant or the acquisition rates a, b and g were obtained from the study of Madden et. al. [5]. We considered f and c as the net retention rates of a virus on the vector. At a rate of f, vectors leave the viruliferous class V_2 and move back to the virus-free vector class V_0 after the retention period of RTBV on the vector lapses. A vector infected with both RTSV and RTBV, the retention period of RTSV is at most 2 days and the retention period of RTBV is 3 days after the retention period of RTSV on the vector lapses [3]. Loss rate of vector is due to death or emigration denoted by μ. Loss rate of host plant is due to crop period and loss of biomass, $q_i, i = 0, 1, 2, 3$.

2.1. Model with Roguing

Plants showing symptoms are removed to prevent spread of the disease and to preserve the quality of rice. As a management control to further virus spread, we examined the effects of roguing ρ, where $0 \leq \rho \leq 1$. The value of ρ represents the percentage of diseased plants uprooted from the field which is represented by the red arrow in Figure 1. Note that when $\rho = 0$, the model reduces to the general model (1). On the other hand, $\rho = 1$ assumes perfect roguing, i.e.

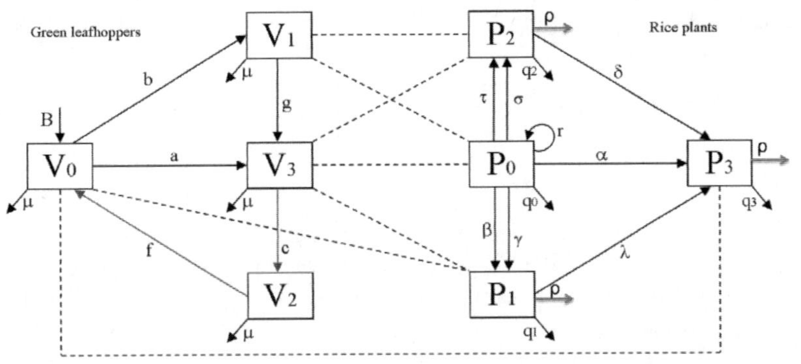

Figure 1. Schematic diagram of transmission of rice tungro disease with roguing.

virus infected rice plants are immediately removed.

$$\begin{aligned}
\frac{dP_0}{dt} &= r(K-N_P) - \frac{\alpha P_0 V_3}{N_P} - \frac{\gamma P_0 V_3}{N_P} - \frac{\tau P_0 V_3}{N_P} - \frac{\beta P_0 V_1}{N_P} - \frac{\sigma P_0 V_2}{N_P} - q_0 P_0,\\
\frac{dP_1}{dt} &= \frac{\beta(1-\rho)P_0V_1}{N_P} + \frac{\gamma(1-\rho)P_0V_3}{N_P} - \frac{\lambda(1-\rho)P_1V_3}{N_P} - q_1(1-\rho)P_1 - \rho P_1,\\
\frac{dP_2}{dt} &= \frac{\tau(1-\rho)P_0V_3}{N_P} + \frac{\sigma(1-\rho)P_0V_2}{N_P} - \frac{\delta(1-\rho)P_2V_3}{N_P} - q_2(1-\rho)P_2 - \rho P_2,\\
\frac{dP_3}{dt} &= \frac{\alpha(1-\rho)P_0V_3}{N_P} + \frac{\lambda(1-\rho)P_1V_3}{N_P} + \frac{\delta(1-\rho)P_2V_3}{N_P} - q_3(1-\rho)P_3 - \rho P_3
\end{aligned} \tag{2}$$

2.2. Numerical Simulation Results

Let $x_0 = (P_0, P_1, P_2, P_3, V_0, V_1, V_2, V_3)$ be the initial value or the population of plant and vector at time $t = 0$. We first consider a typical cropping system, initially all plants are healthy and vectors infected with both RTSV and RTBV arrive in to the system from neighboring fields. The usual time to maturity of TN1 for summer crop lasts 138 days after transplanting [8]. We set the populations $P_0 = 20000$ of plant and $V_3 = 4000$ of vectors. With this initial value, we present simulations using the roguing-free model, and the model with roguing $\rho = 0.40$ and $\rho = 0.80$. This illustrates the situation where vector population is less than the host population. Then, we consider the situation when population of plants and vectors are equal.

The following values of parameters are used in all simulations: $K = 20000$, $V = 5K$,$a = b = g = 0.996$; $c = 0.5, f = 0.33$; $r = 0.001$, $B = 0.033$, $\mu = 0.033$, $q_0 = 0.008$, $q_1 = 0.009$, $q_2 = 0.0125$, $q_3 = 0.025$, $\alpha = 0.035$, $\beta = 0.09$, $\delta = 0.07$, $\gamma = 0.01$, $\tau = 0.06$, $\lambda = 0.03$, $\sigma = 0.08$ [3, 5].

For all simulations, the upper figure displays the dynamics for plant population and the lower figure displays the dynamics for vector population. The red lines represent change in P_0 and V_0. P_1 and V_1 represented by green lines. P_2 and V_2 represented by blue color and the black lines represent P_3 and V_3.

The first case considered was when the initial vector population was less than the plant population. A rice field with 20000 rice plants was used, with an initial vector population of 4000. In the absence of roguing, the number of healthy rice plants decreased over time, and the virus began to spread to both vectors and plants, as shown in Figure 2. On the other hand, applying a roguing rate of $\rho = 0.4$ controlled the spread of the virus in the rice plants, as shown in Figure 3. For the same roguing rate of $\rho = 0.4$ and initial vector population of 1 vector per rice hill, the roguing was not effective in controlling the spread of the disease as shown in Figure

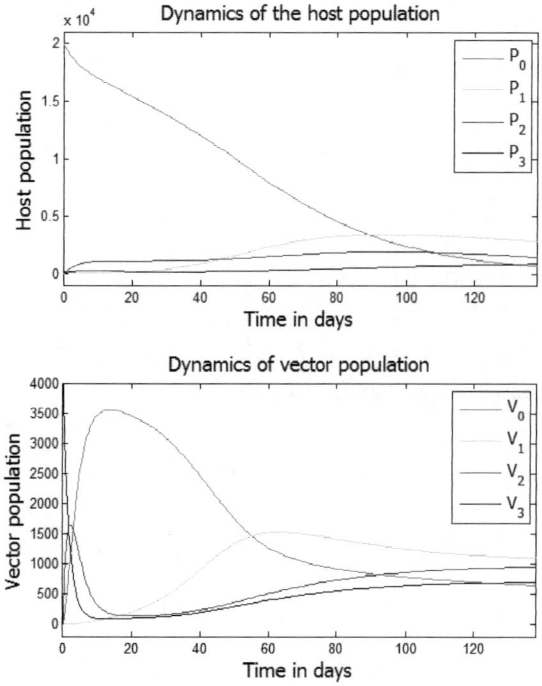

Figure 2. Simulation result for general model with initial $x_0 = (20000, 0, 0, 0, 0, 0, 0, 4000)$.

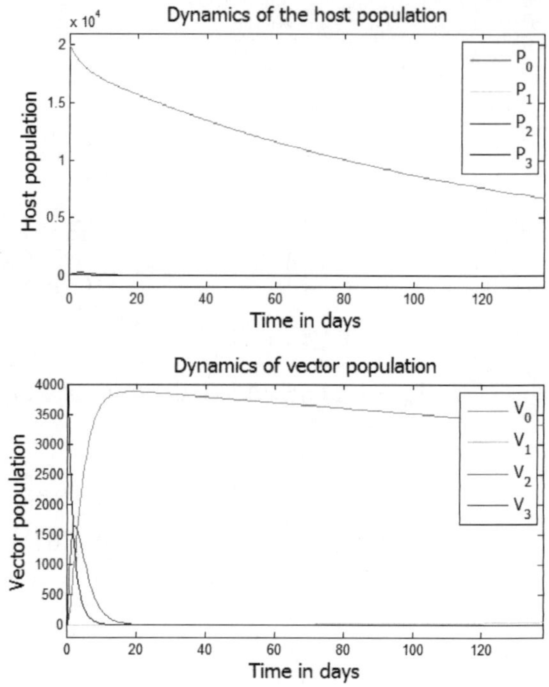

Figure 3. Simulation result for roguing model $\rho = 0.4$ with initial $x_0 = (20000, 0, 0, 0, 0, 0, 0, 4000)$.

4. For the same initial value an almost healthy rice field is attainable at a roguing rate of $\rho = 0.8$ as shown in Figure 5.

3. Conclusion

A rice field was modeled to study the spread of rice tungro disease via green leaf hoppers. Roguing process of removing a percentage of diseased rice plants was incorporated in the model For a roguing rate of $\rho = 0.40$ and low initial vector population, the method was effective in maintaining healthy plant population. For the same roguing rate of $\rho = 0.40$ and a high initial vector population of at least 1 vector per rice hill, the roguing was not effective in controlling the spread of the disease. However, a more aggressive roguing rate of $\rho = 0.8$ was effective in controlling the spread of the virus. The results of the study suggest that roguing is effective in controlling the spread of rice tungro disease in a rice field. The rate of roguing necessary would depend on the population of the green leaf hoppers present in the rice field, which can be estimated by sampling the number of vectors per rice hill. However, a high rate of roguing may not be practical as the process is labor intensive and subject to errors by farmers. Additional studies are needed, particularly those highlighting preventive measures instead of direct control, as suggested by the International Rice Research Institute (IRRI), to model solutions in controlling spread of rice tungro disease in rice fields.

Acknowledgment
The authors would like to thank Department of Science and Technology (DOST) of the Republic of the Philippines for funding the research and Ahmad Shah Olay for the simulations.

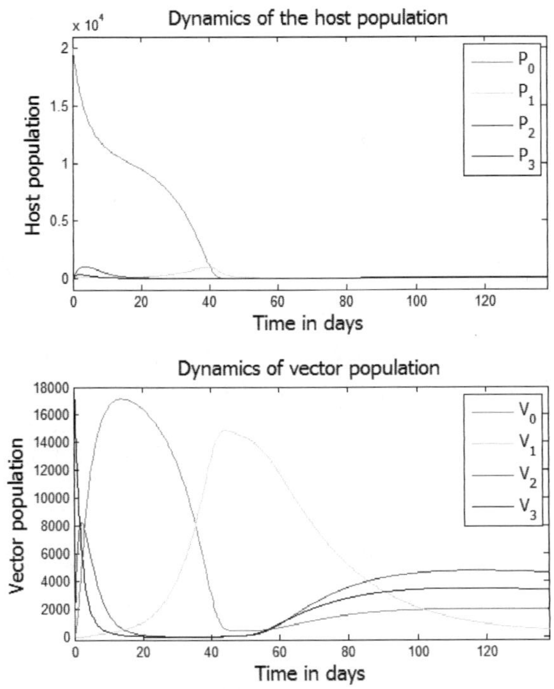

Figure 4. Simulation result for roguing model $\rho = 0.4$ with initial $x_0 = (20000, 0, 0, 0, 0, 0, 0, 20000)$.

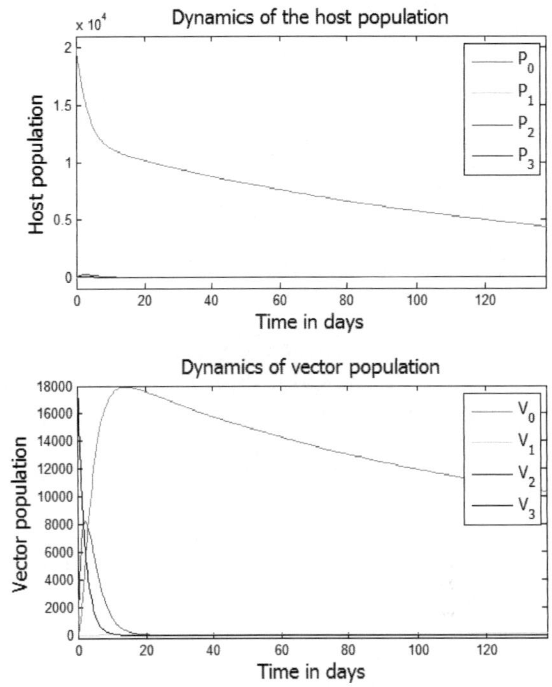

Figure 5. Simulation result for roguing model $\rho = 0.8$ with initial $x_0 = (20000, 0, 0, 0, 0, 0, 0, 20000)$.

4. References

[1] Dahal G, Hibino H, Cabunagan R C et al 1990 Changes in Cultivar Reactions to Tungro Due to Changes in "Virulence" of the Leafhopper Vector *Phytopathology: The Amer. Phytopathological Soc.* **80** pp 659–665

[2] Azzam O and Chancellor T 2002 The Biology, Epidemiology, and Management of Rice Tungro Disease in Asia *Plant Disease: An Int. J. of App. Plant Pathology* **86**(2) pp 88–100

[3] Chowdhury A K, Hibono H and Teng P S 1996 Leafhopper Transmission of Rice Tungro Bacilliform and Rice Tungro Spherical Viruses in Leafhopper Resistant and Susceptible Rice Cultivars *Proc. of the Indian Nat. Sci. Acad.* **B62**(4) pp 287–296

[4] Hibino H, Saleh N and Roechan M 1979 Transmission of Two Kinds of Rice Tungro-Associated Vuruses by Insect Vectors *Phytopathology: The Amer. Phytopathological Soc.* **69**(12) pp 1266-1268

[5] Madden L V, Jeger M J and Bosch F 2000 A Theoretical Assessment of the Effects of Vector-Virus Transmission Mechanism on Plant Virus Disease Epidemics *Phytopathology: The Amer. Phytopathological Soc.* **90**(6) pp 576–594

[6] Zhang X -S, Holt J and Colvin J 2000 Mathematical Models of Host Plant Infection by Helper-Dependent Virus Complexes: Why Are Helper Viruses Always Avirulent? *Phytopathology: The Amer. Phytopathological Soc.* **90**(1) pp 85–93

[7] Nakasuji F, Miyai S, Kawamoto H and Kiritani K 1985 Mathematical Epidemiology of Rice Dwarf Virus Transmitted by Green Rice Leafhoppers: A Differential Equation Model *J. of App. Ecology* **22**(3) pp 839–847

[8] Smith W H and Kumble V 1988 Rice Ratooning *Institute of Rice Research* p 124

[9] Chan M S and Jeger M J 1994 An analytical model of plant virus disease dynamics with roguing and replanting *J. of App. Ecol.* pp 413–427

[10] Holt J and Chancellor T C B 1996 Simulation Modelling of the Spread of of Rice Tungro Virus Disease: the potential for management by roguing *J. of App. Ecol.* **33** pp 927–936

IOP Conf. Series: Journal of Physics: Conf. Series **893** (2017) 012019 doi:10.1088/1742-6596/893/1/012019

Development of the Statistical ARIMA Model: An Application for Predicting the Upcoming of MJO Index

Eddy Hermawan[1], Budi Nurani Ruchjana[2], Atje Setiawan Abdullah[3], I Gede Nyoman Mindra Jaya[4], Sinta Berliana Sipayung[1], Shailla Rustiana[5]

[1] Atmospheric Science and Technology Center National Institute of Aeronautics and Space (LAPAN), Bandung, Indonesia

[2] Department of Mathematics, Faculty of Mathematics, and Natural Sciences Universitas Padjadjaran, Indonesia

[3] Department of Computer Science, Faculty of Mathematics, and Natural Sciences Universitas Padjadjaran, Indonesia

[4] Department of Statistics, Faculty of Mathematics, and Natural Sciences Universitas Padjadjaran, Indonesia

[5] Magister student at Department of Applied Statistics Universitas Padjadjaran and Researcher Assistant at National Institute of Aeronautics and Space (LAPAN) Bandung, Indonesia

E-mail: eddy_lapan@yahoo.com

Abstract. This study is mainly concerned in development one of the most important equatorial atmospheric phenomena that we call as the Madden Julian Oscillation (MJO) which having strong impacts to the extreme rainfall anomalies over the Indonesian Maritime Continent (IMC). In this study, we focused to the big floods over Jakarta and surrounded area that suspecting caused by the impacts of MJO. We concentrated to develop the MJO index using the statistical model that we call as Box-Jenkis (ARIMA) ini 1996, 2002, and 2007, respectively. They are the RMM (Real Multivariate MJO) index as represented by RMM1 and RMM2, respectively. There are some steps to develop that model, starting from identification of data, estimated, determined model, before finally we applied that model for investigation some big floods that occurred at Jakarta in 1996, 2002, and 2007 respectively. We found the best of estimated model for the RMM1 and RMM2 prediction is ARIMA (2,1,2). Detailed steps how that model can be extracted and applying to predict the rainfall anomalies over Jakarta for 3 to 6 months later is discussed at this paper.

1. Introduction

Currently, the *Atmospheric Science and Technology Centre of National Institute of Aeronautics and Space* (LAPAN) in Bandung is developing both models, dynamical and statistical weather and/or climate model. The dynamical weather and/or climate model is designed based on the *Global Circulation Models* (GCM), and *Limited Area Model* (LAM), respectively. While, for the statistical, we still develop the Box-Jenkins (ARIMA) model [1, 5]. As a pre-dominant peak/mode oscillation of the *Intra Seasonal Variability* (ISV) at the equatorial region (including the *Indonesia Maritime Continent*, IMC), the characteristics, mechanism, propagation, and prediction of Madden-Julian Oscillation (MJO) is very important. MJO is the major fluctuation in tropical weather on weekly to monthly timescales. The MJO can be characterized as an eastward moving 'pulse' of cloud and rainfall near the equator that typically recurs every 30 to 60 days [3, 4]. There are many ways to predict the MJO index. One of them is that we call as the Real-time Multivariate MJO (RMM) Index. The MJO phase diagram illustrates the progression of the MJO through different phases, which generally coincide with locations along the equator around the globe. RMM1 and RMM2 are mathematical methods that combine cloud amount and winds at upper and lower levels of the atmosphere to provide a measure of the strength and location of the MJO (http://www.bom.gov.au/climate/mjo/). Based on the above reason, it is necessary to develop an RMM1 and RMM2 using the ARIMA model for predicting the extreme rainfall over Jakarta in 1996, 2002, and 2007.

2. Materials and Methods

The daily of the Real-time Multivariate MJO series 1 and 2 that we call as the RMM1 and RMM2 for period of 1 March 1979 to 1 March 2009 ($\tilde{3}0$ years observation) is taken from http://www.bom.gov.au/climate/mjo/graphics/rmm.74toRealtime.txt. The pentad anomaly of Outgoing Longwave Radiation (OLR) data for the same period observation at position of 100oE, 120oE, and 140oE, respectively is taken from http://www.cpc.ncep.noaa.gov/products/precip/CWlink/daily_mjo_index/proj_norm_order.ascii. We mainly apply the Box-Jenkins (ARIMA) technique to get the best estimated model of RMM1 and RMM2 index prediction. For the reference, we mainly refer to the paper written by [6].

3. Main Result

The results of this study is devided into several steps. They are: identification of data, estimated, predicted, and application of that model for the real problem. Firstly, we are analysing the identification of data as shown below.

3.1. Identification of data

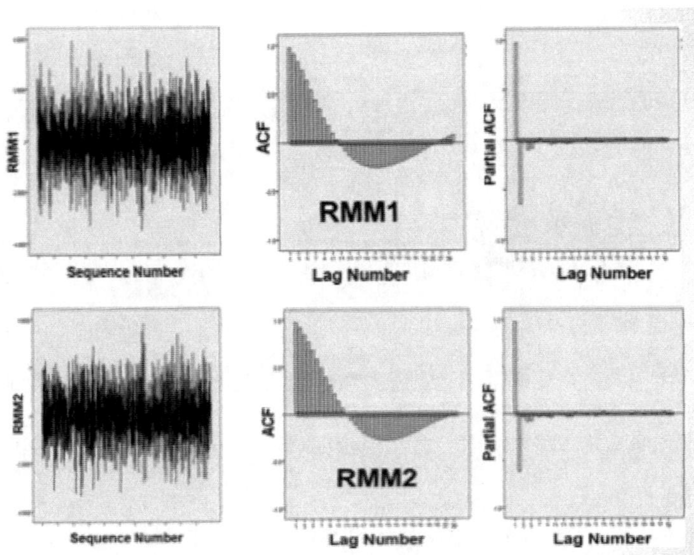

Figure 1. An identification of RMM1 and RMM2 .

Fig. 1 above is showing an identification of RMM1 and RMM2 data. We can see if they looks same each other. To get more information the RMM1 and RMM2 data can be shown in Table 1. While, for data ploting, data, Autocorrelation Function (ACF), and Partial Autocorrelation Function (PACF) as the first differential can be seen at Fig. 2, respectively.

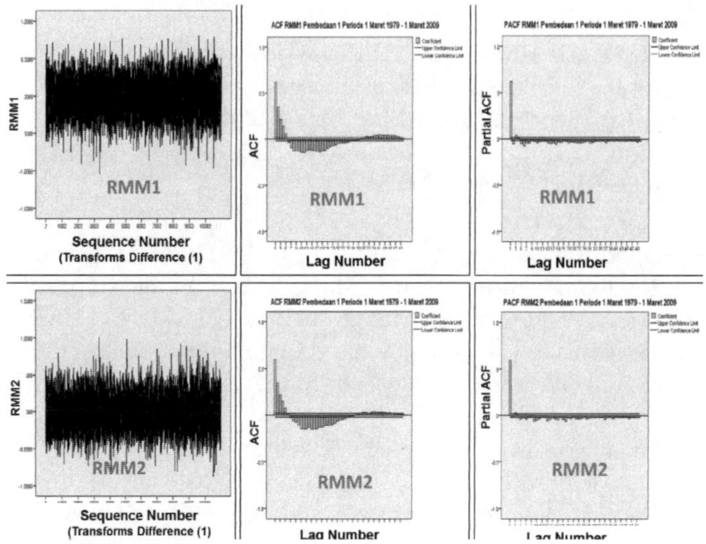

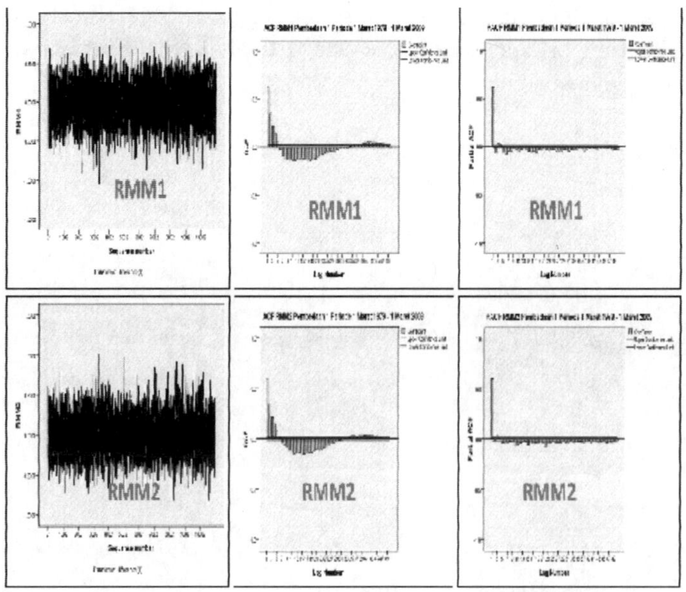

Figure 2. As same as Fig. 1, but for 3 and 6 months later .

As same as Fig. 1, but for Plot data, Autocorrelation Function (ACF), and Partial Autocorrelation Function (PACF) as the first differencing.

After we make the Mean Absolute Deviation (MAD) of the ARIMA model for data RMM1 and RMM2 for period of 1 March 1979 1 March 2009, then after we make Sum of Squares Error for RMM1 and RMM2, respectively for period of 1 March 1979 1 March 2009, we make an estimated model as showing below.

IOP Conf. Series: Journal of Physics: Conf. Series **893** (2017) 012019 doi:10.1088/1742-6596/893/1/012019

3.2. Estimated model

By the Box-Jenkins (ARIMA) technique analysis, we got the best model for the MJO prediction is ARIMA (2,1,2), where for RMM1:

$$\hat{Z}_t = 1,68Z_{t-1} - 0,722Z_{t-2} - 0,02a_{t-1} - 0,05a_{t-2}$$

and for for RMM2 :

$$\hat{Z}_t = 1,714Z_{t-1} - 0,764Z_{t-2} - 0,109a_{t-1} - 0,05a_{t-2}$$

3.3. Predicted model for 3 and 6 months later

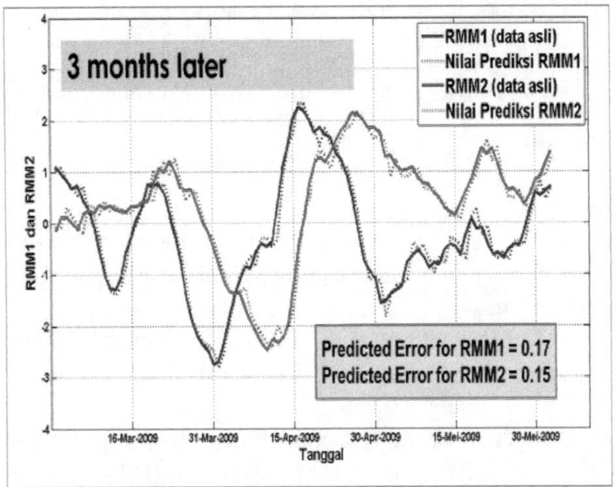

Figure 3. Comparison between original and model of data for 3 months prediction .

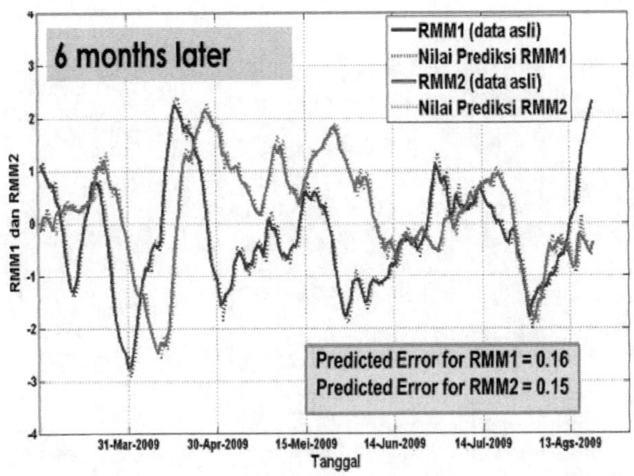

Figure 4. As same as Fig. 3, but for 6 months later .

From Figure 3 and 4, we can clearly how that model can be used well for 3 and 6 months predictions.

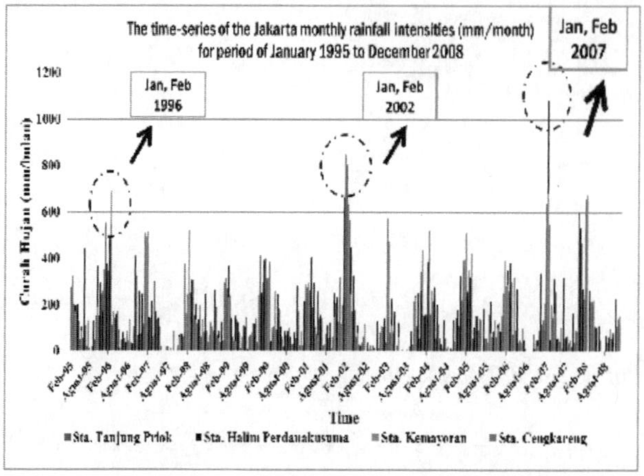

Figure 5. The time-series of the Jakarta monthly rainfall intensities for period Januari 1995 to December 2008.

3.4. Application of model

Figure 5 is showing the rainfall variability over Jakarta for period of January 1995 to December 2008. The variability looks clearly like sinusoidal function. It indicates that Jakarta looks mostly effected by the Monsoon system.

But, please note here, the intensity of that rainfall is increasing gradually with time. We suspect it cause by the other phenomena that we call as MJO as already been studied by [6] and [7].

We are trying now to apply that model to explain the rainfall anomalies prediction over Jakarta using the Real Time Multivariate MJO (RMM1 dan RMM2). We concentrate to analyze during rainy season (December, January, and February) in 1996, 2002, and 2007, respectively as showing at Figs 6 and 7, respectively.

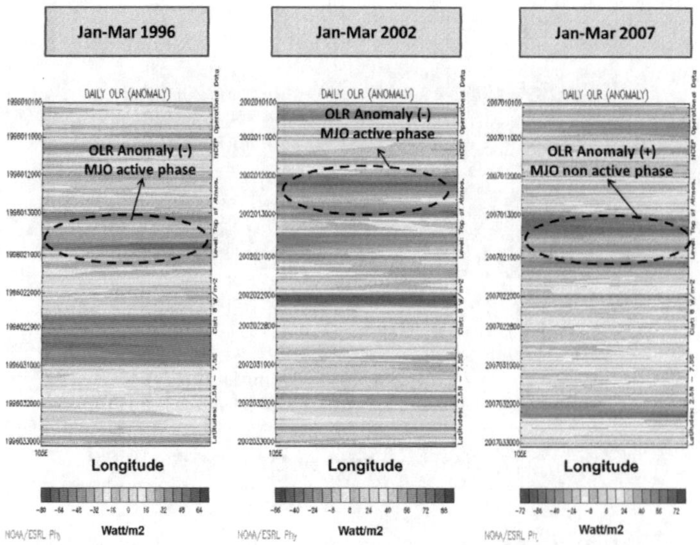

Figure 6. Comparison between RMM1 and RMM2 observed and model for period of 1996 (a), 2002 (b), and 2007 (c), respectively.

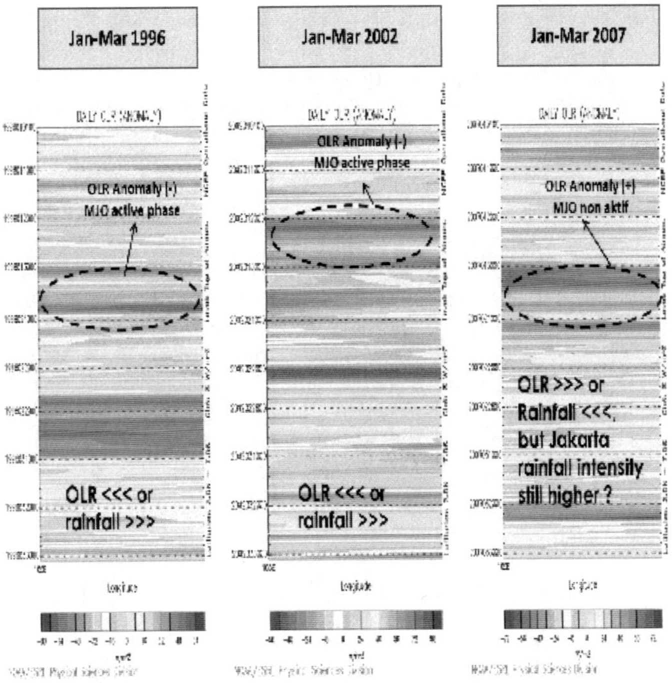

Figure 7. The Hovmoller diagram as more detail explanations of Fig. 6 for period of 1996, 2002 and 2007, respectively

From figure 7, we can see that there are no any differences for the RMM1 and RMM2 prediction both for 3 and 6 months. The higher rainfall intensity over Jakarta (more than 600 mm/month) not always caused by the MJO active condition, although in that time MJO phase is located at phase 4 and 5 (around the Indonesia Maritime Continent, IMC).

We can see, although MJO active phase is located around Africa (very far away from Jakarta), but Jakarta was covered by big floods. It indicates that we need to consider the other phenomena can make significant contribution to increase the rainfall intensity. One of them is that we call as Cold Surge [2].

4. Conclusion

We found the best model for the RMM1 and RMM2 prediction is ARIMA (2,1,2). It means that the RMM1 and RMM2 are influenced by the phenomenon at second order on autoregressive, have the first differencing and also has the second order of moving average. MJO active with active phase not always followed by the extreme rainfall. In 1996 and 2002, we found that MJO with active is responsible for big floods over Jakarta and surrounded area. But, in 2007, big floods at Jakarta was occurred when the MJO with weak phase. It indicates that we need to consider the other phenomena such as Cold Surge.

Acknowledgment

The work reported in this paper is supported by the Directorate General of Higher Education, Ministry of Research-Technology and Higher Education, Indonesia, through Penelitian Kerjasama Luar Negeri (KLN) dan Publikasi Internasional 2016.

References

[1] Chatfield C 1984 *The Analysis of Time Series : An Introduction* London: Chapman and Hall
[2] Lestari D, Cold Surge dan MJO 2009 Pemicu Banjir Bandang *Antara News 14 Februari 2009*
[3] Madden R A, Julian P 1972 Description of global Svale circulation cells in tropics with a 40-50 day period. *J Atmos Sci* **29** pp 1109–1123.
[4] Madden R A, Julian P 1994 Observations of the 40-50 day tropical oscillation *Month Weather Rev* **122** pp 814–837
[5] Makridakis S G, Wheelwright S C, Hyndman R J *Forecasting: Methods and Application* 3rd Edition New York: Wiley
[6] Matthew C, Wheeler and Harry H, Hendon, 2004: An All-Season Real-Time Multivariate MJO Index: Development of an Index for Monitoring and Prediction *Monthly Weather Review* **132** pp 1917–1932
[7] Seto T H 2002 Pengamatan Osilasi Madden Julian dengan radar atmosfer equator (EAR) di Bukittinggi Sumatera Barat *Sains & Teknologi Modifikasi Cuaca* **3** pp 121–124
[8] Wheeler M C, Hendon H H 2004 An all-season real-time multivariate MJO index: development of an index for monitoring and prediction *Month Weather Rev* **132** pp 1917–1932
[9] http://www.bom.gov.au/climate/mjo/, downloaded at November 8, 2016
[10] http://www.bom.gov.au/climate/mjo/graphics/rmm.74toRealtime.txt, downloaded at November 8, 2016
[11] http://www.cpc.ncep.noaa.gov/products/precip/CWlink/daily_mjo_index/proj_norm_order.ascii, downloaded at November 8, 2016

IOP Conf. Series: Journal of Physics: Conf. Series **893** (2017) 012020 doi:10.1088/1742-6596/893/1/012020

The role of ethnomathematics in West Java (a preliminary analysis of case study in Cipatujah)

Dianne Amor Kusuma[1,2], Stanley P. Dewanto[1], Budi Nurani Ruchjana[1] Atje Setiawan Abdullah[3]

[1] Department of Mathematics, Faculty of Mathematics, and Natural Sciences Universitas Padjadjaran, Indonesia
[2] School of Postgraduate Studies-Mathematics Education Indonesia University of Education Bandung, Indonesia
[3] Department of Computer Science, Faculty of Mathematics, and Natural Sciences Universitas Padjadjaran, Indonesia

E-mail: amor@unpad.ac.id

Abstract. Indonesia is a multicultural country of which people usually do anything influenced by their culture. The culture contain of many aspects, one of them is ethnomathematics. It is a study about the connection between culture and mathematics concepts. It also reveals mathematical practices of day life. These mathematical practices could be seen at Cipatujah, West Java, Indonesia. It has several ethnics which implemented of ethnomathematics in their life, for example they implement traditional mathematical concept in the way they determine the time to head seaward for fishing, and the way they construct their houses. The exploration will describe about how deep is the role of ethnomathematics in Cipatujah and state any problem found according to the exploration result. The objectives of this study is to show that ethnomathematics holds an important role in our day life, with the case study of primary school students and for the people of Cipatujah. The method which implemented in this study is exploratory. The result is the people of Cipatujah have been implemented ethnomathematics in their life for many years and believe that ethnomathematics is part of their life, but the teachers of primary school there have not implemented ethnomathematics approach yet in learning process of mathematics. Conclusion of this study is ethnomathematics as the root of culture life in West Java.

1. Introduction

Indonesia is rich of ethnic which influences the lives of its inhabitants. They implement the culture into everyday life. West Java is one of the areas in Indonesia that contain a variety of ethnic characteristic and ethnomathematics is part of the ethnics. Ethnomathematics is the relation between culture and mathematics [3].

Most people (including students) thought that mathematics is simply a set of concepts or formulas that are not related to their daily lives, and this affects the interest of students to learn mathematics more deeply. Actually, their opinion is wrong. The reality is mathematical concepts unwittingly have relevance to everyday life for many years. That means, culture and mathematics have a strong bond. This paper will discuss how the ethnomathematics role in the lives of people from West Java, with the case of the elementary school students and inhabitants in Cipatujah.

1.1. Ethnomathematics
As it grows, there is some definitions of ethnomathematics proposed by the observer of mathematics education in the world, D'Ambrosio [3] suggested that ethnomathematics is mathematics applied in cultural groups. The opinion was supported by Lee [2] that ethnomathematics is a concept to increase knowledge about the development of mathematics in different cultures around the world. Other statement revealed that ethnomathematics is the way in which people of diverse cultures when using mathematics in everyday life [3] and there is another opinion that states that ethnomathematics is a study that relates between mathematics (and math education) with a cultural background, covers the mathematical concepts generated, transferred, and how mathematics to blend into different culture system [15].

From some of those opinions, it can be concluded that ethnomathematics is a concept, knowledge, study, or approaches that associate mathematics with culture. The correlation can be viewed from various aspects, including:

(i) How a society / particular culture use mathematics in everyday life.
(ii) Mathematical concepts contained in a culture.
(iii) How to teach mathematics adapted to the local culture and the unique character of their students.
(iv) How deep mathematics blended to the local culture.
(v) Mathematical activities which have done by locals.

D'Ambrosio [3] suggested that ethnomathematics bring harmony in human behavior, and between humans and nature. It is based on the ethics of diversity:

(i) Respect
(ii) Solidarity
(iii) Cooperation

According to Francois [5], expanded of the use of ethnomathematics corresponding to the cultural diversity of students and with mathematics in their daily practice of bringing mathematics closer to students environment because ethnomathematics implicitly is a program or activity that delivers the values in mathematics and mathematics education.

Adopting ethnomathematics into mathematics learning activities is something that is feasible [4]. In fact it can also be used as an alternative ethnomathematics learning of mathematics [13]. Those opinions could be an inspiration for practitioners in mathematics education to apply ethnomathematics in learning activity.

From the definition above, it can be concluded that ethnomathematics is a concept, knowledge, study, or approaches that associate mathematics with culture.

1.2. Culture in Cipatujah
Cipatujah is sub area of Tasikmalaya, West Java, with a population of 54766 inhabitants. Situated on a plateau and has some beautiful beaches. The name "Cipatujah" comes from the name of the river Cipatujah which means "hit each other". This illustrates that the river flows into the south coast sea. Most residents of Cipatujah are farmers and fishermen. After working all day, they would gather at night in a place called *saung*. The building is made of bamboo and palm trees. In textitsaung, they do discussion between inhabitants and indigenous leaders, and children play traditional instruments from bamboo called textitcalung renteng. There are cultural arts performances and a parade in Cipatujah which held on every December 31th.

1.3. Ethnomathematics in Cipatujah

Inhabitants of Cipatujah still adhere to their ethnic and believing it. The ethnic influence whatever they do in everyday life. With regard to ethnomathematics, they use unique tools that are typical for the dating system called *tunduk (kolenjer)*. Based on their beliefs, they can be used to specify the month, day, and to predict the fate of life. Tunduk made of wood sized 6x25 cm and marked by several lines and dots on its surface. There are three types of tunduk by function, namely:

(i) *Tunduk indit-inditan* is used to determine the good days and directions to guide someone who will do traveling.
(ii) *Tunduk durujana* is used to find the thief or someone who has done bad things to inhabitants of Cipatujah.
(iii) *Tunduk bajo* is used to attack strangers who interfere to inhabitants of Cipatujah.

Tunduk is an important tool for inhabitants of Cipatujah, because the system is calculated based on their culture. In addition to these tools, they use some particular terms in the calculation systems (as unit of measure), as follows: *Jengkal, Jeujeuh, Sasiku, Sadepa, Satuur, Satangtung, Sadampal.*

Those terms are used to calculate the length and inhabitants of Cipatujah never use centimeters, meters, and other units of measure in their daily lives. For measuring the weight, they used to utilize traditional tools made of stone and rope, called *dapon* or *kasungka.*

1.4. Ethnomathematics in Mathematics Learning

Ethnomathematics is an important part of mathematics education, where a math teacher demanded to be more creative and innovative to see or "capture" mathematical ideas contained in a local culture, and apply learning approach in accordance with the unique character of the students based on the particular culture. Ethnomathematics is good to be developed as it has some usefulness as follows:

(i) Sharpen the sensitivity of students.
(ii) Instill a sense of awareness of students.
(iii) Digging math concepts that inherent in a culture.
(iv) Associate mathematical concepts with the culture, therefor it will make the students more appreciate the culture of their nation.

In addition to the aspect of the local language, community activities and objects used in everyday life, ethnomathematics may contain of philosophies believed by certain communities, which can be implemented in mathematical story problems.

Learning ethnomathematics divided into three parts [14], namely:

(i) Learn about the culture, placing culture as a science. The process of learning about the culture has been studied directly by the students through the subjects of arts and crafts, art and literature, painting and drawing. Cultural subjects studied in a particular culture to culture. Product prevailing culture in a society can be used as a method of solving mathematical problems.
(ii) Learn by the culture. Learning by the culture for students include the benefit of various forms of cultural manifestation that becomes learning media or context in learning process in the classroom.
(iii) Learning through the culture. Learning through the culture for students is provided an opportunity to demonstrate achievement of understanding or meaning is created in a subject through a variety of cultural manifestation.

Learning mathematics by using ethnomathematics has some benefits [14], includes:

(i) Their misconceptions in mathematics and is still ongoing, therefore there is a belief that mathematics will be taught effectively and meaningful by relate it with the culture [4].
(ii) To affect how students think about math, not how or what they will learn [2].
(iii) Respect other students who have different cultures with them [6].
(iv) Teaching basic education would be more effective and meaningful if it starts from the socio-cultural context for students, in order to students could recognized it easier [9].
(v) Constructing on students knowledge by bringing culture and their history in order to promote the values and culture [7].
(vi) The development of positive attitudes towards mathematics [12].
(vii) Reduced the abstract nature of mathematics learning [1].
(viii) Bring up the emotion in learning, motivation, spiritual awareness and cultural identity [10].

2. Methodology

The method used in this study is exploratory. The goal is to gather information, to describe circumstances / facts and the specific character of the population. The subjects were 25 students of the 5th grade, 5 teachers at Primary School in Cipatujah, and inhabitants of Cipatujah. The procedures of this study are as follows:

(i) Planning Phase
  (a) Preliminary Study: formulating problems and objectives, it is ethnomathematics in Cipatujah.
  (b) Literature: finalize the concept; they are ethnomathematics and culture of Cipatujah.
  (c) Framing of research instruments: to make the questionnaire for inhabitants of Cipatujah (for information on the application of ethnomathematics in everyday life in Cipatujah) and the questionnaire for the 5th grade students and teachers of primary school in Cipatujah (for information on implementation of ethnomathematics approach in learning process of mathematics).
(ii) Data Collection Phase
  (a) Observing the daily activities that carried out by inhabitants of Cipatujah, which apply ethnomathematics.
  (b) Conducting interview to several inhabitants of Cipatujah, regarding to the implementation of ethnomathematics in their daily lives.
  (c) Conducting interview to the 5th grade students and teachers of primary school in Cipatujah.
(iii) Phase Analysis
  (a) Processing data of observations results and interviews.
  (b) Make a conclusion.

3. Main Result

Cipatujah is one area in West Java that has a unique culture. Inhabitants of Cipatujah do many things in everyday life by adhering to the culture they believe. Unwittingly, they use the basic concepts of mathematics in various activities such as in trading, farming, fishing and building a house. There are some basic concepts of mathematical calculations are usually carried out by inhabitants of Cipatujah for many years, that will be discussed.

3.1. Time Calculation Rules of Farming in Cipatujah (Pranata Mangsa)

Because most inhabitants of Cipatujah are farmers, then they do account when the right time very carefully to start planting and harvesting. Cipatujah society have known the term *prananta mangsa*, means the time calculation rules. According to indigenous leaders, inhabitants of Cipatujah divide it into two types of *mangsa*. The first *mangsa*, divided into 8 *mangsa*, which is used to ngahuma (term in farming, which means it is growing rice in a field or on dry lands). The second *mangsa*, consisting of 12 *mangsa*, which is used to nyawah (term in farming, which means it is growing rice in the fields or in wet lands). Each time of *mangsa* is divided into two periods, namely:

(i) *mangsa* panas bumi (Its not time to grow rice), starting from *mangsa* 1 to *mangsa* 3.
(ii) *mangsa* tiis bumi (it's time to plant), starting from *mangsa* 4 to *mangsa* 8 or *mangsa* 12. *mangsa* 4 is transition *mangsa* from *mangsa* panas bumi to *mangsa* tiis bumi.

Here are the divisions of the first kind of *mangsa*:

(i) *Mangsa hiji* : state of the earth/ soil is getting hottest.
(ii) *Mangsa dua* and *mangsa tilu* : state of the earth is still hot. In these circumstances, any crop planted will die, although it has been raining.
(iii) *Mangsa opat* : state of the earth is getting cold, is characterized by appearance of buds of leaves on each plant.
(iv) *Mangsa lima* : state of the Earth is cold, is characterized by already verdant foliage.
(v) *Mangsa genep* : state of the Earth is still cold.
(vi) *Mangsa tujuh* : this *mangsa* is in extreme cold, and usually characterized by appearance of the grass in the fields.
(vii) *Mangsa dalapan* : on this *mangsa*, the rain rarely falls.

Pranata mangsa 12 times in a year are presented in the following table 1

3.2. Calculation of Repok

In everyday life, especially on marriage, inhabitants of Cipatujah usually do calculation of repok that aims to avoid their life from the altercation which begins from disputes in marriage.

Here is the standard of calculation of *repok* :

(i) Alphabet: H = 1, D = 6, P = 11, M = 16, N = 2, T = 7, D = 12, G = 17, C = 3, S = 8, A = 13, B = 18, R = 4, W = 9, Y = 14, T = 19, K = 5, L = 10, NY =15, NG = 20
(ii) Day of birth: Monday = 4, Tuesday = 3, Wednesday = 7, Thursday = 8, Friday = 6, Saturday = 9, Sunday = 5
(iii) Conclusion of the results of calculation (based on the rest of numbers after deducting 7 / based on modulo 7)
   (a) *Pisang punggel* = less good
   (b) *Lumbung gumuling* = less good
   (c) *Tunggak kasemi* = less good
   (d) *Satria lalaku* = good
   (e) *Sangga waringin* = good.
   (f) *Padaringan kebek* = good
   (g) *Ratu sabdaning pandita* = good

Most of the calculation on these findings has similarity to Mustapa [11]. For example :

A man named Rosadi (born on Sunday) and a woman named Sarinah (born on Wednesday), they come from the village Sindangkerta, Cipatujah, and then the calculation is as follows:

Tabel 1 *Pranata Mangsa* 12 times in a Year

NO.	MONTH OF HIJRIYAH	MONTH OF MASEHI	NAME OF MANGSA	SUM OF DAYS	CHARACTER
1.	*Sawal*	22nd Jun - 2nd Aug	I	41	starts to plant palawija
2.	*Hapit*	2nd Aug - 25th Aug	II	24	to plant second palawija
3.	*Haji*	25th Aug - 18th Sep	III	24	time for harvest of palawija
4.	*Muharam*	18th Sep - 13th Oct	IV	24	time to plant bananas
5.	*Sapar*	13th Oct - 9th Nov	V	27	rainy season
6.	*Mulud*	9th Nov - 22nd Dec	VI	44	working in the fields
7.	*Silih Mulud*	22nd Dec - 3rd Feb	VII	42	windy season
8.	*Jumadil Awal*	3rd Feb - 1st Mar	VIII	26	bulging rice
9.	*Jumadil Ahir*	1st Mar - 26th Mar	IX	25	blooming rice
10.	*Rajab*	26th Mar - 19th Apr	X	23	to plant palawija on the land
11.	*Rewah*	19th Apr - 12th May	XI	24	there are a lot of birds
12.	*Puasa*	12th May - 22nd Jun	XI	41	very cold morning

Hijriyah is tentative, and *masehi* is permanent.

- Ro-sa-di = 4 + 8 + 6 = 18; Sa-ri-nah = 8 + 4 + 2 = 14; So 18 + 14 = 32
- Because based on modulo 7, the rest of 32 is 4, so its included in satria lalaku (at good category)
- Next, added by sum of both peoples days of birth, are 12.
- So 12 + 32 = 44; then 44 - 32 = 12; then 12 - 7 = 5; so its include in sangga waringin (at good category).
- Because Sin-dang-ker-ta = 8 + 6 + 5 + 7 = 26; and based on modulo 7 the rest is 5, so its included sangga waringin.
- The conclusion based on the calculation is Rosadi and Sarinah will avoid the altercation in their marriage.

According to previous discussion, it showed that inhabitants of Cipatujah involve the basic concepts of mathematics in any activities in their daily life. In other words, ethnomathematics hold an important role in cultural life in Cipatujah.

3.3. Primary School Mathematics Learning in Cipatujah

Based on interview result of 25 students of 5th grade and 5 teachers of public primary school in Cipatujah, obtained the following information:

(i) According to the students:

- Mathematics is a difficult subject to understand.
- Learning mathematics in the classroom is less attractive for them.
- Mathematics is not related to their daily lives.
- Mathematics is not related to their culture.
- Mathematics is not useful in real life.

(ii) According to the teachers:

- It is difficult to make students understand mathematics.
- It is difficult to make students interested in mathematics learning.
- Mathematics learning in the classroom usually uses direct learning/expository method.
- Lack of props (i.e. real objects) that can allow students to understand math concepts easier.
- They have never implemented ethnomathematics approach in mathematics learning.

Based on these descriptions can be said that need to do some efforts to make the students can be more interested in mathematics learning, and in order to the students also assume that mathematics is related to their culture and can be applied in everyday life. If ethnomathematics approach implemented in mathematics learning, then the culture will indirectly be maintained.

4. Conclusion

In daily life, inhabitants of Cipatujah still hold on their culture and involving ethnomathematics in performing some rituals based cultures such as the calculation in the use of *tunduk*, prananta mangsa, and calculation of *repok*. Most of Cipatujah cultures have in common with sundanese basic culture on other areas in West Java, but Cipatujah has differences in *prananta mangsa* which dividing of each *mangsa* into 2 parts, *mangsa panas bumi* and *mangsa tiis bum*.

The 5th grade students of primary school in Cipatujah feel that the mathematical concepts taught in the classroom is not related to everyday life and it made them think that mathematics is not interesting (although any of them get good score on math). However, according to the observation, the teachers of primary school in Cipatujah have not yet formally implemented ethnomathematics approach in learning process of mathematics (means, it has not been included in the current curriculum).

Acknowledgment

The authors express their gratitude to Department of Mathematics FMIPA Universitas Padjadjaran and DRPM Universitas Padjadjaran for the support and funding, to the reviewers for carefully reading and valuable inputs that improved the manuscript.

References

[1] Achor E E, Imoko B I, and Uloko E S 2009 Effect of Ethnomathematics Teaching Approach on Senior Secondary Student's Achievement and Retention in Locus *Educational Research and Review* **4**(8) pp 385–390

[2] Adam S 2004 Ethnomathematics Ideas in the Curriculum *Mathematics Education Research Journal* **16** pp 49–68

[3] D'Ambrosio U 1985 Ethnomathematics and Its Place in the History and Pedagogy of Mathematics *For the Learning of Mathematics* **5**(1) pp 44–48

[4] Eduardo 2001 Comparison of the Final of Students in Intermediate Algebra Taught with and without an Ethnomathematical Pedagogy *A Presentation to the Center for Study of Diversity in Teaching and Learning in Higher Education* pp 1–9 Miami Florida

[5] Francois K 2012 Ethnomathematics in a European Context: Towards an Enrichment Meaning of Ethnomathematics *Journal of Mathematics and Culture* **6**(1) pp 191–208

[6] Garegae P 2008 Conceptualizing Ethnomathematics as a Bridge between Peace and Conflict: The Case of Cultural Artefacts *A Presentation to The Fifteenth International Conference on Learning* pp 1–15 University of Illinois Chicago USA

[7] Jenni L 2007 Foundation in Ethnomathematics for Prospective Elementary Teachers *The Journal of Mathematics and Culture* **2**(1) pp 1–19
[8] Lee H S 2004 Ethnomathematics in Taiwan: a Review *Proceedings of the 10th ICME* pp 65–80 Tipografia Editrice Pisana
[9] Matang 2006 Linking Ethnomathematics, Situated Cognition, Social Constructivism and Mathematics Education: An Example from Papua New Guinea *A Presentation to The ICEM-3 Conference* pp 1–9 New Zealand
[10] Massarwe K, et al. 2010 An Ethnomathematics Exercise in Analyzing and Constracting Ornaments in a Geometry Class *Journal of Mathematics and Culture* **5**(1) pp 1–19
[11] Mustapa, Hasan 2010 Adat Istiadat Sunda *PT Alumni, Bandung*
[12] Ng Wee Leng 2006 Effects of an Ancient Chinese Mathematics Enrichment Programme on Secondary School Students Achievement in Mathematics *International Journal of Science and Mathematical Education* **4** pp 485–511
[13] Owens K 2012 Policy and Practices: Indigenous Voices in Education *Journal of Mathematics and Culture* **6**(1) pp 51–75
[14] Supriadi 2014 Mengembangkan Kemampuan dan Disposisi Pemodelan serta Berpikir Kreatif Matematik Mahasiswa PGSD melalui Pembelajaran Kontekstual Berbasis Etnomatematika *Thesis*
[15] Zhang, Weizhong, and Zhang, Qinqiang 2010 Ethnomathematics and Its Integration within the Mathematics Curriculum *Journal of Mathematics Educations* **3**(1) pp 151–157

The Asian Mathematical Conference 2016 (AMC 2016) IOP Publishing
IOP Conf. Series: Journal of Physics: Conf. Series **893** (2017) 012021 doi:10.1088/1742-6596/893/1/012021

Rainfall Prediction of Cimanuk Watershed Regions with Canonical Correlation Analysis (CCA)

Shailla Rustiana[1], Budi Nurani Ruchjana[2], Atje Setiawan Abdullah[3], Eddy Hermawan[4], Sinta Berliana Sipayung[6], I Gede Nyoman Mindra Jaya[6], and Krismianto[7]

[1] Magister Student at Department of Applied Statistics Universitas Padjadjaran and Researcher Assistant at National Institute of Aeronautics and Space (LAPAN) Bandung, Indonesia

[2] Department of Mathematics, Faculty of Mathematics, and Natural Sciences Universitas Padjadjaran, Indonesia

[3] Department of Computer Science, Faculty of Mathematics, and Natural Sciences Universitas Padjadjaran, Indonesia

[4,5,7]Researcher at Center of Atmospheric Science and Technology Center National Institute of Aeronautics and Space (LAPAN)

[6] Department of Statistics, Faculty of Mathematics, and Natural Sciences Universitas Padjadjaran, Indonesia

E-mail: `shailla.rustiana@gmail.com`

Abstract. Rainfall prediction in Indonesia is very influential on various development sectors, such as agriculture, fisheries, water resources, industry, and other sectors. The inaccurate predictions can lead to negative effects. Cimanuk watershed is one of the main pillar of water resources in West Java. This watersheds divided into three parts, which is a headwater of Cimanuk sub-watershed, Middle of Cimanuk sub-watershed and downstream of Cimanuk sub-watershed. The flow of this watershed will flow through the Jatigede reservoir and will supply water to the north-coast area in the next few years. So, the reliable model of rainfall prediction is very needed in this watershed. Rainfall prediction conducted with Canonical Correlation Analysis (CCA) method using Climate Predictability Tool (CPT) software. The prediction is every 3months on 2016 (after January) based on Climate Hazards group Infrared Precipitation with Stations (CHIRPS) data over West Java. Predictors used in CPT were the monthly data index of Nino3.4, Dipole Mode (DMI), and Monsoon Index (AUSMI-ISMI-WNPMI-WYMI) with initial condition January. The initial condition is chosen by the last data update. While, the predictant were monthly rainfall data CHIRPS region of West Java. The results of prediction rainfall showed by skill map from Pearson Correlation. High correlation of skill map are on MAM (Mar-Apr-May), AMJ (Apr-May-Jun), and JJA (Jun-Jul-Aug) which means the model is reliable to forecast rainfall distribution over Cimanuk watersheds region (over West Java) on those seasons. CCA score over those season prediction mostly over 0.7. The accuracy of the model CPT also indicated by the Relative Operating Characteristic (ROC) curve of the results of Pearson correlation 3 representative point of sub-watershed (Sumedang, Majalengka, and Cirebon), were mostly located in the top line of non-skill, and evidenced by the same of rainfall patterns between observation and forecast. So, the model of CPT with CCA method is reliable to use.

IOP Conf. Series: Journal of Physics: Conf. Series **893** (2017) 012021 doi:10.1088/1742-6596/893/1/012021

1. Introduction

Rainfall prediction in Indonesia is very influential on various development sectors. The inaccurate of predictions can lead to negative effects. Indonesia's rainfall has a very large variability. Besides the influence of climate phenomena such as El Nino Southern Oscillation (ENSO) on the Pacific Ocean, Indian Ocean Dipole (IOD) on the Indian Ocean, and the Monsoon (Asian and Australian) causing extreme climate impacts such as the dry and rainy season that is longer than normal conditions [4]. It is certainly affects to various of life sectors in Indonesia. Java Island is the island with the biggest population in Indonesia and make various development sectors center on there. Java is also the largest island affected by climate phenomena compared to other islands in Indonesia, because Java is the center of the Asian-Australian monsoon region [10].

In the climate forecast system the result of research by [9] , the significant common power and coherence between ENSO IOD, ENSO Indian Summer Monsoon Rainfall (ISMR) and IODISMR have clearly brought out the dominant time-period when they interact with each-other. Based on this theory, it needed a fairly accurate prediction method for West Java Province's rainfall, especially on the area of Cimanuk Watershed, because this watershed is the one of the main support of water resources in West Java who can affect agricultural sector [11].

From the various prediction methods commonly used, known a method name as Canonical Correlation Analysis (CCA). In this study, the CCA method was run using Climate Predictability Tool (CPT) software. Climate prediction model results from CPT output is expected will give a fairly accurate prediction for West Javas rainfall.

The observation region in this research were three representative points of Cimanuk Watershed in West Java, which is divided into three parts sub-watershed (Upper, Middle, and Downstream), as shown in Figure 1.

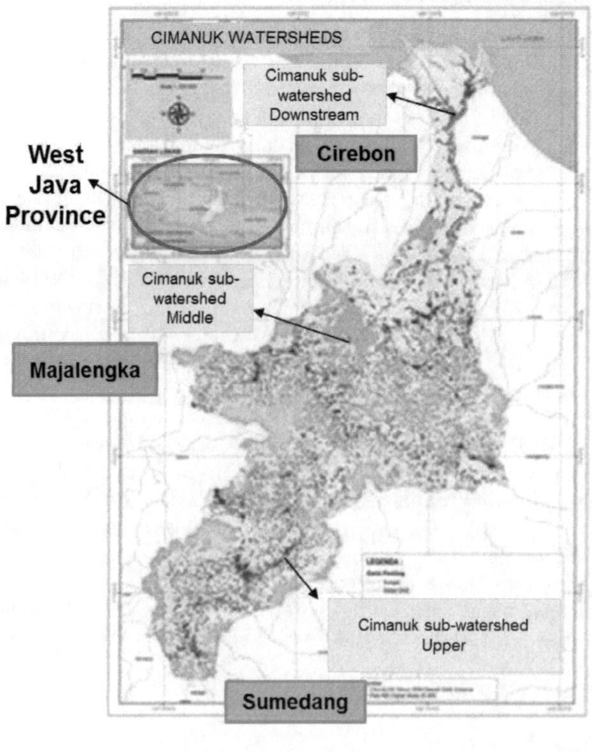

Figure 1. Map of observation region on Cimanuk Watershed

The data used on this research were monthly data from January 2001-2016 (for predictors, as initial condition) consisting of: Nino3.4 index obtained from National Oceanic and Atmospheric Administration (NOAA); Dipole Mode Index (DMI) obtained from the website of the Japan Agency for Marine-Earth Science and Technology (JAMSTEC); Monsoon Index the calculation result of U wind data, from NOAA; and rainfall data of Climate Hazard Group Infrared Precipitation with Station (CHIRPS) over western part of Java Island from January 2001 - December 2015 as predictant.

As already reported by [3], found the lag time of predictor (SST Nino 3.4) is 2 to 3 months later. While it take only one month when we used IOD. We suspect it related well to the location of SST. The SST Nino 3.4 is located a little bit far away from Indonesian Maritime Continent comparing with IOD is located near from Indonesian Maritime Continent. While the Monsoons data dont have lag time because their patter were sinusoidal. The predictant data we used was CHIRPS rainfall data, because the data was satellite rainfall data reanalysis with best resolution, ie $0.05^o \times 0.05^o$ or 50 km [2].

The tool we used were the set of computers with Microsoft Office software, Opening Grid Analysis and Display System (OpenGrADS) version 2.02 could be downloaded at (http://opengrads.org) and the Climate Predictability Tool (CPT) version 14 could be downloaded at (http://iridl.columbia.edu/). The canonical correlation coefficient values between a pair of canonical variables is based on a linear relationship. If the variables are not linearly related, then the relationship can not be explained by the canonical correlation coefficient. Canonical correlation analysis maximized the linear relationship between the set of variables [1].

The canonical correlation equation is:

$$Corr(U,V) = Corr(a^T x, b^T y) = \frac{Cov(U,V)}{\sqrt{Var(U),Var(V)}} = \sqrt{a^T S_{xy}}\sqrt{(a^T s_{xx} a)(b^T S_{yy} b)} \tag{1}$$

with :

U : Predictor =a1Ty + a2Ty + a3Ty + a4Ty + a5Ty + a6Ty
(Nino3.4 +DMI +Monsoon Index: AUSMI, ISMI,
WNPMI, WYMI)

V : Predictant $= b1^T x + b2^T x + ...bn^T x$
(Grided rainfall of West Java)

S : Matriks X,Y

Procedure of this study consisting of 3 step, as shown in Figure 2 below:

2. Main Result

Climate Predictability Tool (CPT) is a software based on windows that was used to develop seasonal climate prediction models based on Model Output Statistics (MOS) [8]. The result for spatial analysis of rainfall prediction map (Fig. 3) shows that the average values of 3monthly rainfall forecast range between 0-1500 mm/3month. Red colours of the map means dry season, and the Blue colours means rainy season. Forecast rainfall on FMA (Feb-Mar-Apr) till MAM (Mar-Apr-May) were the rainy season, and MJJ (May-Jun-Jul) till JJA (Jun-Jul-Aug) season was the dry season, while the AMJ (Apr-May-Jun) and JAS (Jul-Aug-Sep) were the transition season.

The results of Pearson Correlation to explain the skill map of model or the correlation between the predictor and predictant of the model. Pearson correlation values range between -1 to 1. Where the Pearson correlation is the correlation between x variables (predictor) and y variables (predictant) that affect the results of prediction. Pearson correlation directly proportional to

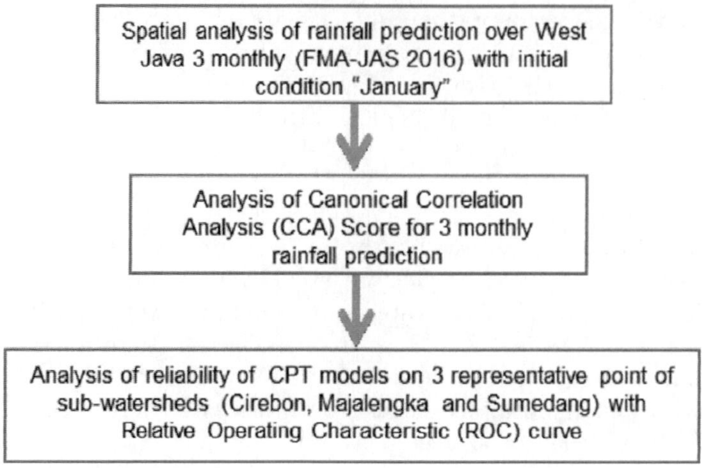

Figure 2. Procedure Research of CCA

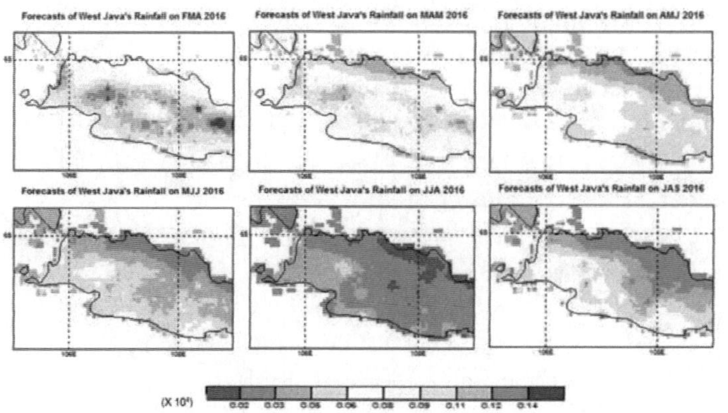

Figure 3. Forecast rainfall map of West Java

the ROC curve. The results in fig.4 showed that Pearson correlation value is mostly positive high (0.3-0.9) for every season on the area of observation Cimanuk sub-watersheds.

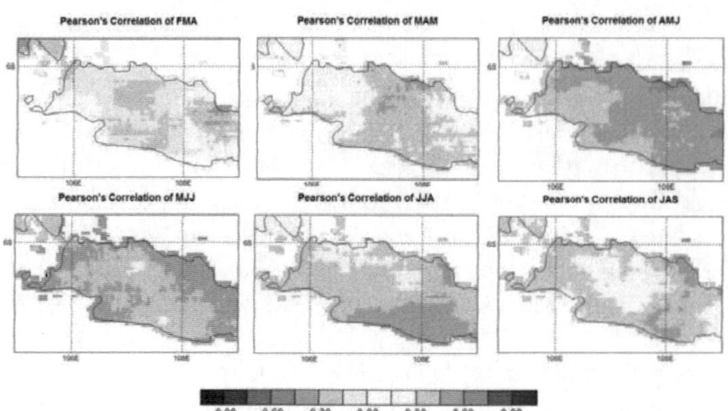

Figure 4. Pearson correlation as skill map of CPT model.

The good prediction model from CPT output indicated by coefficient of CCA (CCA score), shown on Table 1 by the result of formula on equation (1). CCA is a method to find the best linear combination between two multivariate datasets that maximize the correlation coefficient between them. The CCA only focuses on the correlation between a linear combination of the variables in one set and linear combination of the variables in another set. The idea is first to determine the pair of linear combinations having the largest correlation. Next, determine the pair of linear combinations having largest correlation among all pairs uncorrelated with the initially selected pair, and so on. The pairs of linear combinations are called the cannonical's variables, and their correlations are called canonical correlation [6]. This is particularly useful to determine the relationship between criterion measures and the set of reviews their explanatory factors. Involves this technique, first, the reduction of the dimensions of the two multivariate datasets by projection, and second, the calculation of the relationship (measured by the correlation coefficient) between the two projections of the datasets [7]. The basic idea behind forecasting with CCA is straightforward: simple linear regressions are constructed that relate the predictant canonical variables (x) to the predictor canonical variables (y). In this case the entire (I1) field x(t) is used to forecast the (J1) field y(t+), where is the time lag between the two fields in the training data, which becomes the forecast lead time [12]. In applications with atmospheric data it is typical that there are too few observations n relative to the dimensions I and J of the fields for stable sample estimates (which are especially important for out-of-sample forecasting) to be calculated, even if n ¿ max (I, J) so that the calculations can be performed. It is therefore usual for both the x and y fields to be represented by series of separately calculated truncated principal components [12] Based on the equation (1), the canonical correlation coefficient (CCA Score) is the correlation between this pair of canonical variables, X (V) and Y (U), and is the largest possible correlation between pairs of linear combinations of these two data sets (x and y variables).

Table 1. Canonical Correlation Analysis (CCA) Score

Season	CCA Mode					
	1	2	3	4	5	6
FMA	0.94	0.85	0.39	0.87	0.66	0.30
MAM	0.88	0.76	0.49	0.91	0.70	0.24
AMJ	0.77	0.57	1.00	0.95	0.83	0.04
MJJ	0.91	0.73	0.72	0.48	0.80	0.15
JJA	0.68	0.61	1.00	0.79	0.66	0.01
JAS	0.76	0.69	0.37	0.18	0.35	0.10

The CCA score per 3month of the observation season are more than 0.7 average, which means the model was accurate enough to estimate rainfall on the observation region. Time series analysis of Observation rainfall (red line) and Hindcast rainfall from CPT (green line) for 3 representative point of sub-watersheds (Cirebon, Majalengka, and Sumedang) on FMA season (Fig. 5) were close together, and the forecast rainfall for FMA 2016 (cross point) on the normal layer (green layer). On JAS season (Fig. 6), time series analysis between observation and hindcast also close together. That means the model accurate enough to use. However the forecast rainfall for JAS 2016 on the above normal layer (red layer).

To determine the reliability of the output CPT models was carried out by analyzed the Relative Operating Characteristic (ROC) curve. ROC curve is a reliability level of the models. Linear straight lines on the ROC curve is a non-skill line. Curve over non-skill line is a positive skill (reliable), while the curve under non-skill line is a negative skill (unreliable) [5].

The ROC curve for 3 representative point of sub-watersheds (Cirebon, Majalengka, and

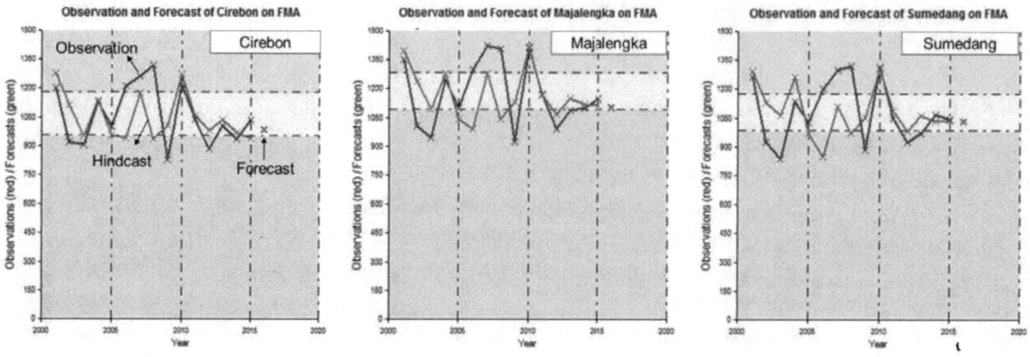

Figure 5. As same as Fig. 3, but for 6 months later .

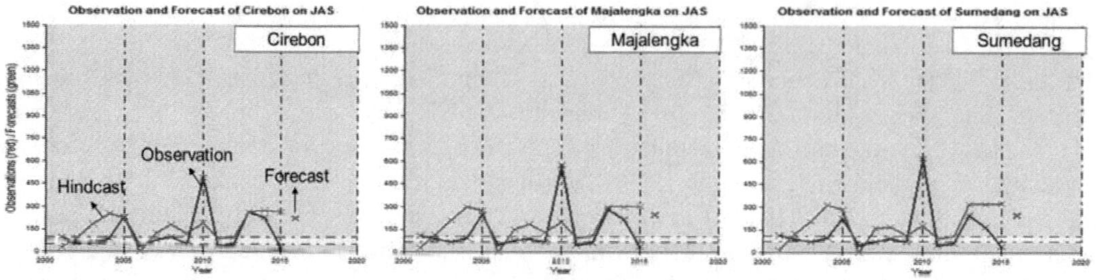

Figure 6. The time-series of the Jakarta monthly rainfall intensities for period January 1995 to December 2008.

Sumedang) on FMA season (Fig. 5) were on the top of non-skill linear line, for above normal curve (red line) and below normal curve (blue line). The ROC curve for on JAS season (Fig. 6) were also on the top of non-skill linear line, for above normal curve and below normal curve. These ROC curve were proportional with Pearson Correlation. So, that means the model predictions from CPT with CCA method were reliable to use.

3. Conclusion

Rainfall prediction with CCA method of CPT software was accurate enough to estimate West Java's rainfall on the area of Cimanuk Watersheds, by the results analysis of Canonical Correlation Analysis (CCA) score. The CCA score per 3month of the observation season were more than 0.7 average, which means the model was accurate enough. Time series analysis of Observation rainfall (red line) and Hindcast rainfall from CPT (green line) for 3 representative points of Cimanuk sub-watersheds (Cirebon, Majalengka, dan Sumedang) on FMA and JAS season were close together. That means the model was also accurate enough. ROC curve at the 3 representative point of Cimanuk sub-watersheds on FMA and JAS season were above the linear line of non-skill, for above normal curve (red line) and below normal curve (blue line). It was clearly that the CPT model was reliable or good enough to used for predict West Javas rainfall.

Acknowledgment

The work reported in this paper is supported by the Directorate General of Higher Education, Ministry of Research-Technology and Higher Education, Indonesia, through Penelitian Kerjasama Luar Negeri (KLN) dan Publikasi Internasional 2016.

References

[1] Asbah M F, Sudarno, and Diah S 2013 Penentuan Koefisien Korelasi Kanonik dan Interpretasi Fungsi Kanonik Multivariat *Jurnal Gaussian* **2**(2) pp 119–128

[2] Funk C, Pete J, Martin F L, Diego H P, James P V, James D R, Bo E R, Gregory J H, Joel C M, and Andrew P V 2014 A Quasi-global Precipitation Time Series for Drought Monitoring *USGS, EROS Data Center, Sioux Falls, SD*

[3] Hermawan E 2012 Model Interkoneksi Kejaidan EL Nino dan Dipole Mode Sebagai indikasi Awal Datangnya Musim Kering/Basah Panjang di Kawasan Barat Indonesia *Orasi Ilmiah Pengukuhan Profesor Riset Bidang Meteorologi*

[4] Jourdain C, Alexander S G, Andrea S T, Caroline C U, Aurel F M, and Karumuri A 2013 The Indo-Australian monsoon and its relationship to ENSO and IOD in reanalysis data the CMIP3/CMIP5 simulations *Clim Dyn* **41** pp 3073 –3102, DOI 10.1007/s00382-013-1676-1

[5] Kadarsah 2010 Aplikasi ROC untuk uji kehandalan model HYBMG *Jurnal Meteorologi dan Geofisika* **11**(1) pp 32–42

[6] Johnson R A and Dean W W 2007 Applied Multivariate Statistical Analysis 6th Edition *United States of America: Pearson Prentice Hall*

[7] Malacarne R L 2014 Canonical Correlation Analysis *The Mathematica Journal* **16** pp 1–22

[8] Mason S J Seasonal forecasting using the Climate Predictability Tool (CPT) *Science and technology infusion Climate Bulletin* In: Mason S.J. editor. 36th NOAA Annual Climate Diagnostics and Prediction Workshop [Internet], 2011, [downloaded April 1, 2014]

[9] Pokhrel S, Chaudhari H S, Subodh K S, Ashish D, Yadav R K, Kiran S, Mahaparta S, and Suryachandra A R 2012 ENSO, IOD, and Indian Summer Monsoon in NCEP climate forecast system *Clim Dyn* **39** pp 2143–2165, DOI 10.1007/s00382-012-1349-5

[10] Qian J H, Robertson A W, and Moron V 2010 Interaction Among ENSO, the Monsoon, and Diurnal Cycle in Rainfall Variability Over Java, Indonesia *Journal of the Atmospheric Sciences* **67** pp 3509 – 3524

[11] Susetyaningsih A 2012 Pengaturan penggunaan lahan di daerah hulu DAS Cimanuk sebagai upaya optimalisasi pemanfaatan sumberdaya air *Jurnal Konstruksi* **10**(01) ISSN: 2302-7312

[12] Wilks D S 2006 Statistical Methods in the Atmospheric Sciences Second Edition *London, United Kingdom: Elsevier Academic Press Publications*

IOP Conf. Series: Journal of Physics: Conf. Series **893** (2017) 012022 doi:10.1088/1742-6596/893/1/012022

Application of universal kriging for prediction pollutant using GStat R

Annisa Nur Falah[1], **Betty Subartini**[2], **and Budi Nurani Ruchjana**[2]

[1] Master Student at Study Program of Mathematics
Faculty of Mathematics and Natural Sciences, Universitas Padjadjaran, Indonesia
[2] Department of Mathematics, Faculty of Mathematics and Natural Sciences
Universitas Padjadjaran, Indonesia

E-mail: annisanurfalah02@gmail.com

Abstract. In the universe, the air and water is a natural resource that is a very big function for living beings. The air is a gas mixture contained in a layer that surrounds the earth and the components of the gas mixture is not always constant. Also in river there is always a pollutant of chemistry concentration more than concentration limit. During the time a lot of air or water pollution caused by industrial waste, coal ash or chemistry pollution is an example of pollution that can pollute the environment and damage the health of humans. To solve this problem we need a method that is able to predict pollutant content in locations that are not observed. In geostatistics, we can use the universal kriging for prediction in a location that unobserved locations. Universal kriging is an interpolation method that has a tendency trend (drift) or a particular valuation method used to deal with non-stationary sample data. GStat R is a program based on open source R software that can be used to predict pollutant in a location that is not observed by the method of universal kriging. In this research, we predicted river pollutant content using trend (drift) equation of first order. GStat R application program in the prediction of river pollutants provides faster computation, more accurate, convenient and can be used as a recommendation for policy makers in the field of environment.

1. Introduction

Geostatistics is a blend of science mining, geology, mathematics, and statistics [1]. The data used in a geostatistics is spatial data based on the observation that the value of the location. One of main geostatistics tools is an interpolation process which initially is used to make predictions of mineral reserves, called as Kriging [6]. Kriging, or spatial Best Linear Unbiased Estimation (BLUE) is coined by Matheron (1962) in honour of D. G. Krige, a South African mining engineer [3]. Kriging is a valuation method that uses spatial data. Kriging is a method to estimate, in a spatial context, the value of a variable of interest at a location where it has not been measured, using data in the neighborhood [10]. Kriging calculation process can be distinguished several kinds, namely: Ordinary Point Kriging, Ordinary Block Kriging, Co-Kriging, Universal Kriging [8]. Universal Kriging is Kriging of data that have a particular tendency to trend [4]. Universal Kriging can also be interpreted as a valuation method used to deal with problems non-stationarity of sample data and can help the problems in real life. One example is the predicted levels of pollutant cadmium in Meuse river floodplain [5]. Thus it is necessary to predict the location that contains cadmium. However there is only few sample data about the cadmium concentration of Meuse river, hence the missing data in unobserved regions need to be generated.

In the calculation of prediction with Universal Kriging Method, it is required gstat and sp library in GStat-R [9]. The current condition of the river has a lot of waste contains pollutants, for example pollutant cadmium that can damage ecosystems and interfere with human health. Thus the need for a method that can predict pollutant of cadmium that pollute the river flood plains ie Universal Kriging Method. GStat R Program is a program for geostatistical model, prediction and simulation in one, two, or three dimensions comprising the sample variogram calculation, cross variogram and match the model. At the beginning of this stand-alone program, but in 2004 were made in the software package R as one package containing spatial data processing functions or command processing program geostatistical applications [2]. Thus, in this study to analyze the Universal Kriging Method can use a software based GStat R Program.

2. Literature Study

In this section, we briefly describe the main theoretical tools that will be used in the analysis of the Universal Kriging Method.

2.1. The Experimental Semivariogram for Universal Kriging

According to [1] develop hypotheses intrinsic meaning that the mean and variance of the shift $[Z(x+h) - Z(x)]$ does not depend on the position of the location x or defined as:

$$E[Z(x+h) - Z(x)] = 0 \tag{1}$$

$$Var[Z(x+h) - Z(x)] \tag{2}$$

According the equation (1) and (2) so the experimental semivariogram is defined as:

$$\hat{\gamma} = \frac{1}{2N(h)} \sum_{i=1}^{N(h)} [(Z(x_i+h) - m(x)) - (Z(x_i) - m(x))]^2 \tag{3}$$

with :

$\hat{\gamma}(h)$	:	experimental semivariogram value with distance h
$Z(x_i)$	:	value of observations in x_i
$Z(x_i+h)$	:	value of observations in $x_i + h$
$m(x)$	:	trend (drift) equation
$N(h)$	:	number of point pairs within h

2.2. The Theoretical Semivariogram Models

Generally the experimental semivariogram is computed by formula (3), we usually observe the distribution of experimental semivariogram and then identify a reasonable theoretical semivariogram model based on the experimental semivariogram distribution or prior knowledge. The most commonly used theoretical semivariogram models are exponential model, gaussian model and spherical model [11]. These theoretical models must be fitted to the experimental semivariogram by determining three parameters: sill c, range a and distance (h)

Spherical Model

$$\gamma(h) = \begin{cases} c\left[\frac{3}{2}\left(\frac{h}{a}\right) - \left(\frac{1}{2}\right)\left(\frac{h}{a}\right)^2\right], & \text{h}< \text{a}; \\ c, & \text{h}\geq \text{a}. \end{cases} \tag{4}$$

Gaussian Model

$$\gamma(h) = \begin{cases} c\left[1 - exp\left(\frac{-3h^2}{a^2}\right)\right], & \text{h}< \text{a}; \\ c, & \text{h}\geq \text{a}. \end{cases} \tag{5}$$

Exponential Model

$$\gamma(h) = \begin{cases} c\left[1 - exp\left(\frac{-3h}{a}\right)\right], & \text{h}< \text{a}; \\ c, & \text{h}\geq \text{a}. \end{cases} \tag{6}$$

2.3. Universal Kriging Method

Universal Kriging Method is one of the Kriging Method which assumes that the average (mean) is not constant. If the equation $m(x)$ is the trend (drift), with the following equation [4] :

$$m(x) = \sum_{l=0}^{n} a_l f_l(x)$$

Because $m(x)$ not constant, then the estimator of universal kriging as follows [8]:

$$\hat{Z}(x_0) - m(x) = \sum_{i=1}^{k} \lambda_i[Z(x_i) - m(x)]$$

$$\hat{Z}(x_0) - m(x) = \sum_{i=1}^{k} \lambda_i[Z(x_i) - m(x)]$$

$$\hat{Z}(x_0) = m(x) + \sum_{i=1}^{k} \lambda_i[Z(x_i) - m(x)]$$

3. Main Result

3.1. Research Data

The data used in this paper is secondary data in the program GStat R . Dataset drawn named Data Meuse is a data set that consists of four heavy metals (Cadmium, Zinc, Lead, Copper) as measured from the top layer of soil along the Meuse River flood plain located around Meers and Maasband, Limburg, Netherlands [2].

Table 1. Table of Cadmium Data

Locations	x(m)	y(m)	Cadmium (ppm)
1	181072	333611	11.7
2	181025	333558	8.6
...	...	...	...
164	180627	330190	2.7

Table 2. Table of Descriptive Statistics

	x	y	Cadmium
Min.	178605	330179	0.7
1st Qu.	179442	331065	1.8
Median	180283	332213	2.9
Mean	180156	332022	4.689
3rd Qu.	180935	332778	7.05
Max.	181390	333611	18.1

From the Table 2 above minimum values obtained coordinate x 178605 meters and a maximum of 181390 meters, the minimum coordinate y value of 329714 meters and a maximum of 333611 meters, while the content of cadmium ie 0.7 ppm to 18.1 ppm and the minimum value to the maximum value.

3.2. Stationarity

In universal kriging necessary assumption that the data is a second order not stationary, while the second order stationary universal kriging not apply, so all the data is done plotting the data in order to determine whether the obtained data is stationary or not. Non-stationary from sample data will look after points in the plot and the content will be visible if the data has a tendency to trend or not. Here is a plot of the distribution of the data based on Table I.

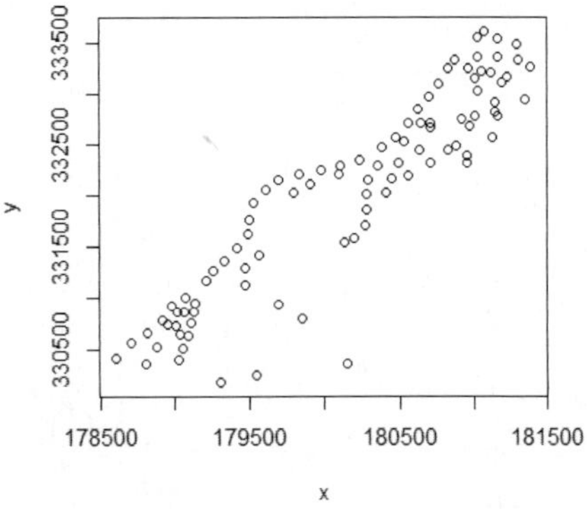

Figure 1. Plot 2D of Cadmium Data

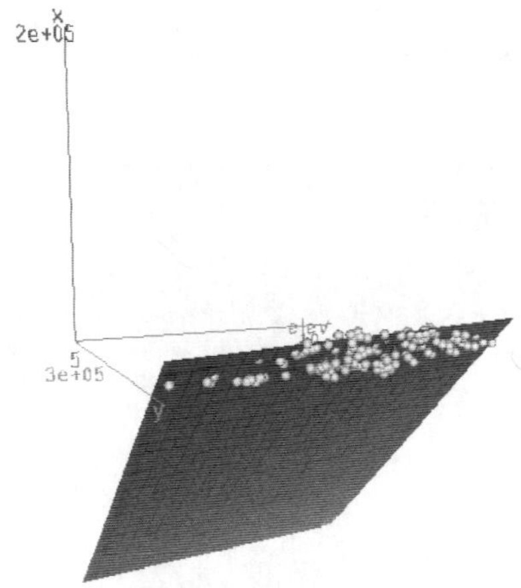

Figure 2. Plot 3D of Cadmium Data

Based from Fig. 1. and Fig. 2. can be seen that the data of pollutant cadmium is not stationary, because there are some outliers or a trend (drift). Therefore the universal kriging method is suitable to use to predict the data of the cadmium.

3.3. Table of The Experimental Semivariogram

The experimental semivariogram is one tool used to get the best of the theoretical semivariogram models, To be able to know the theoretical semivariogram models are suitable for use, first look for the average of the value semivariogram, with the help of R.3.1.1 program average values obtained semivariogram from a pair of data shown in the following table:

Based on the Table 3 the average value obtained table experimental semivariogram as many as 15 locations. To determine the best of the theoretical semivariogram model, can be done by the plot of data from the experimental semivariogram.

Table 3. The Experimental Semivariogram

No	Number of point pairs within h	Distance	Experimental Semivariogram
1	57	79.29244	4.927280
2	299	163.97367	6.470288
3	419	267.36483	6.603915
4	457	372.73542	8.581387
5	547	478.47670	9.326972
6	533	585.34058	10.494006
7	574	693.14526	10.009457
8	564	796.18365	10.610774
9	589	903.14650	11.097955
10	543	1011.29177	12.220606
11	500	1117.86235	14.103238
12	477	1221.32810	13.323896
13	452	1329.16407	15.156499
14	457	1437.25620	10.608018
15	415	1543.20248	11.153523

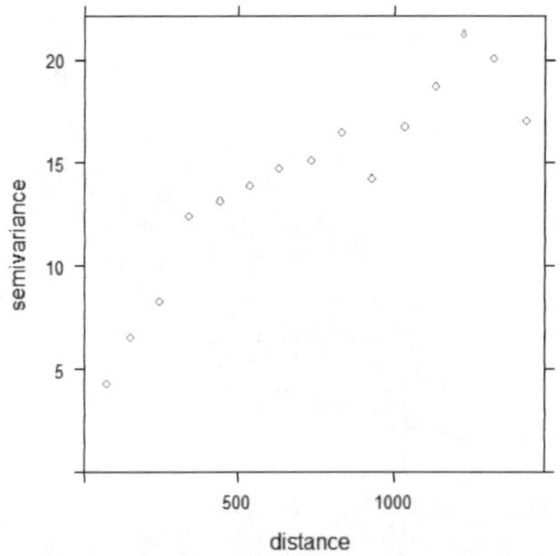

Figure 3. The Experimental Semivariogram

3.4. Fitting The Experimental Semivariogram with The Theoritical Semivariogram Models
Based on the plot of the results of the experimental semivariogram value calculation, the following is the result fitting of the three theoretical semivariogram models.

Table 4. MSE Semivariogram

MSE Of The Theoritical Semivariogram Models		
Spherical	Gaussian	Eksponensial
3.455941	4.195471	18.24626

Based on the Fig. 4. and Table 4, the theoretical semivariogram model is obtained it is best spherical semivariogram model, because it has the most minimum mean square error between gaussian semivariogram model and exponential semivariogram model.

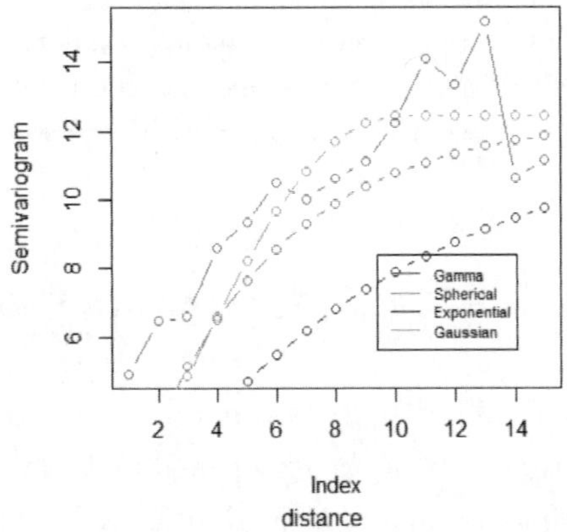

Figure 4. Fitting Theoretical with Experimental Semivariogram

3.5. The Trend (drift)

The universal kriging is kriging of data that have a tendency to trend (drift). In this paper the order of the equation trend (drift) bounded is order one, because the shape is more simple than the other order. Using R.3.1.1 program can be sought from the equation coefficients trend (drift) as follows:

Table 5. Coefficient of the Trend Equation

Coefficients		
Intercept	x	y
-326.65228	-0.00294	0.00259

3.6. Predictions at Unobserved Locations

Table 6. Prediction at Unobserved Locations

Locations	x	y	Prediction	Error Variance
1	181071	333612	8.92048	6.081463
2	180111	331811	0.8250955	6.076976
3	181190	333114	1.611239	5.655549
4	180553	332706	8.903248	5.769673
5	180915	332754	1.727731	5.641941
6	180384	332475	9.630817	5.863041
7	180198	331590	0.9197342	5.727752
8	179472	331124	4.242874	5.987837
9	178813	330665	3.679977	5.90064
10	179302	330178	0.9584527	5.813183

Based on the Table 5, this means that when the value of the absis x is equal to zero, and the value of the ordinate y is equal to zero, then the contant of the pollutant cadmium will be equal to intercept, the value would be decrease by 326.65228. If the value of the absis x shifted one unit then the level will be decrease by 0.00294 whereas if the ordinate y shifted one unit then the content of pollutant cadmium would be increase by 0.00259.

According to the Table 6, it can be seen that the predicted results and the variance of pollutant cadmium in 5 locations at unobserved locations is has the variation value, is not likely to go increase or not likely to go decrease, this is because the type of data that is predictable is the data that is non-stationary or have a tendency trend (drift) so that the results of predictions is very variation.

4. Conclusion

GStat R procedure in calculating prediction at unobserved locations with universal kriging method include the following stages :

a. Testing stationarity by way of plot distribution data.

b. The process of calculating the value of the experimental semivariogram.

c. Fitting the theoretical semivariogram models that best by finding the minimum sum of squared errors.

d. Searching for the coefficients of the equation trend (drift) of the order one.

e. The prediction at unobserved locations using universal kriging method.

Acknowledgment

The work reported in this paper is supported by the Directorate General of Higher Education, Ministry of Research-Technology and Higher Education, Indonesia, through Penelitian Kerjasama Luar Negeri dan Publikasi Internasional 2016.

References

[1] Armstrong M 1998 Basic Linear Geostatistics. New York: Springer
[2] Bivand R, Pebesma E, and Rubio V 2013 Applied Spatial Data Analysis with R (Second Edition) New York: Springer
[3] Cressie N A C 1993 Statistics For Spatial Data. New York: John Wiley and Sons, Inc.
[4] Kambhammettu B, Allena P, and King J 2011 Application and evaluation of universal kriging for optimal contouring of groundwater levels. J.Eart Syst. Sci.
[5] Middelkoop H 2000 Heavy-metal pollution of the river Rhine and Meuse floodplains in the Netherlands. Netherlands Journal of Geosciences **79**(4) pp 411–428
[6] Montero J M, Aviles G F, and Mateu J 2015 Spatial and Spatio-Temporal Geostatistical Modeling and Kriging New York: John Wiley and Sons.
[7] Nerini D, Monestiez P, and Mante C 2010 Cokriging for spatial functional data. Journal of Multivariate Analysis pp 409–418
[8] Olea R A 1999 Geostatistics for engineers and earth scientists. United States of America: Kluwer Academic Publishers
[9] Pebesma E 2004 Multivariable geostatistics in S the gstat package Computers and Geosciences **30**(7) pp 683–691
[10] Wackernagel H 1994 Cokriging versus kriging in regionalized multivariate data analysis Elsevier Science B V pp 83–92
[11] Youkuo C, Yongguo Y, and Wangwen W 2015 Coal Seam Thickness Prediction based on Least Squares Support Vector Machines and Kriging Method EJGE **20** Bund. 1

Estimating the intensity of a cyclic Poisson process in the presence of additive and multiplicative linear trend

I Wayan Mangku

Department of Mathematics, Bogor Agricultural University, Indonesia

E-mail: wayan.mangku@gmail.com

Abstract. In this paper we survey some results on estimation of the intensity function of a cyclic Poisson process in the presence of additive and multiplicative linear trend. We do not assume any parametric form for the cyclic component of the intensity function, except that it is periodic. Moreover, we consider the case when there is only a single realization of the Poisson process is observed in a bounded interval. The considered estimators are weakly and strongly consistent when the size of the observation interval indefinitely expands. Asymptotic approximations to the bias and variance of those estimators are presented.

1. Introduction

In this paper we survey some results on estimation of the intensity function of a cyclic Poisson process in the presence of additive and multiplicative linear trend. The problems of estimating intensity of a cyclic Poisson process using a single realization of the process arises frequently in many diverse areas, including: communications, hydrology, meteorology, insurance, reliability, medical sciences, and seismology.

Some existing methods using parametric approach are as follows. Rathbun and Cressie [14] investigated maximum likelihood estimators and Bayes estimators for regular parametric models with unknown finite dimensional parameter.

Helmers and Zitikis [4] consider a uniform kernel type estimator for $\lambda(s)$ in the case where λ is a parametric function of spatial location. These authors focus their attention to the case that X is a Poisson process on $[0, \infty)$ with intensity function

$$\lambda(s) = \exp\left\{\alpha + \beta s + \gamma s^2 + K_1 \sin(\omega_0 s) + K_2 \cos(\omega_0 s)\right\},$$

$s > 0$, where $\alpha, \beta, \gamma, K_1$, and K_2 are unknown parameters, and ω_0 is a known 'frequency'.

Vere-Jones [16] consider a cyclic Poisson process with intensity function

$$\lambda(s) = A \exp\left\{\rho \cos(\omega_0 s + \phi)\right\},$$

where $A > 0$, $\rho > 0$, $\omega_0 > 0$, and $\phi(0 < \phi < 2\pi)$ are unknown parameters.

In this paper, we consider the problem of estimating the intensity λ of a cyclic Poisson process (with and without trend) without assuming any parametric form of λ, except that it is periodic.

The rest of the paper is organized as follows. In section 2 we rewiew some results on estimation of the intensity of a (purely) cyclic Poisson process. Some results on estimation of the intensity of a cyclic Poisson process in the presence of additive linear trend are presented in section 3. In section 4 we present some results on estimation of the intensity of a cyclic Poisson process in the presence of multiplicative linear trend. Some conclusions are given in section 5.

2. Estimating intensity of a cyclic Poisson process

In this section we rewiew some results on estimation of the intensity of a cyclic Poisson process. Let N_c be a Poisson process on $[0, \infty)$ with (unknown) locally integrable intensity function λ. We assume that λ is a periodic function with (known) period τ. We do not assume any parametric form of λ, except that it is periodic. That is, for each point $s \in [0, \infty)$ and all $k \in \mathbf{Z}$, with $\mathbf{Z}$ denotes the set of integers, we have

$$\lambda(s + k\tau) = \lambda(s).$$

Suppose that, for some $\omega \in \Omega$, a single realization $N_c(\omega)$ of the Poisson process N_c defined on a probability space $(\Omega, \mathcal{F}, \mathbf{P})$ with intensity function λ is observed, though only within a bounded interval $[0, n]$.

Note that, since λ is a periodic function with period τ, the problem of estimating λ at a given point $s \in [0, \infty)$ can be reduced into a problem of estimating λ at a given point $s \in [0, \tau)$. Hence, for the rest of this paper, we will assume that $s \in [0, \tau)$.

We also note that, the meaning of the asymptotic $n \to \infty$ in this paper is somewhat different from the classical one. Here n does not denote our sample size, but it denotes the length of the interval of observations. The size of our samples is a random variable denoted by $N_c([0, n])$.

Let $K : \mathbf{R} \to \mathbf{R}$ be a real valued function, called *kernel*, which satisfies the following conditions: (K1) K is a probability density function, (K2) K is bounded, and (K3) K has (closed) support $[-1, 1]$. Let also h_n be a sequence of positive real numbers converging to 0, that is, $h_n \downarrow 0$, as $n \to \infty$.

The estimator of λ at a given point $s \in [0, \tau)$ is given by

$$\hat{\lambda}_{n,K}(s) = \frac{\tau}{n} \sum_{k=0}^{\infty} \frac{1}{h_n} \int_0^n K\left(\frac{x - (s + k\tau)}{h_n}\right) N_c(dx). \tag{1}$$

For the rest of this section we assume that the intensity function λ is periodic and locally integrable, the kernel K satisfies conditions $(K1), (K2), (K3)$, and $h_n \downarrow 0$ as $n \to \infty$.

Our first result is as follows. If $nh_n \to \infty$, then

$$\hat{\lambda}_{n,K}(s) \xrightarrow{p} \lambda(s), \tag{2}$$

as $n \to \infty$, provided s is a Lebesgue point of λ. In other words, $\hat{\lambda}_{n,K}(s)$ is a consistent estimator of $\lambda(s)$. In addition, the *Mean-Squared-Error* (MSE) of $\hat{\lambda}_{n,K}(s)$ converges to 0, as $n \to \infty$, that is we have

$$MSE(\hat{\lambda}_{n,K}(s)) \rightarrow 0, \tag{3}$$

as $n \to \infty$, provided s is a Lebesgue point of λ.

Under slightly stronger assumption, we also have strong consistency of $\hat{\lambda}_{n,K}(s)$ as follows. If $\sum_{n=1}^{\infty} \exp\{-\epsilon\sqrt{nh_n}\} < \infty$, for each $\epsilon > 0$, then

$$\hat{\lambda}_{n,K}(s) \overset{a.s.}{\rightarrow} \lambda(s), \tag{4}$$

as $n \to \infty$, provided s is a Lebesgue point of λ. In other words, $\hat{\lambda}_{n,K}(s)$ is a strongly consistent estimator of $\lambda(s)$.

Asymptotic approximation to the bias of $\hat{\lambda}_{n,K}(s)$ can be described as follows. If the intensity function λ has finite second derivative λ'' at s, the kernel K is symmetric and $nh_n^2 \to \infty$, then

$$\mathbf{E}\hat{\lambda}_{n,K}(s) = \lambda(s) + \frac{1}{2}\lambda''(s)h_n^2 \int_{-1}^{1} x^2 K(x)dx + o(h_n^2), \tag{5}$$

as $n \to \infty$.

Asymptotic approximation to the variance of $\hat{\lambda}_{n,K}(s)$ is given by

$$Var\left(\hat{\lambda}_{n,K}(s)\right) = \frac{\tau\lambda(s)}{nh_n}\int_{-1}^{1} K^2(x)dx + o\left(\frac{1}{nh_n}\right) \tag{6}$$

as $n \to \infty$, provided s is a Lebesgue point of λ.

From (5) and (6) we obtain asymptotic approximation to the MSE of $\hat{\lambda}_{n,K}(s)$, which is given by

$$\begin{aligned} MSE(\hat{\lambda}_{n,K}(s)) &= \frac{\tau\lambda(s)}{nh_n}\int_{-1}^{1} K^2(x)dx + \frac{1}{4}\left(\lambda''(s)\int_{-1}^{1} x^2 K(x)dx\right)^2 h_n^4 \\ &\quad + o\left(\frac{1}{nh_n}\right) + o(h_n^4), \end{aligned} \tag{7}$$

as $n \to \infty$.

By minimizing the main terms for the variance and the squared bias on the r.h.s. of (7), we then obtain the optimal choice of h_n, which is given by

$$h_n = \left[\frac{\tau\lambda(s)\int_{-1}^{1} K^2(x)dx}{\left(\lambda''(s)\int_{-1}^{1} x^2 K(x)dx\right)^2}\right]^{\frac{1}{5}} n^{-\frac{1}{5}}.$$

With this choice of h_n, the optimal rate of decrease of $MSE(\hat{\lambda}_{n,K}(s))$ is of order $\mathcal{O}(n^{-4/5})$ as $n \to \infty$.

Mathematical details of the results presented in this section can be found in [5] and [6]. Some related work can be found in [3], [10], and [15].

3. Estimating a cyclic intensity plus linear trend

In this section, we survey some results on estimation of the intensity of a cyclic Poisson process in the presence of additive linear trend. Let N be a Poisson point process on $[0,\infty)$ with (unknown) locally integrable intensity function λ which is assumed to consist of two components, namely a periodic or cyclic component with period $\tau > 0$ and a (unknown) linear trend component. In other words, for any point $s \in [0,\infty)$, we can write the intensity function λ as

$$\lambda(s) = \lambda_c(s) + as \tag{8}$$

where $\lambda_c(s)$ is a periodic function with (known) period τ and a denotes the slope of the linear trend.

Suppose now that, for some $\omega \in \Omega$, a single realization $N(\omega)$ of the Poisson process N defined on a probability space $(\Omega, \mathcal{F}, \mathbf{P})$ with intensity function λ (cf. (8)) is observed, though only within a bounded interval $[0,n]$.

An estimator of a is given by

$$\hat{a}_n = \frac{2N([0,n])}{n^2}.$$

The expected value and variance of $\hat{a}_n$ are given by respectively

$$\mathbf{E}\left(\hat{a}_n\right) = a + \frac{2\theta}{n} + \mathcal{O}\left(\frac{1}{n^2}\right)$$

and

$$Var\left(\hat{a}_n\right) = \frac{2a}{n^2} + \mathcal{O}\left(\frac{1}{n^3}\right)$$

as $n \to \infty$, where $\theta = \tau^{-1}\int_0^\tau \lambda_c(s)ds$, the global intensity of the periodic component λ_c. Hence $\hat{a}_n$ is a consistent estimator of a. Its MSE is given by

$$MSE\left(\hat{a}_n\right) = \frac{4\theta^2 + 2a}{n^2} + \mathcal{O}\left(\frac{1}{n^3}\right)$$

as $n \to \infty$.

The estimator of λ_c, the cyclic component of the intensity function λ, is given by

$$\hat{\lambda}_{c,n,K}(s) = \frac{1}{\ln(\frac{n}{\tau})}\sum_{k=1}^{\infty}\frac{1}{kh_n}\int_0^n K\left(\frac{x-(s+k\tau)}{h_n}\right)N(dx) - \hat{a}_n\left(s + \frac{n}{\ln(\frac{n}{\tau})}\right).$$

If we are interested in estimating $\lambda(s)$ at a given point s, then $\lambda(s)$ can be estimated by

$$\hat{\lambda}_{n,K}(s) = \hat{\lambda}_{c,n,K}(s) + \hat{a}_n s.$$

For the rest of this section, we assume that the intensity function λ satisfies (8) and is locally integrable, the kernel K satisfies conditions $(K1), (K2), (K3)$, and $h_n \downarrow 0$ as $n \to \infty$.

Consistency of $\hat{\lambda}_{c,n,K}(s)$ is as follows. If $h_n \ln n \to \infty$, then

$$\hat{\lambda}_{c,n,K}(s) \xrightarrow{p} \lambda_c(s),$$

as $n \to \infty$, provided s is a Lebesgue point of λ_c. Hence, $\hat{\lambda}_{c,n,K}(s)$ is a consistent estimator of $\lambda_c(s)$. In addition, the MSE of $\hat{\lambda}_{c,n,K}(s)$ converges to 0, as $n \to \infty$.

Asymptotic approximation to the bias of $\hat{\lambda}_{c,n,K}(s)$ can be described as follows. If the intensity function λ has finite second derivative λ'' at s and $h_n^2 \ln n \to \infty$, then

$$\mathbf{E}\hat{\lambda}_{c,n,K}(s) = \lambda_c(s) + \frac{\lambda_c''(s)}{2} h_n^2 \int_{-1}^{1} x^2 K(x)dx + o(h_n^2),$$

as $n \to \infty$.

Asymptotic approximation to the variance of $\hat{\lambda}_{c,n,K}(s)$ is given by

$$Var\left(\hat{\lambda}_{c,n,K}(s)\right) = \frac{a\tau}{2h_n \ln n} \int_{-1}^{1} K^2(x)dx + o\left(\frac{1}{h_n \ln n}\right),$$

as $n \to \infty$, provided s is a Lebesgue point of λ_c.

From the asymptotic approximations to the bias and variance of $\hat{\lambda}_{c,n,K}(s)$ we can obtain asymptotic approximations to the MSE of $\hat{\lambda}_{c,n,K}(s)$, which is given by

$$\begin{aligned} MSE\left(\hat{\lambda}_{c,n,K}(s)\right) &= \frac{a\tau}{2h_n \ln n} \int_{-1}^{1} K^2(x)dx + \frac{1}{4}\left(\lambda_c''(s) \int_{-1}^{1} x^2 K(x)dx\right)^2 h_n^4 \\ &\quad + o\left(\frac{1}{h_n \ln n}\right) + o(h_n^4), \end{aligned}$$

as $n \to \infty$.

By minimizing the main terms for the variance and the squared bias of $MSE(\hat{\lambda}_{c,n,K}(s))$, we then obtain the optimal choice of h_n, which is given by

$$h_n = \left(\frac{a\tau \int_{-1}^{1} K^2(x)dx}{2\left(\lambda_c''(s) \int_{-1}^{1} x^2 K(x)dx\right)^2} \right)^{\frac{1}{5}} (\ln n)^{-\frac{1}{5}}.$$

With this choice of h_n, the optimal rate of decrease of $MSE(\hat{\lambda}_{c,n,K}(s))$ is of order $\mathcal{O}((\ln n)^{-4/5})$ as $n \to \infty$.

Mathematical details of the results presented in this section can be found in [2], [11], and [12]. Some related work can be found in [7].

4. Estimating the intensity obtained as the product of a periodic function with the linear trend

In this section, we survey some results on estimation of the intensity of a cyclic Poisson process in the presence of multiplicative linear trend. Let N^* be a Poisson process on $[0, \infty)$ with (unknown) locally integrable intensity function λ. The intensity function λ is assumed to be a periodic function multiplied by the linear trend. That is, for any given point $s \in [0, \infty)$, the intensity function λ can be written as

$$\lambda(s) = (\lambda_c^*(s))(as), \tag{9}$$

where $\lambda_c^*(s)$ is a periodic function with known period τ and $a > 0$ denotes the slope of the linear trend.

Since $a\lambda_c^*$ is also a periodic function with period τ, then, without loss of generality, the intensity function λ given in (9) can also be written as

$$\lambda(s) = (\lambda_c(s))(s), \tag{10}$$

where $\lambda_c(s) = a\lambda_c^*(s)$.

Suppose that, for some $\omega \in \Omega$, a single realization $N^*(\omega)$ of the Poisson process N^* defined on a probability space $(\Omega, \mathcal{F}, \mathbf{P})$ with intensity function λ given by (10) is observed, though only within a bounded interval $[0, n]$.

The estimator of λ_c at a given point $s \in [0, \tau)$ is given by

$$\hat{\lambda}_{c,n,K}^*(s) = \frac{\tau}{n} \sum_{k=0}^{\infty} \frac{1}{h_n(s + k\tau)} \int_0^n K\left(\frac{x - (s + k\tau)}{h_n}\right) N^*(dx).$$

For the rest of this section we assume that the intensity function λ satisfies (10) and is locally integrable, the kernel K satisfies conditions $(K1), (K2), (K3)$, and $h_n \downarrow 0$ as $n \to \infty$.

Our first result of this section is as follows. If $n^2 h_n / \ln n \to \infty$, then

$$\hat{\lambda}_{c,n,K}^*(s) \xrightarrow{p} \lambda_c(s),$$

as $n \to \infty$, provided s is a Lebesgue point of λ. Hence, $\hat{\lambda}_{c,n,K}^*(s)$ is a consistent estimator of $\lambda_c(s)$. In addition, the MSE of $\hat{\lambda}_{c,n,K}^*(s)$ converges to 0, as $n \to \infty$, that is we have

$$MSE(\hat{\lambda}_{c,n,K}^*(s)) \to 0,$$

as $n \to \infty$.

Asymptotic approximation to the bias of $\hat{\lambda}_{c,n,K}^*(s)$ can be described as follows. If λ_c has finite second derivative λ_c'' at s, the kernel K is symmetric and $nh_n^2 \to \infty$, then

$$\mathbf{E}\hat{\lambda}_{c,n,K}^*(s) = \lambda_c(s) + \frac{1}{2}\lambda_c''(s) h_n^2 \int_{-1}^{1} x^2 K(x) dx + o(h_n^2),$$

as $n \to \infty$.

Asymptotic approximation to the variance of $\hat{\lambda}_{c,n,K}^*(s)$ is given by

$$Var\left(\hat{\lambda}_{c,n,K}^*(s)\right) = \frac{\tau^2 \lambda(s) \ln n}{n^2 h_n} \int_{-1}^{1} K^2(x) dx + o\left(\frac{\ln n}{n^2 h_n}\right)$$

as $n \to \infty$, provided s is a Lebesgue point of λ.

From asymptotic approximations to the bias and variance of $\hat{\lambda}^*_{c,n,K}(s)$ we obtain an asymptotic approximation to the MSE of $\hat{\lambda}^*_{c,n,K}(s)$, which is given by

$$\begin{aligned} MSE(\hat{\lambda}^*_{c,n,K}(s)) &= \frac{\tau^2\lambda(s)\ln n}{n^2 h_n}\int_{-1}^{1} K^2(x)dx + \frac{1}{4}\left(\lambda_c''(s)\int_{-1}^{1} x^2K(x)dx\right)^2 h_n^4 \\ &\quad + o\left(\frac{\ln n}{n^2 h_n}\right) + o(h_n^4), \end{aligned}$$

as $n \to \infty$.

By minimizing the sum of the main terms for the variance and the squared bias of $MSE(\hat{\lambda}^*_{c,n,K}(s))$, we then obtain the optimal choice of h_n, which is given by

$$h_n = \left[\frac{\tau^2\lambda(s)\int_{-1}^{1} K^2(x)dx}{\left(\lambda_c''(s)\int_{-1}^{1} x^2K(x)dx\right)^2}\right]^{\frac{1}{5}} \left(\frac{n^2}{\ln n}\right)^{-\frac{1}{5}}.$$

With this choice of h_n, the optimal rate of decrease of $MSE(\hat{\lambda}^*_{c,n,K}(s))$ is of order $\mathcal{O}((n^2/\ln n)^{-4/5})$ as $n \to \infty$.

Mathematical details of the results presented in this section can be found in [8]. Some related work can be found in [1], [9], and [13].

5. Conclusion

In this paper we consider Poisson processes with three models of intensity functions, namely: (i) $\lambda(s) = \lambda_c(s)$ (purely cyclic Poisson process), (ii) $\lambda(s) = \lambda_c(s) + as$ (cyclic Poisson process in the presence of additice linear trend) , and (iii) $\lambda(s) = (\lambda_c(s))as$ (cyclic Poisson process in the presence of multiplicatice linear trend). Estimators of $\lambda_c(s)$ in these three cases are denoted respectively by (i) $\hat{\lambda}_{n,K}(s)$, (ii) $\hat{\lambda}_{c,n,K}(s)$ and (iii) $\hat{\lambda}^*_{c,n,K}(s)$.

First we obtain that all these three estimators are having the same asymptotic approximations to the bias, that is

$$\frac{1}{2}\lambda_c''(s)h_n^2\int_{-1}^{1} x^2K(x)dx + o(h_n^2),$$

as $n \to \infty$. However, those three estimators have different asymptotic approximations to the variance, which are as follows.

$$Var\left(\hat{\lambda}_{n,K}(s)\right) = \frac{\tau\lambda(s)}{nh_n}\int_{-1}^{1} K^2(x)dx + o\left(\frac{1}{nh_n}\right),$$

$$Var\left(\hat{\lambda}_{c,n,K}(s)\right) = \frac{a\tau}{2h_n\ln n}\int_{-1}^{1} K^2(x)dx + o\left(\frac{1}{h_n\ln n}\right),$$

and

$$Var\left(\hat{\lambda}^*_{c,n,K}(s)\right) = \frac{\tau^2\lambda(s)\ln n}{n^2h_n}\int_{-1}^{1} K^2(x)dx + o\left(\frac{\ln n}{n^2h_n}\right)$$

as $n \to \infty$.

The optimal rate of decrease of MSE of those three estimators are as follows: $MSE(\hat{\lambda}_{n,K}(s)) = \mathcal{O}(n^{-4/5})$, $MSE(\hat{\lambda}_{c,n,K}(s)) = \mathcal{O}((\ln n)^{-4/5})$, and $MSE(\hat{\lambda}^*_{c,n,K}(s)) = \mathcal{O}((n^2/\ln n)^{-4/5})$ as $n \to \infty$.

6. References

[1] Erliana W, Mangku I W and Sumarno H 2014 Estimating the intensity obtained as the product of a periodic function with the power function trend of a non-homogeneous Poisson process *Far East J. Math. Sci.* **90** pp 163–72

[2] Helmers R and Mangku I W 2009 Estimating the intensity of a cyclic Poisson process in the presence of linear trend *Ann. Inst. Stat. Math.* **61** pp 599–628

[3] Helmers R and Mangku I W 2012 Predicting a cyclic Poisson process *Ann. Inst. Stat. Math.* **64** pp 1261–1279

[4] Helmers R and Zitikis R 1999 On estimation of Poisson intensity functions *Ann. Inst. Stat. Math.* **51** pp 265–280

[5] Helmers R, Mangku I W and Zitikis R 2003 Consistent estimation of the intensity function of a cyclic Poisson process *J. Multivariate Anal.* **84** pp 19–39

[6] Helmers R, Mangku I W and Zitikis R 2005 Statistical properties of a kernel-type estimator of the intensity function of a cyclic Poisson process *J. Multivariate Anal.* **92** pp 1–23

[7] Mangku I W 2010 Consistent estimation of the distribution function and the density of waiting time of a cyclic Poisson process with linear trend *Far East J. Theor. Stat.* **33** pp 81–91

[8] Mangku I W 2011 Estimating the intensity obtained as the product of a periodic function with the linear trend of a non-homogeneous Poisson process *Far East J. Math. Sci.* **51** pp 141–150

[9] Mangku I W, Budiarti R, Taslim and Casman 2013 Estimating the intensity obtained as the product of a periodic function with the quadratic trend of a non-homogeneous Poisson process *Far East J. Math. Sci.* **82** pp 33–44

[10] Mangku I W, Ruhiyat and Purnaba I G P 2013 Statistical properties of an estimator for the mean function of a compound cyclic Poisson process *Far East J. Math. Sci.* **82** pp 227–237

[11] Mangku I W, Siswadi and Budiarti R 2009 Consistency of a kernel-type estimator of the intensity of the cyclic Poisson process with the linear trend *J. Indones. Math. Soc.* **15** pp 37–48

[12] Mangku I W, Siswadi and Budiarti R 2011 Asymptotic approximations to the bias and variance of a kernel-type estimator of the intensity of the cyclic Poisson process with the linear trend *J. Indones. Math. Soc.* **17** pp 1–9

[13] Maulidi I, Mangku I W and Sumarno H 2015 Strong consistency of a kernel-type estimator for the intensity obtained as the product of a periodic function with the power function trend of a non-homogeneous Poisson process *British J. Appl. Sci. Tech.* **9** pp 383–387

[14] Rathbun S L and Cressie N 1994 Asymptotic properties of estimators for the parameters of spatial inhomogeneous Poisson point processes *Adv. Appl. Probab* **26** pp 122–154

[15] Ruhiyat, Mangku I W and Purnaba I G P 2013 Consistent estimation of the mean function of a compound cyclic Poisson process *Far East J. Math. Sci.* **77** pp 183–194

[16] Vere-Jones D 1982. On the estimation of frequency in point-process data *J. Appl. Probab.* **19A** pp 383–94

IOP Conf. Series: Journal of Physics: Conf. Series **893** (2017) 012024 doi:10.1088/1742-6596/893/1/012024

A multi item probabilistic inventory model

Dharma Lesmono and Taufik Limansyah

Department of Mathematics, Universitas Katolik Parahyangan
Jalan Ciumbuleuit 94 Bandung 40141 Indonesia

E-mail: {jdharma, taufik.limansyah}@unpar.ac.id

Abstract. In this paper, we develop a mathematical model for an inventory problem which consists of several products. Demands for these products are following Gamma distribution. Our objective is to determine the optimal ordering quantity that minimizes the total cost for each product. This is called an individual replenishment policy. The total cost consists of the purchasing cost, ordering cost, handling and shortage costs. Using the same model, we also develop a periodic review model that determines an optimal ordering time for all the products. This is called a joint replenishment policy. We then compare the total cost of the individual and the joint replenishment policy and determine a policy that gives minimum total cost. Our numerical experiments show that in general the joint replenishment policy is preferable than the individual replenishment policy. The joint replenishment policy gives higher shortage cost, but less in ordering and handling costs than the individual replenishment policy.

1. Introduction

Inventory management becomes one important thing that should be considered by the management related to the cash flow of the company especially with the funds contained in the inventory. Having a lot of inventory has a consequence of the unavailability of funds for other sectors in the company since most of the funds are invested in the inventory. On the other hand, having less inventory in hands will increase the probability of the unfilled demand, and this will cost the reputation of the company in the future. Therefore we need a mathematical model to find the optimal inventory to help company dealing with this problem. In order to build a mathematical model, we need to know that there are several costs involved in handling the inventory such as purchasing, ordering, holding and shortage costs. For retailer, two basic questions to be answered related to the inventory are when to order and how many (the optimal quantity). These two questions are the building block in the mathematical model in order to find the optimal replenishment policy.

There have been a lot of mathematical models for inventory problems developed in the last few decades and the first and most simple one is the Economic Order Quantity (EOQ) introduced by Wilson (Tersine [8]). This EOQ model has become the basis for the development of the more advanced and complex inventory models. Some factors that have been incorporated in the models involving multi-item, quantity discount, deterioration rate, deterministic or probabilistic models in terms of demand distributions.

A mathematical model for production system control for one item with uncertain deterioration rate has been developed by Bukhari [1], while Ferguson et.al [2] considers a model for perishable goods by considering time dependent non-linear holding cost. Muckstadt and Sapra [6] and Zhang and Wang [11] have developed a multi-item inventory problem with storage capacity restriction, while a joint inventory model to determine the optimal inventory replenishment product assortment, shelf space and display area is discussed in Hariga et. al [4]. Kasthuri, et.al. [5] developed a fuzzy multi-item inventory model by incorporating storage space and production cost. The case of multi-product newsboy problem for uncertain demand is studied in Zhang and Du [10]. Namit and Chen [7] and Tyworth and Ganeshan [9] discussed the process and method of finding the optimal order quantity and time when demand during lead time is Gamma distributed.

In this paper, we develop a mathematical model for an inventory problem which consists of several products. Demands for these products are following Gamma distribution. Our objective is to determine the optimal ordering quantity that minimizes the total cost for each product. This is called an individual replenishment policy. The total cost consists of the purchasing cost, ordering cost, handling and shortage costs. Using the same model, we also develop a periodic review model that determines an optimal ordering time for all the products. This is called a joint replenishment policy. We then compare the total cost of the individual and the joint replenishment policy and determine a policy that gives minimum total cost. Our numerical experiments show that in general the joint replenishment policy is preferable than the individual replenishment policy. The joint replenishment policy gives higher shortage cost, but less in ordering and handling costs than the individual replenishment policy.

The organization of this paper is as follows. In Section 2 we introduce the mathematical model which consists of the model for individual replenishment policy and joint replenishment policy along with an algorithm to find the optimal solution. Section 3 deals with numerical experiments as an application of the models discussed in Section 2. Here, comparison between joint replenishment policy and individual replenishment policy is performed and also the effect of the joint ordering cost on the optimal policy. Conclusions and further research are relegated in Section 4.

2. The Model

Probabilistic EOQ model is an inventory model that is close to the real situation retailer has to face. In reality, demand will vary from time to time. This probabilistic inventory model will incorporate the variation of the demand and uncertain lead time. Demand variation will cause a shortage especially during lead time when retailer only has a limited amount of goods to cover the demand during lead time and the goods ordered have not been arrived yet. Based on that situation, there are three possibilities that can happen for the probabilistic inventory model. The first one is when demand during lead time is constant but the lead time itself varies. The second is when lead time is constant but demand during lead time is not, and the last possibility is when both lead time and demand during lead time are varies. In this section we will discuss the model for the individual and joint replenishment policy.

2.1. Individual Replenishment Policy

In this subsection we derive a mathematical model for individual replenishment policy by considering that demand during lead time follows a class of Gamma distribution with constant lead time. From this model, we will then determine the optimal order quantity and reorder point for each item. Some assumptions in developing this model are as follows.

(i) Demand during lead time follows Gamma distribution.
(ii) The storage capacity is unlimited.
(iii) Lead time is known and constant.
(iv) Shortages occur when demand exceeds the inventory level during lead time.
(v) Unfilled demand during lead time will be replenished in the next period.

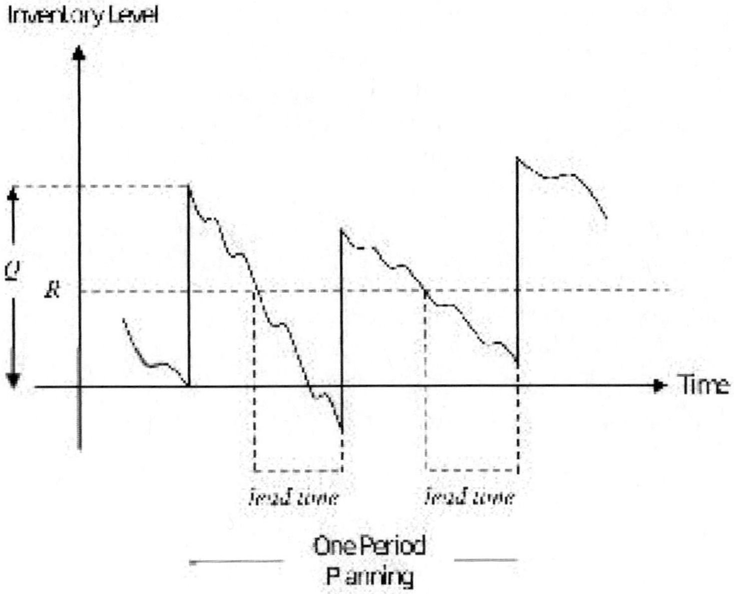

Figure 1. Probabilistic EOQ Model (Taha, 2011)

(vi) The order quantity is always the same in every replenishment.
(vii) Holding cost depends on the average goods that are stored.

We use the following notations for developing our model.

- $D =$ Average demand within one planning period.
- $P =$ Purchase cost/unit.
- $Q =$ Optimal order quantity.
- $S =$ Ordering cost per order.
- $h =$ holding cost fraction per unit per planning period.
- $\pi =$ shortage cost/unit.
- $1 - \omega =$ *service level*, retailer confidence level to fulfill customers' demand. $(0 < \omega < 1)$.
- $R =$ reorder point.
- $f(x) = \frac{1}{\beta^\alpha \Gamma(\alpha)} x^{\alpha-1} e^{-\frac{x}{\beta}}$ is Gamma density function for demand during lead time.
- $TAC =$ Total inventory cost.
- $U =$ Quantity limit where there is price change.

The situation of the problem is depicted in Figure 1 that represents the inventory level from time to time during the planning period. From that figure, we can see that shortage can occur during lead time when the demand during lead time exceeds the reorder point R.

The total inventory cost considered in building the model consists of purchasing cost, ordering cost, holding cost and shortage cost.

(i) Purchasing cost is the cost spent by the retailer to buy goods from the supplier. If the annual demand is D units, then the annual purchasing cost is PD.

(ii) Ordering cost is the cost occurs every time an order is placed. If the ordering cost is S, then the annual ordering cost is $\frac{SD}{Q}$.

(iii) Holding cost usually occurs for storing and maintaining goods, such as rent or insurance cost for goods. It is usually represented as fraction of the purchasing cost per unit, viz. Ph In our model, the annual holding cost is given by $Ph(\frac{Q}{2} + R - E(X))$, where X is random variable for demand during lead time.

(iv) The shortage cost in our model occurs when demand during lead time exceeds the reorder point. The amount of the annual shortage cost can be formulated as $\frac{\pi D}{Q}[\int_R^\infty (x - R)f(x)dx]$.

Therefore, the annual total inventory cost for our model is:

$$TAC(Q, R) = PD + \frac{SD}{Q} + Ph[\frac{Q}{2} + R - E(X)] + \frac{\pi D}{Q}[\int_R^\infty (x - R)f(x)dx] \quad (1)$$

In order to find the minimum total annual inventory cost, conditions $\frac{\partial TAC}{\partial Q} = 0$ and $\frac{\partial TAC}{\partial R} = 0$ must be satisfied. The condition $\frac{\partial TAC}{\partial Q} = 0$ will give:

$$Q = \sqrt{\frac{2D(S + \pi[\int_R^\infty (x - R)f(x)dx]}{Ph}} \quad (2)$$

and from condition $\frac{\partial TAC}{\partial R} = 0$, we have:

$$\int_R^\infty f(x)dx = \frac{PhQ}{\pi D} \quad (3)$$

The right hand side of (3) can be interpreted as the inability of the retailer to fulfill customers' demand. We can write (3) as

$$\int_R^\infty f(x)dx = \omega \quad (4)$$

The procedure to obtain the optimal order quantity and reorder point for our model follows the Hadley-Whitin algorithm (Hadley and Whitin [3]) as follows.

(i) First we start with the EOQ, that is $Q = \sqrt{\frac{2DS}{Ph}}$.

(ii) Calculate the value of R using (3).

(iii) Calculate Q using (2) with the value of R found in step 2.

(iv) Repeat step 2 and 3 until we found the optimal value of Q and R, that is the difference between each value of Q and R in each iteration is less than 1.

(v) Calculate the optimum TAC using (1).

2.2. Joint Replenishment Policy

For the joint replenishment policy, the decision variable in our model is the optimal time of replenishment. This is also called the periodic review model. In this model we assume that all item are ordered at the same time (joint order). The ordering time interval is fixed but the ordering amount for each item varies.

When items are ordered together, some costs such as ordering cost should be cheaper than when they are ordered separately. We denote the ordering cost for joint replenishment policy

Table 1. Data

	Product 1	Product 2	Product 3
Annual Demand	550	400	800
Ordering cost ($)	5	7	4
Holding cost fraction	0.01	0.015	0.01
Shortage cost ($)	1.5	2	1.2
Purchase cost ($)	12	15	8
α	5	5	3
β	2	1	2

by S^* where S^* is less than the summation of all individual ordering cost. Since the decision variable is the optimal time of replenishment, T, we use the relation $Q_i = D_iT$ for each item i in the model for the individual replenishment policy and then find the optimal T. Our model then becomes:

$$\begin{aligned} TAC(T) &= \frac{S^*}{T} + \sum_{i=1}^{n}\{P_iD_i + P_ih_i(\frac{TD_i}{2} + R_i - E(X_i)) \\ &+\frac{\pi_i}{T}\int_{R_i}^{\infty}(x_i - R_i)f(x_i)dx_i\} \end{aligned} \quad (5)$$

where n is the number of items jointly ordered. Taking the first derivative of TAC(T) and equates it to zero will give the optimal time of replenishment as follows.

$$T = \sqrt{\frac{2(S^* + \sum_{i=1}^{n}\pi\int_{R_i}^{\infty}(x_i - R_i)f(x_i)dx_i)}{\sum_{i=1}^{n}P_iD_ih_i}} \quad (6)$$

3. Numerical Experiments

In this section we consider a retailer who sells three different products to customer and orders these products from one supplier. The demand of each product is Gamma distributed with different values of parameters (α and β), and all the data is given in Table 1.

For the individual replenishment policy, using Hadley-Whitin algorithm, we have the following results as in Table 2. The optimal ordering quantity and reorder point occur for each product are also given with the total inventory cost of 19087.54.

For the joint replenishment policy, using the joint ordering cost $S^* = 9$ the results are given in Table 3.

From Table 3, the optimal ordering time T is around 0.29 years, the optimal order quantity for each product are 161, 117 and 234 respectively which are less than the optimal order quantities for the individual replenishment policy. This fact has consequences in the holding cost and shortage cost. The holding cost for the joint replenishment policy is less than the holding cost in the individual joint policy but the shortage cost is larger. These consequences make sense since the more goods we have in hands, the more we have to spend for holding cost but the less we spend for shortage cost. Overall, in this numerical experiments, the total inventory cost for the joint replenishment policy is less than the total inventory cost for the individual replenishment policy.

It is also possible the individual replenishment policy gives lower total inventory cost than the joint replenishment policy in this model. This can happen when the saving from the ordering

Table 2. Results for Individual Replenishment Policy

	Product 1	Product 2	Product 3
Q	218	160	800
R	20	10	15
Total Purchase Cost	6600	6000	6400
Total Ordering Cost	12.67	17.58	11.21
Total Holding Cost	14.2	18.89	12.10
Total Shortage Cost	0.35	0.34	0.20
Total Inventory Cost	6627.22	6036.81	6423.51
TAC		19087.54	

Table 3. Results for Joint Replenishment Policy

	Product 1	Product 2	Product 3
T (year)		0.29	
Q	161	117	234
Total Purchase Cost	6600	6000	6400
Total Ordering Cost		30.86	
Total Holding Cost	12.47	14.38	11.83
Total Shortage Cost	0.48	0.46	0.25
TAC		19070.73	

cost for the joint replenishment policy is not substantial compared with the sum of the individual ordering cost for the individual replenishment policy. In our numerical experiments, the joint ordering cost is 30.84 (Table 3) compared to the sum of the individual ordering cost of 41.46 (Table 2), a saving of 10.62.

4. Conclusion and Further Research

In this paper, we have developed a mathematical model for multi-item probabilistic inventory problem where demand for each product is Gamma distributed with different values of α and β.

We have derived an optimal ordering quantity and reorder point for the individual replenishment policy and an optimal ordering time for the joint replenishment policy. In our numerical experiments, we then compare the individual replenishment policy and joint replenishment policy to find the optimal policy. From our numerical experiment with three products and different distribution for demand during lead time, we find that the joint replenishment policy gives lower total annual inventory cost compared to the individual replenishment policy.

There are some limitations in our model since it has not considered several factors such as deterioration rate of the products, quantity discounts and perhaps non-linear holding cost involved. These factors are interesting to be pursued in further research, analyze their impact on the optimal policy and the optimal ordering quantity, time and reorder point.

Acknowledgments

The authors would like to thank the Institute of Research and Community Service Universitas Katolik Parahyangan for funding this research.

References

[1] Bukhari F 2011 Adaptive control of a production-inventory model with uncertain deterioration rate, *Appl. Math.* **2** pp 1170-74

[2] Ferguson M, Jayaraman V and Souza G C 2007 Note : An application of the EOQ model with nonlinear holding cost to inventory management of perishables, *European J. Oper. Res.* **180**(1) pp 485-490

[3] Hadley G and Whitin T M 1963 *Analysis of Inventory Systems* (Englewood Cliffs, N.J.: Prentice Hall Inc.)

[4] Hariga M A, Al-Ahmari A and Abdel-Rahman A and Mohamed 2007 A joint optimisation model for inventory replenishment product assortment, shelf space, and display area allocation decisions *European J. Oper. Res.* **181** pp 239-251.

[5] Kasthuri R, Vasanthi P, Ranganayaki S and Seshaiah C V 2011 Multi-item Fuzzy inventory model involving three constrains : A Karush-Kuhn-Tucker conditions approach, *Amer. J. Oper. Res.* **1** pp 155-159.

[6] Muckstadt J A and Sapra A 2010 *Principles of Inventory Management* (New York: Springer)

[7] Namit K and Chen J 1999 Solution to the $< Q, r >$ inventory model for Gamma Lead-Time Demand *Int. J. Phys. Distrib. Logist. Manag.* **29**(2) pp 138-151.

[8] Tersine R J 1994 *Principles of Inventory and Material Management* 4th ed (New Jersey: Prentice Hall)

[9] Tyworth J E and Ganeshan R 2000 Research note: A note on solution to the $< Q, r >$ inventory model for Gamma Lead-Time Demand *Inter. J. Phys. Distrib. Logist. Manag.* **30**(6) pp 534-539

[10] Zhang B and Du S 2010 Multi-product newsboy problem with limited capacity and outsourcing, *European J. Oper. Res.* **202** pp 107-113

[11] Zhang B and Wang X 2011 Optimal policy and simple algorithm for a deteriorated multi-item EOQ problem, *Amer. J. Oper. Res.* **1** pp 46-50

IOP Conf. Series: Journal of Physics: Conf. Series **893** (2017) 012025 doi:10.1088/1742-6596/893/1/012025

Parameter estimation and hypothesis testing on geographically weighted gamma regression

Dina Eka Putri, Purhadi, and Dedy Dwi Prastyo
Department of Statistics, Institut Teknologi Sepuluh Nopember, Surabaya, Indonesia
E-mail: msdinaputri@gmail.com, {purhadi, dedy-dp}@statistika.its.ac.id

Abstract. Having applied to observations from different location, the gamma regression yields global parameters which are assumed to be valid in any location. In fact, each location may have different characteristics such that the existence of spatial effect needs to be considered. In such a case the parameters of gamma regression are less representative. Thus, Geographically Weighted Gamma Regression (GWGR) plays into role. This study aims to estimate parameters of GWGR model using Maximum Likelihood Estimation (MLE) method and numerical optimization using Broyden-Fletcher-Goldfarb-Shanno (BFGS) algorithm. Once the parameter estimation done, the hypothesis testing procedure was used to test parameter similarity between gamma regression and GWGR as well as to test the significance of independent variables within the model, both partially using Z-test and simultaneously using Maximum Likelihood Ratio Test (MLRT).

1. Introduction

A popular and broadly used method in spatial proposed by [1, 2] was so-called Geographically Weighted Regression (GWR). The GWR has become establish as a tool in investigation of spatially varying relationship but only valid for normally distributed data. In fact, not all of the data are normally distributed. Some type of response variable might follow other distributions, such as Poisson, exponential, Weibull, Gamma, and Inverse-Gaussian.

Gamma regression is a nonlinear regression used to model relationship between one or more independent variables and a positive continuous dependent variable that follows gamma distribution. Data are generally related with the geographical location where they observed, which is called spatial data [3]. Having applied to observations from different location, gamma regression yields global parameters which are assumed to be valid in any location. In fact, each location may have different characteristics such that the existence of spatial effect needs to be considered. In such a case the parameters of gamma regression are less representative.

Thus, Geographically Weighted Gamma Regression (GWGR) plays into role. GWGR is a local form of gamma regression that considers the spatial effect, i.e. longitude and latitude of point observations. Unlike conventional regression which produces a single equation to summarize global relationship among the explanatory variable and response variable, local form regression such as GWGR generates spatial data that express the spatial variation in the relationship among variables.

This study aimed to estimate parameters of GWGR model using Maximum Likelihood Estimation (MLE) method along with to test the significance of those estimated parameters. As the MLE method could not produce solution analytically, a numerical optimization method using

Broyden-Fletcher-Goldfarb-Shanno (BFGS) algorithm was employed to estimate the parameters of GWGR model. In addition, a Generalized Cross Validation (GCV) was used to determine the optimum bandwidth in Kernel function used as the weight function. GWGR model was expected to report better estimation than the gamma regression did. Therefore a similarity model test using F-test was employed to evaluate the performance of those two models. Once the parameter estimation done, the hypothesis testing procedure was used to test the significance of independent variables within the model, both partially using Z-test and simultaneously using Maximum Likelihood Ratio Test (MLRT).

2. Material and Methodology

2.1. Gamma Distribution

Gamma distribution is a general form of exponential distribution [4] firstly introduced by the Swiss mathematician, Leonhard Euler in the 18-*th* century. The gamma distribution is positive continuous distribution which is often applied in many practical areas such as meteorology, business, finance, and health. The probability density function (pdf) of gamma distribution with two parameters is given by:

$$f(y,\alpha,\theta)=\frac{1}{\Gamma(\alpha)\theta^{\alpha}}y^{\alpha-1}e^{-\frac{y}{\theta}}, \tag{1}$$

where $\alpha,\theta > 0$ and $0 < y < 0$. Thus, $Y \sim Gamma(\alpha,\theta)$. The expected value of Y according to [5] is $\mu = \alpha\theta$.

2.2. Geographically Weighted Gamma Regression (GWGR)

Gamma regression is a non-linear regression used to model relationship between one or more independent variables and a positive continuous dependent variable that follows gamma distribution. The gamma regression had been developed by [6]. Given univariate response variable Y and several predictors X_1, X_2, ..., X_k, the gamma regression modeling is supposed to use log linear model so that expected value (μ) is only able to be positive-valued [7]:

$$\mu_i = E(Y_i) = \exp(\mathbf{x}_i^T\beta) = \alpha\theta_i \tag{2}$$

such that $\theta_i = \frac{\exp(\mathbf{X}_i^T\beta)}{\alpha}$. Therefore, the pdf of gamma regression is:

$$f(y_i,\alpha,\beta)=\frac{y_i^{\alpha-1}}{\Gamma(\alpha)\left(\frac{\exp(\mathbf{X}_i^T\beta)}{\alpha}\right)^{\alpha}}\exp\left(\frac{-y_i}{\left(\frac{\exp(\mathbf{X}_i^T\beta)}{\alpha}\right)}\right), \tag{3}$$

with $i = 1,2,\ldots,n$. Instead of calibrating a single equation that valid globally to all of observation, the GWGR generates separate regression for each observation using different weight for each data set. A response variable Y follows gamma distribution with different parameter of α and θ for each location, i.e. $Y \sim Gamma(\alpha(u_i,v_i),\theta(u_i,v_i))$, where (u_i,v_i) captures the coordinates location i [8]. Thus, the pdf of GWGR is given by:

$$f(y_i)=\frac{1}{\Gamma(\alpha(u_i,v_i))\left(\frac{\exp(\mathbf{X}_i^T\beta(u_i,v_i))}{\alpha(u_i,v_i)}\right)}\times A_i \tag{4}$$

in which

$$A_i = y_i^{\alpha-1}\exp\left(\frac{-y_i}{\left(\frac{\exp(\mathbf{X}_i^T\beta(u_i,v_i))}{\alpha(u_i,v_i)}\right)}\right).$$

The assumption is that observation nearby one another have greater influence on one another's parameter estimates than observation farther apart [9].

2.3. Spatial Effects

Spatial homogeneity caused by condition of unit in observation which is basically not homogenous. The hypothesis of Breusch-Pagan (BP) test [10] is: $H_0 : \sigma_1^2 = \sigma_2^2 = \cdots = \sigma_n^2 = \sigma^2$ against H_1 : there is at least one i in which $\sigma_i^2 \neq \sigma^2$. The BP statistics test is:

$$BP = \frac{1}{2}\mathbf{f}^T\mathbf{Z}(\mathbf{Z}^T\mathbf{Z})^{-1}\mathbf{f} \sim \chi^2_{(k+1)},$$

where $\mathbf{f} = (f_1, f_2, \dots, f_n)^T$ with $f_i = \left(\frac{e_i^2}{\widehat{\sigma}^2} - 1\right)$, $e_i = y_i - \widehat{y}_i$, $\widehat{\sigma}^2$ is the estimated variance of y, $\mathbf{Z}$ is a standardized matrix with size $n \times (k+1)$. The H_0 is rejected if $BP > \chi^2_{(k)}$.

Spatial dependence indicates the effect of one location to other locations. The hypothesis being tested using Moran's I test is $H_0 : I = 0$ (There is no spatial dependence) against $H_1 : I \neq 0$ (There is spatial dependence). The statistics test employed is:

$$Z_I = \frac{\widehat{I} - E(\widehat{I})}{\sqrt{Var(\widehat{I})}},$$

where $\widehat{I} = \frac{n\sum\limits_{i=1}^{n}\sum\limits_{i=1}^{n} w_{il}(y_i-\overline{y})(y_l-\overline{y})}{\left(\sum\limits_{i=1}^{n}\sum\limits_{i=1}^{n} w_{il}\right)\sum\limits_{l=1}^{n}(y_i-\overline{y})^2}$, $E(\widehat{I}) = \frac{-1}{N-1}$, and $Var(\widehat{I}) = \frac{NS_1 - S_2S_3}{(N-1)(N-2)(N-3)\left(\sum\limits_{i=1}^{n}\sum\limits_{j=1}^{n} w_{ij}\right)}$,

with

$$\begin{aligned}
S_1 &= (N^2 - 3N + 3)\left(\frac{1}{2}\sum_{i=1}^{n}\sum_{j=1}^{n}(w_{ij} + w_{ji})^2\right) \\
&\quad -N\sum_{j=1}^{n}\left(\sum_{j=1}^{n} w_{ij} + \sum_{j=1}^{n} w_{ji}\right)^2 \\
&\quad +3\left(\sum_{j=1}^{n} w_{ij}\sum_{j=1}^{n} w_{ji}\right)^2 \\
S_2 &= \frac{N^{-1}\sum\limits_{i=1}^{n}(y_i - \overline{y})^4}{\left(N^{-1}\sum\limits_{i=1}^{n}(y_i - \overline{y})^2\right)^2} \\
S_3 &= \left(\frac{1}{2}\sum_{j=1}^{n}\sum_{i=1}^{n}(w_{ij} + w_{ij})^2\right) - 2N\left(\frac{1}{2}\sum_{j=1}^{n}\sum_{i=1}^{n}(w_{ij} + w_{ij})^2\right) + 6\left(\sum_{j=1}^{n}\sum_{i=1}^{n} w_{ij}\right)^2,
\end{aligned} \quad (5)$$

2.4. Spatial Weighting

Before choosing spatial weight, distance decay and the bandwidth need to be selected. Distance calculated using Euclidean distance, while the optimal bandwidth selection might be chosen by researcher or optimized using an algorithm that seeks to maximize cross-validation score. Due to its importance, the bandwidth selection procedure has been being the researchers' focus over the time but there is still no method claimed to have the best performance. This study used GCV considering the method is asymptotically optimum and do not include the parameter of population.

Suppose d_{il} is the Euclidean distance from location (u_i, v_i) to (u_l, v_l) and w_{il} is the geographical weight at *l-th* location for coefficient estimate at *i-th* location. Some of spatial weighting which are frequently used for geographically weighted regression [8] are as follows:

- *Fixed Gaussian*: $w_{il}(u_i, v_i) = \exp\left[-\frac{1}{2}\left(\frac{d_{il}}{h}\right)^2\right]$
- *Fixed Bisquare*: $w_{il}(u_i, v_i) = \begin{cases} \left(1 - \left(\frac{d_{il}}{h}\right)^2\right)^2, & d_{il} \leq h \\ 0, & d_{il} > h \end{cases}$
- *Adaptive Gaussian*: $w_{il}(u_i, v_i) = \exp\left[-\frac{1}{2}\left(\frac{d_{il}}{h_i}\right)^2\right]$
- *Adaptive Bisquare*: $w_{il}(u_i, v_i) = \begin{cases} \left(1 - \left(\frac{d_{il}}{h_i}\right)^2\right)^2, & d_{il} \leq h_i \\ 0, & d_{il} > h_i \end{cases}$

The spatial weight employed is the kernel commonly used in nonparametric regression, see [11], and also in statistical learning method, see [12].

2.5. Procedure

(i) Estimating parameter on GWGR model using following algorithm:
  (a) Determine parameter set under H_0 (ω) and under population (Ω).
  (b) Determine likelihood and ln-likelihood function of GWGR using MLE method.
  (c) If the equation yielded is not close form, then BFGS method should be employed. The detail of BFGS algorithm can be seen on [13].

(ii) Testing on parameter estimation both simultaneously using Maximum Likelihood Ratio Test (MLRT) and partially using Z test.

3. Results and Discussion

3.1. Parameter Estimation

The GWGR parameter estimation is conducted using MLE which aims to maximize the log likelihood function as follows:

$$\ln L(\alpha(u_i, v_i), \beta(u_i, v_i) i = 1, 2, \ldots, n) = B + C + D, \quad (6)$$

where

$$\begin{aligned} B &= -\ln(\Gamma(\alpha(u_i, v_i))) \sum_{i=1}^{n} w_{il} + \alpha(u_i, v_i) \ln \alpha(u_i, v_i) \sum_{i=1}^{n} w_{il} \\ C &= -\alpha(u_i, v_i) \sum_{i=1}^{n} (\mathbf{x}_i^T \beta(u_i, v_i)) w_{il} + (\alpha(u_i, v_i) - 1) \sum_{i=1}^{n} \ln y_i w_{il} \\ D &= -\alpha(u_i, v_i) \sum_{i=1}^{n} y_i w_{il} \exp(-(\mathbf{x}_i^T \beta(u_i, v_i))), \end{aligned}$$

Since it does not generate close form solution then numerical optimization using BFGS should be performed. The detail of the algorithm is not described here because the limited space available. A note necessarily to be mentioned is that the result of BFGS algorithm depends on initial values of parameters, therefore the way how to choose the good initial point is a challenge. An evolutionary algorithm, for example a Genetic algorithm [12], can be the alternative solution for future research.

3.2. Hyphotesis Testing

Hypothesis testing consists of three kind of tests: (i) parameters similarity test between GWGR and gamma regression, (ii) simultaneous test of GWGR parameters, and (iii) partial test of GWGR parameters.

3.2.1. Parameter Similarity Test GWGR method provides a feasible way for testing global gamma regression relationship for spatial data. Hypothesis being tested is the null H_0 : $\beta_{ji}(u_i, v_i) = \beta_{ji}$ and $H_0 : \alpha(u_i, v_i) = \alpha$, with $j = 0, 1, 2, \ldots, k; i = 1, 2, \ldots, n$, against the alternative H_1: at least one of $\beta_{ji}(u_i, v_i) \neq \beta_{ji}$ and $H_1 : \alpha(u_i, v_i) \neq \alpha$. The statistics test is then constructed as:

$$F = \frac{G_1^2/df_1}{G_2^2/df_2}, \tag{7}$$

with G_1^2 is the deviance of gamma regression and G_2^2 is the deviance of GWGR, whereas df_1 is the degree of freedom of gamma regression and df_2 is the degree of freedom of GWGR. The null hypothesis H_0 is rejected if $F > F_{(\alpha, df_1, df_2)}$.

Given the parameters set under population is $\Omega = \{\beta_0, \beta_1, \ldots, \beta_k, \alpha\} = \{\beta, \alpha\}$ and parameters set under the null hypothesis $H_0(\omega)$ is $\omega = \{\beta_0, \alpha_0\}$, the deviance of gamma regression is calculated by $G_1^2 = -2 \ln \Lambda = -2 \ln \left\{ L(\widehat{\omega} - L(\widehat{\Omega})) \right\}$ resulted in:

$$G_1^2 = E + F + G + H, \tag{8}$$

with

$$\begin{aligned}
E &= 2 \ln \Gamma(\widehat{\alpha}) - 2\widehat{\alpha} \sum_{i=1}^{n} \mathbf{x}_i^T \widehat{\beta} + 2\widehat{\alpha}(n \ln \widehat{\alpha}) \\
F &= 2(\widehat{\alpha} - 1) \sum_{i=1}^{n} \ln y_i - 2 \sum_{i=1}^{n} \frac{y_i}{\left(\frac{\exp(\mathbf{x}_i^T \widehat{\beta})}{\widehat{\alpha}} \right)} \\
G &= 2n \ln \Gamma(\alpha_0) + 2n\widehat{\alpha}_0 \widehat{\beta}_0 - 2\widehat{\alpha}_0 (n \ln \alpha_0) \\
H &= -2(\widehat{\alpha}_0 - 1) \sum_{i=1}^{n} \ln y_i + 2 \sum_{i=1}^{n} \frac{y_i}{\left(\frac{\exp(\mathbf{x}_i^T \widehat{\beta}_0)}{\widehat{\alpha}_0} \right)}
\end{aligned}$$

The G_1^2 follows *Chi-square* distribution with degree of freedom equal to k, i.e. the number of predictors in GWGR model.

For GWGR, the deviance is obtained by letting the parameters set under $H_0(\omega)$ is $\omega = \{\beta_0(u_i, v_i), \alpha_0(u_i, v_i), i = 1, 2, \ldots, n\}$ whereas the parameter set under population is $\Omega = \{\beta(u_i, v_i), \alpha_0(u_i, v_i), i = 1, 2, \ldots, n\}$.

The log likelihood under H_0 of GWGR model is:

$$lnL^*(\omega) = I + J + K, \tag{9}$$

where

$$\begin{aligned}
I &= -\sum_{i=1}^{n} w_{il} \ln(\Gamma(\alpha_0(u_i, v_i))) + \sum_{i=1}^{n} \alpha_0(u_i, v_i) w_{il} \ln \alpha_0(u_i, v_i) \\
J &= -\sum_{i=1}^{n} \alpha_0(u_i, v_i) w_i l \beta_0(u_i, v_i) + \sum_{i=1}^{n} (\alpha_0(u_i, v_i) - 1) w_{il} \ln y_i \\
K &= -\sum_{i=1}^{n} y_i w_{il} \exp(-\beta_0(u_i, v_i)) \alpha_0(u_i, v_i)
\end{aligned}$$

The log likelihood under population of GWGR model becomes:

$$lnL^*(\Omega) = L + M + N, \tag{10}$$

where

$$\begin{aligned}
L &= -\sum_{i=1}^{n} w_{il} \ln\Gamma(\alpha(u_i,v_i)) + \sum_{i=1}^{n} \alpha(u_i,v_i) w_{il} \ln\alpha(u_i,v_i) \\
M &= -\sum_{i=1}^{n} \alpha(u_i,v_i)(\mathbf{x}_i^T \beta(u_i,v_i)) w_{il} + \sum_{i=1}^{n} (\alpha(u_i,v_i)-1) w_{il} (\ln y_i) \\
N &= -\sum_{i=1}^{n} y_i \alpha(u_i,v_i) w_{il} \exp(-\mathbf{x}_i^T \beta(u_i,v_i))
\end{aligned}$$

The parameters in log likelihood under H_0 and population are then maximized using MLE by calculating first derivation and make it equal to zero. If the equation yielded is not close form, BFGS once again needs to be performed. Thus, the statistics obtained is:

$$G_2^2 = 2(L(\widehat{\Omega}) - (L(\widehat{\omega})) = O + P, \tag{11}$$

with

$$\begin{aligned}
O = &-2\sum \ln(\Gamma(\widehat{\alpha}(u_i,v_i))) + 2\sum_{i=1}^{n} \widehat{\alpha}(u_i,v_i) \ln\widehat{\alpha}(u_i,v_i) - \\
&2\sum_{i=1}^{n} \widehat{\alpha}(u_i,v_i)(\mathbf{x}_i^T \widehat{\beta}(u_i,v_i)) + 2\sum_{i=1} (\widehat{\alpha}(u_i,v_i) - 1)\ln y_i - \\
&2\sum_{i=1} y_i \widehat{\alpha}(u_i,v_i) exp(-\mathbf{x}_i^T \widehat{\beta}(u_i,v_i))
\end{aligned}$$

and

$$\begin{aligned}
P = &\ 2\sum_{i=1}^{n} \ln(\Gamma(\widehat{\alpha_0}(u_i,v_i))) - 2\sum_{i=1}^{n} \widehat{\alpha}_0(u_i,v_i) \ln\widehat{\alpha_0}(u_i,v_i) + \\
&2\sum_{i=1}^{n} \widehat{\alpha_0}(u_i,v_i)\widehat{\beta_0}(u_i,v_i) - 2\sum_{i=1}^{n} (\widehat{\alpha_0}(u_i,v_i) - 1)\ln y_i + \\
&2\sum_{i=1}^{n} y_i \exp(-\widehat{\beta_0}(u_i,v_i))\widehat{\alpha_0}(u_i,v_i)
\end{aligned}$$

The G_2^2 follows *Chi-square* distribution with the degree of freedom equal to $tr(S)$, where S is the matrix hat described in [8].

3.2.2. Simultaneous Test The hypothesis being tested is the null $H_0 : \beta_{11}(u_i,v_i) = \ldots = \beta_k(u_i,v_i) = \beta(u_i,v_i) = 0, \quad j = 1,2,\ldots,k; i = 1,2,\ldots,n$ against the alternative H_1: at least one of $\beta_j i(u_i,v_i) \neq 0$. The test performed in simultaneous test similar to previous procedure in finding the deviance of GWGR (G_2^2) in parameter similarity test. The null H_0 is rejected if $G_2^2 > \chi_{(\alpha,df_2)}$

3.2.3. Partial Test Hypothesis being tested for partial test is $H_0 : \beta_j(u_i, v_j) = 0$ against $H_1 : \beta_j(u_i, v_j) \neq 0, j = 0, 1, 2, \ldots, k$. The statistics test for partial test of GWGR is:

$$Z = \frac{\widehat{\beta}_j(u_i, v_i)}{SE(\widehat{\beta}_j(u_i, v_i))} \tag{12}$$

If $|Z_{statistic}| > Z_{\alpha/2}$, then the H_0 is rejected.

4. Conclusion

Based on this study, it is concluded that the GWGR is a local form of gamma regression model considering the geographical spatial effects, i.e. latitude and longitude coordinates. The GWGR model produces estimates that are local to each observation's location. The parameter estimation in GWGR is performed using Maximum Likelihood Estimation (MLE) with quasi-Newton numerical optimization using Broyden-Fletcher-Goldfarb-Shanno (BFGS) algorithm. Once the parameters estimation was done, the hypothesis testing procedure was used to test the significance of independent variables within the model, both partially using Z-test and simultaneously using Maximum Likelihood Ratio Test (MLRT).

References

[1] Brudson C, Fotheringham A S, and Charton M 1996 Geographically Weighted Regression: A Method for Exploring Spatial Nonstationarity, *Geographical Analysis* **28**(4) pp 281–298

[2] Brudson C, Fotheringham A S, and Charton M 1998 Geographically Weighted Regression: Modelling Spatial Nonstationarity *The Statistician* **47**(3) pp 431–443

[3] Mei C L 2004 *Geographically Weighted Regression Technique for Spatial Data Analysis* (China: Working paper, School of Science, Xi'an Jiaotong University)

[4] Tripathi R C, Gupta C R, and Pair K P 1993 Statistical Test Involving Several Independent Gamma Distribution *Ann. Inst. Statist. Math.* **45**(4) pp 773–786

[5] Casella G and Berger R L 2002 *Statistical Inference, Second Edition* (Duxburry: Pacific Grove)

[6] Al-Abood A, Bakir S, and Young D 1986 *Improved Estimator for The Shape Parameter in Gamma Regression* (London: Brunel University Maths Technical Papers) pp 1–27

[7] Davidian M 2005 *Applied Longitudinal Analysis* (Raleigh: North Carolina State University)

[8] Fotheringham A, Brundson C, and Charlton M 2002 *Geographically Weighted Regression: The Analysis of Spatially Varying Relationships* (New York: John Wiley and Son)

[9] Mennis J 2006 Mapping the Result of Geographically Weighted Regression *The Cartographic Journal* **43** pp 171–179

[10] Anselin L 1988 *Spatial Econometrics: Methods and Models* (Dordrecht: Kluwer Academic)

[11] Rismal, Budiantara I N, and Prastyo D D 2016 Mixture model of spline truncated and kernel in multivariable nonparametric regression, *AIP Conference Proceedings* **1739** (020085)

[12] Haerdle W K, Prastyo D D, and Hafner C M 2014 Support vector machines with evolutionary model selection for default prediction, In Racine J S, Su L, and Ullah A, ed. *The Oxford Handbook of Applied Nonparametric and Semiparametric Econometric and Statistics* (New York: Oxford University Press) pp 346–373

[13] Nocedal J and Wright S 1999 *Numerical Optimization* (New York: Springer) pp 194–201

IOP Conf. Series: Journal of Physics: Conf. Series **893** (2017) 012026 doi:10.1088/1742-6596/893/1/012026

Classification of customer lifetime value models using Markov chain

Dony Permana[1], Udjianna S. Pasaribu[2], Sapto W. Indratno[2], and Suprayogi[3]

[1] Statistics Study Program, Faculty of Mathematics and Natural Sciencies
Universitas Negeri Padang,Indonesia
[2] Statistics Research Division, Faculty of Mathematics and Natural Sciencies
Institut Teknologi Bandung, Indonesia
[3] Industrial System and Techno-Economics, Research Group, Faculty of Industrial Technology
Institut Teknologi Bandung, Indonesia

E-mail: donypermana27@gmail.com, {udjianna,sapto, yogi}@math.itb.ac.id, yogi@mail.ti.itb.ac.id

Abstract. A firm's potential reward in future time from a customer can be determined by customer lifetime value (CLV). There are some mathematic methods to calculate it. One method is using Markov chain stochastic model. Here, a customer is assumed through some states. Transition inter the states follow Markovian properties. If we are given some states for a customer and the relationships inter states, then we can make some Markov models to describe the properties of the customer. As Markov models, CLV is defined as a vector contains CLV for a customer in the first state. In this paper we make a classification of Markov Models to calculate CLV. Start from two states of customer model, we make develop in many states models. The development a model is based on weaknesses in previous model. Some last models can be expected to describe how real characters of customers in a firm.

1. Introduction

In a marketing system, a consumer of a goods or services can be seen to has some states that passed. Transition a consumer from one state to another is assumed to follow Markovian [4, 9]. Relationship inter states have meaning characteristics of the consumer. Consumers who are loyal to a product of goods or services are marked with the transition from customer state to itself. Such transition is called retention. The Retention need a kind of marketing cost. A consumer is called former customer if he or she don't buy again the product. But transition from former customer state to customer state is possible and it is called acquisition of former. The acquisition also need another kind of marketing cost. Model describes characteristics of a consumer based on relationship inter states can be made by Markov Model.

In previous research, the Markov model is used to determine customer lifetime value (CLV) [6, 8, 10]. It is called models of CLV. Another research, it is used to determine an inverse problem of CLV model [5]. Model of CLV is used to expenditure decision management [3], segmentation of consumer, data mining in an auto-repair and maintenance company [2], etc. There are some models of CLV, but not yet classified.

The objective of this paper are to (1) classify Markov models, create a naming system and describing characters of a customer in a marketing system, and (2) make transition chart for every model. Procedures of classifying are (1) analyzing the number of state of customer in a firm that is needed, (2) analyzing relationships inter the states, (3) constructing a transition chart, (4) setting a transition matrix, (5) developing another model, and (6) creating a naming system.

2. Result

We make two types of model of CLV. The first is CF (Customer-Former) model, contains customer state and former customer state. The second is PCF (Prospect-Customer-Former) model, contains three states are prospect, customer, and former customer. Indexes number in brackets have meaning the number of each state. Example, $CF(1,2)$ have one customer state and two levels former customer states, $PCF(1,2,1)$ have one prospect state, two levels customer state, and one former customer state. Here is diagram of classify of model of CLV and the naming system of it.

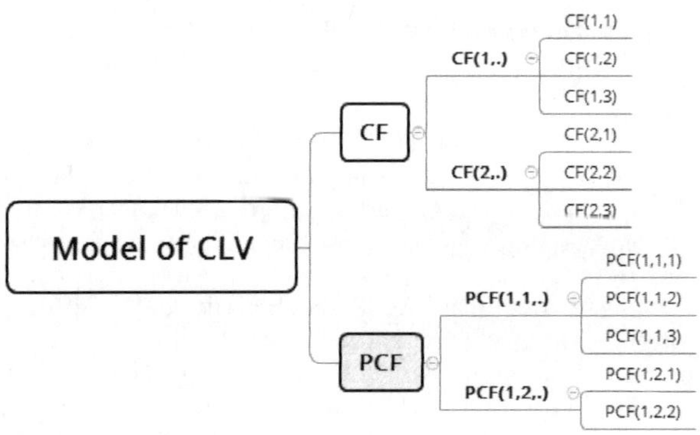

Figure 1. Diagram of classify of model CLV and its naming system

Look at Figure 1. Actually, we can make some another model from CF, example $CF(3,.)$, $CF(4,.)$ and so on. For PCF model, we can also make $PCF(1,3,.)$, $PCF(1,4,.)$ and so on. It depends on the reality of the characteristics of customer in a firm or needs in the field. Figure 1 show how we classify the model and make a naming system of it. The next figures show diagram transition of every model that written in Figure 1.

Figure 2 shows transition diagram of models of $CF(1,.)$. There are three models are $CF(1,1)$, $CF(1,2)$, and $CF(1,3)$. The $CF(1,1)$ model has no absorbent state. It isn't realistic in the field of a product marketing. Because models without absorbent will take a lot of marketing costs. The $CF(1,2)$ is better than $CF(1,1)$. Here, absorbent state is saved in former customer level 2. But the weakness of this model is former customer level 1 only have one offer to come back to the customer state.

Thus, we make $CF(1,3)$ that is better than $CF(1,2)$. We may think that $CF(1,4)$ is better than $CF(1,3)$. That may be true, but it must be remembered that the more of former customer state will need a lot of cost for marketing.

Figure 3 show transition diagram of models of $CF(2,.)$. There are three models too, $CF(2,1)$, $CF(2,2)$, and $CF(2,3)$. Model of $CF(2,.)$ is proposed for a firm whose two level customer. Example in health insurance, level 1's customer has low premium price and level 2'customer

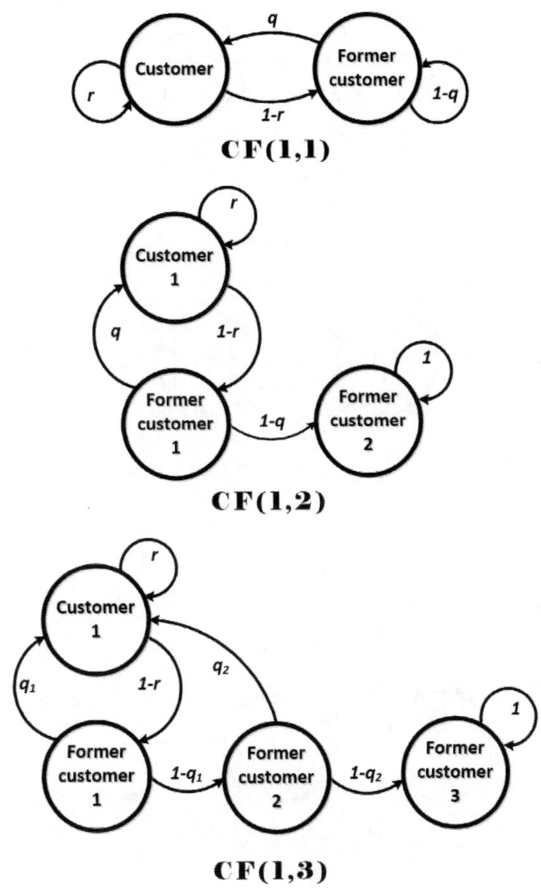

Figure 2. Transition diagram of models of $CF(1,.)$

has high premium price. In a service product, level 1's customer is for member and level 2's customer is for non member.

Figure 4 show transition diagram of models of $PCF(1,1,.)$. There are three models, $PCF(1,1,1)$, $PCF(1,1,2)$, and $PCF(1,1,3)$. In this model, there is a prospect customer state. It is a state for a people that is offered to be a customer. But if he or she declined the offer, then he or she moved to former customer state. Such as $CF(1,1)$, $PCF(1,1,1)$ isn't realistic in the field of a product marketing. So, we construct $PCF(1,1,2)$ and $PCF(1,1,3)$.

Based on transition diagrams, we can set matrices of transition. Size of matrix depends on the number of states. Model of $CF(1,2)$ have three states, thus size of transition matrix is 3×3. The following are the transition matrix of the models.

$$P_{CF(1,1)} = \begin{pmatrix} r & 1-r \\ q & 1-q \end{pmatrix} \tag{1}$$

$$P_{CF(1,2)} = \begin{pmatrix} r & 1-r & 0 \\ q & 0 & 1-q \\ 0 & 0 & 1 \end{pmatrix} \tag{2}$$

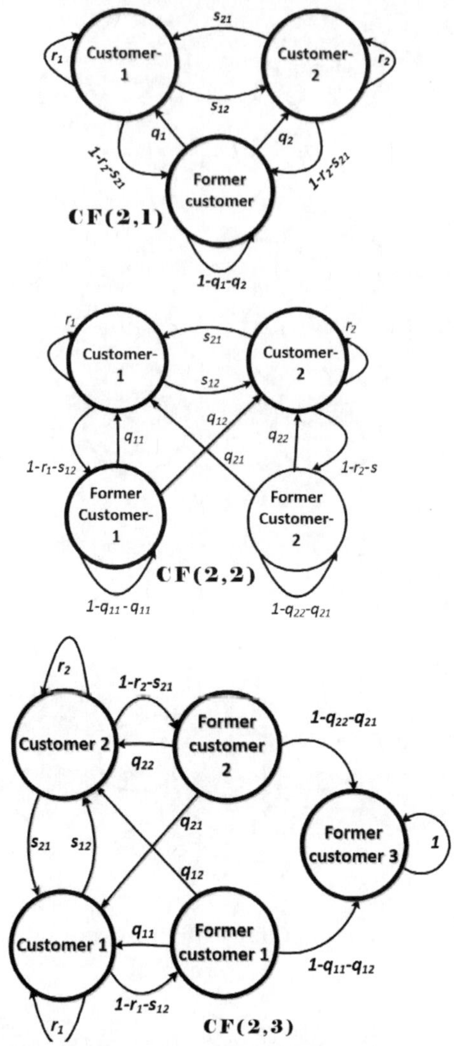

Figure 3. Transition diagram of models of $CF(2,.)$

$$P_{CF(1,3)} = \begin{pmatrix} r & 1-r & 0 & 0 \\ q_1 & 0 & 1-q_1 & 0 \\ q_2 & 0 & 0 & 1-q_2 \\ 0 & 0 & 0 & 1 \end{pmatrix} \tag{3}$$

$$P_{CF(2,1)} = \begin{pmatrix} r_1 & s_{12} & 1-r_1-s_{12} \\ s_{21} & r_2 & 1-r_2-s_{21} \\ q_1 & q_2 & 1-q_1-q_2 \end{pmatrix} \tag{4}$$

$$P_{CF(2,2)} = \begin{pmatrix} r_1 & s_{12} & 1-r_1-s_{12} & 0 \\ s_{21} & r_2 & 0 & 1-r_2-s_{21} \\ q_{11} & q_{12} & 1-q_{11}-q_{12} & 0 \\ q_{21} & q_{22} & 0 & 1-q_{21}-q_{22} \end{pmatrix} \tag{5}$$

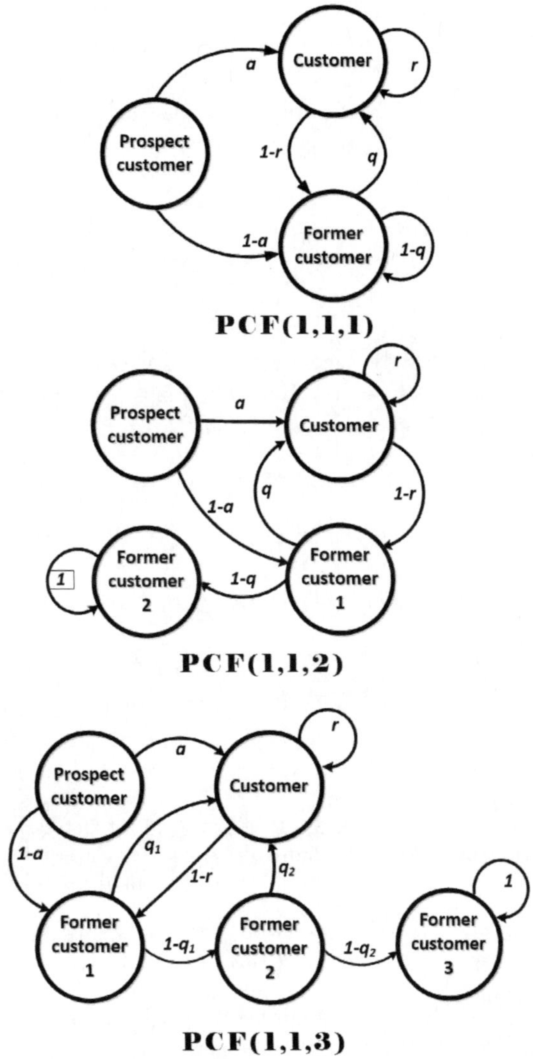

Figure 4. Transition diagram of models of $PCF(1,1,.)$

$$P_{CF(2,3)}=\begin{pmatrix} r_1 & s_{12} & 1-r_1-s_{12} & 0 & 0 \\ s_{21} & r_2 & 0 & 1-r_2-s_{21} & 0 \\ q_{11} & q_{12} & 0 & 0 & 1-q_{11}-q_{12} \\ q_{21} & q_{22} & 0 & 0 & 1-q_{21}-q_{22} \\ 0 & 0 & 0 & 0 & 1 \end{pmatrix} \tag{6}$$

$$P_{PCF(1,1,1)} = \begin{pmatrix} 0 & a & 1-a \\ 0 & r & 1-r \\ 0 & 0 & 1 \end{pmatrix} \tag{7}$$

$$P_{PCF(1,1,2)} = \begin{pmatrix} 0 & a & 1-a & 0 \\ 0 & r & 1-r & 0 \\ 0 & q & 0 & 1-q \\ 0 & 0 & 0 & 1 \end{pmatrix} \tag{8}$$

$$P_{PCF(1,1,3)} = \begin{pmatrix} 0 & a & 1-a & 0 & 0 \\ 0 & r & 1-r & 0 & 0 \\ 0 & q_1 & 0 & 1-q_1 & 0 \\ 0 & q_2 & 0 & 0 & 1-q_2 \\ 0 & 0 & 0 & 0 & 1 \end{pmatrix} \tag{9}$$

Furthermore, the transition matrix is used to calculate CLV[7]or will be estimated by inverse problem[1].

3. Conclusion

There is the naming system to classify Markov models for describing characters of a customer in a firm. The models are used to calculate CLV. The classification is made based on a number of state of customer that followed by a customer in a firm.

Acknowledgments

We would like to thank the Penprinas MP3EI 2016 Program which has supported funding of this research.

References

[1] Al-Ma'shumah F, Permana D, and Sidarto K A 2015 *Solving Inverse Problem for Markov Chain Model (MCM) of Customer Lifetime Value (CLV) Using Flower Pollination Algorithm (AIP Conference. Proceedings) Conf. 1692* pp. 020030.1-020030.3

[2] Cheng C J, Chiu S, Cheng C B, and Wu J Y 2012 Customer Lifetime Value Prediction by a Markov Chain Based Data Mining Model: Application to an Auto Repair and Maintenance Company in Taiwan *Scientia Iranica* **19**(3) pp 849–855

[3] Ekinci Y, Ulengin F, Uray N, and Ulengin B 2014 Analysis of customer lifetime value and marketing expenditur decisions through a Markovian-based model *European Journal of Operation Research* **237** pp 402–411

[4] Gupta S, Hanssens D, Hardie B, Kahn W, Kumar V, Lin N, et al 2006 Modelling Customer Lifetime Value *Journal of Service Research* **9**(2) pp 139–155

[5] Pasaribu U S, Al-Ma'shumah F, and Permana D 2015 Estimation of the Transition Matrix in Markov Chain Model of Customer Lifetime Value (CLV) Using Flower Pollination Algorithm *Applied Mathematical Sciences* **9**(69) pp 3409–3418

[6] Permana D, Effendie A R, Pasaribu U S, and Indratno S W 2015 Modelling of Dynamical Interest Rates using Binomial and Trinomial Tree *Far East Journal of Mathematics and Sciences* **99**(6) pp 909–920

[7] Permana D, Pasaribu U S, Indratno S W, and Suprayogi 2015 The Probability of First-time and Estimation of the Customer Lifetime Value in Health Insurance Data Using Markov Chain Model *International Journal of Applied Mathematics and Statistics* **53**(3) pp 174–18

[8] Permana D, Pasaribu U S, and Indratno S W 2014 *Analysis Study Of Behavior and Determination of Customer Lifetime Value (CLV) Using Markov Chain Model(AIP Conference. Proceedings) Conf. 1589* pp 456–459

[9] Pfeifer P and Carraway R 2000 Modeling Customer Relationships As Marcov Chain *Journal of Interactive Marketing* **14**(2) pp 43–55

[10] Widyawan A, Pasaribu U S, Henintyas, and Permana D 2015 *Estimation of Customer Lifetime Value on The Health Insurance with Assumption Uniform Distribution of Interest Rates (AIP Conference. Proceedings) Conf. 1692* pp. 020015.1-020015.4

IOP Conf. Series: Journal of Physics: Conf. Series **893** (2017) 012027 doi:10.1088/1742-6596/893/1/012027

A survey of kernel-type estimators for copula and their applications

I W Sumarjaya
Department of Mathematics, Udayana University, Bali, Indonesia
E-mail: sumarjaya@unud.ac.id

Abstract. Copulas have been widely used to model nonlinear dependence structure. Main applications of copulas include areas such as finance, insurance, hydrology, rainfall to name but a few. The flexibility of copula allows researchers to model dependence structure beyond Gaussian distribution. Basically, a copula is a function that couples multivariate distribution functions to their one-dimensional marginal distribution functions. In general, there are three methods to estimate copula. These are parametric, nonparametric, and semiparametric method. In this article we survey kernel-type estimators for copula such as mirror reflection kernel, beta kernel, transformation method and local likelihood transformation method. Then, we apply these kernel methods to three stock indexes in Asia. The results of our analysis suggest that, albeit variation in information criterion values, the local likelihood transformation method performs better than the other kernel methods.

1. Introduction

Data in fields such as finance, insurance, health and hydrology are often nonlinear in nature. Researchers often interested in seeing dependence structure in these data. A commonly used distribution to model this dependence is multivariate normal. However, the multivariate normal distribution has serious shortcomings. First, it cannot capture tail dependence which plays important role in risk management. Second, the dependence for multivariate normal is summarized in correlation matrix. However, correlation is not the best measure of dependence; it is a measure of linear dependence [1]. Thus, a tool that can capture this nonlinearity and invariant under transformation is required. Copula offers flexibility in modelling this nonlinearity by combining various margins to form new distributions.

In general, copula is a function that couples multivariate distribution with their univariate margins [2]. This means that copula can model rich dependence structure. The flexibility of copula demands invariant dependence measures. Few dependence measures that are invariant under transformation are Kendall's τ, Spearman's ρ, Gini's γ and Blomqvist's β. For a review of some of these dependence measures see [1], [2], [3] and references therein.

Modelling using copula means modelling with many parameters. Basically these parameters consist of two parameters: the copula parameters and the marginal parameters. Methods for estimating copulas include parametric, nonparametric and semiparametric methods. For a review of these methods see [4] and [5].

This article is organized as follows. In section one we give a motivation for copula as a flexible dependence modelling. Section two introduces basic concepts of copula. A general copula estimation is briefly discussed in section three. Section four discusses kernel-type estimators of copulas such as mirror reflection kernel, beta kernel, transformation method and local likelihood transformation method. Application of these kernel-type estimators to Asian stock indexes data is discussed in section five. Section six concludes the article.

2. Basic concepts of copula

The concept of copula can be traced back to the seminal paper of [6]. Let X and Y be random variables with distribution functions $F(x)$ dan $G(y)$, respectively. Let $H(x,y)$ be the joint distribution function. Now, Sklar's Theorem (see [2]) guarantees that there exits a function C such that

$$H(x,y) = C(F(x), G(y)). \tag{1}$$

The proof of Sklar's Theorem in (1) and its extension to multivariate case can be found in [2].

An example of copula is Gumbel copula which is often used to model tail dependence and has the form

$$C(u,v) = \exp\{-[(-\ln u)^{\theta} + (-\ln v)^{\theta}]^{\theta}\}, \theta \in [1,\infty).$$

Let $C(u,v)$ be a copula. The copula density $c(u,v)$ that is associated with copula $C(u,v)$ is defined as

$$c(u,v) = \frac{\partial^2 C(u,v)}{\partial u\, \partial v}$$

provided that the derivative exists. In the next section we will discuss briefly how to estimate the copula $C(\cdot,\cdot)$ and put more emphasis on using kernel method for estimating the copula density $c(\cdot,\cdot)$.

3. Estimation of copula

In general, there are three methods that are commonly used to estimate copula. They are parametric, nonparametric and semiparametric. See [4] and [5] for details of these methods. Parametric estimation can be properly used if the margins are explicitly known. However, without prior information about the margins the parametric estimation method may lead to severe underestimation [7]. In this case, nonparametric estimation procedure is preferred.

Let $(X_1,Y_1)^T, \ldots, (X_n,Y_n)^T$ be independent and identically distributed samples from bivariate random vector $(X,Y)^T$ with joint distribution function H and marginal distribution functions F and G. Based on this random sample we can define empirical distribution function of the form

$$H_n(x,y) = \frac{1}{n}\sum_{i=1}^{n} I_{\{X_i \le x, Y_i \le y\}}, -\infty < x,y < \infty$$

where $I_{\{A\}}$ is an indicator function. Let $F_n(x)$ and $G_n(x)$ be marginal distribution functions that associated with H_n that is $F_n(x) = H_n(x,+\infty)$ and $G_n(y) = H_n(+\infty,y)$ for $-\infty < x,y < +\infty$. Empirical copula C_n is defined as [8]

$$C_n(u,v) = H_n(F_n^{-1}(u), G_n^{-1}(v)), 0 \le u,v \le 1$$

and empirical copula process is defined as

$$Z_n(u,v) = \sqrt{n}(C_n - C)(u,v), 0 \le u,v \le 1. \tag{2}$$

Nonparametric estimator for copula in (2) is proposed by [9] and one popular estimator has the form

$$\hat{C}_n(u,v) = \frac{1}{n}\sum_{i=1}^{n} I_{\{\hat{U}_i \le u, \hat{V}_i \le v\}}$$

with $\hat{U}_i = F_n(x_i)$ and $\hat{V}_i = G_n(y_i)$. Here F_n and G_n are empirical cumulative distribution functions of the marginals.

Another approach of estimating copula is proposed by [10] by inversion of marginal cumulative distribution functions and probability density functions.

4. Kernel-type estimators for copula density

In the previous section we discussed estimation of copula distribution $\hat{C}_n$. However, copula distribution cannot capture invaluable information such as mode, tail, or mode near boundary [11]. This information can be captured by copula density $\hat{c}_n$. Nonparametric estimation of copula density can be done via kernel, wavelets, spline, or Bernstein polynomials. In this section we survey some kernel-type estimators for copula density. The use of kernel is motivated by the fact that a density should be estimated by a density and kernel provides a natural way of estimating a density.

4.1. Mirror reflection kernel

Kernel estimation for copula is known to suffer from boundary bias. One way of overcoming this problem is by reflecting all data points with respect to all corners and edges of the unit square (see [7] and [12]). This idea was introduced by [13] and the technique is known as mirror reflection. The mirror reflection (MR) kernel has the form [13]

$$\hat{c}_n^{(\mathrm{MR})} = \frac{1}{n}\sum_{i=1}^{n}\sum_{l=1}^{9}\left[K\left(\frac{u-\hat{U}_i^{(l)}}{h_n}\right) - K\left(\frac{-\hat{U}_i^{(l)}}{h_n}\right)\right]\left[K\left(\frac{v-\hat{V}_i^{(l)}}{h_n}\right) - K\left(\frac{-\hat{V}_i^{(l)}}{h_n}\right)\right] \tag{3}$$

with $\{(\hat{U}_i^{(l)}),(\hat{V}_i^{(l)}), i = 1,\ldots,n; l = 1,\ldots,9\} = \{(\pm\hat{U}_i,\pm\hat{V}_i),(\pm\hat{U}_i, 2-V_i),(2-\hat{U}_i,\pm\hat{V}_i),(2-\hat{U}_i, 2-\hat{V}_i), i = 1,\ldots,n\}$. An improved version of the MR kernel in (3) is suggested by [14] by adding a quantity $r(w) = \min(\sqrt{w},\sqrt{1-w})$ to the kernel so that it will shrink bandwidth to zero near the corner of the unit rectangle. The bias and the variance of the MR kernel can be found in [12].

4.2. Beta kernel

Another approach of estimating the density of a copula is by using kernel whose support matches the support of the density we wish to estimate [15]. Beta kernel is proposed by [16] for univariate density estimation. Kernel estimator for copula density $\hat{c}_n$ with bandwidth parameter b_n is defined as (see [7] and [12])

$$\hat{c}^{(\mathrm{Beta})}(u,v) = \tfrac{1}{n}\textstyle\sum_{i=1}^{n} K\left(U_i, \tfrac{u}{b_n}+1, \tfrac{1-u}{b_n}+1\right) K\left(V_i, \tfrac{v}{b_n}+1, \tfrac{1-v}{b_n}+1\right)$$

for all $(u,v) \in [0,1]^2$. Here $K(x,a,b)$ is the density of Beta(a,b) random variable. Derivation of asymptotic integrated mean square error (AMISE) and bandwidth selection can be found in detail in [12].

4.3. Transformation method

The idea behind the transformation method is as follows (see [12]). Rather than having supports on the unit square, the data is transformed such that it has full support on R^2. Then, standard kernel techniques can be applied to the transformed domain. Finally, apply back-transformation to estimate the copula density. One popular choice is by using standard normal cumulative distribution function. Let Φ be the standard normal distribution function and ϕ be the standard normal density function. The random vector $(X,Y) = (\Phi^{-1}(U), \Phi^{-1}(V))$ has normal margin and full support on R^2. By using the Sklar's Theorem the joint density function can be written as

$$f(x,y) = c(\Phi(x), \Phi(y))\phi(x)\phi(y).$$

Now, using the kernel estimator of the form

$$\hat{f}_n = \frac{1}{n}\sum_{i=1}^{n} K_{b_n}(x - X_i)K_{b_n}(y - Y_i)$$

for all $(x,y) \in R^2$, we obtain the following transformation estimator

$$\hat{c}_n^{(\mathrm{T})} = \frac{1}{n\phi(\Phi^{-1}(u))\phi(\Phi^{-1}(v))} \sum_{i=1}^{n} K_{b_n}(\Phi^{-1}(u) - \Phi^{-1}(U_i))K_{b_n}(\Phi^{-1}(v) - \Phi^{-1}(V_i))$$

for all $(u,v) \in [0,1]^2$. Asymptotic properties and bandwidth selection for this estimator can be found in [12]. Another extension is proposed by [17] by tapering the back-transformation in the tails. Recent extension using logit transformation kernel estimator is proposed by [11].

4.4. Local likelihood transformation method

Extension of transformation method is proposed by [18] by fitting polynomial locally to the log-density (TLL1) and by quadratic polynomials (TLL2). Let $Z_i = (U_i, V_i)$ and let Δ_{k_n} be Euclidean distance between (x,y) and the kth closest observation between $(\Phi^{-1}(Z_i))_{i=1,\ldots,n} = (\Phi^{-1}(U_i), \Phi^{-1}(V_i))_{i=1,\ldots,n}$. For all $(u,v) \in [0,1]^2$, the local likelihood transformation of the copula density $c(u,v)$ with nearest-neighbour factor Δ_{k_n} and bandwidth matrix B has the form

$$\hat{c}_n^{(\mathrm{TLL})}(u,v) = \frac{\exp[\hat{\alpha}(\Phi^{-1}(u), \Phi^{-1}(v))]}{\phi(\Phi^{-1}(u))\phi(\Phi^{-1}(v))}$$

where the expression of $\hat{\alpha}(\cdot,\cdot)$ can be found in [18] or [12].

5. Application

Various applications of the kernel-type estimators in previous section can be found in [11], [18], [12] and [17]. In this section we apply the kernel estimators to three weekly log return of stock indexes in Asia (STI, HSI and AORD) for period 3 January 2000 to 7 March 2016. We use the R package kdecopula of [15].

Figure 1 shows the kernel density estimation for empirical copula of stock indexes pairs using local likelihood transformation kernel. AIC values can be seen in Table 1. Here we use the following abbreviation: A = AORD, H = HSI and S = STI.

6. Conclusion

In this article we have reviewed kernel-type estimators for copula density estimation. We conclude that the transformation likelihood (TLL2) performs better than the other estimators since its AIC values are the smallest among the other four kernel estimators (highlighted in bold in Table 1).

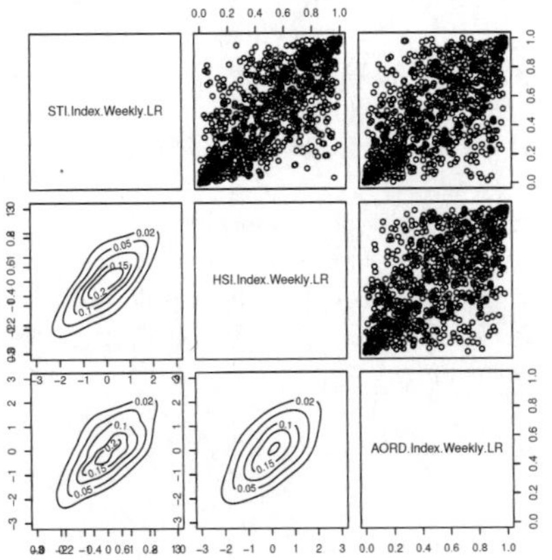

Figure 1. Kernel density estimation of stock indexes using local likelihood transformation kernel (TLL2).

Table 1. AIC values for kernel estimators for stock indexes data

	MR	Beta	T	TLL1	TLL2
S vs H	-586.12	-614.50	-654.03	-641.20	**-667.94**
S vs A	-398.33	-437.68	-471.11	-486.86	**-489.93**
H vs A	-352.01	-379.03	-416.23	-400.61	**-434.49**

Acknowledgment
The author would like to thank three anonymous reviewers for their comments that lead to the improvement of this paper.

References
[1] Joe H 2015 *Dependence Modeling with Copulas* (Boca Raton: CRC Press)
[2] Nelsen R B 2006 *An Introduction to Copulas* (Berlin: Springer)
[3] Liebscher E 2014 Copula-based dependence measures *Depend. Model.* **2** pp 49–64
[4] Trivedi P K and Zimmer D M 2005 Copula modeling: an introduction for practitioners *Found. Trends Econom.* **1** pp 1–111
[5] Choroś B, Ibragimov R and Permiakova E 2010 *Copula Theory and Its Applications* (Berlin: Springer) ch. Copula Estimation pp 77–91
[6] Sklar A 1959 Fonctions de répartition à n dimensions et leurs marges *Publ Inst Statist Univ Paris* **8** pp 229–231
[7] Charpentier A, Fermanian J D and Scaillet O 2007 *Copulas: From Theory to Application in Finance*(Risk Publications) ch. The Estimation of Copulas: Theory and Practice pp 35–60
[8] Fermanian J D, Radulovic D and Wegkamp M 2004 Weak convergence of empirical copula processes *Bernoulli* **10** pp 847–860
[9] Deheuvels P 1979 La fonction de dépendance empirique et ses propriétés un test non-paramétrique d'indépendance *Acad. Royale Belgique. Bulletin de la classe des sciences 5e Sér* **65** pp 274–292
[10] Racine J S 2015 Mixed data kernel copulas *Empir. Econ.* **48** pp 37–59
[11] Li S and Silvapulle P 2015 Kernel estimation of copula densities (*Preprint* http://dx.doi.org/10.2139/ssrn.2620511)
[12] Nagler T 2014 *Kernel methods for vine copula estimation Master's thesis* (Munich: Technische Universitat Munchen)

[13] Gijbels I and Mielniczuk J 1990 Estimating the density of a copula function *Comm. Statist. Theory Methods* **19** pp 445–464
[14] Omelka M, Gijbels I and Veraverbeke N 2009 Improved kernel estimation of copula: weak convergence and goodness-of-fit testing *Ann. Statist.* **37** pp 3023–3058
[15] Nagler T 2016 kdecopula: an R package for the kernel estimation of copula densities (*Preprint ArXiv*: 1603.04229v2)
[16] Chen S X 1999 Beta kernel estimators for density functions *Comput. Statist. Data Anal.* **31** pp 131–145
[17] Wen K and Wu X 2015 Transformation-kernel estimation of the copula density (*Preprint* http://agecon2. tamu. edu/people/faculty/wu-ximing/agecon2/public/copula.pdf)
[18] Geenens G, Charpentier A and Paindaveine D 2014 Probit transformation for nonparametric kernel estimation of the copula density (*Preprint ArXiv*: 1404.4414)

IOP Conf. Series: Journal of Physics: Conf. Series **893** (2017) 012028 doi:10.1088/1742-6596/893/1/012028

Estimation of truncated spline in nonlinear structural equation modeling using weighted least squares method

Ruliana[1], **I. N. Budiantara**[2], **B. W. Otok**[2] **and W. Wibowo**[2]

[1] Department of Mathematics, Universitas Negeri Makassar, Indonesia
[2] Department of Statistics, Faculty of Mathematics and Natural Sciences
Institut Teknologi Sepuluh Nopember (ITS), Surabaya, Indonesia

E-mail: ruliana.t@gmail.com, ruliana.t@unm.ac.id

Abstract. Structural equation modeling (SEM) consists of two parts, the measurement model and the structural model. In this paper, we develop a SEM model that the measurement model is linear and the structural model is nonlinear. Structural model explain about relationship between latent variables. Truncated spline is used for modeling the nonlinear relationship between the latent variables. Trought weighted least squares (WLS) method, It is found that estimation of parameter truncated spline in nonlinear SEM for two exogen and an endogen latent variables is resulted by optimizing

$$\min_{\gamma}\{\|\Sigma^{-\frac{1}{2}}(\omega_\eta - \Gamma_{\xi\kappa}\gamma)\|^2\}$$

and it gave an estimator

$$\hat{\gamma} = (\mathbf{\Gamma'}_{\xi\kappa}\mathbf{\Sigma}^{-1}\mathbf{\Gamma}_{\xi\kappa})^{-1}\mathbf{\Gamma'}_{\xi\kappa}\mathbf{\Sigma}^{-1}\omega_\eta$$

The estimator influenced by the location and some knot points κ_1 and κ_2.

1. Introduction

Structural equation modeling (SEM), is one of the statistical approaches used for research that uses a variable that cannot be measured directly (latent variable) but it is conducted through indicators. In general, studies with SEM approach which assumes that the relationship between the variables is linear, while it is known that the application in the field there is a phenomenon that can not be enforced because the actual relationship is nonlinear.

For certain cases, linearity in structural models can be modified into nonlinear in order to clarify the relationship between the latent variables that are not linear [1], [2], [3], [4], and [5]. Therefore, this work aims to investigating the parameter estimation of nonlinear structural model by using weighted least squares method (WLS). The nonlinear is indicated with a change the relation pattern between latent variables, which differs in the certain range.

2. Structural Equation Modelling (SEM)

Research variable in SEM is a vector $\mathbf{h}$ dimension p with $\mathbf{h} = (\eta_j, \xi_j)'$ that can not be measured directly, but through an indicator $\mathbf{z} = (z_1, z_2, ..., z_p)'$, it can be stated with the relationship $\mathbf{z} =$

$\tau + \mathbf{\Lambda h} + \mathbf{e}$, a vector for the intercept τ and factor loading matrix $\mathbf{\Lambda}$, as well as measurement error models $\mathbf{e}$. Wall and Amemiya [2] formulated the measurement model in general form i.e.

$$\mathbf{z} = \Lambda \mathbf{h} + \mathbf{e} \tag{1}$$

where $\mathbf{z}$ was a vector $p + q$ of observed variable which was called indicator, $\mathbf{\Lambda}$ was a loading factor matrix, $\mathbf{h} = (\eta, \xi)'$ was the vector of latent variables and $\mathbf{e}$ was error, with $\mathbf{e}\ \tilde{\mathbf{N}}(\mathbf{0},\Theta_e)$. Bollen [6] defined the general structural model following

$$\eta = \Gamma\xi + \zeta \tag{2}$$

where η and ξ were the vectors of endogen and exogen latent variables respectively, Γ was coefficient matrix that measured the linear relation of latent variables.
In many cases vector $\mathbf{z}$ as described in Equation (1) is divided in two parts, the exogen indicators $\mathbf{x}$ and the endogen indicators $\mathbf{y}$. Figure 1 shows the path diagram for two exogen latent variables and one endogen latent variable. Measument model for two eksogenous dan one endogenous latent variable respectively are written as follow:

$$\begin{bmatrix} x_{11} \\ x_{21} \\ \vdots \\ x_{p_1 1} \\ x_{12} \\ x_{22} \\ \vdots \\ x_{p_2 2} \end{bmatrix} = \begin{bmatrix} \lambda_{x_{11}1} I_n & 0 \\ \lambda_{x_{21}1} I_n & 0 \\ \vdots & \vdots \\ \lambda_{x_{p_1}1} I_n & 0 \\ 0 & \lambda_{x_{12}2} I_n \\ 0 & \lambda_{x_{22}2} I_n \\ \vdots & \vdots \\ 0 & \lambda_{x_{p_2}2} I_n \end{bmatrix} \begin{bmatrix} \xi_1 \\ \xi_2 \end{bmatrix} + \begin{bmatrix} \delta_{11} \\ \delta_{21} \\ \vdots \\ \delta_{p_1 1} \\ \delta_{12} \\ \delta_{22} \\ \vdots \\ \delta_{p_2 2} \end{bmatrix} \tag{3}$$

and

$$\begin{bmatrix} y_1 \\ y_2 \\ \vdots \\ y_q \end{bmatrix} = \begin{bmatrix} \lambda_{y_1} I_n \\ \lambda_{y_2} I_n \\ \vdots \\ \lambda_q I_n \end{bmatrix} \eta_1 + \begin{bmatrix} \varepsilon_1 \\ \varepsilon_2 \\ \vdots \\ \varepsilon_q \end{bmatrix} \tag{4}$$

In matrix notation the Confirmatory Factor Analysis (CFA) model can be written as:

a. variance-covariance matrix among the indicators $\mathbf{x}$

$$\begin{aligned} \mathbf{\Sigma_{xx}}(\theta) &= \mathbf{E}[(\mathbf{\Lambda_x}\xi + \delta)(\mathbf{\Lambda_x}\xi + \delta)'] \\ &= \mathbf{E}(\mathbf{\Lambda_x}\xi\xi'\mathbf{\Lambda_x'} + \mathbf{\Lambda_x}\xi\delta' + \delta\xi'\mathbf{\Lambda_x'} + \delta\delta') \\ &= \mathbf{\Lambda_x\Phi\Lambda_x'} + \mathbf{\Theta}_\delta \end{aligned} \tag{5}$$

where the matrix $\mathbf{\Phi}$ represents the correlation matrix between the factors.

b. variance-covariance matrix among the indicators $\mathbf{y}$

$$\begin{aligned} \mathbf{\Sigma_{yy}}(\theta) &= ((\mathbf{\Lambda_y}\eta + \varepsilon)(\mathbf{\Lambda_y}\eta + \varepsilon)') \\ &= \mathbf{E}(\mathbf{\Lambda_y}\eta\eta'\mathbf{\Lambda_y'} + \mathbf{\Lambda_y}\eta\varepsilon' + \varepsilon\eta'\mathbf{\Lambda_y'} + \varepsilon\varepsilon') \\ &= \mathbf{\Lambda_y E}(\eta\eta')\mathbf{\Lambda_y'} + \mathbf{\Theta}\varepsilon \end{aligned} \tag{6}$$

In order to determine the covariance matrix for the endogenous variables η it is necessary to use the reduce form $\eta = (\mathbf{I} - \mathbf{B})^{-1}(\mathbf{\Gamma}\xi + \zeta)$

$$\begin{aligned} \mathbf{E}(\eta\eta') &= \mathbf{E}[[(\mathbf{I} - \mathbf{B})^{-1}(\mathbf{\Gamma}\xi + \zeta)][(\mathbf{I} - \mathbf{B})^{-1}(\mathbf{\Gamma}\xi + \zeta)]'] \\ &= (\mathbf{I} - \mathbf{B})^{-1}\mathbf{E}[\mathbf{\Gamma}\xi\xi'\mathbf{\Gamma}' + \mathbf{\Gamma}\xi\zeta' + \zeta\xi'\mathbf{\Gamma}' + \zeta\zeta'](\mathbf{I} - \mathbf{B})^{-1'} \\ &= (\mathbf{I} - \mathbf{B})^{-1}(\mathbf{\Gamma\Phi\Gamma}' + \mathbf{\Psi})(\mathbf{I} - \mathbf{B})^{-1'} \end{aligned} \tag{7}$$

So, the covariance matrix among the $\mathbf{y}$'s is

$$\mathbf{E}(\mathbf{yy}') = \mathbf{\Lambda_y}[(\mathbf{I}-\mathbf{B})^{-1}(\mathbf{\Gamma\Phi\Gamma}' + \mathbf{\Psi})(\mathbf{I}-\mathbf{B})^{-1'}]\mathbf{\Lambda}_\mathbf{y}' + \mathbf{\Theta}_\varepsilon \tag{8}$$

c. covariance matrix between $\mathbf{x}$'s and $\mathbf{y}$'s

$$\begin{aligned}\mathbf{\Sigma_{xy}}(\theta) &= \mathbf{E}[(\mathbf{\Lambda_x}\xi + \delta)(\mathbf{\Lambda_y}\eta + \varepsilon)'] \\ &= \mathbf{E}((\mathbf{\Lambda_x}\xi + \delta)(\eta'\mathbf{\Lambda}_\mathbf{y}' + \varepsilon)) \\ &= \mathbf{\Lambda_x}\mathbf{E}(\xi\eta')\mathbf{\Lambda}_\mathbf{y}' + \mathbf{\Lambda_x}\mathbf{E}(\xi\varepsilon') + \mathbf{E}(\delta\eta')\mathbf{\Lambda}_\mathbf{y}' + \mathbf{E}(\delta\varepsilon')\end{aligned} \tag{9}$$

Subtituting η by reduced form we have

$$\begin{aligned}\mathbf{E}(\mathbf{xy}') &= \mathbf{\Lambda_x}\mathbf{E}[\xi[(\mathbf{I}-\mathbf{B})^{-1}(\mathbf{\Gamma}\xi + \zeta)]']\mathbf{\Lambda}_\mathbf{y}' + \mathbf{\Lambda_x}\mathbf{E}(\xi\varepsilon') + \\ &\quad + \mathbf{E}(\delta[(\mathbf{I}-\mathbf{B})^{-1}(\mathbf{\Gamma}\xi + \zeta)]')\mathbf{\Lambda}_\mathbf{y}' + \mathbf{E}(\delta\varepsilon') \\ &= \mathbf{\Lambda_x}[\mathbf{E}(\xi\xi')\mathbf{\Gamma}'(\mathbf{I}-\mathbf{B})^{-1'} + \mathbf{E}[(\xi\zeta')(\mathbf{I}-\mathbf{B})^{-1'}]]\mathbf{\Lambda}_\mathbf{y}' \\ &= \mathbf{\Lambda_x}\mathbf{\Phi\Gamma}'(\mathbf{I}-\mathbf{B})^{-1'}\mathbf{\Lambda}_\mathbf{y}'\end{aligned} \tag{10}$$

d. covariance matrix between $\mathbf{y}$'s and $\mathbf{x}$'s

$$\begin{aligned}\mathbf{\Sigma_{yx}}(\theta) &= \mathbf{E}[(\mathbf{\Lambda_y}\eta + \varepsilon)(\mathbf{\Lambda_x}\xi + \delta)'] \\ &= \mathbf{E}[\mathbf{\Lambda_y}\eta\xi''\mathbf{\Lambda}_\mathbf{x}' + \mathbf{\Lambda_y}\eta\delta' + \varepsilon\xi'\mathbf{\Lambda}_\mathbf{x}' + \varepsilon\delta'] \\ &= \mathbf{\Lambda_y}\mathbf{E}[\eta\xi']\mathbf{\Lambda}_\mathbf{x}' + \mathbf{\Lambda_y}\mathbf{E}[\eta\delta'] + \mathbf{E}[\varepsilon\xi']\mathbf{\Lambda}_\mathbf{x}' + \mathbf{E}[\varepsilon\delta']\end{aligned} \tag{11}$$

Substituting η by reduced form we have

$$\begin{aligned}\mathbf{E}(\mathbf{yx}') &= \mathbf{\Lambda_y}\mathbf{E}[[(\mathbf{I}-\mathbf{B})^{-1}(\mathbf{\Gamma}\xi + \zeta)]\xi']\mathbf{\Lambda}_\mathbf{x}' + \mathbf{\Lambda_y}\mathbf{E}[[(\mathbf{I}-\mathbf{B})^{-1}(\mathbf{\Gamma}\xi + \zeta)\delta']] + \\ &\quad + \mathbf{E}[\varepsilon\xi']\mathbf{\Lambda}_\mathbf{x}' + \mathbf{E}(\varepsilon\delta') \\ &= \mathbf{\Lambda_y}(\mathbf{I}-\mathbf{B})^{-1}[\mathbf{\Gamma}\mathbf{E}[\xi\xi'] + \mathbf{E}[\zeta\xi']]\mathbf{\Lambda}_\mathbf{x}' \\ &= \mathbf{\Lambda_y}(\mathbf{I}-\mathbf{B})^{-1}\mathbf{\Gamma\Phi\Lambda}_\mathbf{x}'\end{aligned} \tag{12}$$

If structured equation of (5), (8), (10) and (12) in a matrix form $\mathbf{\Sigma} = \mathbf{\Sigma}(\theta)$ than we will obtain the covariance matrix for the observation variables $\mathbf{x}$ and $\mathbf{y}$ as a function of the model parameters

$$\mathbf{\Sigma}(\theta) = \begin{bmatrix}\mathbf{\Lambda_y}(\mathbf{I}-\mathbf{B})^{-1}\mathbf{F} & \mathbf{\Lambda_y}(\mathbf{I}-\mathbf{B})^{-1}\mathbf{\Gamma\Phi\Lambda}_\mathbf{x}' \\ \mathbf{\Lambda_x}((\mathbf{I}-\mathbf{B})^{-1})'\mathbf{G} & \mathbf{\Lambda_x\Phi\Lambda}_\mathbf{x}' + \mathbf{\Theta}_\delta\end{bmatrix} \tag{13}$$

where
$\mathbf{F} = (\mathbf{\Gamma\Phi\Gamma}' + \mathbf{\Psi})((\mathbf{I}-\mathbf{B})^{-1})'\mathbf{\Lambda}_{\mathbf{y}'} + \mathbf{\Theta}_\varepsilon$;
$\mathbf{G} = \mathbf{\Phi\Gamma}'\mathbf{\Lambda}_\mathbf{y}'$
For null B model (one endogenous latent variable), matrix in equation (13) became

$$= \begin{bmatrix}\mathbf{\Lambda_y}(\mathbf{\Gamma\Phi\Gamma}' + \mathbf{\Psi})\mathbf{\Lambda}_\mathbf{y}' + \mathbf{\Theta}_\varepsilon & \mathbf{\Lambda_y\Gamma\Phi\Lambda}_\mathbf{x}' \\ \mathbf{\Lambda_x\Phi\Gamma}'\mathbf{\Lambda}_\mathbf{y}' & \mathbf{\Lambda_x\Phi\Lambda}_\mathbf{x}' + \mathbf{\Theta}_\delta\end{bmatrix}$$

Furthermore,

$$\eta = \mathbf{f}(\xi_1, \xi_2) + \zeta \tag{14}$$

η is vector $n \times 1$ of endogen latent variable, $\mathbf{f}(\xi_1, \xi_2)$ is functions of exogen latent variables ξ_1 and ξ_1 and then ζ is vector $n \times 1$ which contains the error structural model.

3. Nonlinear SEM

The lives of children with Cystic Fibrosis by using quadratic and interaction SEM approach can be found in [2]. Nonlinear SEM suggest by dramatically increased feelings of dejection come at a sort of tipping point or breaking point. That is, after a certain level of stress or a certain lack of self-esteem, the feelings of dejection are much higher. The main investigation of this work is to estimate parameters and functions in structural model as mentioned in Equation (14) using WLS. Ruliana, et al. [1] had written down the factor scores for ξ is ω_{ξ_1} and ω_{ξ_2} and the score factor for η is ω_η . Model nonlinear SEM for two variables of latent exogenous is written down the following:

$$\omega_{\eta \mathbf{i}} = \sum_{\mathbf{r}=\mathbf{0}}^{\mathbf{m_1}} \alpha_{\mathbf{1r}} \omega_{\xi_1 \mathbf{i}}^{\mathbf{r}} + \sum_{\mathbf{s}=\mathbf{1}}^{\mathbf{K_1}} \beta_{\mathbf{1s}} (\omega_{\xi_1 \mathbf{i}} - \kappa_{\mathbf{s}})_+^{\mathbf{m_1}} + \sum_{\mathbf{t}=\mathbf{0}}^{\mathbf{m_2}} \alpha_{\mathbf{2t}} \omega_{\xi_2 \mathbf{i}}^{\mathbf{t}} + \sum_{\mathbf{u}=\mathbf{1}}^{\mathbf{K_2}} \beta_{\mathbf{2u}} (\omega_{\xi_2 \mathbf{i}} - \kappa_{\mathbf{u}})_+^{\mathbf{m_2}} + \zeta_{\mathbf{i}} \tag{15}$$

Model in Equation (15) can be wtitten as

$$\zeta_i = \omega_{\eta i} - A \tag{16}$$

where
$A = \sum\limits_{r=0}^{m_1} \alpha_{1r} \omega_{\xi_1 i}^r + \sum\limits_{s=1}^{K=1} \beta_{1s} (\omega_{\xi_1 i} - \kappa_s)_+^{m_1} + \sum\limits_{t=0}^{m_2} \alpha_{2t} \omega_{\xi_2 i}^t + \sum\limits_{u=1}^{K_2} \beta_{2u} (\omega_{\xi_2 i} - \kappa_u)_+^{m_2}$
In the form of matrix model (16) can be written down such as the following:

$$\zeta = \omega_\eta - \mathbf{\Gamma}_{\xi\kappa} \gamma \tag{17}$$

Furthermore, given the function

$$\begin{aligned} \mathbf{Q}(\gamma) &= \zeta' \mathbf{\Sigma}^{-1} \zeta \\ &= (\omega_\eta - \mathbf{\Gamma}_{\xi\kappa}\gamma)' \mathbf{\Sigma}^{-1} (\omega_\eta - \mathbf{\Gamma}_{\xi\kappa}\gamma) \\ &= \omega_\eta' \mathbf{\Sigma}^{-1} \omega_\eta - \omega_\eta' \mathbf{\Sigma}^{-1} \mathbf{\Gamma}_{\xi\kappa}\gamma + \\ &= -\gamma' \mathbf{\Gamma}_{\xi\kappa} \mathbf{\Sigma}^{-1} \omega_\eta + \gamma' \mathbf{\Gamma}_{\xi\kappa}' \mathbf{\Sigma}^{-1} \mathbf{\Gamma}_{\xi\kappa} \gamma \end{aligned} \tag{18}$$

where $\mathbf{\Sigma}^{-1}$ is variance covariance matrix or weighted matrix, $\mathbf{\Gamma}_{\xi}\kappa$ is the function-based containing exogenous latent variable and knot [1].
In order to modelling nonlinear relationship between latent variable in structural equation modeling (SEM), it is necessary to find the parameter estimation. The method applied for the estimation parameter was a weighted least square (WLS) that minimized the sum of quadratics error, which was weighted by a matrix inverse of error variance. Ordinary least square (OLS) method, was assumed that all elements of $\mathbf{\Theta}^{-1}$ should be similar. In WLS It was weighted by matrix inverse of error variance whose the objective was to solve if the assume was not fulfilled. A theorem of parameter estimation was given with WLS method for latent variables following:

Theorem 1 *If model is given as described in Equation(15), and the error term in Equation (17) then estimator of parameter* γ *with method of weighted least square (WLS) is given by* $\hat{\gamma} = (\mathbf{\Gamma}_\xi' \kappa \mathbf{\Sigma}^{-1} \mathbf{\Gamma}_\xi \kappa)^{-1} \mathbf{\Gamma}_\xi' \kappa \mathbf{\Sigma}^{-1} \omega_\eta$ *with*
$\gamma = [\gamma_1', \gamma_2']'$, *with*

$$\gamma_1' = \begin{bmatrix} \alpha_{10} & \alpha_{11} & \cdots & \alpha_{1m_1} & \beta_{11} \cdot & \beta_{1K_1} \end{bmatrix}' \tag{19}$$

$$\gamma_2' = \begin{bmatrix} \alpha_{20} & \alpha_{21} & \cdots & \alpha_{2m_2} & \beta_{21} & \cdots & \beta_{2K_2} \end{bmatrix}' \tag{20}$$

and
$\mathbf{\Gamma}_\xi \kappa = [\mathbf{\Gamma}_{\xi\kappa(\mathbf{1})} \mathbf{\Gamma}_{\xi\kappa(\mathbf{2})}]$

Proof:
Based on model as mentioned in Equation (15) is obtained a measurement $\zeta = \omega_\eta - \mathbf{\Gamma}_{\xi\kappa}\gamma$ with using WLS method, then estimator for γ was resulted by optimizing:

$$\begin{aligned}\min_\gamma\{\|\mathbf{\Sigma}^{-\frac{1}{2}}(\omega_\eta - \mathbf{\Gamma}_{\xi\kappa}\gamma)\|\}^{\mathbf{2}} &= \min_\gamma\{(\omega_\eta - \mathbf{\Gamma}_{\xi\kappa}\gamma)'\mathbf{\Sigma}^{-\mathbf{1}}(\omega_\eta - \mathbf{\Gamma}_{\xi\kappa}\gamma)\}\\ &= \min_\gamma\{(\omega'_\eta - \gamma'\mathbf{\Gamma}'_{\xi\kappa})'\mathbf{\Sigma}^{-\mathbf{1}}(\omega_\eta - \mathbf{\Gamma}_{\xi\kappa}\gamma)\}\\ &= \min_\gamma\{(\omega'_\eta\mathbf{\Sigma}^{-\mathbf{1}}\omega_\eta - \omega'_\eta\mathbf{\Sigma}^{-\mathbf{1}}\mathbf{\Gamma}_{\xi\kappa}\gamma - \gamma'\mathbf{\Gamma}'_{\xi\kappa}\mathbf{\Sigma}^{-\mathbf{1}}\omega_\eta + \\ &\quad +\gamma'\mathbf{\Gamma}'_{\xi\kappa}\mathbf{\Sigma}^{-\mathbf{1}}\mathbf{\Gamma}_{\xi\kappa}\gamma)\}\\ &= \min_\gamma\{\mathbf{Q}(\gamma)\}\end{aligned} \tag{21}$$

where

$$\mathbf{Q}(\gamma) = \{(\omega'_\eta\mathbf{\Sigma}^{-\mathbf{1}}\omega_\eta - \omega'_\eta\mathbf{\Sigma}^{-\mathbf{1}}\mathbf{\Gamma}_{\xi\kappa}\gamma - \gamma'\mathbf{\Gamma}'_{\xi\kappa}\mathbf{\Sigma}^{-\mathbf{1}}\omega_\eta + \gamma'\mathbf{\Gamma}'_{\xi\kappa}\mathbf{\Sigma}^{-\mathbf{1}}\mathbf{\Gamma}_{\xi\kappa}\gamma)\} \tag{22}$$

Derivation of Equation (21) with respect to γ is obtained

$$\begin{aligned}\partial\mathbf{Q}(\gamma)\partial\gamma &= \frac{\partial}{\partial\gamma}\{\{(\omega'_\eta\mathbf{\Sigma}^{-\mathbf{1}}\omega_\eta - \omega'_\eta\mathbf{\Sigma}^{-\mathbf{1}}\mathbf{\Gamma}_{\xi\kappa}\gamma - \gamma'\mathbf{\Gamma}'_{\xi\kappa}\mathbf{\Sigma}^{-\mathbf{1}}\omega_\eta + \gamma'\mathbf{\Gamma}'_{\xi\kappa}\mathbf{\Sigma}^{-\mathbf{1}}\mathbf{\Gamma}_{\xi\kappa}\gamma)\}\}\\ &= \frac{\partial}{\partial\gamma}\{\{(\omega'_\eta\mathbf{\Sigma}^{-\mathbf{1}}\omega_\eta - (\omega'_\eta\mathbf{\Sigma}^{-\mathbf{1}}\mathbf{\Gamma}_{\xi\kappa}\gamma)' - \gamma'\mathbf{\Gamma}'_{\xi\kappa}\mathbf{\Sigma}^{-\mathbf{1}}\omega_\eta + \gamma'\mathbf{\Gamma}'_{\xi\kappa}\mathbf{\Sigma}^{-\mathbf{1}}\mathbf{\Gamma}_{\xi\kappa}\gamma)\}\}\\ &= \frac{\partial}{\partial\gamma}\{\{(\omega'_\eta\mathbf{\Sigma}^{-\mathbf{1}}\omega_\eta - \mathbf{2}\gamma'\mathbf{\Gamma}'_{\xi\kappa}\mathbf{\Sigma}^{-\mathbf{1}}\omega_\eta + \gamma'\mathbf{\Gamma}'_{\xi\kappa}\mathbf{\Sigma}^{-\mathbf{1}}\mathbf{\Gamma}_{\xi\kappa}\gamma)\}\} - \mathbf{2}\mathbf{\Gamma}'_{\xi\kappa}\mathbf{\Sigma}^{-\mathbf{1}}\omega_\eta\\ &\quad +\mathbf{2}\gamma\mathbf{\Gamma}'_{\xi\kappa}\mathbf{\Sigma}^{-\mathbf{1}}\mathbf{\Gamma}_{\xi\kappa}\end{aligned} \tag{23}$$

Equation (23) is similar to zero and then the estimation for parameter γ is gave $\hat{\gamma}\mathbf{\Gamma}'_{\xi\kappa}\mathbf{\Sigma}^{-\mathbf{1}}\mathbf{\Gamma}_{\xi\kappa} = \mathbf{\Gamma}'_{\xi\kappa}\mathbf{\Sigma}^{-\mathbf{1}}\omega_\eta$. Because $\mathbf{\Gamma}_{\xi\kappa}$ is full matrix rank, we have

$$\hat{\gamma} = (\mathbf{\Gamma}'_{\xi\kappa}\mathbf{\Sigma}^{-\mathbf{1}}\mathbf{\Gamma}_{\xi\kappa})^{-\mathbf{1}}\mathbf{\Gamma}'_{\xi\kappa}\mathbf{\Sigma}^{-\mathbf{1}}\omega_\eta \tag{24}$$

4. Data Applications

The application of this estimator in data concentration CO_2 in rainy season (January, 2013) of ambient air in Surabaya, Indonesia. The concentration of CO_2 influenced by two latent variables, they are meteorology and environment. The indicators of meteorology are wind speed (Kec_A), wind direction (Ar_A), solar radiation (Rad_M), temperature (Temp), humidity (Kelemb), and the indicators of environment are: elevation (elev), green space (GS), and non green space (NGS).

The result of the measurement model is given in the Table 1 below. From the table, it clear that the indicators to measure latent variables meteorological i.e. wind speed, wind direction, solar radiation, temperature and humidity are significant at P value < 0.05. The indicators that measure the latent variables environment i.e. green space and non green space are significant at P values < 0.05. CO_2 variable is measured directly by concentration CO_2 (kon_CO_2).

Table 1. Combined loadings and cross-loadings

Indicator	Meteo	Ling	CO_2	Type	SE	P-value
kec_A	-0.709	-0.33	-0.03	reflect	0.068	< 0.001
Ar_A	0.34	-0.286	0.458	reflect	0.068	< 0.001
Rad_M	-0.159	0.132	-0.22	reflect	0.068	< 0.001
Temp	0.757	0.047	-0.217	reflect	0.068	< 0.001
kelemb	0.789	-0.192	-0.059	reflect	0.068	< 0.001
elev	-0.284	0.538	-0.319	reflect	0.068	< 0.001
GS	0.065	0.971	0.067	reflect	0.068	< 0.001
NGS	-0.093	-0.963	-0.11	reflect	0.068	< 0.001
kon_CO_2	0	0	1	reflect	0.068	< 0.001

Plots of score factor are shown in Fig. 1 and Fig. 2.

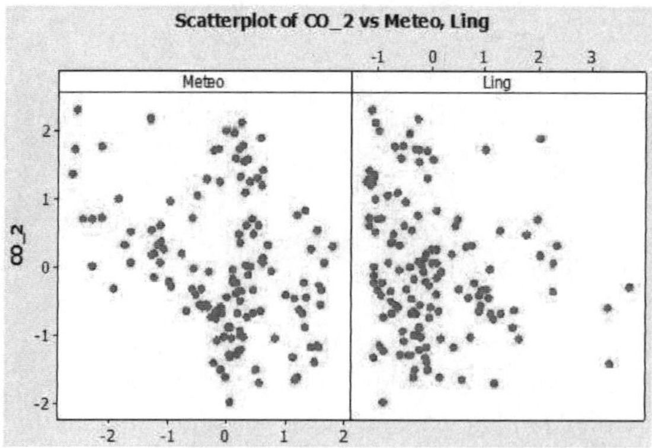

Figure 1. a. Scatterplot of Score Factor CO_2 and Meteo, b. Scatterplot of Score Factor CO_2 and Ling

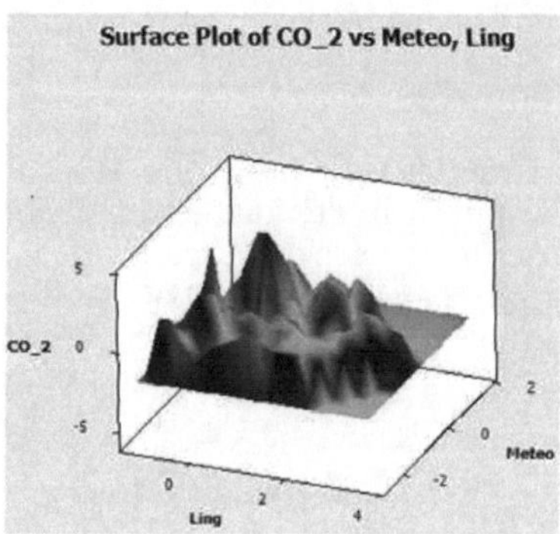

Figure 2. Surface Plot of CO_2 vs Meteo and Ling

From the Figure 1 and Figure 2, we can see that the relation between laten variable is nonlinear, and by using the estimator in the Theorem 1 for $\Sigma = \mathbf{I}$, we get the structural model estimation in the Table 2 and Path Diagram 3 as folows:

Table 2. Structural Model

Coef	SE Coef	T	Pvalue
2.2990	0.9240	2.4881	0.0141
-3.0147	0.9319	-3.2351	0.0015
4.1882	1.4418	2.9048	0.0043
-3.8396	1.4810	-2.5925	0.0106
-6.8345	2.1936	-3.1157	0.0023
6.7444	2.2190	3.0393	0.0029
Minimum GCV	:	0.0000417	
Knot Optimum	:	-2.453; 0.0324; 0.21; -1.014	

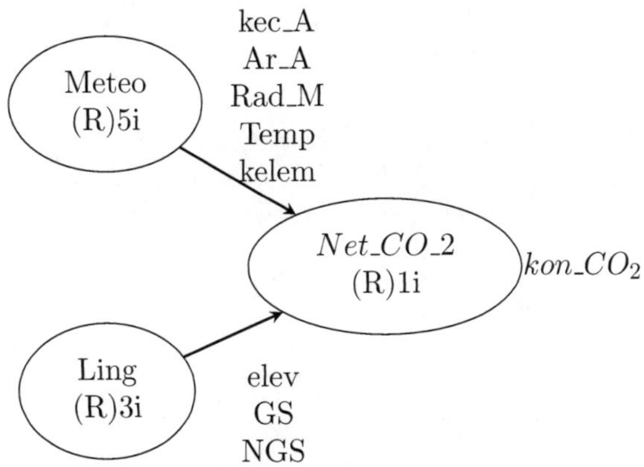

Figure 3. Path Diagram of CO_2, Meteorology, and Environment

From the table above, the estimation of structural model of nonlinear SEM by using spline truncated is:

$$CO_2 = 2.2990\,Meteo - 3.0147(Meteo + 2.453)_+ + 4.1882(Meteo - 0.0324)_+ - 3.8396(Meteo + \\ - 0.21)_+ - 6.8345\,Ling + 6.7444(Ling + 1.014)_+$$

All of parameter in structural model estimation are significant for P value < 0.05 [7], [8] with minimum GCV is 0.0000417. The plot of CO_2 model with the estimator are given in Figure 4.

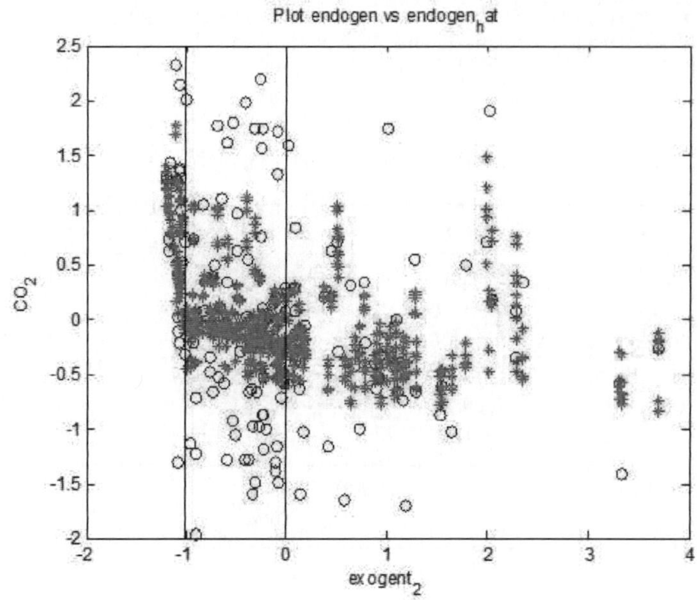

Figure 4. Plot of CO_2 and Estimator Spline Truncated

5. Conclusion

By using confirmatory factor analysis for measurement model and weighted least squares ((WLS) method for the estimation parameter of spline truncated in nonlinear SEM $\omega_\eta = \mathbf{\Gamma}_{\xi\kappa}\gamma + \zeta$ with respect to $\mathbf{Q}(\gamma) = \zeta'\mathbf{\Sigma}^{-1}\zeta$ can be concluded as follow:

a. variance-covariance matrix among the indicators $\mathbf{x}$ is $\mathbf{\Lambda_x \Phi \Lambda_x'} + \mathbf{\Theta}_\delta$, among the indicators $\mathbf{y}$ is $\mathbf{\Lambda_y}[(\mathbf{I}-\mathbf{B})^{-1}(\mathbf{\Gamma\Phi\Gamma'}+\mathbf{\Psi})(\mathbf{I}-\mathbf{B})^{-1'}]\mathbf{\Lambda_y'} + \mathbf{\Phi}_\varepsilon$, covariance matrix between $\mathbf{x}$ and $\mathbf{y}$ is $\mathbf{\Lambda_x\Phi\Gamma'}(\mathbf{I}-\mathbf{B})^{-1'}\mathbf{\Lambda_y'}$ and matrix between $\mathbf{y}$ and $\mathbf{x}$ is $\mathbf{\Lambda_y}(\mathbf{I}-\mathbf{B})^{-1}\mathbf{\Gamma\Phi\Lambda_x'}$.

b. Estimation parameter γ of spline truncated in nonlinear SEM $\omega_\eta = \mathbf{\Gamma}_{\xi\kappa}\gamma + \zeta$ is $\hat{\gamma} = (\mathbf{\Gamma}'_{\xi\kappa}\mathbf{\Sigma}^{-1}\mathbf{\Gamma}_{\xi\kappa})^{-1}\mathbf{\Gamma}'_{\xi\kappa}\mathbf{\Sigma}^{-1}\omega_\eta$

5.1. Acknowledgments

Many thanks to Kemenristek Dikti Indonesia for the Ph.D Grand in cooperation with Universitas Negeri Makassar (UNM) and Institut Teknologi Sepuluh Nopember (ITS) Surabaya.

References

[1] Ruliana, Budiantara I N, Otok B W and Wibowo W 2015 *Applied Mathematical Sciences* **9** 7439–7451
[2] Wall M M and Amemiya Y 2007 *Handbook of latent variable and related models* 321–343
[3] Lee S Y and Zhu H T 2000 *British Journal of Mathematical and Statistical Psychology* **53** 209–232
[4] Lee S Y and Song X Y 2003 *Computational statistics & data analysis* **44** 125–142
[5] Lee S Y, Song X Y and Lee J C 2003 *Journal of Educational and Behavioral Statistics* **28** 111–134
[6] Bollen K A 2009 *Structural equations with latent variables* (Wiley)
[7] Ruliana, Budiantara I N, Otok B W, Wibowo W *et al.* 2016 *Journal of Theoretical and Applied Information Technology* **89** 372
[8] Ruliana, Budiantara I N, Otok B W and Wibowo W 2016 *Global Journal of Pure and Applied Mathematics (GJPAM)* **12** 3775–3786
[9] Bollen K A and Curran P J 2006 *Latent curve models: A structural equation perspective* vol 467 (John Wiley & Sons)
[10] Wall M M and Amemiya Y 2000 *Journal of the American statistical association* **95** 929–940

IOP Conf. Series: Journal of Physics: Conf. Series **893** (2017) 012029 doi:10.1088/1742-6596/893/1/012029

Calculation of benefit reserves based on true m-thly benefit premiums

Riaman, Dwi Susanti, Agus Supriatna, and Budi Nurani Ruchjana

Department of Mathematics, Faculty of Mathematics and Natural Sciences
Universitas Padjadjaran, Indonesia

E-mail: riaman02@yahoo.co.id

Abstract. Life insurance is a form of insurance that provides risk mitigation in life or death of a human. One of its advantages is measured life insurance. Insurance companies ought to give a sum of money as reserves to the customers. The benefit reserves are an alternative calculation which involves net and cost premiums. An insured may pay a series of benefit premiums to an insurer equivalent, at the date of policy issue, to the sum of to be paid on the death of the insured, or on survival of the insured to the maturity date. A balancing item is required and this item is a liability for one of the parties and the other is an asset. The balancing item, in loan, is the outstanding principle, an asset for the lender and the liability for the borrower. In this paper we examined the benefit reserves formulas corresponding to the formulas for true m-thly benefit premiums by the prospective method. This method specifies that, the reserves at the end of the first year are zero. Several principles can be used for the determined of benefit premiums, an equivalence relation is established in our discussion.

1. Introduction

Risk was very close to the soul of man is death. Because death is the end of human life in this world, then people need to plan things that will be accomplished by themselves. One of plan is economic planning. Especially for someone who is already married, matters relating to the economy highly preferred. Such as paying bills, pay for school children, provide for the family, and so forth. Therefore, to reduce the economic burden of the family when someone dies is to buy a life insurance policy. With a life insurance policy, the insurance company (insurer) will compensate financial losses (compensation) experienced by the insured (the policyholder). The insurance product which provides death protection within a specified period is term life insurance. Unfortunately, if a person lives longer until the insurance expires, the premium money will be forfeited.

In life insurance, the insurance company will receive the premiums of the insured person. Premiums received by insurance companies, not just used to provide compensation to policyholders, but there are costs that are required in carrying out their duties. The costs would be borne by the policyholder paid together with net premium and so-called gross premiums. In fact, premiums in the early years were not enough to cover the cost, but the shortfall will be covered by premiums recent years. This situation force insurance companies looking for sources of additional funding to cover the costs of the early years which will then be repaid from premiums a year later. The insurance company has a duty to provide insurance money that would be taken as a reserve. Reserves calculation involving net premiums and fees, called reserves are adjusted. There are several methods used to calculate the adjusted reserves, one of which is the method the New Jersey. This paper will discuss the calculation of the adjusted reserves in term life insurance by using New Jersey [1].

2. Main Result

2.1. Mortality Table

Insurance companies use Mortality Table to calculate insurance premiums. This table contains the person's chances of dying by age of a group of people who are insured (policyholders) and are expected to describe the actual probability of dying from a group of people who are insured.

The number of people who were born at the same time is denoted by l_0, of a number of a number of these people will be l_0 of this person, there are l_x the people who will reach the age of x years old at the same time. The number of people who died from l_x people before reaching age $x+1$ is denoted by d_x thus:

$$d_x = l_x - l_{x+1} \tag{1}$$

and $l_x = d_x + d_{x+1} + d_{x+2} + d_{x+3} + \ldots + d_{xn-1} + d_{x+n}$ where $n \geq 1$. Probability of person aged x will die before reaching age $x+1$ year, is denoted by q_x , then:

$$q_x = \frac{l_x - l_{x+1}}{l_x} = \frac{d_x}{l_x} \tag{2}$$

There are some mortality tables are used in the calculation of insurance. In this paper, the Indonesian Mortality Table 1993 is used.

2.2. Annuity

Annuity is a payment of a certain amount, which is conducted every certain time interval and, on an ongoing basis. Life annuity in which the payment is made at a period of time called the term life annuity. Life annuity futures end with a period of n years denoted by $a_{x:\bar{n}|}$ whereas the initial term life annuity is denoted by $\ddot{a}_{x:\bar{n}|}$, the calculation is as follows:

$$a_{x:\bar{n}|} = \frac{N_{x+1} - N_{x+n+1}}{D_x} \tag{3}$$

$$\ddot{a}_{x:\bar{n}|} = \frac{N_x - N_{x+n}}{D_x} \tag{4}$$

2.3. Discrete Term Life Insurance

Discrete term life insurance is term life insurance, which is repaid at the end of the policy year, the intent of this policy is the final year of the policy at the time of death. Single premium life insurance discrete futures to age x, n-year term insurance, a sum amount of 1 paid at the end of policy year, denoted by $A_{x:\bar{n}|}$ [1]

$$A^1_{x:\bar{n}|} = \frac{M_x - M_{x+n}}{D_x} \tag{5}$$

2.4. The Annual Premium

The annual premium is the amount of premiums payment is the same every year . The annual premium n-year term life insurance, the sum insured 1, payable at the end of the policy year is $P_{x:\bar{n}|}$ [5]

$$P^1_{x:\bar{n}|} = \frac{M_x - M_{x+n}}{N_x - N_{x+n}} \tag{6}$$

The annual premium whole life insurance with insurance money paid out by the end of the policy and the payment period m years ${}_mP_x$ is

$$ {}_mP_x = \frac{M_x}{N_x - N_{x+n}} \tag{7} $$

2.5. *Benefit Reserves*

Reserves are the amount of money available on the company within the coverage. The calculation of the reserve fund is based on the assumption of annual premium. Reserve fund formed due to excess funds from the annual premium on the value of insurance between the preceding year and the year. Insurance companies sometimes have to pay compensation because the insured dies before the payment period expires. To overcome this problem, the insurer must pay compensation for the use of reserves [3].

Prospective reserves calculation at year t is the value of benefits that will come with the cash value is reduced premiums to come. Mathematically, prospective reserves for term life insurance with a sum insured 1 is

$$ {}_tV^1_{x:\overline{n}|} = A^1_{x+t:\overline{n-t}|} - {}_tP^1_{x:\overline{n}|} \cdot \ddot{a}_{x+t:\overline{n-t}|} \tag{8} $$

2.6. *Adjusted Reserves*

Reserves are adjusted premium reserve calculation that uses the assumptions adjusted premiums. Sources of additional funds to cover the initial costs can be obtained by adjusting the premium reserve. Let P stated net premium for a particular type of insurance. The premium will be replaced with α in the first year followed by a β in subsequent years. α and β are adjusted premiums. Policyholders pay only net premiums equal each year, the P + fees. Value α and β exist only in the actuarial calculations and had nothing to do with the policyholder. P on the one hand, and α and β on the other hand are connected by [5]

$$ P = \alpha + \beta \tag{9} $$

2.7. *Benefit Reserves based on True m-thly Benefit Premiums*

The benefit reserve formulas corresponding to the formulas for true m-thly benefit premiums. The method is based on prospective. [1]

$$ {}^h_kV^{(m)}_{x:\overline{n}|} = A_{x+k:\overline{n-k}|} - {}_hP^{(m)}_{x:x:\overline{n}|}\,\ddot{a}^{(m)}_{x+k:\overline{h-k}|}, \quad h < k \tag{10} $$

The difference between ${}^h_kV^{(m)}_{x:\overline{n}|}$ and ${}^h_kV_{x:\overline{n}|}$ of limited payment endowment insurance, for $h < k$,

$$ \begin{aligned} {}^h_kV^{(m)}_{x:\overline{n}|} - {}^h_kV_{x:\overline{n}|} &= {}_hP_{x:x:\overline{n}|}\ddot{a}^{(m)}_{x+k:\overline{h-k}|} - {}_hP^{(m)}_{x:x:\overline{n}|}\ddot{a}^{(m)}_{x+k:\overline{h-k}|} \\ &= {}_hP^{(m)}_{x:x:\overline{n}|}\frac{\ddot{a}^{(m)}_{x:\overline{h}|}}{\ddot{a}_{x:\overline{n}|}}\ddot{a}^{(m)}_{x+k:\overline{h-k}|} \end{aligned} \tag{11} $$

Based on the assumption of a uniform distribution of deaths in each years of age, equation (15), becomes

$$ {}^h_kV^{(m)}_{x:\overline{n}} = \beta(m)\,{}_hP^{(m)}_{x:\overline{n}|}\;V^1_{x:\overline{h}|} $$

In other words, the benefit reserve for an insurance with true m-thly benefit premiums equals to the corresponding fully discrete benefit reserve plus a fully discrete benefit reserve over the premium paying period for a fraction, $\beta(m)$, of the true m-thly benefit premium for the plan of insurance. [2]

Table 1. Reserves are adjusted by the method of New Jersey to the age of 30 years on a 30 years term life insurance

Prospective Method	New Jersey Method
IDR 86.991,41	IDR -
IDR 178.165,18	IDR 257.578,13
IDR 273.196,01	IDR 350.046,92
IDR 372.649,75	IDR 446.783,58
IDR 477.094,52	IDR 548.345,66
IDR 586.518,27	IDR 654.711,36
IDR 699.811,36	IDR 764.763,01
IDR 814.897,52	IDR 876.417,18
IDR 928.997,30	IDR 986.887,12
IDR 1.040.669,67	IDR 1.094.719,97
IDR 1.148.952,05	IDR 1.198.938,87
IDR 1.253.449,30	IDR 1.299.131,82
IDR 1.354.272,08	IDR 1.395.390,47
IDR 1.451.234,91	IDR 1.487.509,27
IDR 1.544.190,16	IDR 1.575.318,85
IDR 1.631.770,71	IDR 1.657.429,72
IDR 1.712.003,66	IDR 1.731.844,44
IDR 1.782.478,85	IDR 1.796.125,29
IDR 1.840.068,41	IDR 1.847.113,21
IDR 1.879.459,35	IDR 1.879.459,35
IDR 1.894.318,08	IDR 1.894.318,08
IDR 1.878.182,09	IDR 1.878.182,09
IDR 1.825.924,22	IDR 1.825.924,22
IDR 1.735.628,19	IDR 1.735.628,19
IDR 1.601.393,37	IDR 1.601.393,37
IDR 1.416.615,10	IDR 1.416.615,10
IDR 1.173.685,80	IDR 1.173.685,80
IDR 863.666,84	IDR 863.666,84
IDR 476.462,83	IDR 476.462,83
IDR -	IDR -

2.8. Problems

On the basis of the illustrative Life Table under the UDD of each year of age and i = 6%, calculate the following for a 20-year endowment insurance issued to (50) with a unit benefit and true semiannual benefit premiums :

- The benefit reserve at the end of the tenth year if the benefit is payable at the end of the year of death.
- The benefit reserve at the end of the tenth year if the benefit is payable at the moment of the death.

And the solution of the problems, require
$A^{1}_{60:\overline{10}|} = 0.136789, A_{60:\overline{10}|} = 0.587984, \ddot{a}_{60:\overline{10}|}$
${}_{10}V^{1}_{50:\overline{20}|} = A^{1}_{60:\overline{10}|} - P^{1}_{50:\overline{20}|} \times \ddot{a}_{60:\overline{10}|} = 0.0525752$
${}_{10}V_{50:\overline{20}|} = A_{60:\overline{10}|} - P_{50:\overline{20}|} \times \ddot{a}_{60:\overline{10}|} = 0.355380$
Under the assumption of a uniform distribution of deaths over each year age, we have

$\ddot{a}^{(2)}_{60:\bar{10}|} = \alpha(2)\ddot{a}_{60:\bar{10}|} - \beta(2)(1 -_{10} E_{60}) = 7.139240$
Then the benefit reserve, ${}_{10}V^{2}_{50:\bar{20}|}$,
$A_{60:\bar{10}|} - P^{(2)}_{50:\bar{20}|} \times \ddot{a}^{(2)}_{60:\bar{10}|} = 0.355822$ or
${}_{10}V_{50:\bar{20}|} + \beta(2)P^{(2)}_{50:\bar{20}|}\ {}_{10}V^{1}_{50:\bar{20}|} = 0.355822$
And the second one need additional calculated value:
$\frac{i}{\delta}A^{1}_{50:\bar{20}|} = 0.13423835$
$P^{(2)}\ (\bar{A}_{50:\bar{20}|}) = \frac{\bar{A}_{50:\bar{20}|}}{\ddot{a}^{(2)}_{50:\bar{20}|}} = 0.0328683$
$\frac{i}{\delta}A^{1}_{60:\bar{10}|} = 0.14085233$
$\bar{A}_{50:\bar{20}|} = 0.36471188$
$P\ (\bar{A}_{50:\bar{20}|}) = \frac{\bar{A}_{50:\bar{20}|}}{\ddot{a}_{50:\bar{20}|}} = 0.03229873$
$\bar{A}_{60:\bar{10}|} = 0.59204806$
${}_{10}V(\bar{A}_{50:\bar{20}}) = \bar{A}_{60:\bar{10}|} - P\ (\bar{A}_{50:\bar{20}|})\ddot{a}_{60:\bar{10}|} = 0.3569475$
${}_{10}V^{(2)}(\bar{A}_{50:\bar{20}}) = \bar{A}_{60:\bar{10}|} - P\ (\bar{A}^{(2)}_{50:\bar{20}|})\ddot{a}^{(2)}_{60:\bar{10}|} = 0.3573937$
$\beta(2)P^{(2)}\ (\bar{A}^{(2)}_{50:\bar{20}|})\ {}_{10}V^{1}_{50:\bar{20}|} = 0.000446$

3. Conclusion

The amount of the reserve fund with an annual premium paid will be greater than the reserves for premiums paid m times a year. Or in other words the amount of the reserve fund will be more flexible if the premiums payable m times a year .

Acknowledgment

The authors thank to Rector Universitas Padjadjaran and Director of Directorate of Research and Service Community Unpad, who gave funding for dissemination of this paper through Academic Leadership Grant 2016.

References

[1] Bowers et al. 1997 Actuarial Mathematics 2nd Edition *Schaumberg Illinois: The Society of Actuaries*
[2] Futami T 1993 Life Insurance Mathematics Part 1 *Tokyo: Incorporated Foundation Oriental Life Insurance Cultural Development Center*
[3] Futami T 1994 Life Insurance Mathematics Part 2 *Tokyo: Incorporated Foundation Oriental Life Insurance Cultural Development Center*
[4] Irfani Azis, Agus Supriatna, Riaman 2015 Perhitungan Cadangan Yang Disesuaikan Dengan Metode New Jersey Pada Asuransi Jiwa Berjangka
[5] Larson, Robert E, Gaumnitz, Erwin A 1962 Life Insurance Mathematics *New York John Wiley & Sons, Inc. London*
[6] Riaman et al. 2016 Calculation Of Adjusted Reserves By Using New Jersey Method In Term Life Insurance *Proceedings Of The National Seminar of Mathematics and Mathematics Education (SENDIKA 2016) University of Muhammadiyah Purworejo*
[7] Sukma, Rizky P 2014 Perhitungan Cadangan Yang Disesuaikan Pada Produk Asuransi Jiwa Endowment Semikontinu Dengan Menggunakan Metode Illinois Dan Metode Canadian *Skripsi, Tidak Dipublikasikan, Jurusan Matematika, FMIPA, UNPAD*

Structural model to evaluate the effect of participation and satisfaction on ecotourism sustainability

Eka N Kencana[1] **and T Manutami**[2]

[1] Tourism Research Consortium, Udayana University – Denpasar 80234, Indonesia
[2] Department of Mathematics, Udayana University – Jimbaran, Badung 80361 - Indonesia

E-mail: `i.putu.enk@unud.ac.id`, `tresianimanutami@yahoo.com`

Abstract. This paper is directed to study the effect of local community participation and visitors' satisfaction on ecotourism sustainability at Badung regency of Bali province, Indonesia. Two important aspects regarding ecotourism sustainability had been studied, i.e. (a) economic benefits for local people and (b) tourists' satisfaction. Applying variance-based structural equation modeling, data were collected in July 2015 from local community leaders of Kiadan Village at Badung regency and tourists whom visited this village, were analysed. Four latent variables, namely (a) community participation, (b) economic benefits, (c) tourists' satisfaction, and (d) ecotourism sustainability, were used to build structural model. The results showed sustainability of Kiadan's ecotourism was significantly affected by local community participation and visitors' satisfaction although community participation's effect slightly greater than tourists' satisfaction with path values for participation and satisfaction as much as 0.651 and 0.627, respectively.

1. Introduction

Many scholars argue sustainability of tourist destinations depends on the involvement and participation of local community on the tourism system in their villages [1][2][3]. By taking participation in tourists activities, local people have chances to initiate and manage their small businesses that take focus on servicing and supplying visitors needs. Furthermore, for people with lack of abilities to start their own business, they might be employed by tourist industries located around their village. In general, tourism give many economic benefits for local people whom participated in tourism activities at their village by direct or indirect participation [1].

Despite of the economies beneficiaries, undoubtedly Bali's tourism also causes some negative potentials especially on socio-cultural and environmental aspects. Recently, Balinese people experienced water scarcity that is never happened before tourism so intensive in this small island. Another example is the disruption of the traffic's flow as consequences of the concentration of tourism industries and people's activities. These conditions lead to the inquiry of 'the friendly-formed of tourism' that is argued has less negative impacts. Ecotourism is proposed to be an alternative.

This research is motivated to elaborate and to study the relationship between local community participation and sustainability of ecotourism, especially in highland region of Bali. We focused our study on three main issues, namely (a) the effect of community participation on economies' benefit gained by local people, (b) the effect of community participation and tourists satisfaction on ecotourism sustainability, and (c) the effect of ecotourism sustainability on economies' benefit for local people. Variance-based structural equation modeling (PLS-SEM) is applied to study these issues. This technique is chosen with following considerations:

(i) Classified to soft modeling technique, PLS-SEM is not too strict regarding the fulfillment of normality assumption as found on covariance-based SEM techniques such as AMOS and LISREL [4];

(ii) PLS-SEM can be used without loss of power test for small sample. This technique has been proven its power even for sample only sized 6 [5].

In brief, SEM is a statistical method that is grouped to multivariate statistical analysis as well as factor analysis, and is commonly used by social researcher to study the relationship among latent variables [5][6]. According to Bollen [6] and Tenenhaus et al. [4], SEM is a powerful technique to study latent variable or construct, a variable that can not be measured directly. Latent variable is usually measured by its indicators or manifests.

2. Research Method

The data in this research is obtained from local people and community leaders who participate in the management of Kiadan Village Ecotourism, and from tourists whom visiting Kiadan Village ecotourism on May–July 2015. Kiadan is located at northern part of Badung regency, Province of Bali. Self-administered questionnaires were distributed for 120 respondents, 60 samples were taken from the local people and their leaders, and another 60 were taken from the tourists whom visited Kiadan village. Prior to use for collecting data, the questionnaire were tested to evaluate its validity and reliability.

We developed our SEM's model that consists of six first-order latent variables, namely (a) direct participation, (b) indirect participation, (c) economies' benefit, (d) tourists' attraction, (e) accessibility, and (f) amenities; and three second-order latent variables, i.e. (a) ecotourism sustainability, (b) local participation, and (c) visitors satisfaction. The conceptual model of this research is depicted in Fig. 1.

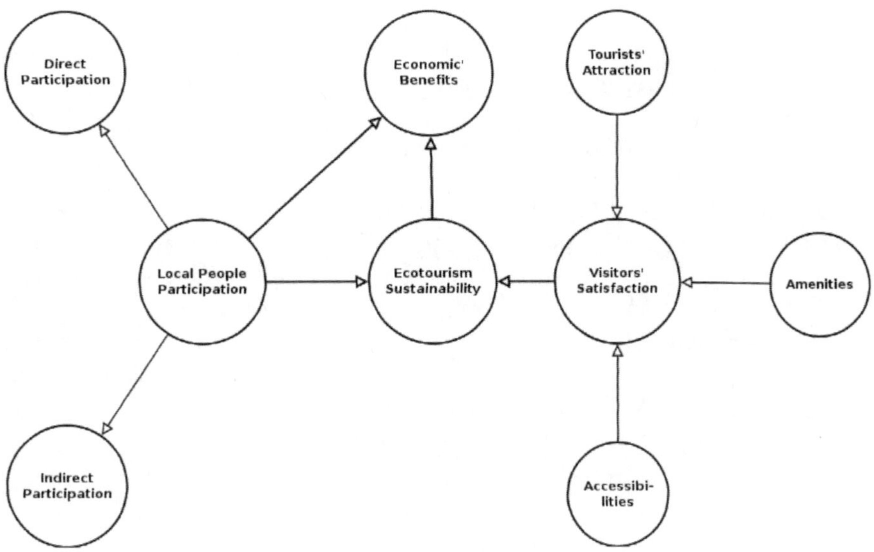

Figure 1. Conceptual Research Model

IOP Conf. Series: Journal of Physics: Conf. Series **893** (2017) 012030 doi:10.1088/1742-6596/893/1/012030

3. Results and Discussion

3.1. The Assessment of Questionnaire

Prior to data collection, we tested the validity and reliability of the questionnaire. For reflective construct, an item is declared invalid, if its correlation with total items score on the same construct is < 0.30 [8]. Furthermore, a reflective construct is called reliable if the value of its Cronbach's alpha $(\alpha) \geq 0.60$ [9]. The assessment results is listed in Table 1.

Table 1. The Assessment of Reflective Constructs on Instrument

Construct	Total item	Valid item	α value
Direct Participation	4	4	0.727
Indirect Participation	3	3	0.652
Tourists' Attraction	11	9	0.771
Accessibilities	2	2	0.612
Amenities	2	2	0.636
Economic Benefit	3	2	0.602

Based on the assessment results, three items that are proved insignificant to represent their constructs (tourists' attraction and economic benefit) were eliminated on subsequent analysis. We used SmartPLS 3.2.3 from [10] to analyse the model.

3.2. Outer Model Analysis

As described previously, a latent variable ξ is measured through its manifests x_h. For reflective measurement – a latent variable is reflected into p manifest (observable) variables – the causal relationship can be expressed in equation (1) as follow:

$$x_h = \pi_{h0} + \pi_h \xi + \epsilon_h \tag{1}$$

In (1), ξ has mean m and standard of deviation equals to 1. According to H. Wold (cited in [4, p.163]), the *predictor specification* must satisfy

$$E(x_h|\xi) = \pi_{h0} + \pi_h \xi \tag{2}$$

Outer model analysis is aimed to elaborate the relationship between construct and its items. This assessment is conducted by observing the individual loading factor of the manifests, evaluating constructs' internal consistency by observing the composite reliability (CR) and checking constructs' convergent validity through its average variance extracted (AVE) [7]. A construct is argued has internal consistency if its CR's ≥ 0.708 [10] and achieved convergent validity if its AVE's ≥ 0.50 [10][11]. In addition, the outer loading has to be greater than 0.60 [9] and significant [11]. The AVE for one contruct that is reflected into p manifest variables can be counted using formula (3) that is introduced in [4] and [16]:

$$AVE = \frac{1}{p} \sum_{h=1}^{p} Cor^2(x_h, \xi) \tag{3}$$

By examining the AVE's values for all of first-order reflective constructs in our model, suggested all of constructs achieved convergent validity. The smallest and the largest AVE's were found for tourists' attraction and for tourists' amenities as much as 0.508 and 0.683, respectively. In addition, the examination of CR's values showed all constructs also achieved internal consistency and accessibilities and tourists' attraction constructs with the smallest and the largest CR's values as much as 0.741 and 0.849, respectively. Refers to this outer assessment, we concluded that the inner model analysis is worth to be conducted.

3.3. Inner Model Analysis

Inner or structural model analysis relates to causal relationships among constructs [9]. The assessment of inner model is conducted by examining the path coefficients that relate one or more exogenous constructs to an endogenous construct. The path coefficients for inner relationship is depicted in Fig. 2. We got eight out of nine paths are significant at 10 percent significance level, while one path coefficient is insignificant.

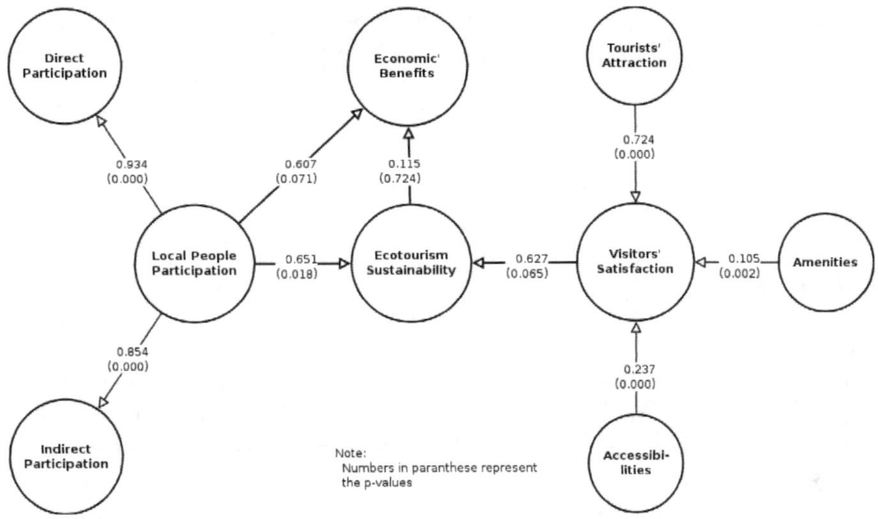

Figure 2. Path Coefficients for Inner Relationships

Furthermore, the coefficient of determination R^2 that represents the amount of endogenous' variance is explained by its respective exogenous constructs has to be noted. According to [13], an endogenous construct is claimed has a weak, moderate, or substantial predictive accuracy whenever its R^2 greater than the threshold values as much as 0.19, 0.33, and 0.67, respectively. In addition to examine the R^2, one should check the predictive relevance of endogenous constructs by noting the Stone-Geisser's Q^2 value. The Q^2 value > 0.0 suggests the construct is relevantly predicted by its respective exogenous constructs [11]. The R^2s and Q^2s for endogenous constructs in our model is listed in Table 2. Refers to Table 2, except for economies' benefit construct that is moderately predicted by its exogenous constructs, the rest are substantially predicted by their respective constructs.

3.4. Discussion

Despite the community participation has been acknowledged and promoted since 1950s as an important agent to assure the successful of development process, its implementation and mechanism are still debated among economic development scholars [14]. Community involvement is argued has a central position in the development process as well as an indispensable tool to educate people [1]. Our work showed people of Kiadan village significantly affect ecotourism sustainability at this area through their participation. The more local

Table 2. The Assessment of the Inner Model

Construct	R^2	Accuracy?	Q^2	Relevance?
Direct Participation	0.871	Yes	0.484	Yes
Indirect Participation	0.729	Yes	0.306	Yes
Tourists' Satisfaction	0.919	Yes	0.298	Yes
Economic Benefit	0.492	Yes	0.184	Yes
Sustainable Ecotourism	0.995	Yes	0.171	Yes

people get involved on ecotourism at Kiadan, the more its sustainability could be expected. Furthermore, their participation on ecotourism also positively affect the economic' benefits for the host. This finding is inline with the similar research is done by [15].

From demand perspective, the sustainability of ecotourism at Kiadan is significantly affected by its visitors' satisfaction, although this effect is slightly less than the effect arise from local participation. Visitors felt satisfied with attraction, accessibilities, and amenities in this destination. Observing the path values of these constructs, it is clear that natural attraction outperformed accessibilities and amenities factors in forming tourists' satisfaction whom visited this ecotourism destination.

4. Conclusion

The results of this work gave several explanations regarding the causal relationships between community participation, ecotourism sustainability, visitors' satisfaction, and the economic' benefit for local people. The sustainability of ecotourism at Kiadan village, Badung regency of Bali, is significantly affected by local community participation and tourists whom visited the destination. The more local people get involved in ecotourism activities at this village, the more economic benefit they will get. In addition, we concluded tourists' attraction is the ultimate factor to keep ecotourism at this village sustain. From this finding, we suggest the tourism stakeholders at Badung regency to maintain natural landscape and to expand local activities that support ecotourism at Kiadan Village.

Acknowledgments

The authors would like to thank people and leaders of Kiadan Village for their support in this work by supplying valuable information.

References

[1] Cevat T and Timothy D J 2003 Community participation in the tourism development process *J. of Tourism Studies* **14** (2) pp 1–15

[2] Hwansuk C C and Ercan S 2006 Sustainability indicators for managing community tourism *Tourism Management* **27** pp 1275–1289

[3] Pham H L 2012 Tourism impacts and support for tourism development in Ha Long Bay, Vietnam: An Examination of resident's perceptions *Asian Social Science* **8** 8 pp 28–39

[4] Michel T, Vincenzo E V, Yves-Marie C and Carlo L 2005 PLS path modeling *Computational Statistics and Data Analysis* **48** pp 159–205

[5] Hox J J and Bechger T M 1998 An Introduction to structural equation modeling *Family Science Review* **11** pp 354–373

[6] Kenneth A B 2002 Latent variables in psychology and the social sciences *Annual Review of Psychological* **53** pp 605–634

[7] Joe F H, Marko S, Christian M R and Jeannette A M 2012 An assessment of the use of partial least squares structural equation modeling in marketing research *J. of Academic Marketing Science* **40** pp 414–433

[8] Gilbert A C 1979 A Paradigm for developing better measures of marketing constructs *J. of Marketing Research* **16** 1 pp 64–73
[9] Hair J F, Anderson R E, Tatham R L and Black W C 1995 *Multivariate Data Analysis with Readings 4th edition* (New Jersey: Prentice-Hall Inc.)
[10] Christian M R, Sven W and Alexander W 2014 *SmartPLS 3.0* (http://www.smartpls.de)
[11] Joseph F H Jr, Tomas M H, Christian M R and Marko S 2014 *A Primer on Partial Least Square Equation Modeling (PLS-SEM)* (California, USA: SAGE Publications, Inc.)
[12] David X P and Fujun L 2012 Using partial least squares in operations management research: A practical guideline and summary of past research *J. of Operation Management* **30** pp 467–480
[13] Wynne W C 1998 *The partial least squares approach to structural equation modeling* (Modern Methods for Business Research) ed G A Marcoulides (Mahwah, NJ: Lawrence Erlbaum Associates) pp 295–358
[14] Kadt E D 1982 Community participation for health: The case of Latin America *World Development* **10** 7 pp 573–584
[15] Sebele L S 2010 Community-based tourism ventures, benefits and challenges: Khama Rhino Sanctuary Trust, Central District, Botswana *Tourism Management* **31** pp 136–146
[16] Vinzi V E et al 2010 *PLS Path Modeling: From Foundations to Recent Developments and Open Issues for Model Assessment and Improvement* (Handbook of Partial Least Squares, Springer Handbooks of Computational Statistics) ed V E Vinzi et al (Heidelberg, Berlin: Springer-Verlag) pp 47–82

IOP Conf. Series: Journal of Physics: Conf. Series **893** (2017) 012031 doi:10.1088/1742-6596/893/1/012031

Bootstrap resampling to detect active zone for extreme rainfall in Indonesia

Heri Kuswanto, Sri Hidayati, Mutiah Salamah, and Brodjol Sutijo Ulama

Department of Statistics, Faculty of Mathematics and Natural Science
Insitut Teknologi Sepuluh Nopember, Indonesia

E-mail: heri_k@statistika.its.ac.id

Abstract. This research aims to develop a methodology to detect active zones related to the extreme rainfall event in one of regions in Indonesia. Active zone is defined as regions or zones on the atmospheric level which have significantly different pattern with other regions prior to the occurrence of extreme event happen in the study area. The detection will be useful for the forecasters to predict the extreme rainfall events, and hence, the risk of the disaster caused by the events can be minimized. In order to predict the active zone, this paper examines statistical procedure that is able to test the significant difference between weather phenomena at the atmospheric level onset with prior to the occurrence.

1. Introduction

Climate change happens within the last decade has lead to the increasing of extreme events intensity. Pall et al.[7] showed that the global gas emission anthropogenic greenhouse increases the risk of flood in England. Using the case of Northern Hemisphere data, Hansen et al.[5] showed that the human behavior towards the greenhouse gas has contributed to the increasing of heavy precipitation intensity. Dai[1] also found that the percentage of dry global area increased about 1.74% per decade spanning from 1950 top 2008 caused by the global warming. Min et al.[6] argues that extreme anomaly such as happen in Texas and Oklahoma in 2011 and Moscow 2010 were the consequence of global warming. Another research discussed the impact of climate change to the extreme weather is Rahmstorf and Coumou[11] among others.

Indonesia is also one of the most vulnerable countries towards the impact of climate change. Especially in East Java, extreme precipitation has led to severe flood and caused significant losses in term of infrastructure and human victim. To deal with this, actions to reduce disaster risk is necessarily to be carried out. One of the feasible mitigation strategies is to study the meteorological pattern of some weather variables prior to the occurrence of the extreme events.

This research proposes composite maps showing the atmospheric dynamic prior to the occurrence of extreme precipitation in Indramayu, East Java. Similar maps has been developed by Grotjahn and Faure [4] to predict several types of extreme events in Sacramento, California. In fact, the maps has a very good performance used as a guidance for the forecaster in the Meteorological Office. The composite maps developed in this research differ from Grotjahn and Faure [4] in the sense that the grid spaces is smaller, meaning that the proposed map has higher resolution. Grotjahn and Faure used 2.5 x 2.5 degree as the grid space, while the proposed maps is developed with 0.25 x 0.25 degree. As a consequent of using more precise grid space, the predictability of the event will increase.

The map is composed from reforecast data treated with statistical resampling procedure namely bootstrap. The bootstrap will generate artificial data to calculate the climatological mean as the basis to identify regions or zones over Indonesia which show significant different pattern on one and two days prior to the heavy precipitation in Indramayu. Bootstrap is used due to the fact that the underlying distribution of weather variable is unknown or maybe not normal, and hence, bootstrap nonparametric approach can be a powerful method to solve the problem of lack information about distribution. As in illustration, the composite maps will be developed for Mean Sea Level Pressure (MSLP).

2. Research Methodology

In order to develop the composite maps, this research uses two kind of dataset as follow:

(i) Field observed Precipitation data
This is daily rainfall data collected from one of the meteorological stations in Indramayu. The dataset span from 1979 to 2006.

(ii) Atmospheric data (Reforecast Era-Interim data)

The steps of analysis are described below:

- identify the date of extreme events from observation data
- describe the MSLP on the Indonesian regions (specified by latitude and longitude)
- for all grids, analyze the MSLP at 1 day and 2 day prior to the occurrence of extreme rainfall in Indramayu using bootstrap
- apply bootstrap resampling to create artificial data for calculating the climatological mean
- estimate the percentile bootstrap to create the confidence level with certain significant level
- test the significant different between mean of bootstrap with climatological mean (both day-1 and day-2). The zone is said to be active if the mean of extreme events lies outside the interval.

3. Bootstrap

Bootstrap is used to estimate the sample mean, variance and other measures. The method estimates the sampling distribution of statistical measure using a simple procedure i.e. resampling (Efron and Tibshirani [2]; Efron [3]). Bootstrap estimates the properties (statistical measures) of estimator by by measuring the property from the closest sampling distribution. One of the standard choices is through the empirical distribution of the analyzed data. In case where sequence of observations are assumed to be identical and independent, the empirical distribution can be developed by generating number of resampling from the observed data (with the number of sample same as the raw data), by which each of them are drawn randomly from the raw data. Bootstrap can also be used to test the hypothesis. In this case, bootstrap is usually used as alternative for statistical inference that need assumption about underlying distribution, or to calculate a very complex standard error.

Below are the steps of bootstrap resampling:

- given the sample with the size of n, do resampling with or without replacement and get R replication. The number of replication R has to be sufficient enough to estimate the sampling distribution.
- calculate the mean estimator from each resampling
- calculate the bootstrap mean by averaging mean values of all resampling.

Bootstrap has been applied to solve many problems in meteorology and climatology for instance in Lall and Sharma [8], Orlowsky et al.[10] as well as Mudelsee [11] who discusses bootstrap application for risk climate analysis.

4. Results and Discussion

The descriptive statistic of the daily rainfall in Indramayu for the annual period within 1979 to 2006 observed from Bangkir station is performed in order to study the characteristic of the rainfall to be used for further analysis.

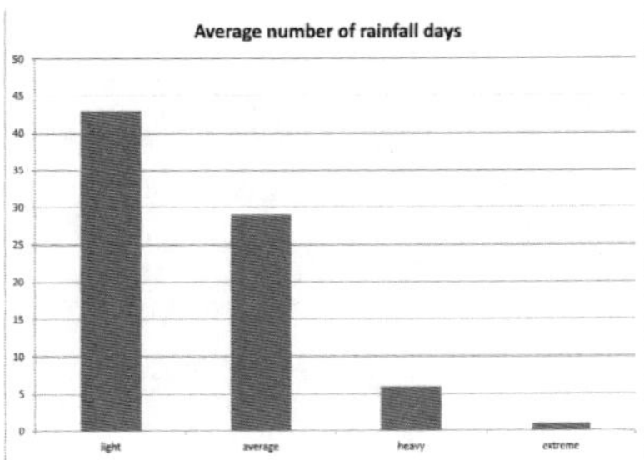

Figure 1. Number of rainfall based on each category.

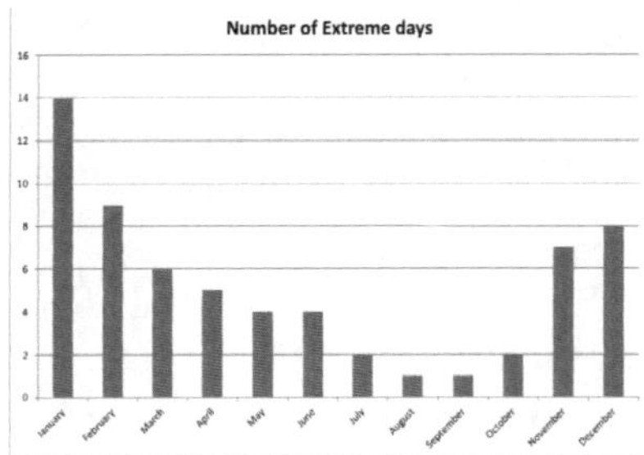

Figure 2. Average number of rainfall every month.

Figure 1 shows that on the period of one year, the average days of light rainfall is 43 days, average or normal rainfall is 29 days and heavy rainfall in about 6 days, with 1 day is considered to be extreme. The information about daily rainfall on monthly period is depicted in Figure 2, and it shows that the rainy season in Indamayu happens on November to April, while the dry season happens on May to October.

4.1. Extreme Events Detection in Indramayu

Identification of extreme events can be seen from normal probability plot.

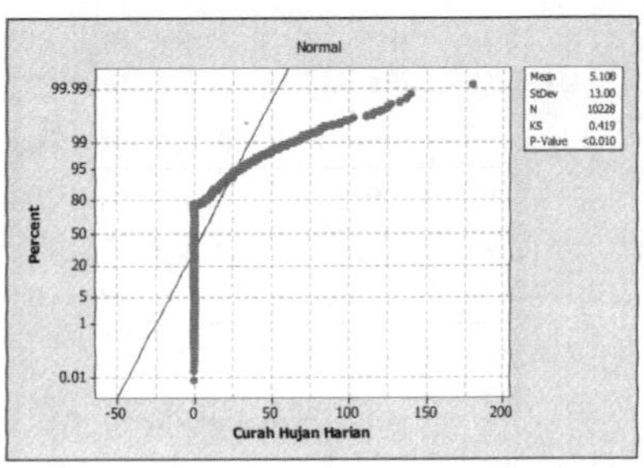

Figure 3. Normal Probability Plot.

Figure 3 shows that the pattern of daily rainfall consists of tail distribution (extreme value), and hence the probability distribution function is away from normal distribution. This fact shows nonlinear pattern of the distribution which strongly indicates that the distribution is not normal. It is confirmed also by the p-value of the statistical test less than 0.01.

Another popular method to identify the extreme value statistically is by using Peak Over Threshold (POT). It uses a threshold value u as the cut off for determining extreme. This research uses Mean Residual Life Plot (MRLP) to find the threshold.

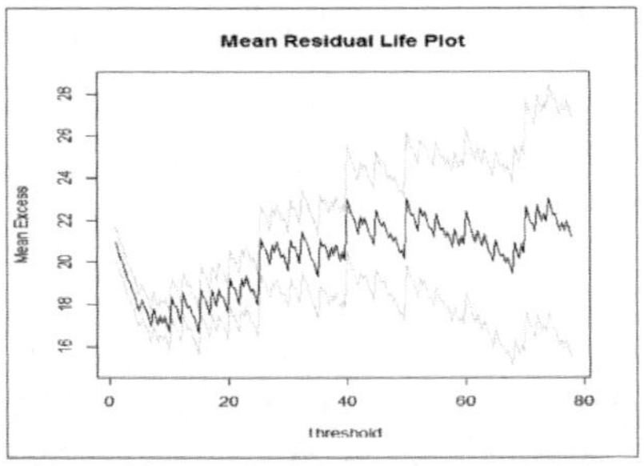

Figure 4. Mean Residual Life Plot.

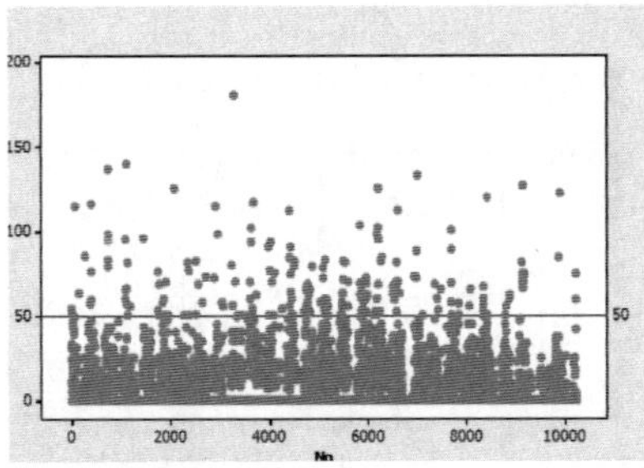

Figure 5. Rainfall with the threshold.

From the MRLP plot, the threshold is chosen by identifying the magnitude of the mean excess in which it shows a linear pattern. From Figure 4, the MRLP start to be stable at the point of 50, and hence, the threshold will be set as 50. This value is consistent with the extreme rainfall definition specified by the Agency for Meteorology, Climatology and Geophysics (BMKG) Indonesia. The extreme rainfall data are observations above the threshold as shown in Figure 5.

Another statistic reveals that extreme rainfall mostly happen in November, December, January, February and March. These months are exactly the period of rainy season in Indonesia. January is the month with the highest intensity of extreme rainfall with about 60 events. Further analysis will use only extreme rainfall detected within the rainy season due to the fact that heavy rainfall happening in dry season might be the result of convolution.

4.2. Analysis of Mean Sea Level Pressure over Indonesia Region

In this research, the MSLP data that will be analyzed covers the area of over Indonesia region. The region is sliced into grid space with the size of of 0.25 degree x 0.25 degree. The dataset are reforecast era interim data published by European Center for Medium Range Weather Forecasting Model (ECMWF) and available at http://ecmwf.int/.

4.3. Bootstrap to identify Active Zones for Mean Sea Level Pressure

We listed the date of extreme rainfall based on the threshold specified on the previous section. The idea of bootstrapping is to create artificial data corresponding to one and two days prior to the occurrence of the events.The analysis is done for each grid coordinate. Detail procedures of bootstrap resampling for the considered case are as follows:

- Dataset to be sampled span from 1979 to 2006
- The number of random sample drawn in each resampling process is 146 days (i.e. number of extreme rainfall events) both for day-1 and day-2
- The resampling process is done with replacement
- The number of replication is set to be 1000 times, and calculate the estimate mean of sample, and hence, there will be 1000 mean values
- For all estimated mean, sort the value and find the percentile with confidence level of 99%. Note that the active zones may consist of two condition i.e. very low level of pressure and very high level of pressure. Therefore, the test will be based on 1 % and 99% data.
- Test the mean of MSLP on day-1 and day-2 with mean of bootstrap sample. If the sample mean of MSLP is less than the 1 percent-th data, then the MSLP is significantly low at this zone. Moreover, if the mean of MSLP is above the the 99th percent of the bootstrap

data, then the MSLP is significantly high. The zone is said to be inactive zone (normal condition) if the mean of MSLP lies between 1

An illustration of bootstrap result is given on Figure 6. The figure performs histogram of mean bootstrap of MSLP in Indonesia on one coordinate, together with the vertical line showing the lower and upper limit as the basis for the significant test. From the figure, the lower bound is 92000 and the upper zone is 100500. If the mean sample of MSLP on day-1 lies outside the interval 1(let say lower than 92000), then the region is active zone. It simply means that on one day prior to the occurrence of extreme rainfall in Indramayu, this region will show MSLP that is significantly different pattern compared to other regions/zones. Basically we test for the following hypothesis:
H0 : $\mu = \mu_0$ (the MSLP on the corresponding coordinate is not active zone)
H1 : $\mu \neq \mu_0$ (the MSLP on the corresponding coordinate is active zone)

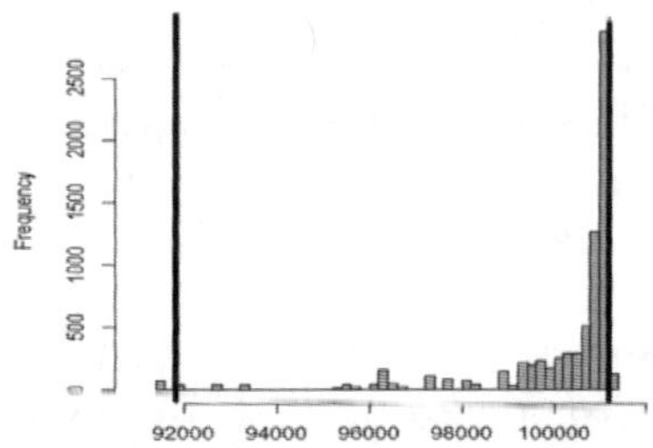

Figure 6. Illustration of bootstrap result in one grid point.

Figure 7. Composite maps showing active zone on day-1.

Using bootstrap resampling method, the active zones can be plotted into a composite map as performed in Figure 7.The Figures revealed that most of the regions in Sumatra and few part of Kalimantan have plot with red color, which show that the MSLP on day-1 is significantly high on that region, while over Java and Bali, the color is black showing that the MSLP on that region is significantly low. We can see also that the Sea Pressure at the atmospheric level on most of the area above Java Sea does not show any significant different patterns with usual days. To deal with this, the forecaster can only focus on observing the active zones to predict the occurrence of extreme rainfall in Indramayu.

5. Conclusion

Indramayu is a region in Indonesia with low intensity of rainfall, however the varians of the rainfall intensity is very high, in a year, the average of light rainfall event is 43 days, normal rainfall is 29 days and heavy rainfall is 6 days. The threshold to characterize extreme rainfall event is 50 mm, which fits the definition of BMKG Indonesia. On day-1 and day-2 prior to the occurrence of the extreme rainfall in Indramayu, the MSLP on Sumatra and some Kalimantan regions will show a very high level of MSLP. Meanwhile, the MSLP in Bali and east part of Java will be very low.

References

[1] Dai A 2011 Drought under global warming: a review *Climate Change* **2**(1) pp 45–65
[2] Efron B 1979 A Note on Bootstrapping the Sample Median *Ann. Statist* **7**(1)
[3] Efron B and Tibshirani R J 1993 *An Introduction to the Bootstrap* Chapman and Hall, London
[4] Grotjahn R and Faure G 2008 Composite Predictor Maps of Extraordinary Weather Events in the Sacramento California Region *Weather and Forecasting* **23** pp 313–335

[5] Hansen J, Sato M and Ruedy R 2012 Perception of climate change *Proceeding National Academic of Sciences US*
[6] Min S, Zhang X and Hegerl G C 2011 Human contribution to more-intense precipitation extremes *Nature* **470** pp 378–381
[7] Pall P, Aina T, Stone D A, Stott P A, Nozawa T, Hilberts A G 2000 Anthropogenic greenhouse gas contribution to flood risk in England and Wales in autumn *Nature* **470** pp 328–385
[8] Lall U and A Sharma 1996 A Nearest Neighbor Bootstrap For Resampling Hydrologic Time Series *Water Resour. Res.* **32(3)** pp 679–693
[9] Mudelsee M 2011 The Bootstrap in Climate Risk Analysis *Extremis* pp 44–58
[10] Orlowsky B, Bothe O, Fraedrich K, Gerstengarbe F W, Zhu X 2010 Future climates from bias-bootstrapped weather analogues: an application to the Yangtze river basin *Journal of Climate* **23**(13) pp 3509–3524
[11] Rahmstorf S and Coumou D 2011 Increase of extreme events in a warming world. *Proc. Natl Acad. Sci. USA* **108** pp 17905–17909

IOP Conf. Series: Journal of Physics: Conf. Series **893** (2017) 012033 doi:10.1088/1742-6596/893/1/012033

Hierarchical time series bottom-up approach for forecast the export value in Central Java

D A Mahkya, B S Ulama, and Suhartono

Department of Statistics, Faculty of Mathematics and Natural Sciences
Sepuluh Nopember Institute of Technology, Surabaya, Indonesia

E-mail: danialmahkya@gmail.com, {brodjol_su, suhartono}@statistika.its.ac.id

Abstract. The purpose of this study is Getting the best modeling and predicting the export value of Central Java using a Hierarchical Time Series. The export value is one variable injection in the economy of a country, meaning that if the export value of the country increases, the country's economy will increase even more. Therefore, it is necessary appropriate modeling to predict the export value especially in Central Java. Export Value in Central Java are grouped into 21 commodities with each commodity has a different pattern. One approach that can be used time series is a hierarchical approach. Hierarchical Time Series is used Buttom-up. To Forecast the individual series at all levels using Autoregressive Integrated Moving Average (ARIMA), Radial Basis Function Neural Network (RBFNN), and Hybrid ARIMA-RBFNN For the selection of the best models used Symmetric Mean Absolute Percentage Error (sMAPE). Results of the analysis showed that for the Export Value of Central Java, Bottom-up approach with Hybrid ARIMA-RBFNN modeling can be used for long-term predictions. As for the short and medium-term predictions, it can be used a bottom-up approach RBFNN modeling. Overall bottom-up approach with RBFNN modeling give the best result.

1. Introduction

Export is a method or activity to market goods produced domestically and sold or marketed abroad. If a country open international trade and become an exporter of goods, the domestic producers of goods that will benefit and domestic consumer goods will be harmed. The value of exports is one variable injection in the economy of a country, meaning that if a country's exports increases, the country's economy will increase even more [1]. The economic growth of Central Java province after 1999 have always had a positive growth is also determined by several indicators that tend to increase. Foreign sector plays an important role in the economy of Central Java province. During the period 1997 to 2010 the value of exports and imports increased. [2] Every year in Central Java province exports continued to rise, means the economy in Central Java will continue to grow and will certainly have an impact on Indonesia's export sector in general. Predicting the export value, especially in Central Java province is certainly much needed for policy development. Because with an approximate value in the future, will help in the future careful planning. Moreover, the value of exports in the province of Central Java, which tends to increase from year to year, so it needs a correct policy that these conditions persist. If policymakers about how the development of exports has predictive value of exports in Central Java province in the future, the policy created will be able to anticipate the phenomena that occur in the pattern of export Central Java province. Therefore, it is necessary an appropriate

IOP Conf. Series: Journal of Physics: Conf. Series **893** (2017) 012033 doi:10.1088/1742-6596/893/1/012033

predictive models to determine whether export in Central Java province will continue to rise or it will go down. Research on the value of exports of Central Java never been done before, among other things by Rejekiningsih in 2012 studied the concentration of exports in Central Java province. Data of export value in Central Java has a hierarchical structure or data that have levels. The value of exports in total as a data hierarchy level 0 or level up while the value of exports in commodities were 21 commodity as a data hierarchy level 1 or lower levels. In the data analysis used hierarchical level 1 series as much as 19 because there are some data that can not be modeled. Data of export value in Central Java for each commodity has a pattern of stable or unstable. It is therefore we model and forecast all series in hierarchy individually at all level using ARIMA, RBFNN and Hybrid ARIMA-RBFNN which is expected to capture the phenomena that occurs in the data pattern. Selection of the best model in this study using sMAPE.

2. Methods

According to Wei (2006) time series is a sequence of observations. Usually sorted by time, at a certain moment a few sorted by time interval. One of the important things in building a time series model is able to predict the value of a time series for the future [4]. By predicting the value of a time series that will come, may be obtained new stage in an analysis, which is the stage of policy making. Policy in terms of economic, political and social.

2.1. Hierarchical Time Series

The structure of Hierarchical Time Series can be illustrated in Fig.1 [5]:

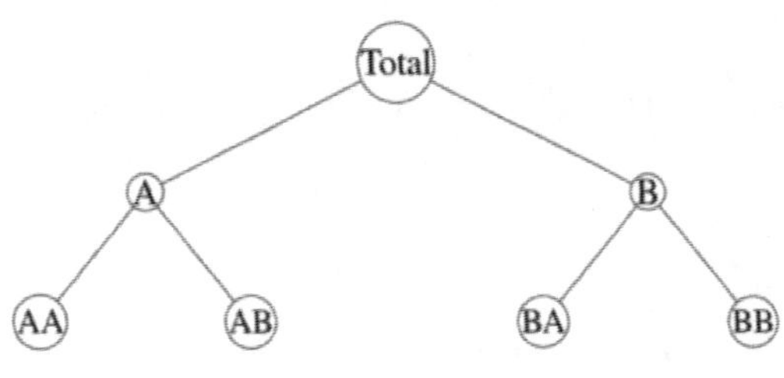

Figure 1. Hierarchical Time Series Structure

Based on Fig.1 variable Total is the time series data level 0 which is the sum of time series data at level 1. Likewise for time series data level 1 which is the sum of the time series data at level 2 and so on until level to K, In general notation for Hierarchical Time Series approaches, used Yi,t consisting of all observations to a level i and time t. All observations on all time t is denoted by $Y_t = [Y_t, Y'_{1,t}, ..., Y'_{K,t}]$' is a column vector [6]. So it can be written as follows: $Y_t = SY_{K,t}$ the S is the sum of the order m x mK matrix which is the sum of all the lower level time series data. The concept of Buttom-up approach is to generate the prediction result data for each of the time series data at a lower level and then summing to obtain total predictions or predictions of top level [5]. The advantages of the approach Buttom-up is no information is lost during the process of incorporation for each level. But on the other hand, Buttom-up approach probably would not be accurate to predict if there is missing data or noisy at lower levels [6].

2.2. Autoregressive Integrated Moving Average (ARIMA)

There are 3 stages that are used to build an ARIMA model such as [7] ARIMA model tentatively identified by analyzing the data of the past, estimating the model parameters are unknown, checked diagnostic residuals generated to determine the adequacy of the model or indicate whether or not to do repair. Once those steps are completed, then the best ARIMA model is obtained. General model of ARIMA (p, d, q) is as follows [3]:

$$(1-\phi_1 B - .. - \phi_p B^p)(1-B)^d Y_t = \theta_0 + (1-\theta_1 B - .. - \theta_q B^q)a_t \qquad (1)$$

2.3. Radial Basis Function Neural Network (RBFNN)

RBFNN appear as one type of neural network around the 80s. Construction of RBFNN contains three distinct layers. Input Layer The first layer is derived from the sensor units. The second layer is the Hidden Layer with high dimensional RBF which aims to process many layer Perceptron. The third layer is the output layer of the network provides a response to the activation pattern is applied to the Input Layer [8]. RBFNN have architecture as follows [9]:

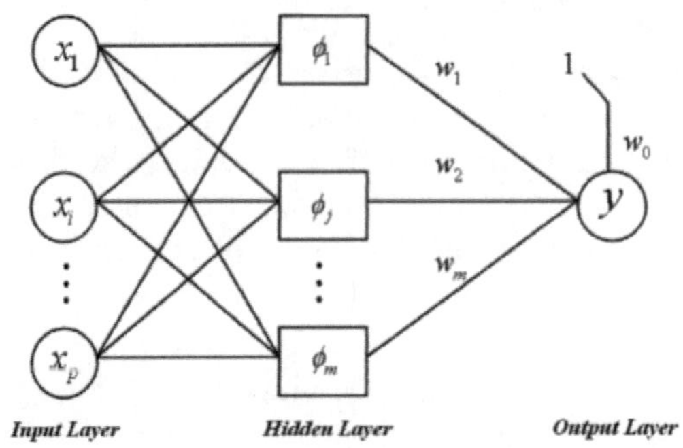

Figure 2. Architecture of RBFNN with input p and m neurons in the hidden layer

where $\phi_j(x) = \exp\left(-\frac{\|x-\mu\|^2}{2\sigma_j^2}\right)$
Based on Fig. 2 there are p neurons in the input layer and nonlinear processed with gaussian activation function. Inputs are grouped into several groups so that the average and variance of each group can be determined.

2.4. Hybrid Model

Time Series forecasting is a very important aspect of Forcasting in which past observations of the same variable are collected and analyzed to a develop a model describing the underlying relationship [10]. Various models have been developed time series analysis, from the univariate models to multivariate models. Therefore models to the analysis of time series is very popular and is often used for forecasting. Meanwhile Neural Network (NN) is one method that can be used in forecasting. NN is a highly flexible modeling. By NN, no special model in its establishment. The resulting model adapting the patterns generated from the data [10]. Both models are very much used for forecasting, but not between the two models are suited to the circumstances of various phenomena. Therefore [10] proposed a Hybrid model for forecasting by considering the autocorrelation linear structure and nonlinear components with the following equation:

$$Y_t = L_t + N_t \qquad (2)$$

where L_t is a component of linear and nonlinear component is N_t. The two components are estimated from the data using each method. For the first model representing linear components, the model will generate residuals from linear models containing nonlinear relationships.

2.5. Accuracy of Prediction

In measuring the accuracy of prediction, often used Mean Absolute Percentage Error (MAPE). But the value of MAPE would be problematic if the data has a value close to zero or equal to zero [11]. Therefore, in this study used Symmetric Mean Absolute Percentage Error to measure the accuracy of the model by the formula:

$$sMAPE = \frac{1}{n}\sum_{t=1}^{n}\frac{\left|Y_t - \hat{Y}_t\right|}{(|Y_t| + |\hat{Y}_t|)/2} \text{ x } 100 \tag{3}$$

At the stage of checking the accuracy of prediction used the Rolling Forecast Origin approach with illustration shown in Fig 3 and Fig 4.
Here are the stages of Rolling Forecast Origin:
- Determine the best model is used based on the training data.
- Estimate based on the best model and re-estimate based on best model in observation n+i with i = 0,1,2 .., N-n where n is the number of observations in which the training data and N is the number of overall data.
- Calculate the accuracy of the predictions for each i.

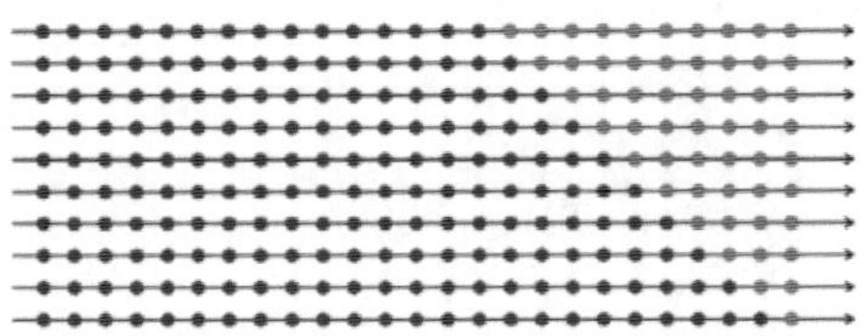

Figure 3. Data partition of Rolling Forecast Origin

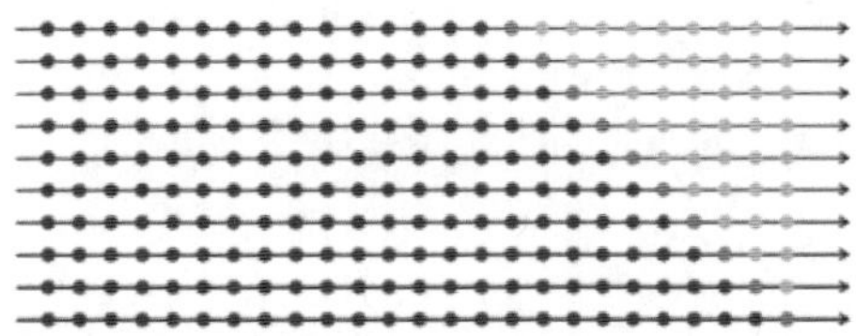

Figure 4. Process of Rolling Forecast Origin at h=1

Based on Fig 3 and Fig 4, the blue dot is re-estimated data, while the red dot is the predicted outcome for each h. Where h is the prediction of the future.

3. Methodology

The data used in this study is secondary data obtained from the Berita Resmi Statistik (BRS) Badan Pusat Statistik province of Central Java with the period from January 2007 until December 2015. In the analysis of this study, there are several steps that must be completed are as following:

(i) ARIMA modeling for each variable in level 0 and level 1 with the following steps:
  (a) Transforming or differencing if not stationary time series data.
  (b) Identify the ARIMA model based on ACF and PACF pattern.
  (c) Estimate the parameter.
  (d) Check the white noise assumption for residuals.
  (e) Getting the best model.
  (f) Calculate forecast value

(ii) RBFNN modeling for each variable in level 0 and level 1 with the following steps:
  (a) Decide the input to the model RBFNN based on ACF and PACF patterns.
  (b) Decide the mean an variance using K-means Cluster.
  (c) Getting the best criteria of some mean and variance.
  (d) Calculate forecast value.

IOP Conf. Series: Journal of Physics: Conf. Series **893** (2017) 012033 doi:10.1088/1742-6596/893/1/012033

(iii) Hybrid ARIMA-RBFNN modeling for each variable in level 0 and level 1 with the following steps:

(a) ARIMA Modeling for each variable in level 0 and level 1 and get the predicted value .
(b) Produce the Residuals ARIMA of each variable in level 0 and level 1.
(c) RBFNN Modeling for every Residuals ARIMA in level 0 dan level 1.
(d) Calculate forecast value with $\hat{Y}_t = \hat{L}_t + \hat{N}_t$

(iv) Calculating a prediction hierarchy bottom-up approach with ARIMA, RBFNN and Hybrid modeling.

4. Result and Discussion

The result of individual modelling with ARIMA, RBFNN and Hybrid ARIMA-RBFNN show in Table 1, Table 2 and Table 3. On selecting the best model, used Hierarchical Time Series Bottom-up approach with ARIMA, RBFNN and Hybrid ARIMA-RBFNN modeling. using Origin Rolling Forecast. The result is shown in Fig 5 and Table 4 is based on criteria of sMAPE.

Table 1. The Result of ARIMA Modeling

Series	ARIMA Order	Series	ARIMA Order
Total	ARIMA (2,0,0)	11	ARIMA (2,1,0)
1	ARIMA (0,1,1)	12	ARIMA (1,0,1)(0,1,1)6
2	ARIMA (1,1,1)	13	ARIMA (0,1,1)
3	ARIMA (0,1,1)	14	ARIMA (0,1,1)
4	ARIMA (0,1,1)	15	ARIMA (0,1,1)
5	ARIMA (0,1,1)	16	ARIMA (0,1,1)
6	ARIMA (0,1,1)	17	ARIMA (1[20],0,0)
7	ARIMA (1[10],1,0)	18	ARIMA (0,1,1[23])
8	ARIMA (2,0,0)	19	ARIMA (0,1,1)
9	ARIMA (0,1,1)		
10	ARIMA (1,1,1)		

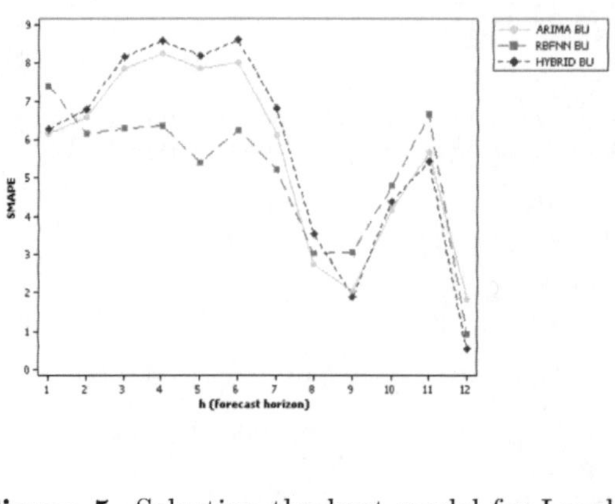

Figure 5. Selection the best model for Level 0

Table 2. The Result of RBFNN Modeling

Series	RBFNN Model	Series	RBFNN Model
Total	2 Input 5 Neuron*	11	3 Input 4 Neuron*
1	2 Input 6 Neuron*	12	2 Input 3 Neuron*
2	2 Input 2 Neuron*	13	2 Input 4 Neuron*
3	2 Input 6 Neuron*	14	2 Input 4 Neuron*
4	2 Input 4 Neuron*	15	2 Input 6 Neuron*
5	2 Input 5 Neuron*	16	2 Input 3 Neuron*
6	2 Input 3 Neuron*	17	2 Input 3 Neuron*
7	2 Input 6 Neuron*	18	2 Input 2 Neuron*
8	2 Input 3 Neuron*	19	2 Input 3 Neuron*
9	3 Input 6 Neuron*		
10	2 Input 6 Neuron*		

* Neuron in Hidden Layer

Table 3. The Result of Hybrid in RBFNN Process

Series	RBFNN Model	Series	RBFNN Model
Total	2 Input 6 Neuron*	11	2 Input 4 Neuron*
1	2 Input 6 Neuron*	12	2 Input 2 Neuron*
2	2 Input 6 Neuron*	13	2 Input 4 Neuron*
3	2 Input 6 Neuron*	14	2 Input 2 Neuron*
4	2 Input 3 Neuron*	15	2 Input 5 Neuron*
5	2 Input 2 Neuron*	16	2 Input 3 Neuron*
6	2 Input 5 Neuron*	17	2 Input 6 Neuron*
7	2 Input 4 Neuron*	18	2 Input 4 Neuron*
8	2 Input 2 Neuron*	19	2 Input 3 Neuron*
9	2 Input 6 Neuron*		
10	2 Input 3 Neuron*		

* Neuron in Hidden Layer

Based on Fig 5 and Table 4 it can be concluded that the level 0, i.e. Total Export Value, Bottom-up approach with RBFNN modeling can be used for prediction of short and medium term. While the bottom-up approach with hybrid ARIMA-RBFNN modeling can be used for long-term predictions. Overall bottom-up approach with RBFNN modeling give the best result. The next step is to predict the value of exports in Central Java province with the chosen model. The results of the predictive value of total exports Central Java is shown in Fig 6.

Based on Fig 6. it can be explained that the forcasting result coincide with the actual data. But in 12-step ahead forcast shows the result with slight fluctuations.

5. Conclussion

Based on the results and discussion, then obtained some conclusions. First based of Fig 5 and Table 4, for a total export value bottom-up approach with hybrid ARIMA-RBFNN modeling

Table 4. sMAPE value for total export value (level 0)

h	Bottom-up			Individual		
	ARIMA	RBFNN	HYBRID	ARIMA	RBFNN	HYBRID
1	6,15	7,39	6,28	6,73	9,75	6,31
2	6,58	6,15	6,78	7,50	10,86	7,31
3	7,87	6,29	8,18	8,50	13,01	8,55
4	8,25	6,37	8,59	9,73	13,01	10,03
5	7,87	5,41	8,20	9,64	12,47	9,77
6	8,03	6,24	8,61	9,81	10,12	9,81
7	6,12	5,23	6,80	7,62	8,45	7,69
8	2,71	3,03	3,54	4,47	10,72	4,39
9	2,04	3,06	1,88	2,33	10,96	1,97
10	4,16	4,79	4,39	4,76	11,88	4,46
11	5,67	6,66	5,43	5,50	13,11	5,68
12	1,82	0,91	0,54	1,51	17,43	2,06
Mean	5,61	5,13	5,77	6,51	11,81	6,50

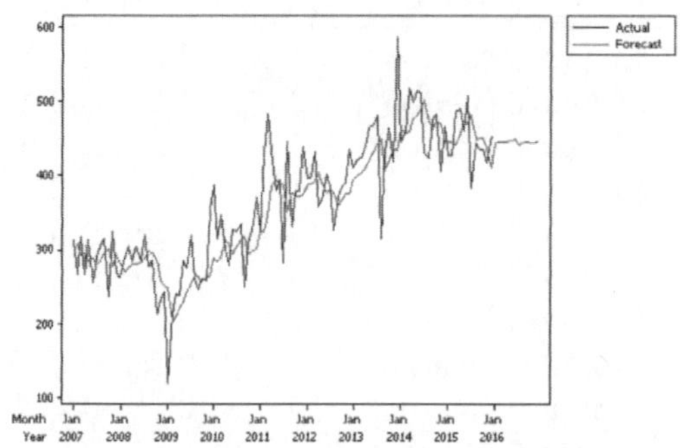

Figure 6. Predictied total export value (level 0) for 12 step ahead using Bottom-up with ARIMA modeling

provides the best results for long-term predictions, while for the short and medium term may use a bottom-up approach with RBFNN modeling. Overall bottom-up approach with RBFNN modeling give the best result. The actual data and forecast values for 12 step ahead show in Fig 6.

References

[1] Nursiah C 2011 Peranan ekspor dalam perekonomian Indonesia *Jurnal Ekonomi* **19** 1
[2] Tri W R 2012 Konsentrasi ekspor provinsi Jawa Tengah *Jurnal Ekonomi Kuantitatif Terapan* **5** pp 109–118
[3] Wei W W S 2006 *Time Series Analysis Univariate and Multivariate Method 2nd Edition* (New York: Pearson Education)
[4] Jonathan D C and Kung-Sik C 2008 *Time Series Analysisi with Applications in R* (Lowa City, Lowa: Springer)

[5] George A, Ahmed, Roman A & Rob J H 2007 *Hierarchical Forcasting for Australian Domestic Tourism*(Monash University)

[6] Lin H S and Peter W F S 2013 Grouped time-series forecasting with an application to regional infant mortality counts *Centre for Population Change Working Paper* 40

[7] Douglas C M, Cheryl J and Murat K 2008 *Introduction to Time Series Analysis and Forecasting* (Hoboken, New Jersey: Willey)

[8] Zhigang Z and Jun W 2010 *Advances in Neural Network Research and Applications* (Berlin: Springer)

[9] Grady H 2011 *Artificial Neural Networks in Biological and Environmental Analysis* (New York: CRC Press)

[10] Zhang G P 2003 Time series forecasting using a hybrid ARIMA and neural network model *Neurocomputing* **50** pp 159–175

[11] Jan G D G and Rob J H 2006 25 years of time series forecasting *Int. J. of Forescating* **22** pp 443–473

Automatics adjusment on private pension fund for Asian Mathematics Conferences

J Purwadi

Faculty of Mathematics and Sciences, Departement of Mathematics
University of Ahmad Dahlan, Janturan - Yogyakarta

E-mail: joko@math.uad.ac.id

Abstract. This paper discussed about how the automatic adjustment mechanism in the pension fund with defined benefits in case conditions beyond assumptions - assumptions that have been determined. Automatic adjustment referred to in this experiment is intended to anticipate changes in economic and demographic conditions. The method discuss in this paper are indexing life expectancy. In this paper discussed about how the methods on private pension fund and how's the impact of the change of life expectancy on benefit.

1. Introduction

Defined benefit retirement systems are often designed with mechanisms to self-adjust for many factors. For example, plans have adjusted for inflation and real wage growth over the working lifetime with a myriad of design factors. Final average defined benefit plans adjust by basing the final benefit on final pay; cash balance defined benefit plans accrue over time with interest.

However, most systems do not have factors to take into account other risks. One key risk, improvement in average longevity, will be discussed at length as it has been the focus of many self-adjusting mechanisms put into place in recent years. But other factors will be discussed as well, including mechanisms to adjust for solvency in social insurance systems in ways other than raising or lowering taxes.

Within defined benefit retirement income systems both employer provided systems and social security systems much work has been done recently to put self adjusting mechanisms in place to deal with improvements in life expectancy. Improvements in life expectancy raise the lifetime generosity of these retirement income systems, raising their costs, and placing strains on their financing. As many of these systems have wrestled with significant improvements in life expectancy in the second half of the 20th century, we have seen several models emerge for how to deal with this systematic risk.

The failure to adjust social security systems for improvements in life expectancy is one of the causes of the financing problems they are facing [3]. Traditional social security systems generally are not self-sustaining in the face of increasing life expectancy at older ages because retirees receive benefits for more years. By contrast, in defined contribution systems, life-expectancy indexing of benefits occurs automatically when benefits are annuitized using current mortality tables because increases in life expectancy reduce benefit levels. Eventually, traditional social security systems require adjustments to deal with the increased benefit cost caused by rising life expectancy. With a pay-as-you-go system, solvency can be maintained three ways: cutting benefits, raising taxes, or raising the early retirement age. Raising the early retirement age will help Social Security finances if benefits are not raised to adjust for the increase in age of eligibility [4].

Increased life expectancy raises the cost to employers of providing defined benefit pension plans. While in the short run, changes in interest rates normally dominate changes in life expectancy in their effect on pension liabilities, over a period of decades, the effect on pension costs of the increase in longevity can be considerable, while changes in interest rates generally have much less effect. The short run and long run importance of the effects on pension cost of changes in life expectancy are thus quite different. The life expectancy coefficient is a mechanism affecting the pension amount and used to hold in check the pension expenditure due to a prolonged life expectancy. The coefficient is based on the average length in life expectancy calculated on the basis of the mortality rates for the preceding five years. The coefficient is determined annually for those who have turned 65 years.

The purpose of index security for pensions is to ensure that a person retiring may receive a starting pension that is reasonable considering the income level while still working, and to ensure the future purchasing power of the pension in payment.

When determining the starting earnings-related pension, the pension provider adjusts the insureds wages and income from work during his or her career to the level of the first year of pension by the wage coefficient. Adjustment by the wage coefficient, i.e. indexing, ensures that a person who is retiring will receive a pension that is proportionate to his or her income level while still working.

Following retirement, the pension provider annually adjusts the earnings-related pension in payment at the beginning of January according to changes in the earnings-related pension index. Indexation with the earnings-related pension index secures and even improves the pensions purchasing power. The Social Insurance Institution annually adjusts national pensions in payment based on changes in the national pension index at the beginning of January. Indexation preserves the purchasing power of national pensions in relation to changes in consumer prices.

The life expectancy coefficient is defined so that the capital value of the pension adjusted with the coeficient is the same when calculated with the available mortality rates of TMI. When calculating the capital value, a two-per-cent interest rate is used.

1.1. Method

The life expectancy coefficient is calculated by dividing the longevity indicator (with 6 decimals) for the base year 2010 with the longevity indicator for year 2015 (with 6 decimals). The life expectancy coefficient is given with five (5) decimals. [1]

$$EAK = EAL_{2010}/EAL_{2010} \tag{1}$$

The longevity indicator (EAL) is calculated using the formula [1]

$$EAL = \sum(1.02^{-x+0.5-65})(L_x \div l_{65}) \tag{2}$$

in which,

$x =$ the age use in the calculation, $x = 65, 66, ..., 100$

$l_x =$ the number of pension alive at age x, $l_{65} = 1$; the number of pension alive at age $x+1$ is obtain using formula

$l_x = (1 - q_x) \cdot l_x$

$q_x =$ based on mortality table indonesia

$L_x =$ the average number of person alive at age cohorts x, $x+1$

$L_x = (l_x + l_{x+1}) \div 2$ the calculation of the life expectancy coefficient for the years 2015

Therefore, the life expectancy coefficient for 2015 is 0.93969

Table 1. Life Expectancy Coefficient 2015

x	q_x	l_x	L_x	EAL
65	0.021	1	0.98950000	0.923241
66	0.02288	0.97900000	0.96780024	0.885289
67	0.02486	0.95660048	0.944709936	0.84723
68	0.02702	0.932819392	0.92021702	0.809076
69	0.02921	0.90761461	0.894358901	0.770922
⋮	⋮	⋮	⋮	⋮
96	0.34662	0.01615405	0.013354393	0.006744
97	0.3677	0.010554734	0.008614246	0.004265
98	0.39016	0.006673758	0.005371842	0.002607
99	0.41413	0.004069925	0.003227186	0.001536
100	0.43974	0.002384447	0.001860179	0.000868

which result the EAL = 11.979234

The calculation of the life expectancy coefficient for the years 2010

Table 2. Life Expectancy Coefficient 2010

x	q_x	l_x	L_x	EAL
65	0.02415	1	0.987925	0.921772
66	0.02653	0.97585	0.96290535	0.880811
67	0.02915	0.9499607	0.936115022	0.839515
68	0.03203	0.922269345	0.907499202	0.797894
69	0.03518	0.892729058	0.877025954	0.755981
⋮	⋮	⋮	⋮	⋮
96	0.37269	0.01148307	0.009322014	0.004708
97	0.40523	0.007160957	0.00571004	0.002827
98	0.43542	0.004259123	0.003331869	0.001617
99	0.46687	0.002404615	0.001843294	0.000877
100	0.49945	0.001281973	0.000640986	0.000299

which result the EAL = 11.256785

1.2. Pension Fund

In this section will be the notion of Pension Funds, various pension funds, Benefit Functions is a function that is used to determine the amount of benefit that an employee will receive on retirement, resignation, disability, and death during the active period. The formula for the benefit is;

$$B_x = \sum b_x \qquad (3)$$

Normal Cost is an annual fee imposed on those who are still active which is determined using an actuarial cost method. The formula for the normal cost is;

$$^{\tau}(NC)_x = b_x p_x^{(T)} v^{(r-x)} \qquad (4)$$

some actuarial funding methods used in the pension program, has a different actuarial liability. In general Actuarial Liability can be expressed as follows :

$$^{\tau}(AL)_x = B_x p_x^{(T)} v^{(r-x)} \ddot{a} \tag{5}$$

Next will be discuss about the total unfunded Liability plans in the year t is defined as the difference between the asset and the actuarial liabilities denoted as follows:

$$ul(t) = AL(t) - f(t) \tag{6}$$

Annuity function is a series of periodic payments, there are two types of functions, namely annuities Life Annuity function which is a series of periodic payments in which each payment is only made if the designated person is still alive at the time payment is due.

$$\ddot{a} = \sum p_x^m v^t \tag{7}$$

And the fixed annuity function which is a series of periodic payments that are carried out within a certain period. There are two types of fixed annuity, immediate and due.[2]

$$a_{\bar{n}|} = (1 - v^n) \div i \tag{8}$$

and

$$\ddot{a}_{\bar{n}|} = (1 - v^n) \div d \tag{9}$$

1.2.1. Muhammadiyah Pension Fund The Muhammadiyah pension fund uses the defined benefit scheme, the result of life coefficient indexation on benefit in muhammadiyah pension fund.

Table 3. Life Expectancy indexation on benefit

Birth	Retire	$\ddot{a}_{65}$	B	R*	E(R)
1950	2015	16.1	-1%	65	18 y, 6 m
1955	2020	16.4	-2%	65 and 3 m	18 y, 6 m
1960	2025	16.8	-4%	65 and 4 m	18 y, 8 m
1965	2030	16.9	-6%	65 and 8 m	19 y, 2 m
1970	2035	17.1	-9%	66 and 1 m	19 y, 3 m
1975	2040	17.6	-11%	66 and 4 m	19 y, 4 m
1980	2045	17.8	-12%	66 and 7 m	19 y, 5 m
1985	2050	17.9	-12%	66 and 11 m	19 y, 6 m
1990	2055	18.0	-13%	67	19 y, 8 m
1995	2060	18.1	-13%	67 and 2 m	19 y, 10 m
2000	2065	18.4	-14%	67 and 3 m	19 y, 11 m

note : y = years ; m = months

As we can see on Table 3, the annuity factor at age 65 differ according to the year of birth. The change in Benefit (B) because of the the change of life expectancy also can be seen here. The R* on Table 3 is the retirement age required to neutralize the effect of increased life expectancy and the E(R) is the implying the expected length of retirement.

2. Conclusion

Automatics adjustment on private pension sector due to the change of demographics and economics can be solved by adjusting the life expectancy coefficient. The research result produce the life expectancy coefficient for 2015 is 0.93969, which mean that the capital value of the pension adjusted with the coefficient. The consequences because of change in the life expectance coefficient made the benefit also change according to the year of birth, the retirement age needs to be neutralize and also the expected retirement length the result can be seen on Table 3.

Acknowledgments

The authors would like to thanks to Ahmad Dahlan University and Muhammadiyah Pension Fund

References

[1] Jukka A 2014 *The Life Expectancy Coefficient For 2015 According To The Employees Pensions Act* (Finnish Centre For Pension)
[2] Stephen G K 1991 *The Teory of Interest 2nd Edition* (USA: Irwin McGraw-Hill)
[3] Edward P 2000 The Swedish pension reform model: Framework and issues *World Bank Social Protection Discussion Paper* 0012
[4] Turner J A 2007 Autopilot: Self-adjusting mechanisms for sustainable retirement systems *Pension Research Council J.*

IOP Conf. Series: Journal of Physics: Conf. Series **893** (2017) 012035 doi:10.1088/1742-6596/893/1/012035

An application of almost marginal conditional stochastic dominance (AMCSD) on forming efficient portfolios

Isnandar Slamet, Siska Mardiana Putri Carissa, and Hasih Pratiwi

Study Program of Statistics, Sebelas Maret University, Surakarta, Indonesia 57126

E-mail: isnandar06@yahoo.com, siskacarissa@gmail.com, hasihpratiwi@gmail.com

Abstract. Investors always seek an efficient portfolio which is a portfolio that has a maximum return on specific risk or minimal risk on specific return. Almost marginal conditional stochastic dominance (AMCSD) criteria can be used to form the efficient portfolio. The aim of this research is to apply the AMCSD criteria to form an efficient portfolio of bank shares listed in the LQ-45. This criteria is used when there are areas that do not meet the criteria of marginal conditional stochastic dominance (MCSD). On the other words, this criteria can be derived from quotient of areas that violate the MCSD criteria with the area that violate and not violate the MCSD criteria. Based on the data bank stocks listed on LQ-45, it can be stated that there are 38 efficient portfolios of 420 portfolios where each portfolio comprises of 4 stocks and 315 efficient portfolios of 1710 portfolios with each of portfolio has 3 stocks.

1. Introduction

Investments are made on investment objects or assets, one of which is stock. A collection of the assets is called portfolio. In fact, investors prefer an efficient portfolio. The portfolio is said to be efficient if it has maximum return on a specific risk or minimal risk in specific return. Therefore, many methods have been developed to obtain the efficient portfolio.

In 1994, Shalit and Yitzhaki [6] introduced the marginal criteria conditional stochastic dominance (MCSD) to solve the problem of second order stochastic criteria in identifying efficiency portfolio. MCSD criteria states the probabilistic conditions under which all risk averse individuals. Given a portfolio of asset, investors prefer to increase the share of one risky asset over the other. Furthermore, Denuit et al. [3] introduced criteria almost marginal conditional stochastic dominance (AMCSD) by replacing the utility function of second order stochastic dominance used in MCSD with utility functions of almost second order stochastic dominance. AMCSD criteria is used when there is violation area which violates the rule of MCSD. The value of AMCSD criteria is ratio of area which violates the rule of MCSD with area which violates and/or not violates the rule of MCSD. AMCSD criteria is therefore used to determine the efficiency portfolio and this provides solutions if portfolio is not efficient (Denuit et al. [3]).

The banking sector is the foundation of economy. Some stocks of banking sector are listed in the LQ−45. These stocks are the stocks of Bank BRI, Bank BTN, Bank Mandiri, Bank BCA, and Bank BNI. Stocks listed in the LQ−45 is the best stocks because they have good liquidity (BEI [1]). In fact, investors are interested in investing the stocks with good liquidity (Levy [4]).

2. Research Method

In this research, we apply the AMCSD criteria to form efficient portfolios which consist of banking stocks in the LQ−45. The steps are as follows. Firstly, we apply the AMCSD criteria to determine the area which violates the rule of MCSD. Next, we determine the value of ε which is ratio of area which violates the rule of MCSD with area which violates and/or not violates the rule of MCSD. For the application of AMCSD, the first step is to calculate the stock returns and the portfolio return. After that, we calculate the value of Absolute Concentration Curve (ACC) for every stocks. The next step is to calculate the value of ε of any comparison of two stocks in a portfolio. Finally, we conclude whether the portfolio is efficient or not.

3. Almost Marginal Conditional Stochastic Dominance (AMCSD)

Almost marginal conditional stochastic dominance (AMCSD) is a decision-making criteria when there is violation of MCSD criteria. Utility functions used in AMCSD is utility function $U_2{}^*(\varepsilon_2)$, defined by

$$U_2{}^*(\varepsilon_2) = \{u : u^{'}(x) \geq 0, u^{''}(x) \leq 0, u^{'}(x) \leq inf\{-u^{''}(x)\}[\tfrac{1}{\varepsilon_2} - 1], \forall x \epsilon S\}.$$

Conditional expected return of R_i when the portfolio's return equals r_+ is defined by $E[R_i|R_+ = r_+]$. Furthermore, this is denoted by $\mu_i(r_+)$. Suppose we take two assets in the portfolio i.e asset k and asset j. Asset k dominates asset j by MCSD if $ACC_k{}^\alpha(\xi) \geq ACC_j{}^\alpha(\xi)$. The value of $ACC_i{}^\alpha(\xi)$ is denoted by $ACC_i{}^\alpha(\xi) = \int_{-\infty}^{F_{R_+}^{-1}(\xi)} \mu_i(r_+)f(r_+)dr$. Then,

$$ACC_i^\alpha(\xi) = E[R_i|R_+ \leq F_{R_+}^{-1}(\xi)]f(r_+), \tag{1}$$

where $F_{R_+}^{-1}(\xi) = inf\{r_+ \epsilon R_+ | F_{R_+}(r_+) \geq (\xi)\}$. If return is bounded in interval $[a, b]$, the difference of ACC's value of asset k and asset j is defined by

$$B(t) = \int_a^t (\mu_k(r_+) - \mu_j(r_+)).f(r_+)dr_+. \tag{2}$$

If asset k dominate asset j by MCSD then we get $B(t) \geq 0$.

Given the condition that asset k does not dominate asset j. because for the area L_2, it can be found that $ACC_k{}^\alpha(\xi) \leq ACC_j{}^\alpha(\xi)$ so that the criria of $MCSD$ can not be used. Suppose Ω is defined as a set of areas that violate the rules of MCSD as $\Omega = \{t\epsilon[a, b]|B(t) < 0\}$. The area L_2 is an area which violate the MCSD's criteria because $B(t) < 0$. The area of L_2 can be obtained as

$$L_2 = \int_\Omega -B(t)dt. \tag{3}$$

Furthermore, the total area that lies between the curve of ACC's asset k and the curve of ACC's asset j is written as

$$(L_1 + L_3) + L_2 = \int_a^b |B(t)|dt. \tag{4}$$

The ratio of area which violates the rule of MCSD with area which violates or not violates the rule of MCSD is denoted by ε. Therefore, ε can be expressed as

$$\varepsilon = \frac{\int_\Omega -B(t)dt}{\int_a^b |B(t)|dt}. \tag{5}$$

In the economy, the critical areas of AMCSD's criteria is 0,032 (Levy *et al.* [5]).

Based on Shalit and Yitzhaki [6], if there is no assets dominates in portfolio, then the portfolio is an efficient portfolio. However, if there is asset k which dominates asset j by AMCSD in the portfolio with $k \neq j$, then the portfolio is not an efficient potfolio.

4. APPLICATION OF AMCSD ON FORMING EFFICIENT PORTFOLIOS

The data used in this researh are the monthly stocks price data of the stocks of Bank BMRI, BBCA, BBRI, BBNI, and BBTN respectively. The data taken are data from January 2011 to October 2015. In this research, the portfolios formed consists of 5 stocks, 4 stocks and 3 stocks. Then, AMCSD will be used to analyze portfolio with any portion of stocks. For the illustration of calculation, we calculated the BBRI's stock returns in February 2011 using Brigham and Houston [2]. The return is

$$R(t) = \frac{5.800 - 5.850{,}66}{5.850{,}66} \times 100\% = 0{,}8659\%.$$

In this example, the BMRI's stock price in January 2011 is $5.850, 66$ and BMRI 's stock price in February 2011 is 5.800. Random test was performed according to Wackerly et al. [7]. With a significance level of $0, 05$, it was obtained that stock returns of BMRI, BBCA, BBRI, BBNI, and BBTN are random. Jarque-Bera normality test was performed with a significance level of 0, 05 based on the stock returns data and it obtained that the stock returns data of BMRI, BBCA, BBRI, BBNI, and BBTN are not normally distributed. Therefore, AMCSD criteria can be used.

In this research, the lowest portion of stocks in a portfolio consists of 5 stocks, 4 stocks, and 3 stocks were at 10%, 10%, and 5%, respectively. The portfolios formed are 126 portfolios that consist of 5 stocks with different portion of each stock. There are 5 combinations stocks that can be formed which consist of 4 stocks. Each combination consist of 84 portfolios with different portions, so that the number of portfolios which has 4 stocks is 420 portfolios. There are 10 combinations of portfolios that consists of 3 stocks. Each combination consists of 171 portfolios with different portions, so that the number of portfolios which has 3 stocks is 1710 portfolios. For example, portfolio that has 3 stocks is a portfolio consisting of 70% BMRI's stock, 5% of BBRI's stock, and 25% of BBNI's stock. Furthermore, these portfolios will be analyzed whether the portfolios are efficient or not. Returns of the portfolio for February 2011 can be calculated, and the result is

$$R_+ = 70\%.(-0.8659\%) + 5\%.(-3.0928\%) + 25\%.(10.0775\%) = 1.7586\%.$$

To calculate the ACC of BBRI when the cumulative probability of portfolio's return is $0, 0702$, we use the equation of (1), and we obtained $-0, 8951$.

It is known that $E[R_{BBRI}] \geq E[R_{BBNI}] \geq E[R_{BMRI}]$. Therefore, we just need to compare between stocks of BBRI with BBNI, BBRI with BMRI, and BBNI with BMRI. Then, $B(t)$ was calculated using equation (2). In comparison of stock BBRI with BBNI, the value of $B(t)$ is negative which means that there is violation of MCSD's criteria. Furthermore, the value of $\varepsilon_{BBRI.BBNI}$ was calculated using equation (5) and the result is $0, 1580$. The same way, the value of $\varepsilon_{BBRI.BMRI}$ and $\varepsilon_{BBNI.BMRI}$ are obtained at $0, 0181$ and $0, 1266$, respectively. The value of $\varepsilon_{BBRI.BMRI}$ is less than $0, 032$ so that it can be said that BBRI's stock dominates BMRI's stock by AMCSD. The portfolio can be said to be an inefficient portfolio.

Table 1 shows five examples of portfolios with each portfolio consists of stock BMRI, BBRI, and BBNI. In similar way, there are no efficient portfolios out of 126 portfolios which each portfolio has 5 stocks. Furthermore, there are 38 efficient portfolios out of 420 portfolios which each portfolio of 38 portfolios consists of BMRI, BBRI, BBNI, and BBTN. Then, there are 315 efficient portfolios out of 1710 portfolios. Out of 315 efficient portfolios, there are 119 efficient portfolios with each portfolio consist of BMRI's stock, BBRI's stock,and BBNI's stock. There are 614 efficient portfolios with each portfolio consists of BMRI's stock, BBRI's stock, and

Table 1. The Analysis of Portfolio.

No	BMRI	BBRI	BBNI	$\varepsilon_{BBRI.}$ BBNI	$\varepsilon_{BBRI.}$ BMRI	$\varepsilon_{BBNI.}$ BMRI	Result
1	90	5	5	0, 3892	0, 0066	0, 0524	not
2	85	10	5	0, 4704	0, 0177	0, 0526	not
3	80	15	5	0, 6120	0, 0210	0, 0526	not
4	75	20	5	0, 6467	0, 0392	0, 0576	Eficient
5	70	25	5	0, 7018	0, 0823	0, 0786	Eficient

BBTN's stock, and there are 76 efficient portfolios with each portfolio consists of BMRI's stock, BBNI's stock, BBTN's stock. There are 59 efficient portfolios with each portfolio consists of BBRI's stock, BBNI's stock, and BBTN's stock.

5. Conclusion

Based on the discussion, it can be concluded that:

(i) The criteria of almost marginal conditional stochastic dominance is given by

$\int_\Omega -B(t)dt \leq \varepsilon \int_a^b |B(t)|dt$, where $0 < \varepsilon < 0.032$.

(ii) Based on the banking stocks data listed on the LQ−45, there are 38 efficient portfolios out of 420 portfolios with each portfolio has 4 stocks and 315 efficient portfolios out of 1710 portfolios with each portfolio has 3 stocks.

References

[1] BEI 2010 Panduan Index 2010 (Jakarta: Bursa Efek Indonesia)
[2] Brigham E F and Houston J F 2007 *Fundamental of Financial Management* (USA: Thomson Rights)
[3] Denuit M M, Huang R J, Tzeng L Y and Wang C W 2014 Almost marginal conditional stochastic dominance *J. Bank. Financ.* **41** pp 57–56
[4] Levy H 2012 Almost tochastic dominance and efficient investment sets *Amer.J.Oper.Res.* **2**
[5] Levy H 2010 Economically relevant preferences for all observed epsilon *An.Oper.Res.* **176**
[6] Shalit H and Yitzhaki S 1994 Marginal conditional stochastic dominance *Manag.Sci.* **40** pp 670–684
[7] Wackerly D D, Mandenhall III W and R L Sceaffer 2008 *Mathemathical Statistics with Applictions* (USA : Thomson Brooks/Cole)

Bayesian mixture modeling for blood sugar levels of diabetes mellitus patients (case study in RSUD Saiful Anwar Malang Indonesia)

Ani Budi Astuti[1], **Nur Iriawan**[2], **Irhamah**[2], **Heri Kuswanto**[2], **and Laksmi Sasiarini**[3]

[1] Department of Mathematics, Statistics Study Program, Faculty of Mathematics and Natural Sciences, University of Brawijaya Malang and Department of Statistics, Faculty of Mathematics and Natural Sciences, Institut Teknologi Sepuluh Nopember Surabaya

[2] Department of Statistics, Faculty of Mathematics and Natural Sciences, Institut Teknologi Sepuluh Nopember Surabaya

[3] Faculty of Medicine, University of Brawijaya Malang

E-mail: ani_budi@ub.ac.id

Abstract. Bayesian statistics proposes an approach that is very flexible in the number of samples and distribution of data. Bayesian Mixture Model (BMM) is a Bayesian approach for multimodal models. Diabetes Mellitus (DM) is more commonly known in the Indonesian community as sweet pee. This disease is one type of chronic non-communicable diseases but it is very dangerous to humans because of the effects of other diseases complications caused. WHO reports in 2013 showed DM disease was ranked 6th in the world as the leading causes of human death. In Indonesia, DM disease continues to increase over time. These research would be studied patterns and would be built the BMM models of the DM data through simulation studies where the simulation data built on cases of blood sugar levels of DM patients in RSUD Saiful Anwar Malang. The results have been successfully demonstrated pattern of distribution of the DM data which has a normal mixture distribution. The BMM models have succeed to accommodate the real condition of the DM data based on the data driven concept.

1. Introduction

When a set of real data observed have a very small amount and a multimodal distribution then it is necessary need specially handling. The Bayesian statistics proposes an approach that is very flexible in the number of samples and distribution of data. These approach is based directly on the posterior distribution of the data. Therefore Bayesian approach can accommodate the real condition of the data based on the data driven concept ([1], [2], [3], [4], and [5]). Bayesian Mixture Model (BMM) is one model that uses a Bayesian approach for multimodal models and its parameter models is seen as random variables in the model parameter space [6].

Various studies have previously been carried out by BMM modeling. The modeling of BMM for the DM data in Surabaya is the research that has been done by [7] and the research of [8] has been conducted for the DM data in Malang by BMMA modeling. Some of the BMM research in the other field, among others [9] and [10]. Modeling of cases is very important in order to know the exact model and can be used for decision making in accordance with the cases observed [11].

One disease that is not contagious but it is very dangerous and deadly to humans is Diabetes Mellitus (DM). These disease can cause complications of other very dangerous diseases so the handling costs are very expensive. According to the WHO report in 2013, DM was ranked as the world's sixth largest as the leading causes of death [12]. The impact of the DM disease is very dangerous because it can lead to death of the patient and the number of patients continues to increase sharply over time. A person is said to have the disease if the blood sugar levels had reached more than 200 mg/dL for the condition is not fasting, more than 126 mg/dL for fasting conditions and the average blood sugar levels exceeding 150 mg/dL ([7], [13], and [14]).

A study of the blood sugar levels in DM patients data modeling is very important because it can be known the exact model for policy making in the management of disease control and prevention of DM diseases. Challenge of the DM data is to have a mixture distribution. The studies of [8] and [15] indicates that the DM data have a mixture normal distribution. These paper would be studied patterns of DM data and would be built the BMM models of the DM data through simulation studies where the simulation data built on cases of blood sugar levels of DM patients in RSUD Saiful Anwar Malang.

2. Materials and Methods

The data used in this study is simulation data which is constructed based on the cases of blood sugar levels of DM patients in RSUD Saiful Anwar Malang for Lowokwaru District in 2015. In this study, there is 7 villages of Lowokwaru District analyzed. To construct simulation data, we need information on real data distribution and value ranges of data from each village in the Lowokwaru District. Based on the information that has been obtained, the data generation is done as much as the number of real data and is repeated 100 times. The detection of the number of mixture components using RJMCMC algorithms and to build BMM models using MCMC with Gibbs Sampler approach by Winbugs program. Finally, to detect feasibility of this model with the Kolmogorov-Smirnov criterion. In this study hypothesized that normal mixture models (BMM-Normal) proposed as a model for the DM cases in Malang. In building a BMM-Normal model using the R software. The following are descriptions of the DM data DM in the Lowokwaru district of Malang.

3. Bayesian Analysis, BMM Analysis, and Kolmogorov-Smirnov

3.1. Bayesian Analysis

The posterior probability distribution model has become the basis of Bayesian statistical analysis method. These distribution is a blend of the prior information and the likelihood of the data. Thomas Bayes had found this concept in 1702-1761, where in model parameter, $\theta \in \Omega$, is treated as a random variable ([16], [8], [16], [17], [18] and [19]).

If $f(x \mid \theta)$ is a likelihood function of observational data x and $p(\theta)$ is the prior θ then $p(\theta \mid x)$ is posterior probability distribution of θ is, determined by the rules of probability according to Bayes' theorem [1] as in Equation 1.

$$
\begin{aligned}
p(\theta \mid x) &= \frac{f(x \mid \theta)p(\theta)}{f(x)} \text{ where,} \qquad (1)\\
f(x) &= E[f(x \mid \theta)] = \int_{\theta \in R} f(x \mid \theta) f(\theta) d\theta \text{ if } \\
&\qquad \theta \text{ continuous and} \\
f(x) &= E[f(x \mid \theta)] = \sum_{\theta \in B} f(x \mid \theta) p(\theta) \text{ if } \theta \text{ discrete,}
\end{aligned}
$$

where $f(x)$ is a normalized constant [6]. So that 1 can be written as:

$$p(\theta \mid x) \propto f(x \mid \theta)p(\theta). \qquad (2)$$

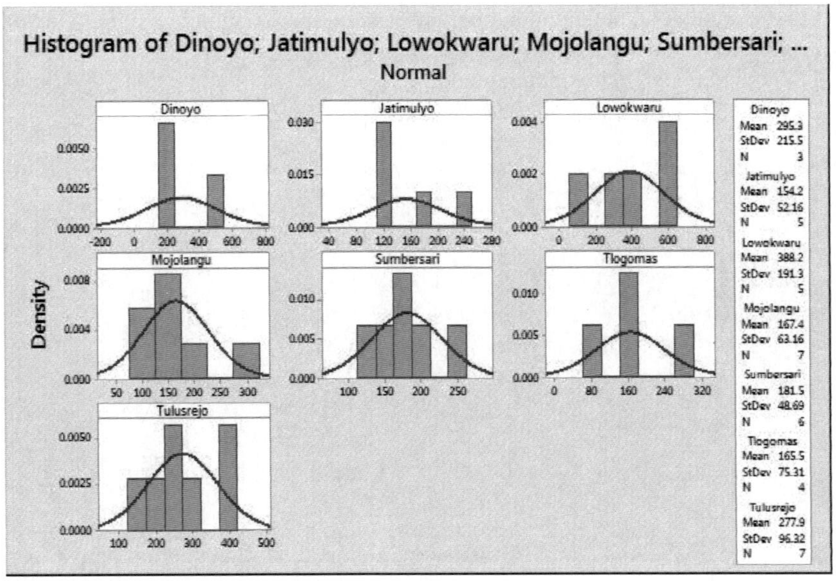

Figure 1. Descriptions of blood sugar levels of DM patients data in Lowokwaru District of Malang

Derived from the Equation Equation 2, the posterior probability that will be used for decision making is proportional to the product of the likelihood function and the prior probability of the model parameters ([1], [20], [21], [22], [23], and [24]).

3.2. BMM Analysis

3.2.1. Mixture Models Mixture model is a special model for the data that have multimodal distribution, where each proportion of sub-group is a constituent component of the mixture models. The mixture model called the particular model because the model is able to combine some different data but still retains the characteristics of the original data ([6], [8], [24], [25], and [26]).

Mixture probability function of an observation $\mathbf{x} = (x_1, x_2, \ldots, x_n)$ taken from a number of k-subgroup can be expressed as in Equation 3 ([6], [8], [18], [24], and [26]).

$$f(\mathbf{x} \mid \theta, \mathbf{w}) = \sum_{j=1}^{k} w_j g_j(\mathbf{x} \mid \theta_j), \tag{3}$$

where $f(\mathbf{x} \mid \theta, \mathbf{w})$ is a function of the probability mixture, $g_j(\mathbf{x} \mid \theta_j)$ is a jth probability function of k number of sub-group, and θ is a mixture model parameters containing of $(\theta_1, \theta_2, \ldots, \theta_k)$. Parameter θ_j, $j = 1, 2, \ldots, k$ represents the characteristic distribution of g_j on each component in mixture models. While $\mathbf{w}$ is the parameter vector of proportions (weighted) mixture model containing of $(w_1, w_2, \ldots, w_k)$, where $0 < w_j < 1$, $\forall j$ and

$$\textstyle\sum_{j=1}^{k} = 1 \text{ for each model parameter} \theta_j.$$

3.2.2. BMM Models In the mixture model, each observation x_i would be classified on each unknown number of sub-group. If the allocation of each observation on each sub-group in

Equation 3 is denoted by z, then the allocation of each observation $z_i, i = 1, 2, \ldots, n$ could be determined based on Equation Equation 4 ([8], [24], and [27]).

$$p(z_i = j) = w_j, j = 1, 2, \ldots, k. \tag{4}$$

Given the value of z_i then the observation data x_i can be derived from the sub-group as in 5.

$$x_i \mid z_i \sim f(x \mid \theta_{z_i}), i = 1, 2, \ldots, n. \tag{5}$$

Thus the resulting joint posterior distribution of all parameter in the mixture model can be expressed as in Equation 6.

$$\begin{aligned} p(k, w, z, \theta, x) &= p(k)p(w \mid k)p(z \mid w, k)p(\theta \mid z, w, k) \\ & \quad p(x \mid \theta, z, w, k) \end{aligned} \tag{6}$$

Furthermore, to estimate each parameter in Equation 6 by employing the full conditional distribution of each parameter [27].

Suppose $p(\theta)$ is a stationary distribution of parameter θ. The full conditional distribution of parameter θ is constructed by making a partition of θ as shown in Equation 7 ([8], [10], and [24]).

$$\theta = (\theta_s, \theta_{-s}), \tag{7}$$

where θ_s denote the-s^{th} parameter to be estimated and θ_{-s} denote the complement of θ_s that is parameter θ without the-s^{th} component. Therefore the full conditional distribution can be established based on the joint distribution of all the parameters as in Equation 8 ([8], [10], and [28]).

$$p(\theta_s \mid \theta_{-s}) = \frac{p(\theta_s, \theta_{-s})}{\int p(\theta_s, \theta_{-s}) d\theta_s}. \tag{8}$$

3.2.3. RJMCMC Algorithm Reversible Jump Markov Chain Monte Carlo (RJMCMC) concept can be used to identify the number of mixture components model in which the number of mixture components is unknown. In its application, RJMCMC algorithm uses the concept of birth/death and split/merge with 6 types of movement, that is $\mathbf{w}$ updating, θ updating, $\mathbf{z}$ updating, hyperparameter β updating, split/merge components, and birth/death of an empty component. The movement split/merge is a random choice between split $(k \to k+1)$ or merge $(k+1 \to k)$ ([8], [10], and [27]). This method is more flexible to use than the Bayes Factor method.

3.2.4. MCMC Alogarithm The posterior distribution in the Bayesian analysis is often very complicate and requires a difficult integration process in determining the marginal posterior of a model parameter. Therefore it takes a numerical approximation with Markov Chain Monte Carlo (MCMC) algorithm ([3], [8], [23], and [29]). MCMC algorithms with Gibbs Sampler approach can be described as follows:

MCMC Algorithm with Gibbs Sampler Approach

3.3. Kolmogorov-Smirnov (KS).

The concept of Kolmogorov-Smirnov test is to compare the empirical cumulative distribution function (CDF), $F_n(x_i)$, and the hypothesis cumulative distribution function, $\hat{F}(x)$. If statistic

Step 1. We will provide initial value.
$\theta^{(0)} = \left(\theta_1^{(0)}, \ldots, \theta_r^{(0)}\right)$

Step 2. We will do the sampling of the parameter.
Generate the value of $\theta_j, j = 1, \ldots, r$ from their conditional distribution as follows:

Step 2.1. Sampling $\theta_1^{(k+1)}$ from $p\left(\theta_1 \mid \mathbf{x}, \theta_2^{(k)}, \ldots, \theta_r^{(k)}\right)$

Step 2.2. Sampling $\theta_2^{(k+1)}$ from $p\left(\theta_2 \mid \mathbf{x}, \theta_1^{(k+1)}, \theta_3^{(k)}, \ldots, \theta_r^{(k)}\right)$

$\vdots$

Step 2.r. Sampling $\theta_r^{(k+1)}$ from $p\left(\theta_r \mid \mathbf{x}, \theta_1^{(k+1)}, \theta_2^{(k+1)}, \ldots, \theta_{r-1}^{(k+1)}\right)$

Step 3. We will do iteration.
Execute step 2 as K times with $K \to \infty$

order $X_{(1)}, X_{(2)}, \ldots, X_{(n)}$ is the independent random variable with hypothesis distribution, $\hat{F}(x)$ and $F_n(x_i)$ is the empirical distribution as in Equation 9,

$$F_n(x_i) = \frac{\text{the number of data } X_i \leq x_i}{n} \quad \text{for i = 1, 2, 3, \ldots, k ¡ n,} \tag{9}$$

where $F_n(x_i)$ is right continuous step function then the formula of test statistic D_n can be written as in Equation 10.

$$D_n = \sup\left\{|F_n(x) - \hat{F}(x)|\right\} \tag{10}$$

In this case, D_n is the greatest distance between $F_n(x) - \hat{F}(x)$. If the value of D_n smaller than the tested models is better ([8], [23] and [30]).

4. Result and Discussion

4.1. Description of the Data Analysis

Description of blood sugar levels of DM patients data for Lowokwaru District in Malang 2015 can be seen that in the Fig. 2.

According to Fig. 2 can be indicated that the data have a mixture distribution (multimodal distribution). These can be seen by the existence of different groupings of the data. The next step will be identified distribution data from each village. In this step, the proposed hypothesis is as follows:

H_0 : data has a normal distribution
versus
H_1 : data don't have a normal distribution
More identification test results are shown in TABLE I.

Derived from TABLE I can be known that the data have a normal distribution for all 7 villages in Lowokwaru District Malang because P-value for each village is larger than 0.05, so that the null hypothesis is accepted. Therefore based on Fig. 2 and TABLE I can be indicated that the data have a normal mixture distribution.

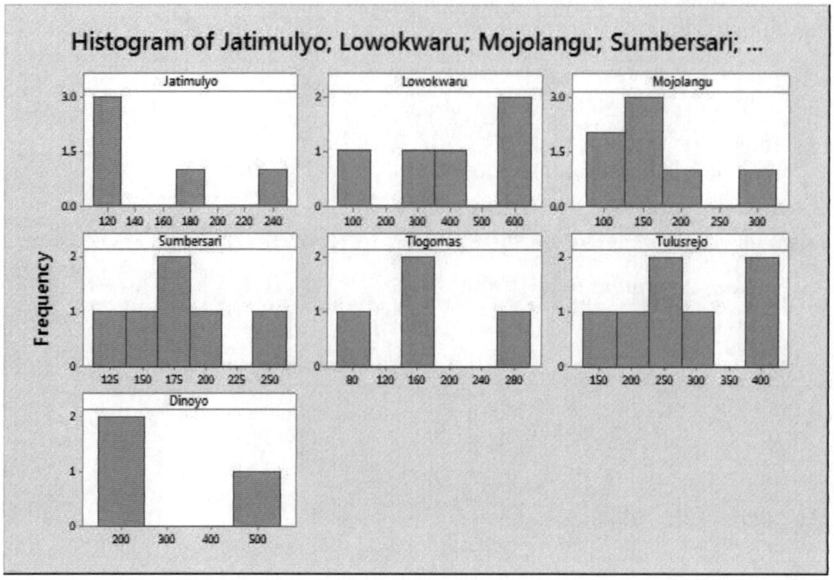

Figure 2. Histogram of blood sugar levels of DM patients data for Lowokwaru District in Malang

4.2. Identification for the Number of Components Mixture by RJMCMC Algorithm.

The identification results for the number of components mixture in the normal mixture distribution by RJMCMC algorithm can be seen that in the TABLE I and TABLE II.

Based on Table I and Table II can be seen that the mixture normal distribution with 2 components for all villages in Lowokwaru District is the best model because P-value of k2 (with k-Maks= 2) from RJMCMC algorithm is the biggest. This means that the exact model to be built for the DM cases in the Lowokwaru District of Malang is a normal mixture model with 2 components. There are two groups of DM patients in the Lowokwaru district of Malang that each group has a normal distribution.

Table 1. Normality test for DM data in the Lowokwaru District

No	Villages	Data	P-value	Result
1.	Tulusrejo	Original	0.9222	Normal Distribution
		100 times Simulation		Normal Distribution
2.	Tlogomas	Original	0.9139	Normal Distribution
		100 times Simulation		Normal Distribution
3.	Sumbersari	Original	0.9704	Normal Distribution
		100 times Simulation		Normal Distribution
4.	Mojolangu	Original	0.7532	Normal Distribution
		100 times Simulation		Normal Distribution
5.	Lowokwaru	Original	0.9551	Normal Distribution
		100 times Simulation		Normal Distribution
6.	Jatimulyo	Original	0.6418	Normal Distribution
		100 times Simulation		Normal Distribution
7.	Dinoyo	Original	0.7304	Normal Distribution
		100 times Simulation		Normal Distribution

4.3. Weighting of the BMM Models

In TABLE III are presented the results of the identification for weighted (w) values in each of mixture components on the BMM modeling. The weighting values of the first mixture component are w1 and the weighting values of the second mixture component are w2. This weighting is obtained by the use of MCMC algorithms with Gibbs Sampler approach for normal mixture distribution through Winbugs program with 3 steps execution.

Based on the weighting values shown in TABLE III it can be seen that the villages of Sumbersari, Lowokwaru, Tulusrejo, and Mojolangu have the second weighting value of mixture components (w2) is greater than the first weighting value of mixture components. Its means that in the four villages have had the proportion of DM patients with higher blood sugar levels more than the number of DM patients with lower blood sugar levels. As for the three other villages namely Tlogomas, Jatimulyo, and Dinoyo have w2 lower than w1. However, the average weighted values, Lowokwaru District has a weighted value of the second mixture components (w2) higher than a weighted value of the first mixture components (w1).

4.4. BMM Models and Goodness of Fits BMM Models by KS

BMM models and Goodness of Fits BMM models by KS for all villages in Lowokwaru District Malang can be shown in the Fig. 3 and TABLE IV.

Derived from Fig. 2, it can be shown that the BMM models for DM cases in the Lowokwaru

Table 2. RJMCMC Algorithm for DM data in the Lowokwaru District

No	Villages	Data	P-value of k2	k-Maks	Result
1.	Tulusrejo	Original	0.3839	2	Mixture Normal Distribution with 2 Components
		100 times Simulation (average)	0.381437		Mixture Normal Distribution with 2 Components
2.	Tlogomas	Original	0.4015	2	Mixture Normal Distribution with 2 Components
		100 times Simulation (average)	0.394409		Mixture Normal Distribution with 2 Components
3.	Sumbersari	Original	0.3216	2	Mixture Normal Distribution with 2 Components
		100 times Simulation (average)	0.334833		Mixture Normal Distribution with 2 Components
4.	Mojolangu	Original	0.347	2	Mixture Normal Distribution with 2 Components
		100 times Simulation (average)	0.348434		Mixture Normal Distribution with 2 Components
5.	Lowokwaru	Original	0.3727	2	Mixture Normal Distribution with 2 Components
		100 times Simulation (average)	0.373755		Mixture Normal Distribution with 2 Components
6.	Jatimulyo	Original	0.3649	2	Mixture Normal Distribution with 2 Components
		100 times Simulation (average)	0.367506		Mixture Normal Distribution with 2 Components
7.	Dinoyo	Original	0.4868	2	Mixture Normal Distribution with 2 Components
		100 times Simulation (average)	0.488151		Mixture Normal Distribution with 2 Components

IOP Conf. Series: Journal of Physics: Conf. Series **893** (2017) 012036 doi:10.1088/1742-6596/893/1/012036

Table 3. The Weighting Values of the BMM Models

No	Villages	Data	w1-value	w2-value
1	Tulusrejo	Original	0.3384	0.6616
		100 times simulation (Average)	0.3384	0.6616
2	Tlogomas	Original	0.5041	0.4959
		100 times simulation (Average)	0.504	0.496
3	Sumbersari	Original	0.5041	0.4959
		100 times simulation (Average)	0.2544	0.7456
4	Mojolangu	Original	0.4527	0.5473
		100 times simulation (Average)	0.4527	0.5473
5	Lowokwaru	Original	0.2935	0.7065
		100 times simulation (Average)	0.2935	0.7065
6	Jatimulyo	Original	0.5687	0.4313
		100 times simulation (Average)	0.5687	0.4313
7	Dinoyo	Original	0.6058	0.3942
		100 times simulation (Average)	0.6058	0.3942
		Average of w	**0.448914286**	**0.551085714**

District in Malang have a normal mixture distribution with two components, namely DM patients with average blood sugar levels less than 200 mg/dL and greater than or equal to 200 mg/dL. Each village has a different characteristic of the normal mixture distribution that vary according to the condition data. Based on the shape of the distribution and the value of the blood sugar levels of DM patients in the Lowokwaru District, the villages that need attention from the Local Government in Malang respectively are Lowokwaru, Tulusrejo, Mojolangu, Sumbersari, Dinoyo, Tlogomas, and Jatimulyo. From TABLE IV it can be seen that the average value of KS as Goodness of Fits BMM models is 0.35739667. Its means that the BMM models that have been built have high accuracy and the BMM models are able to explain the condition of the real data with data driven concept. Derived from the exact model that has been constructed for DM cases, it can be utilized for the Department of Health and related agencies in the prevention and control of the DM diseases.

5. Conslusion

The BMM models for cases of the blood sugar levels of DM patients in RSUD Saiful Anwar Malang in the Lowokwaru District have normal mixture distribution with 2 components, namely DM patients with average blood sugar levels less than 200 mg/dL and greater than or equal to

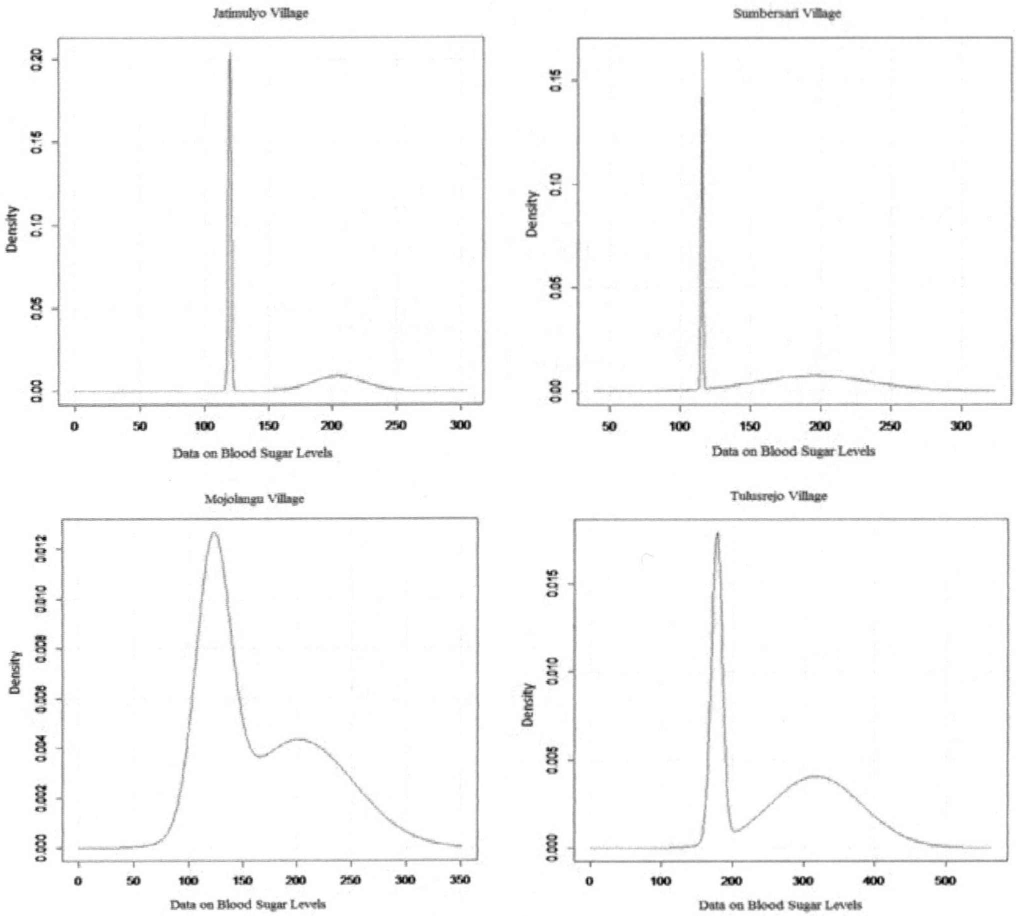

Figure 3. The BMM Models for Some Villages at Lowokwaru District in Malang

200 mg/dL with the proportion of the second group is greater than the first group, respectively. Villages that need attention from Local Government in Malang are Lowokwaru, Tulusrejo, Mojolangu, Sumbersari, Dinoyo, Tlogomas, and Jatimulyo, respectively. The BMM-Normal model for Lowokwaru District has an average KS value of 0.35739667 so that these models have high accuracy and able to accommodate the real condition of the DM data with driven data concept.

Acknowledgement
This paper is part of research Grants PUPT/BOPTN University of Brawijaya in 2015-2016 and the part of Doctoral Research at Department of Statistics, Institut Teknologi Sepuluh Nopember (ITS), Surabaya, Indonesia. We would like the first thank to Directorate General of Higher Education and University of Brawijaya which have financed this research, and the second thank to RSUD Saiful Anwar Malang for allowing use of research data and thank to anonymous reviewer to this paper.

Table 4. The Goodness of Fits BMM Models by KS

No	Villages	Data	KS-value
1.	Tulusrejo	Original	0.226512714
		100 times simulation (Average)	0.223355145
2.	Tlogomas	Original	0.300381216
		100 times simulation (Average)	0.292200553
3.	Sumbersari	Original	0.396896742
		100 times simulation (Average)	0.40181093
4.	Mojolangu	Original	0.214107821
		100 times simulation (Average)	0.221674647
5.	Lowokwaru	Original	0.518545269
		100 times simulation (Average)	0.527494706
6.	Jatimulyo	Original	0.35239443
		100 times simulation (Average)	0.365280553
7.	Dinoyo	Original	0.381005007
		100 times simulation (Average)	0.381131639
	Average KS Simulation		**0.35739667**

References

[1] Gosh J K, Delampady M and Samanta T, 2006 *An Introduction to Bayesian Analysis Theory and Method* (New York: Springer)

[2] Hoeting J A, Madigan D, Raftery A E and Volinsky C T 1999 Bayesian model averaging: tutorial (with discussion), *Stat.Sci.*, **14** pp 382–401

[3] Iriawan N 2000 *Computationally Intensive Approaches to Inference in Neo-Normal Linear Models* (Australia: CUT).

[4] Iriawan N 2012 *Modeling and Analysis of Data-Driven* (in Indonesian) Vol.1 (Surabaya: ITS)

[5] Sebastiani P, Xie H and Ramoni M F 2006 Bayesian analysis of comparative microarray experiments by model averaging, *Internat.Soc.Bayesian.Anal.* **1** pp 707–732

[6] Gelman A, Carlin J B, Stern H S and Rubin D B 1995 *Bayesian Data Analysis* (London: Chapman & Hall)

[7] Suyono S, *Textbook of Medicine* (in Indonesian) Vol.3 4ed (Jakarta: FKUI) (2006)

[8] Astuti A B, Iriawan N, Irhamah, Kuswanto H, and Sasiarini L 2016 Blood sugar levels of diabetes mellitus patients modeling with bayesian mixture model averaging *Glob. J. Pure Appl. Math.* **12** pp 3143–3158

[9] Do K A, Muller P and Tang F 2005 *A Bayesian Mixture Model For Differential Gene Expression* (Texas: Research in Department of Biostatistics, The University of Texas)

[10] Wati D M 2006 *Analysis of Bayesian Markov Chain Monte Carlo in Normal Mixture Modeling with Unknown Number of Components (Case Study: Rate of Inflation in Indonesia)* (in Indonesian) (Surabaya: Statistics Department, ITS)

[11] Astuti A B 2014 Cox regression model for lifetime of a newborn baby (neonates) in Dr. Saiful Anwar Malang Hospital *Glob. J. Pure Appl. Math.* **10** pp 591–600.

[12] http://www.tempo.co/read/news/2013/11/14/060529766/ Indonesian-Log-10-Big-State-with- Diabetes-Sufferers, (in Indonesian).
[13] http://medicastore.com/diabetes/terapidia, Treatment of Diabetes Mellitus. (in Indonesian).
[14] Smeltzer S and Bare B 2008 *Textbook of MedicalSurgical Nursing* 11ed (Philadelphia: Lippincott)
[15] Suharjo B 2003 *Implementation of Bayesian Markov Chain Monte Carlo in Normal Mixture Distributions* (in Indonesian) (Surabaya: Statistics Department, ITS)
[16] Aitkin M 2001 Likelihood and bayesian analysis of mixtures *Stat.Model.* **1** (2001) pp 287–304
[17] Box G E P and Tiao 1973 *Bayesian Inference in Statistical Analysis* (Boston MA: Addison-Wesley, Massachusetts)
[18] Iriawan N 2001 Univariable normal mixture model estimation: a bayesian methods approach with MCMC (in Indonesian) *Proc. of National Seminar and Konferda VII on Mathematics in DIY & Central Java Region* pp 105–110
[19] Iriawan N 2003 *Teaching Modules of Simulation Technique* (in Indonesian) (Surabaya: ITS)
[20] Astuti A B, Iriawan N, Irhamah and Kuswanto H 2014 Model components selection in bayesian model averaging using occams window for microarray data, *Proc. of* 4th *Annual Basic Science International Conference 2014 in conjunction with The* 5th *International Conference on Global Resource Conservation 2014*
[21] Astuti A B, Iriawan N, Irhamah and Kuswanto H 2014 Kolmogorov-smirnov and continuous ranked probability score validation on the bayesian model averaging for microarray data *Appl. Math. Sci.* **146** pp 7277–7287
[22] Astuti A B, Iriawan N, Irhamah, and Kuswanto H 2015 Occams window selection in bayesian model averaging modeling for gene expression data from chickpea plant *Int.J.Appl.Mat.Stat.* **53** pp 160–165.
[23] Astuti A B, Iriawan N and Sasiarini L 2015 Bayesian model averaging with occams window selection components model on diabetes mellitus data (in Indonesia) *SEMIRATA 2015 MIPA BKS-PTN West* University of Tanjungpura, Pontianak.
[24] Astuti A B, Iriawan N, Irhamah, and Kuswanto H 2015 An algorithm for determining the number of mixture components on the bayesian mixture model averaging for microarray data *J.Mat.Stat.* **11** pp 45–51
[25] Astuti E Y 2006 *Implementation of the Bayesian Markov Chain Monte Carlo in Optimal Portfolio Modeling with Mixture Model Approach of Multiple Mixture* (in Indonesian) (Surabaya: Statistics Department, ITS)
[26] McLachlan G J and Basford K E 1988 *Mixture Models Inference and Applications to Clustering* (New York: Marcel Dekker)
[27] Richardson S and Green P J 1997 On Bayesian analysis of mixtures with an unknown number of components (with discussion) *J. Roy. Statist. Soc* **59** pp 731–792
[28] Gilks W R 1995 *Full Conditional Distribution, in Markov Chain Monte Carlo in Practise.* (eds) (London: Gilks, W. R., Richardson, S. and Spiegelhalter, D. J. Chapman & Hall) pp 75–88
[29] Gamerman D 1997 *Markov Chain Monte Carlo* (London: Chapman & Hall)
[30] Hustianda V F 2012 *Comparison of Bayesian Model Averaging and Multiple Linear Regression in Predicting Factors Affecting Number of Infant Death in East Java, (in Indonesian)* (Surabaya: Statistics Department, ITS)

The application of cat swarm optimisation algorithm in classifying small loan performance

Eka N Kencana[1], Nyoman Kiswanti[2], and Kartika Sari[2]

[1] Tourism Research Consortium, University of Udayana – Denpasar 80234, Indonesia
[2] Department of Mathematics, Udayana University – Jimbaran, Badung 80361 - Indonesia

E-mail: i.putu.enk@unud.ac.id, mangkisz@gmail.com, sari_kaartika@yahoo.co.id

Abstract. It is common for banking system to analyse the feasibility of credit application before its approval. Although this process has been carefully done, there is no warranty that all credits will be repaid smoothly. This study aimed to know the accuracy of Cat Swarm Optimisation (CSO) algorithm in classifying small loans' performance that is approved by Bank Rakyat Indonesia (BRI), one of several public banks in Indonesia. Data collected from 200 lenders were used in this work. The data matrix consists of 9 independent variables that represent profile of the credit, and one categorical dependent variable reflects credit's performance. Prior to the analyses, data was divided into two data subset with equal size. Ordinal logistic regression (OLR) procedure is applied for the first subset and gave 3 out of 9 independent variables i.e. the amount of credit, credit's period, and income per month of lender proved significantly affect credit performance. By using significantly estimated parameters from OLR procedure as the initial values for observations at the second subset, CSO procedure started. This procedure gave 76 percent of classification accuracy of credit performance, slightly better compared to 64 percent resulted from OLR procedure.

1. Introduction

Extracting valuable information such as recognising pattern from hidden data is one aim of data mining technique. This technique belongs to multidisciplinary field that intersects statistical methods, machine learning, pattern recognition, and data visualisation. One purpose in data mining technique is classifying objects based on their attribute(s). The classification problem can be veiewed as a problem to classify objects into a set of clusters without any *priori* knowledge [1].

To classify an object into its appropriate group can be done by several methods. Among these methods, K-means algorithm is very popular. Despite of its popularity, refers to [2], the major drawback of this algorithm is frequently found it gets stuck at local minima and classification results is heavily depended on the choice of initial cluster center. To overcome this limitation, some alternative techniques in classifying objects were developed. Metaheuristic algorithms such as genetic algorithm as well as particle swarm optimisation had been operationalised to classify objects regarding their attribute(s) [3][4].

This work implements Cat Swarm Optimisation (CSO) algorithm in classifying small loans performance were approved by Bank Rakyat Indonesia, one of several public bank in Indonesia. CSO is belonged to swarm intelligence method that models cats behavior in solving optimisation problem [5]. Researchers had been applied this technique to solve various problem dimensions, but to our best knowledge, CSO is rarely used to classify financial data. In addition, to elaborate its performance, we used Ordinal Logistic Regression (OLR) to predicts loan's classification from the loan's attributes.

Chu and Tsai [5] propose CSO as a new metaheuristic technique in optimisation problems by observing the behavior of cats. Basically, CSO algorithm consists of two states of cats, namely (a) seeking mode, represents cats during a resting period but being alert, and (b) tracing mode, represents mode of cats while tracing their target. According to [5], cats which are awake spend most of their time resting and spend very little time to chase target. The proportion on cats in tracing mode is defined as mixture ratio (MR).

2. Research Method

The small loan data from 200 lenders for this work were acquired from Bank Rakyat Indonesia, Melati branch office at Denpasar. The data consists of 9 independent variables represent credit's profile and one categorical dependent variable represents the performance of credit. The operational definition of the variables is listed on Table 1.

Prior to analysis stage, the data is randomly divided into two subset with equal size. In order to make fair comparison, before OLR and CSO algorithm were applied, we check the proportion of loans are classified as smoothly paid, less smoothly, and doubtful in each of data subset. The maximum proportion differences is allowed for same classification between subset is five percent. The first subset is analysed with OLR technique to identify influential variables in determining performance of credit, and the second subset is analysed using CSO algorithm with number of dimensions to move is determined by the number of influential variables that is identified with the OLR. The last stage is to compare the classification accuracy is resulted from both techniques. Classification accuracy of both methods were counted through the Apparent Correct Classification Rate (ACCR) measure [6], which is stated by equation

$$ACCR = \frac{\sum_{i=1}^{c} n_i}{N} \tag{1}$$

In equation (1), N represents total observation numbers, n_i is the number of observations were correctly classified in class i, and c represents number of class.

3. Results and Discussion

3.1. OLR Analysis

OLR is an extension of binary logistic regression (BLR). OLR as well as BLR utilised maximum likelihood method to estimate the parameters of regression model [5]. Using STATA to analyse the first subset data, we obtained three out of nine predictor variables significantly affect small loan performance. The OLR's result is listed on Table 2. The Nagelkerke *pseudo* R^2 showed the variance of loan's performance can be explained by its predictors as much as 61.5 per cent while the rest is unexplained.

The number of small loan performance were predicted correctly by using this OLR model as smoothly, less smoothly, and doubtful credits are 34, 24, and 6 observation respectively. By applying eq. (1), the classification accuracy is resulted from OLR method as much as 64 per cent.

3.2. CSO Analysis

Arbitrary, we create 30 cats in our work. In initialisation stage, these cats were randomly assigned between tracing and seeking mode according to MR value. As aforementioned explain, in seeking mode where cats are in resting period, four factors have to be considered and defined as follows [6]:

Table 1. List of Small Loan Variables

Code	*Description*	*Measurement*	*Values*
PER	Loan performance	Categorical	1: Smoothly paid 2: Less smoothly 3: Doubtful
X1	Amount of loan	Continuous	-
X2	Business type	Categorical	1: Services 2: Trading
X3	Loan's period	Continuous	-
X4	Lender's education	Categorical	1: Elementary 2: Junior school 3: High school 4: Undergraduate
X5	Income per month	Continuous	-
X6	Business' duration	Continuous	-
X7	Number of children	Continuous	-
X8	Home ownership	Categorical	1: Owned 2: Family 3: Rent
X9	Lender's age	Continuous	-

Table 2. The OLR's Estimators and Their Significances

Code	*Estimate*	*Wald's Statistic*	*p-Values*
Constant 1	-1.901	0.505	0.478
Constant 2	0.572	0.046	0.830
X1	0.972	6.874	0.009
X2	-0.406	0.441	0.507
X3	0.025	0.872	0.351
X4	0.035	0.020	0.887
X5	-5.743	10.955	0.001
X6	-0.589	16.171	0.000
X7	0.196	1.299	0.254
X8	0.044	0.038	0.845
X9	0.001	0.000	0.985

(i) Seeking Memory Pool (SMP), represents the seek's memory size of each cat, and denotes the number of neighboring positions are considered by a cat;

(ii) Count of Dimension to Change (CDC), represents number of dimension will be mutated;

(iii) Seeking Range (SR), represents mutative ratios for the selected dimension; and

(iv) Self-position Consideration (SPC), is a Boolean variable which its value indicates the current position of cat is considered as a new neighbor position. If SPC $= 1$ then current position is viewed as a new neighbor position, otherwise cat should consider a different position.

The computational steps of cat in seeking mode can be listed as follows [5][6]:

(i) Create n copies of current position of cat_i, n = SMP. If SPC = 1 then $n = n - 1$;

(ii) For each of dimension according to CDC's value, replace the SR's old value for each cat with randomly plus or minus x percent of the old value;

(iii) Randomly select point to move to from the candidates points and replace the position of cat_i.

Once cat_i goes into tracing mode, its steps can be listed as follows:

(i) Update its velocities for every dimension to consider (CDC) according to equation (2)

$$v_{i,d} = v_{i,d} + r_1.c_1(x_{best,d} - x_{i,d}) \tag{2}$$

where $x_{best,d}$ and $x_{i,d}$ are the position of cat with best fitness value and position of cat_i on dimension d, respectively; c_1 is an arbitrary constant and r_1 is a random value in the range [0, 1].

(ii) Update the position of cat_i at dimension d ($x_{i,d}$) according to equation (3)

$$x_{i,d} = x_{i,d} + v_{i,d} \tag{3}$$

3.3. CSO Result

To make fair comparison with OLR technique, we position each cat in five dimensional space (R^5). These dimensions represents two constants and three significant variables according to OLR analysis. The values from OLS analysis were applied as initial positions in each dimensions of cat. In applying CSO algorithm to classify the loan performance, we set the minimum, increment, and maximum values for CSO's parameters as follows:

Parameters	*Value(Min; Increment; Max)*
MR	(0; 0.03; 1)
SR	[0; 0.10; 1]
CDC	[0; 0.20; 1]
SMP	[1; 1; 10]

After 2000 iterations, for 30 cats and using $c_1 = 2$ as suggested by [6] we found the best parameter for MR, SR, SMP, and CDC values in our work are 0.20, 0.50, 3, and 0.20 respectively. Refers to Table 3, the classification accuracy result from CSO method as much as 76 percent, slightly greater than 64 percent was obtained from OLR technique.

4. Conclusion

The application of CSO algorithm in classifying small loan performance by utilizing the significantly estimated parameters from OLR technique proved can increase the classification accuracy. This finding is inline with similar experiment result conducted by [7] in classifying gene expression data. Kumar and Mishra utilised CSO method to update the gene's weight values resulted from Principal Component Analysis (PCA) and Factor Analysis (FA) for high dimensionality of the datasets.

As a metaheuristic algorithm, CSO and other swarm optimisation techniques were known as an efficient and effective global search tehnique. However, CSO has never been investigated its

Table 3. Classification Accuracy

Actual Group	*Predicted Group*			*Total*
	Smoothly	Less Smoothly	Doubtful	
Smoothly	36	1	0	37
Less Smoothly	11	24	3	38
Doubtful	1	8	16	25
Total	48	33	19	100

performance according to the characteristics of object under studied. We showed by eliminating unsignificant variables in small loan classification process, the classification accuracy increase.

Despite of its performance to classify small loan data is better than OLR method, CSO as well as other particle swarm optimisation techniques, its theoretical background is rather weak. Refers to [8], PSO algorithm has a limited condition for the particle(s) move stable from the point of math's.

References

[1] Liu Y and Shen Y 2010 Data clustering with cat swarm optimization *J. Convergence Information Technology* **5** (8) pp 21-28.

[2] Selim S Z and Ismail M A 1984 K-means-type algorithm: generalized convergence theorem and characterization of local optimality *IEEE Transactions on Pattern Analysis and Machine Intelligence* **6** (1) pp.81-87.

[3] Laszlo M and Mukherjee S 2006 A genetic algorithm using hyper-quadtrees for low-dimensional k-means clustering *IEEE Transactions on Pattern Analysis and Machine Intelligence* **28** (4) pp. 533-543.

[4] Kao Y T, Zahara E and Kao I W 2008 A hybridized approach to data clustering *Expert Systems with Applications* **34** (3) pp. 1754-62.

[5] Chu S C and Tsai P W 2007 Computational intelligence based on the behavior of cats *International Journal of Innovative Computing, Information and Control* **3** (1) pp. 163-173.

[6] Hosmer D and Lemeshow S 2000 *Applied Logistic Regression* 2nd ed (New York, USA: John Wiley & Sons Inc.).

[7] Kumar A and Mishra D 2013 Cat swarm based optimization of gene expression data classification international *Journal of Computer Trends and Technology* **4** (5) pp. 1185-90.

[8] Clerc M and Kennedy J 2002 The particle swarm - explosion, stability, and convergence in a multidimensional complex space *IEEE Transactions on Evolutionary Computation* **6** (1) pp. 58-73.

IOP Conf. Series: Journal of Physics: Conf. Series **893** (2017) 012038 doi:10.1088/1742-6596/893/1/012038

Generalized space time autoregressive with exogenous variable model and its application

Dewi Astuti[1], Budi Nurani Ruchjana[2], and Soemartini[3]

[1] Central Bureau of Statistics, West Sumatera, Indonesia
[2] Department of Mathematics, Faculty of Mathematics and Natural Sciences Universitas Padjadjaran, Indonesia
[3] Department of Statistics, Faculty of Mathematics and Natural Sciences Universitas Padjadjaran, Indonesia

E-mail: dewiastuti@bps.go.id

Abstract. In this paper we proposed the Generalized Space Time Autoregressive with variable Exogenous, abbreviated GSTARX as GSTAR development with the addition of exogenous variables. GSTARX not only involves the element of time and location, but also the influence of exogenous variables in the model. GSTARX equation can be written as a linear model, so we can estimate parameters of GSTARX model using Ordinary Least Squares (OLS) method. For our case study, we use GSTARX model with uniform and inverse distance weights to predict an export volume of Crude Palm Oil (CPO) in several locations on the island of Sumatera, where X is the international CPO prices.

1. Introduction

Time series data were recorded simultaneously in multiple locations called space-time data, data with time and location simultaneously observation. We can use multivariate time series model [5] for space time phenomena, such as VAR, STAR, and GSTAR models [1]. Nowadays, space-time model, not only influenced by previous observations at the same location and previous observations in different location, or there are not only time and location dependencies, but also there are some other things that affect, which can be expressed as an exogenous variable. The development of multivariate time series model for the space time data involving exogenous variables are VARX, STARX and GSTARX. GSTARX model itself is a special form of VARX model that combine time and location and involves an exogenous variables [3].

CPO export volume data recorded simultaneously in multiple locations, especially at province in Sumatera island, Indonesia. In this study, the space time data, CPO export volume in North Sumatera, Riau and West Sumatera, are not only influenced by the previous data of CPO export volume in the same province, and the previous ones in different provinces, but there are also other factors that influence, expressed in an exogenous variable, that is international CPO prices, we called it X.

2. Literature Study

2.1. GSTARX Model

Generalized Space Time Autoregressive with Exogenous Variable (GSTARX) model is the development of GSTAR model [4] with the addition of an exogenous variables in the model.

GSTARX model with order time p and order location λ_k, GSTARX(p, λ_k) for k = 1,2, ..., N could be written as:

$$Z_{(N\times1)}(t) = \Big(\sum_{k=1}^{p}\sum_{l=0}^{\lambda_k} \Phi_{kl(N\times N)} W^{(l)}_{(N\times N)} Z_{(N\times1)}(t-k)\Big) + \gamma X(t) + a_{(Nx1)}(t)$$

where $\Phi_{kl(N\times N)}$ and $\gamma_{(N\times N)}$ where are the parameters of GSTARX model, and $a_{(N\times1)}(t) \sim$ iid $N(0, \sigma^2 I)$.

For example, GSTARX model with Autoregressive's order is 1 and spatial's order is also 1, GSTARX(1;1), with location $N = 3$, in matrix notation could be written as:

$$\begin{bmatrix}\dot{Z}_1(t)\\ \dot{Z}_2(t)\\ \dot{Z}_3(t)\end{bmatrix} = \begin{bmatrix}\phi_{10} & 0 & 0\\ 0 & \phi_{20} & 0\\ 0 & 0 & \phi_{30}\end{bmatrix}\begin{bmatrix}\dot{Z}_1(t-1)\\ \dot{Z}_2(t-1)\\ \dot{Z}_3(t-1)\end{bmatrix} + \begin{bmatrix}\phi_{11} & 0 & 0\\ 0 & \phi_{21} & 0\\ 0 & 0 & \phi_{31}\end{bmatrix}\begin{bmatrix}0 & w_{12} & w_{13}\\ w_{21} & 0 & w_{23}\\ w_{31} & w_{32} & 0\end{bmatrix}\begin{bmatrix}\dot{Z}_1(t-1)\\ \dot{Z}_2(t-1)\\ \dot{Z}_3(t-1)\end{bmatrix}$$
$$+ \begin{bmatrix}\gamma_1 & 0 & 0\\ 0 & \gamma_2 & 0\\ 0 & 0 & \gamma_3\end{bmatrix}\begin{bmatrix}X_1(t)\\ X_2(t)\\ X_3(t)\end{bmatrix} + \begin{bmatrix}a_1(t)\\ a_2(t)\\ a_3(t)\end{bmatrix} \tag{1}$$

The parameter of GSTARX(1,1) model could be estimated using ordinary least square method (OLS). In order to satisfy the equation of a linear model, so the GSTARX(1,1) restructured in to the equation[4]:

$$\begin{bmatrix}\dot{Z}_1(t)\\ \dot{Z}_2(t)\\ \dot{Z}_3(t)\end{bmatrix} = \begin{bmatrix}\dot{Z}_1(t) & 0 & 0 & V_1(t-1) & 0 & 0\dot{X}_1(t) & 0 & 0\\ 0 & \dot{Z}_2(t) & 0 & 0 & V_2(t-1) & 0 & 0\dot{X}_2(t) & 0\\ 0 & 0 & \dot{Z}_3(t) & 0 & 0 & V_3(t-1) & 0 & 0\dot{X}_3(t)\end{bmatrix}\begin{bmatrix}\phi_{10}\\ \phi_{20}\\ \phi_{30}\\ \phi_{11}\\ \phi_{21}\\ \phi_{31}\\ \gamma_1\\ \gamma_2\\ \gamma_3\end{bmatrix} + \begin{bmatrix}a_1(t)\\ a_2(t)\\ a_3(t)\end{bmatrix}$$

2.2. Uniform Weight for GSTARX (1,1) Model

Uniform weight gives equal weight to each location. Therefore, this weight often used on data that have same distance between locations. The uniform weight is defined as:

$$w_{ij} = \frac{1}{n_i} \tag{2}$$

where n_i is the number of the location where are located near to location i [4].

For example, for three different locations uniform weight matrix could be written as:

$$\begin{bmatrix}0 & w_{12} & w_{13}\\ w_{21} & 0 & w_{23}\\ w_{31} & w_{32} & 0\end{bmatrix}$$

Because there are three locations, so the number of location where are located near to location i are two locations. Thus, the uniform weighting matrix w_{ij} could be elaborated into:

$$\begin{bmatrix}0 & \frac{1}{2} & \frac{1}{2}\\ \frac{1}{2} & 0 & \frac{1}{2}\\ \frac{1}{2} & \frac{1}{2} & 0\end{bmatrix}$$

IOP Conf. Series: Journal of Physics: Conf. Series **893** (2017) 012038 doi:10.1088/1742-6596/893/1/012038

2.3. Inverse Distance Weight for GSTARX (1,1) Model

Inverse Distance weight for GSTARX model refers to the actual distance between location at the same region, and then normalized. Normalization is conducted to satisfy the condition the condition $\sum_{i \neq j} w_{ij}^{(1)} = 1$. Suppose the distance between three locations are defined as follows:

r_1 : the distance between location 1 and 2
r_2 : the distance between location 1 and 3
r_3 : the distance between location 2 and 3

so that the matrix of inverse distance weight is obtained [1]:

$$w_{ij} = \begin{bmatrix} 0 & w_{12} & w_{13} \\ w_{21} & 0 & w_{23} \\ w_{31} & w_{32} & 0 \end{bmatrix} = \begin{bmatrix} 0 & \frac{1/r_1}{1/r_1+1/r_2} & \frac{1/r_2}{1/r_1+1/r_2} \\ \frac{1/r_1}{1/r_1+1/r_3} & 0 & \frac{1/r_3}{1/r_1+1/r_3} \\ \frac{1/r_2}{1/r_2+1/r_3} & \frac{1/r_3}{1/r_2+1/r_3} & 0 \end{bmatrix} = \begin{bmatrix} 0 & \frac{r_2}{r_1+r_2} & \frac{r_1}{r_1+r_2} \\ \frac{r_3}{r_1+r_3} & 0 & \frac{r_1}{r_1+r_3} \\ \frac{r_3}{r_2+r_1} & \frac{r_2}{r_2+r_3} & 0 \end{bmatrix}$$

3. Main Result

3.1. Descriptive Analysis

In case study, we use CPO export volume (HS 1511100000) in three locations in Sumatra, namely North Sumatera, Riau, and West Sumatra, from January 2004 until August 2015. From the plot of time series data [2], the CPO export volume movement in these three locations tend to be similar.

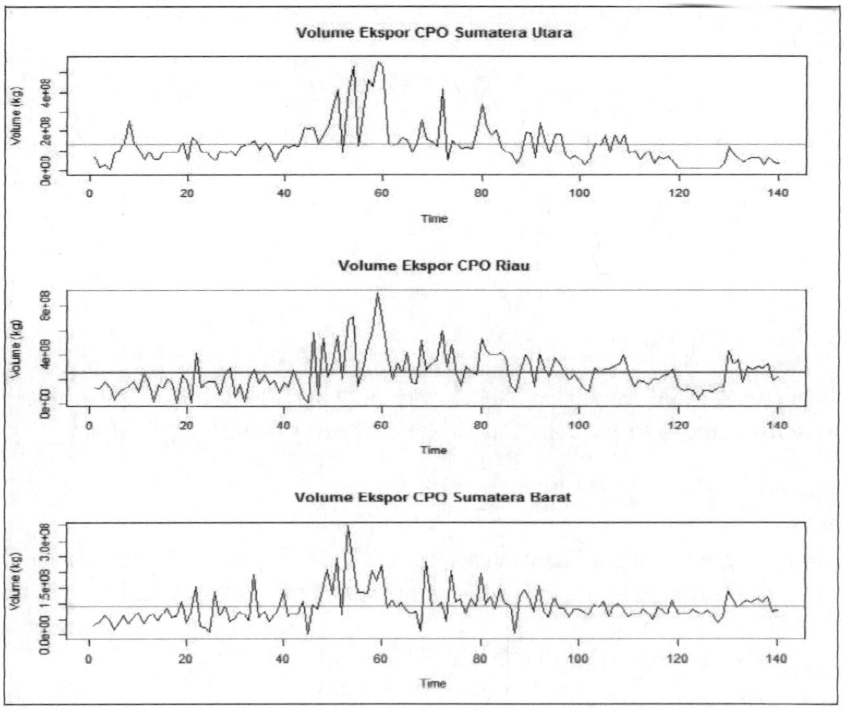

Figure 1. Plot of time series data, CPO export volume, January 2004 - August 2015

Identification process of CPO export volume data in these three locations by using the Pearson correlation between observation at three locations. Table 1 presents the correlation between three of location of the CPO expert volume. We can see in Table 1 that the CPO export volume data between locations are correlated. The correlation coefficient were high enough.

IOP Conf. Series: Journal of Physics: Conf. Series **893** (2017) 012038 doi:10.1088/1742-6596/893/1/012038

Table 1. Correlation of CPO Export Volume between Locations

	North Sumatera	Riau	West Sumatera
North Sumatera	1	0.740 (0.000)	0.620 (0.000)
Riau	0.740 (0.000)	1	0.663 (0.000)
West Sumatera	0.620 (0.000)	0.663 (0.000)	1

Time series data analysis performed by dividing the data into two part, in sample data and out sample data. For each province, the data from January 2004 until April 2013 is used as the in sample data, 112 observations. Meanwhile the out sample data is the monthly data from May 2013 until August 2015, 28 observations.

In further analysis, to get the stationary data in mean level, so the data has been centralized, it means that the initial data is reduced by the average $(Z_t - \mu)$ for each variable.

3.2. The GSTARX Model for CPO Export Volume

Before performing the analysis of time series data, both univariate and multivariate, the main requirements that must be fulfilled is stationarity. Visually, the stationarity of the data in the mean can be seen from ACF plot. ACF plot rapidly towards zero indicates that the data is already stationary. To make sure of the assessment stationarity data visually, we perform stationarity test with Augmented Dickey Fuller (ADF) test for CPO export volume data at each location. The ADF test results of the CPO export volume data in three locations, in Table II, shows that the p-value is less than 0.01. So it was decided to reject H_0. It can be concluded that the data did not contain a unit root, or the data is already stationary.

Table 2. ADF Test of Centralized CPO Export Volume Data

	Location	**ADF**	***p*-value**
1.	North Sumatera	-6.1278	< 0.01
2.	Riau	-7.7209	< 0.01
3.	West Sumatera	-8.9654	< 0.01

For determining the order of GSTARX model can be performed by the smallest AIC value. The AIC minimum criteria in order selection are generally used in multivariate time series model, because the order of each univariate variables can be different. The AIC minimum criteria gives AIC value of each order so that it can be compared, and then determined the lag time that is appropriate for the model GSTARX.

Table 3. AIC Value of CPO Export Volume Data

Lag	MA (0)	MA (1)	MA (2)	MA (3)	MA (4)	MA (5)
AR(0)	25.9649	26.1296	26.1759	26.0679	26.1022	25.9553
AR(1)	25.6925	25.6900	25.7019	25.7312	25.8558	25.7579
AR(2)	25.7295	25.7103	25.8613	25.8804	25.9711	25.8974
AR(3)	**25.6517**	25.7339	25.8986	26.0571	26.1685	26.0727
AR(4)	25.8115	25.8399	25.9797	26.1059	26.2075	26.1999
AR(5)	25.8379	25.8427	26.0064	26.2842	26.1022	25.9553

Table 3 shows that the smallest AIC value lies in AR (3) and MA (0). The identification results of AR order is used to model GSTARX is the order of 3. For the spatial order, order one is chosen, because all of the three of locations are in one area at Sumatera island. So the GSTARX model can be formed as GSTARX(3,1).

3.3. GSTARX(3,1) Model with Uniform Weight Matrix

The uniform weight matrix in modeling GSTARX on CPO export volume data assumes that the CPO export volume in one location have the same effect on the CPO export volume in other locations. In this study, the weight applied uniformly, because the three locations are in one area. Uniform weight matrix in this study can be written as follows:

$$W_{ij} = \begin{bmatrix} 0 & 0.5000 & 0.5000 \\ 0.5000 & 0 & 0.5000 \\ 0.5000 & 0.5000 & 0 \end{bmatrix}$$

GSTARX(3,1) model for CPO export volume data in the three locations has the equation as follows:

$$\begin{bmatrix} \hat{Z}_1(t) \\ \hat{Z}_2(t) \\ \hat{Z}_3(t) \end{bmatrix} = \sum_{k=1}^{2} \left[\begin{bmatrix} \phi_{10}^k & 0 & 0 \\ 0 & \phi_{20}^k & 0 \\ 0 & 0 & \phi_{30}^k \end{bmatrix} + \begin{bmatrix} \phi_{11}^k & 0 & 0 \\ 0 & \phi_{21}^k & 0 \\ 0 & 0 & \phi_{31}^k \end{bmatrix} \begin{bmatrix} 0 & 0.5386 & 0.4614 \\ 0.3051 & 0 & 0.6949 \\ 0.2734 & 0.7266 & 0 \end{bmatrix} \right] \begin{bmatrix} Z_1(t-k) \\ Z_2(t-k) \\ Z_3(t-k) \end{bmatrix}$$

$$+ \begin{bmatrix} \gamma_1 & 0 & 0 \\ 0 & \gamma_2 & 0 \\ 0 & 0 & \gamma_3 \end{bmatrix} \begin{bmatrix} X_1(t) \\ X_2(t) \\ X_3(t) \end{bmatrix} + \begin{bmatrix} a_1(t) \\ a_2(t) \\ a_3(t) \end{bmatrix}$$

Parameter estimation of GSTARX(3,1) model with uniform weight on CPO export volume data in three locations with OLS method produces 21 parameters (Table 4).

Table 4. Parameter Estimation of GSTARX(3,1) Model with Uniform Weight Matrix

Location	Parameter	Estimation	Standard Error	t-value	p-value	Variable
North Sumatera	ϕ_{10}^1	0.5173	0.1334	3.88	**0.0002***	$Z_1(t-1)$
	ϕ_{11}^1	-0.2100	0.1451	-,1.45	0.1508	$V_1(t-1)$
	ϕ_{10}^2	0.1998	0.1422	1.41	0.1630	$Z_1(t-2)$
	ϕ_{11}^2	-0.2391	0.1432	-,1.67	**0.0983***	$V_1(t-2)$
	ϕ_{10}^3	0.3583	0.1414	2.53	**0.0128***	$Z_1(t-3)$
	ϕ_{11}^3	-0.0618	0.1417	-,0.44	0.6640	$V_1(t-3)$
	γ_1	0.0637	0.0378	1.68	**0.0952***	$X(t)$
Riau	ϕ_{20}^1	-0.1268	0.1340	-,0.94	0.3487	$Z_2(t-1)$
	ϕ_{21}^1	0.7628	0.2993	2.55	**0.0123***	$V_2(t-1)$
	ϕ_{20}^2	-0.1133	0.1346	-,0.84	0.4021	$Z_2(t-2)$
	ϕ_{21}^2	0.4608	0.3076	1.50	0.1372	$V_2(t-2)$
	ϕ_{20}^3	-0.0887	0.1376	-,0.65	0.5195	$Z_2(t-3)$
	ϕ_{21}^3	0.6985	0.3191	2.19	**0.0309***	$V_2(t-3)$
	γ_2	0.1221	0.0540	2.26	**0.0259***	$X(t)$
West Sumatera	ϕ_{30}^1	-0.1236	0.1250	-,0.99	0.3252	$Z_3(t-1)$
	ϕ_{31}^1	0.1216	0.0562	2.17	**0.0326***	$V_3(t-1)$
	ϕ_{30}^2	0.1441	0.1249	1.15	0.2514	$Z_3(t-2)$
	ϕ_{31}^2	0.0210	0.0578	0.36	0.7171	$V_3(t-2)$
	ϕ_{30}^3	-0.0206	0.1229	-,0.17	0.8671	$Z_3(t-3)$
	ϕ_{31}^3	0.0554	0.0554	1.00	0.3194	$V_3(t-3)$
	γ_3	0.0451	0.0198	2.28	**0.0247***	$X(t)$

Note: * Significant with $\alpha = 10\%$

Based on the parameters in Table VI, the GSTARX(3,1) with uniform weight on the CPO export volume data can be written in matrix form as follows:

$$\begin{bmatrix} \hat{Z}_1(t) \\ \hat{Z}_2(t) \\ \hat{Z}_3(t) \end{bmatrix} = \left[\begin{bmatrix} 0.5173 & 0 & 0 \\ 0 & -0.1268 & 0 \\ 0 & 0 & -0.1236 \end{bmatrix} + \begin{bmatrix} 0 & -0.1050 & -0.1050 \\ 0.3814 & 0 & 0.3814 \\ 0.0618 & 0.0618 & 0 \end{bmatrix} \right] \begin{bmatrix} \hat{Z}_1(t-1) \\ \hat{Z}_2(t-1) \\ \hat{Z}_3(t-1) \end{bmatrix}$$

$$+\left[\begin{bmatrix}0.1998 & 0 & 0\\ 0 & -0.1132 & 0\\ 0 & 0 & -0.1441\end{bmatrix}+\begin{bmatrix}0 & -0.1195 & -0.1195\\ 0.2304 & 0 & 0.2304\\ 0.0105 & 0.0105 & 0\end{bmatrix}\right]\begin{bmatrix}\hat{Z}_1(t-2)\\ \hat{Z}_2(t-2)\\ \hat{Z}_3(t-2)\end{bmatrix}$$

$$+\left[\begin{bmatrix}0.3583 & 0 & 0\\ 0 & -0.0887 & 0\\ 0 & 0 & -0.0206\end{bmatrix}+\begin{bmatrix}0 & -0.0309 & -0.0309\\ 0.3492 & 0 & 0.3492\\ 0.0277 & 0.0277 & 0\end{bmatrix}\right]\begin{bmatrix}\hat{Z}_1(t-3)\\ \hat{Z}_2(t-3)\\ \hat{Z}_3(t-3)\end{bmatrix}$$

$$+\left[\begin{bmatrix}0.0637 & 0 & 0\\ 0 & 0.1221 & 0\\ 0 & 0 & 0.0451\end{bmatrix}\begin{bmatrix}X_1(t-1)\\ X_2(t-1)\\ X_3(t-1)\end{bmatrix}+\begin{bmatrix}a_1(t)\\ a_2(t)\\ a_3(t)\end{bmatrix}\right]$$

Furthermore, we interpret the result of GSTARX(3,1) with uniform space weight. It indicates that the CPO export volume in North Sumatera province in the same province at time lag 1 month and 3 months before, the CPO export volume in Riau and West Sumatera 2 months before as well as the international CPO price in the same month. CPO export volume in Riau province significantly affected CPO export volume in North Sumatera and West Sumatera at time lag 1 month and 3 months before and international CPO price in the same month. And the CPO export volume in West Sumatera is significantly affected by the CPO export volume in North Sumatera and Riau at tie lag 1 month before, as well as international CPO price in the same month.

3.4. GSTARX(3,1) Model with Inverse Distance Weight Matrix

GSTARX(3,1) model with an inverse distance weight is another way to see the relationship between the CPO export volume in three locations based on the distance between locations. This model assumes that the CPO export volume in one province is affected by distance or the closeness with other locations.

Table 5. Coordinate of Location Research

Location	Latitude	Longitude
North Sumatera	3.580592	98.671687
Riau	0.517062	101.445981
West Sumatera	-0.937687	100.361094

We calculate the distances between locations using the Euclidean distance formula. The coordinates of the location of each study are presented in Table V. Based on the distance of each of these locations, then the inverse distance weight matrix can be notified as follows:

$$W_{ij}=\begin{bmatrix}0 & 0.5386 & 0.4614\\ 0.3051 & 0 & 0.6949\\ 0.2734 & 0.7266 & 0\end{bmatrix}$$

$$\begin{bmatrix}\hat{Z}_1(t)\\ \hat{Z}_2(t)\\ \hat{Z}_3(t)\end{bmatrix}=\sum_{k=1}^{2}\left[\begin{bmatrix}\phi_{10}^k & 0 & 0\\ 0 & \phi_{20}^k & 0\\ 0 & 0 & \phi_{30}^k\end{bmatrix}+\begin{bmatrix}\phi_{11}^k & 0 & 0\\ 0 & \phi_{21}^k & 0\\ 0 & 0 & \phi_{31}^k\end{bmatrix}\begin{bmatrix}0 & 0.5386 & 0.4614\\ 0.3051 & 0 & 0.6949\\ 0.2734 & 0.7266 & 0\end{bmatrix}\right]\begin{bmatrix}Z_1(t-k)\\ Z_2(t-k)\\ Z_3(t-k)\end{bmatrix}+$$

$$\begin{bmatrix}\gamma_1 & 0 & 0\\ 0 & \gamma_2 & 0\\ 0 & 0 & \gamma_3\end{bmatrix}\begin{bmatrix}X_1(t)\\ X_2(t)\\ X_3(t)\end{bmatrix}+\begin{bmatrix}a_1(t)\\ a_2(t)\\ a_3(t)\end{bmatrix}$$

Parameter estimation GSTARX(3,1) model with the inverse distance weight on the CPO export volume data in three locations with OLS method produces 21 parameters (Table 6).

IOP Conf. Series: Journal of Physics: Conf. Series **893** (2017) 012038 doi:10.1088/1742-6596/893/1/012038

Table 6. Parameter Estimation of GSTARX(3,1) Model with Inverse Distance Weight Matrix

Location	Parameter	Estimation	Standard Error	t-value	p-value	Variable
North Sumatera	ϕ^1_{10}	0.5186	0.1327	3.91	**0.0002***	$Z_1(t-1)$
	ϕ^1_{11}	-,0.2075	0.1380	-,1.50	0.1358	$V_1(t-1)$
	ϕ^2_{10}	0.2053	0.1417	1.45	0.1505	$Z_1(t-2)$
	ϕ^2_{11}	-,0.2354	0.1365	-,1.72	**0.0876***	$V_1(t-2)$
	ϕ^3_{10}	0.3695	0.1410	2.62	**0.0101***	$Z_1(t-3)$
	ϕ^3_{11}	-,0.0714	0.1352	-,0.53	0.5986	$V_1(t-3)$
	γ_1	0.0654	0.0376	1.74	**0.0847***	$X(t)$
Riau	ϕ^1_{20}	-,0.0370	0.1330	-,0.28	0.7815	$Z_2(t-1)$
	ϕ^1_{21}	0.6636	0.3393	1.96	**0.0532***	$V_2(t-1)$
	ϕ^2_{20}	-,0.0865	0.1324	-,0.65	0.5151	$Z_2(t-2)$
	ϕ^2_{21}	0.5384	0.3377	1.59	0.1140	$V_2(t-2)$
	ϕ^3_{20}	-,0.0752	0.1345	-,0.56	0.5776	$Z_2(t-3)$
	ϕ^3_{21}	0.7530	0.3583	2.13	**0.0357***	$V_2(t-3)$
	γ_2	0.0944	0.0544	1.74	**0.0855***	$X(t)$
West Sumatera	ϕ^1_{30}	-,0.0939	0.1233	-,0.76	0.4479	$Z_3(t-1)$
	ϕ^1_{31}	0.1017	0.0489	2.08	**0.0401***	$V_3(t-1)$
	ϕ^2_{30}	0.1941	0.1227	1.58	0.1168	$Z_3(t-2)$
	ϕ^2_{31}	0.0032	0.00499	0.06	0.9489	$V_3(t-2)$
	ϕ^3_{30}	0.0235	0.1219	0.19	0.8476	$Z_3(t-3)$
	ϕ^3_{31}	0.0289	0.0488	0.59	0.5549	$V_3(t-3)$
	γ_3	0.0412	0.0199	2.07	**0.0408***	$X(t)$

Note: * Significant with $\alpha = 10\%$

Based on the parameters in Table 6, it can be formed matrix equation of the GSTARX(3,1) model with the distance inverse space weight on the CPO export volume data as follows

$$\begin{bmatrix}\hat{Z}_1(t)\\ \hat{Z}_2(t)\\ \hat{Z}_3(t)\end{bmatrix} = \left[\begin{bmatrix}0.5186 & 0 & 0\\ 0 & -0.0370 & 0\\ 0 & 0 & -0.0939\end{bmatrix} + \begin{bmatrix}0 & -0.1118 & -0.0957 0.2025 & 0 & 0.4612 0.0278 & 0.0739 & 0\end{bmatrix}\right]\begin{bmatrix}\hat{Z}_1(t-1)\\ \hat{Z}_2(t-1)\\ \hat{Z}_3(t-1)\end{bmatrix}$$

$$+\left[\begin{bmatrix}0.2052 & 0 & 0\\ 0 & -0.0865 & 0\\ 0 & 0 & -0.1941\end{bmatrix} + \begin{bmatrix}0 & -0.1268 & -0.1086\\ 0.1643 & 0 & 0.3741\\ 0.0009 & 0.0023 & 0\end{bmatrix}\right]\begin{bmatrix}\hat{Z}_1(t-2)\\ \hat{Z}_2(t-2)\\ \hat{Z}_3(t-2)\end{bmatrix}$$

$$+\left[\begin{bmatrix}0.3695 & 0 & 0\\ 0 & -0.0752 & 0\\ 0 & 0 & -0.0235\end{bmatrix} + \begin{bmatrix}0 & -0.0385 & -0.0329\\ 0.2297 & 0 & 0.5232\\ 0.0079 & 0.0210 & 0\end{bmatrix}\right]\begin{bmatrix}\hat{Z}_1(t-3)\\ \hat{Z}_2(t-3)\\ \hat{Z}_3(t-3)\end{bmatrix}$$

$$+\begin{bmatrix}0.0654 & 0 & 0\\ 0 & 0.0944 & 0\\ 0 & 0 & 0.0412\end{bmatrix}\begin{bmatrix}X_1(t-1)\\ X_2(t-1)\\ X_3(t-1)\end{bmatrix} + \begin{bmatrix}a_1(t)\\ a_2(t)\\ a_3(t)\end{bmatrix}$$

Furthermore, from GSTARX(3,1) model with the distance inverse space weight also can be interpreted CPO export volume in each province. Model GSTARX(3,1) with the inverse distance weight indicates that the CPO export volume in North Sumatera province is significantly affected by the CPO export volume in the same province at 1 month and 3 months before, the CPO export volume in Riau and West Sumatera 2 months before as well as the international CPO price in the same month. CPO export volume in Riau province significantly affected CPO export volume in North Sumatera and West Sumatera at 1 month and 3 months before and international CPO price in the same month. As for the CPO export volume in West Sumatera is significantly affected by the CPO export volume in North Sumatera and Riau at 1 month before, as well as international CPO price in the same month.

3.5. Diagnostic Checking Model

Diagnostic check model is an examination of whether the model assumptions fulfilled. The basic assumption of the model is residual $\{a_t\}$ white noise. The detection of the residual white noise can be done by modeling the residual obtained from the model for each of the space weights. If the value of the smallest AIC currently on lag-0, it can be said that there is no correlation between residual, meaning that the residuals are white noise.

Table 7. AIC Value of Residual GSTARX(3,1) Model (Uniform Weight)

Lag	MA(0)	MA(1)	MA(2)	MA(3)	MA(4)	MA(5)
AR (0)	25,4792	25,6175	25,6701	25,6860	25,7968	25,7726
AR (1)	25,5244	25,5518	25,6927	25,7414	25,8696	25,9529
AR (2)	25,6053	25,6666	25,8021	25,8808	26,0431	26,1375
AR (3)	25,5857	25,7238	25,8497	26,0139	26,1734	26,2829
AR (4)	25,7084	25,8408	25,9724	26,1571	26,3440	26,2949
AR (5)	25,7851	26,9475	26,1439	26,3272	26,5118	26,5626

Table 8. AIC Value of Residual GSTARX(3,1) Model (Inverse Distance Weight)

Lag	MA(0)	MA(1)	MA(2)	MA(3)	MA(4)	MA(5)
AR (0)	25,5732	25,6866	25,7474	25,7355	25,8268	25,8096
AR (1)	25,6327	25,6504	25,7993	25,8109	25,9136	26,0083
AR (2)	25,7237	25,7457	25,8724	25,9412	26,0955	26,1861
AR (3)	25,6809	25,7886	25,9266	26,0918	26,2084	26,2946
AR (4)	25,7642	25,8831	26,0208	26,1935	26,3639	26,2789
AR (5)	25,8398	26,0061	26,1982	26,3587	26,5330	26,5687

Furthermore, from GSTARX(3,1) model with the distance inverse space weight also can be interpreted CPO export volume in each province. Model GSTARX(3,1) with the distance inverse space weight indicates that the CPO export volume in North Sumatera province is significantly affected by the CPO export volume in the same province at 1 month and 3 months before, the CPO export volume in Riau and West Sumatera 2 months before as well as the international CPO price in the same month. CPO export volume in Riau province significantly affected CPO export volume in North Sumatera and West Sumatera at 1 month and 3 months before and international CPO price in the same month. As for the CPO export volume in West Sumatera is significantly affected by the CPO export volume in North Sumatera and Riau at 1 month before, as well as international CPO price in the same month.

3.6. Diagnostic Checking Model

Diagnostic check model is an examination of whether the model assumptions fulfilled. The basic assumption of the model is residual a_t white noise. The detection of the residual white noise can be done by modeling the residual obtained from the model for each of the space weights. If the value of the smallest AIC currently on lag-0, it can be said that there is no correlation between residual, meaning that the residuals are white noise.

Table VII and VIII presents the AIC value of residual GSTARX(3,1) model with uniform and inverse distance weight. The table shows that for each weight, the smallest AIC value are in lag-0, or in AR (0) and MA (0) to model GSTARX (3,1) with uniform and inverse distance weight. This indicates that the residuals of the model GSTARX (3.1) with two weights have fulfilled the assumption of white noise.

3.7. The Accuracy of Forecasting GSTARX(3,1) Model

Comparison of the accuracy of the estimation is done through the value of Mean Square Error (MSE) in-sample data from the model GSTARX (3,1) using two weight matrix. The result are presented in Table IX. Table IX shows that the smallest MSE value for in-sample data GSTARX(3,1) model is on uniform weight. So, it can conclude that GSTARX(3,1) model with uniform weight matrix is the fit model for CPO export volume data in three location at Sumatera island.

Table 9. The Comparison of MSE Value in Sample Data of GSTARX(3,1) Model using Two Weight Matrix

Space Weight	MSE Value for In-Sample Data			
	North Sumatera	Riau	West Sumatera	Three Location
Uniform Weight	7.96×10^3	17.33×10^3	2.62×10^3	9.30×10^3
Inverse Distance Weight	7.92×10^3	18.34×10^3	2.71×10^3	9.66×10^3

Otherwise, in order to determine the most appropriate space weight for forecasting the CPO export volume in the three locations, the forecast data should validate with the actual one in out sample data. The comparison of the accuracy of forecasting can be shown through value Mean Square Error (MSE) of out-sample data. Table 9 shows that the smallest MSE value in the out-sample data GSTARX (3.1) models on the average obtained from the inverse distance weighting locations.

4. Conclusion

Based on the analysis and the discussion above, it can be concluded that the best model for CPO export volume data in three provinces above are GSTARX(3,1) model. CPO export volume in one location affected by the CPO export volume in the past in the same location and the volume of exports of CPO in the past in other location as well as in international CPO prices. GSTARX(3,1) model with uniform space weight is the fit model for CPO export volume data in three location. Otherwise, for forecasting the CPO export volume data in three location, the GSTARX(3,1) model with inverse distance space weight is the most appropriate model, which the smallest MSE value in out sample data.

Acknowledgment

The work reported in this paper is supported by the Directorate General of Higher Education, Ministry of Research-Technology and Higher Education, Indonesia, through Penelitian Kerjasama Luar Negeri dan Publikasi Internasional 2016. The authors thank to the Reviewer of Proceedings AMC 2016 for valuable suggestion to improve this paper.

References

[1] Borovkova S A, Lopuhaa, and Ruchjana B N 2012 Least Squares Estimation of Generalized Space Time Auto Regressive (GSTAR) Model and Its Properties. *AIP Conf Proc* pp 61–64

[2] Box G E P, Jenkins G M, dan Reinsel G C 1994 *Time Series Analysis: Forecasting and Control* Third Edition, Englewood Cliffs: Prentice Hall.

[3] Enders W 2010 *Applied Econometric Time Series* Third Edition New York: John Wiley & Sons Inc.

[4] Ruchjana B N 2002 *Space Time Autoregressive Model and Its Application to Oil Production Data*, Ph.D. Dissertation, Unpublished, Bandung: ITB

[5] Wei W W S 2006 *Time series analysis univariate and multivariate methods* Second Edition Pearson

IOP Conf. Series: Journal of Physics: Conf. Series **893** (2017) 012039 doi:10.1088/1742-6596/893/1/012039

Characterization of completely k-magic regular graphs

A A Eniego[1] and I J L Garces[2]

[1] Science and Mathematics Department, National University, Manila, The Philippines
[2] Department of Mathematics, Ateneo de Manila University, Quezon City, The Philippines

E-mail: `aaeniego@national-u.edu.ph`, `ijlgarces@ateneo.edu`

Abstract. Let $k \in \mathbb{N}$ and $c \in \mathbb{Z}_k$. A graph G is said to be c-sum k-magic if there is a labeling $\ell : E(G) \to \mathbb{Z}_k \setminus \{0\}$ such that $\sum_{u \in N(v)} \ell(uv) \equiv c \pmod{k}$ for every vertex v of G, where $N(v)$ is the neighborhood of v in G. We say that G is completely k-magic whenever it is c-sum k-magic for every $c \in \mathbb{Z}_k$. In this paper, we characterize all completely k-magic regular graphs.

1. Introduction

Let $G = (V(G), E(G))$ be a finite, simple (unless otherwise stated) graph with vertex set $V(G)$ and edge set $E(G)$. A *factor* of G is a subgraph H with $V(H) = V(G)$. In particular, if a factor H of G is h-regular, then we say that H is an *h-factor* of G. An *h-factorization* of G is a partition of $E(G)$ into disjoint h-factors. If such factorization of G exists, then we say that G is *h-factorable.*

The following theorem is attributed to Petersen [7], which we state using the versions of Akiyama and Kano [2] and Wang and Hu [10].

Theorem 1.1 ([2, Theorem 3.1], [7], [10, Theorem 10]). *Let G be a $2r$-regular connected general graph (not necessarily simple), where $r \geq 1$. Then G is 2-factorable, and it has a $2k$-factor for every k, $1 \leq k \leq r$. Moreover, if G is of even order, then it is r-factorable.*

A graph G is *λ-edge connected* if it remains connected whenever fewer than λ edges are removed.

Theorem 1.2. [6] *Let r and k be integers such that $1 \leq k < r$, and G be a λ-edge connected r-regular general graph, where $\lambda \geq 1$. If one of the following conditions holds:*

(1) *r is even, k is odd, $|G|$ is even, and $\frac{r}{\lambda} \leq k \leq r(1 - \frac{1}{\lambda})$,*

(2) *r is odd, k is even, and $2 \leq k \leq r(1 - \frac{1}{\lambda})$, or*

(3) *r and k are both odd and $\frac{r}{\lambda} \leq k$,*

then G has a k-regular factor.

Let k be a positive integer. A finite simple graph $G = (V(G), E(G))$ is said to be *k-magic* if there exists an edge labeling $\ell : E(G) \to \mathbb{Z}_k \setminus \{0\}$, where $\mathbb{Z}_1 = \mathbb{Z}$ the group of integers, and $\mathbb{Z}_k = \{0, 1, 2, \ldots, k-1\}$ the group of integers modulo $k \geq 2$, such that the induced vertex labeling $\ell^+ : V(G) \to \mathbb{Z}_k$, defined by $\ell^+(v) = \sum_{uv \in E(G)} \ell(uv)$, is a constant map. If $c \in \mathbb{Z}_k$

and $\ell^+(v) = c$ for all $v \in V(G)$, then we call c is a *magic sum* of G. In particular, if G is k-magic with magic sum c, then we say that G is *c-sum k-magic.* If G is c-sum k-magic for all $c \in \mathbb{Z}_k$, then it is said to be *completely k-magic.* The set of all magic sums $c \in \mathbb{Z}_k$ of G is the *sum spectrum of G with respect to k* and is denoted by $\Sigma_k(G)$. If $c = 0$, then we say that G is *zero-sum k-magic.* The *null set* of G, denoted by $N(G)$, is the set of all positive integers k such that G is a zero-sum k-magic graph.

Remark 1.3. *If $c \in \mathbb{Z}_k$ and ℓ is a c-sum k-magic labeling of G, then the labeling ℓ', defined by $\ell'(e) = k - \ell(e)$, is a $(k - c)$-sum k-magic labeling of G.*

Remark 1.4. *Any 2-magic graph is not completely 2-magic.*

The concept of A-magic graphs is due to Sedlacek [9]. Over the years, many papers have been published in connection with magic graphs. Akbari, Rahmati, and Zare [1] investigated the zero-sum k-magic labelings and null sets of regular graphs. Dong and Wang [4] solved affirmatively a conjecture posed in [1] on the existence of a zero-sum 3-magic labeling of 5-regular graphs. Salehi [8] determined the integer-magic spectra of certain classes of cycle-related graphs.

Using the term "index set," Wang and Hu [10] initially studied the concept of completely k-magic graphs. They gave a partial list of completely 1-magic regular graphs. Eniego and Garces [5] completely added the remaining cases in this list. They also presented the sum spectra of some regular graphs that are not completely k-magic.

Theorem 1.5 ([1, Theorem 13]). *Let G be an r-regular graph, where $r \geq 3$ and $r \neq 5$. If r is even, then $N(G) = \mathbb{N}$ (the set of positive integers); otherwise, $\mathbb{N} \setminus \{2, 4\} \subseteq N(G)$.*

Theorem 1.6 ([4, Theorem 2.1]). *Every 5-regular graph admits a zero-sum 3-magic labeling.*

Theorem 1.7 ([5, Theorem 3.3]). *Let $n \geq 3$ and $k \geq 3$ be integers, and C_n the cycle with n vertices.*

(1) *If n is even, then C_n is completely k-magic for all k.*

(2) *If n is odd, then C_n is not completely k-magic for any k. Moreover, we have*

$$\Sigma_k(C_n) = \begin{cases} \mathbb{Z}_k \setminus \{0\} & \text{if } k \text{ is odd,} \\ \{0, 2, \ldots, k-2\} & \text{if } k \text{ is even.} \end{cases}$$

Theorem 1.8 ([5, Lemma 3.4]). *Let $k \geq 4$ be an even integer. Then there exists no k-magic graph of odd order that is completely k-magic. In particular, if c is a magic sum of a k-magic graph of odd order, then c must be even.*

Theorem 1.9 ([5, Theorem 3.6]). *Let $k, r \geq 3$ be integers, and G an r-regular graph. If $\gcd(r, k) = 1$, then $\{1, 2, \ldots, k-1\} \subseteq \Sigma_k(G)$.*

Theorem 1.10 ([5, Theorem 3.7]). *Let G be a zero-sum k-magic r-regular graph, where $k \geq 3$ and $r \geq 3$. If G has a 1-factor, then G is completely k-magic.*

Theorem 1.11 ([10, Theorem 13], [5, Theorem 2.1]). *Let G be an r-regular graph of order n. Then*

$$\Sigma_1(G) = \begin{cases} \mathbb{Z} \setminus \{0\} & \text{if } r = 1, \\ \mathbb{Z} & \text{if } r = 2 \text{ and } G \text{ contains even cycles only,} \\ 2\mathbb{Z} \setminus \{0\} & \text{if } r = 2 \text{ and } G \text{ contains an odd cycle,} \\ 2\mathbb{Z} & \text{if } r \geq 3,\ r \text{ even, and } n \text{ odd,} \\ \mathbb{Z} & \text{if } r \geq 3 \text{ and } n \text{ even,} \end{cases}$$

where $2\mathbb{Z}$ is the set of all even integers.

With Remark 1.4 and Theorem 1.11, it remains to characterize all completely k-magic regular graphs for $k \geq 3$. This characterization is the main theorem of this paper, which we state as follows.

Theorem 1.12 (Main Theorem). *Let $r \geq 2$ and $k \geq 3$ be integers, and G an r-regular graph of order $n \geq 3$. Then G is completely k-magic if and only if one of the following properties holds:*

(1) $k \geq 3$, $r = 2$, *and G contains even cycles only,*
(2) $k \geq 5$ *and* $r \geq 3$ *odd,*
(3) $k \geq 5$, $r \geq 4$ *even, and* n *even,*
(4) $k \geq 5$ *odd,* $r \geq 4$ *even, and* n *odd,*
(5) $k = 4$, $r \geq 3$, n *even, and G zero-sum 4-magic, or*
(6) $k = 3$ *and any one of the following conditions holds:*
  (i) $r \not\equiv 0 \pmod 3$,
  (ii) $r \equiv 0 \pmod 6$, *or*
  (iii) $r \equiv 0 \pmod 3$, r *odd, and G has a factor H such that* $d_H(v) \equiv 1 \pmod 3$ *for all* $v \in V(H)$.

For convenience, we only consider graphs that are finite and simple (unless otherwise stated). We also write $\mathbb{Z}_k^*$ to mean $\mathbb{Z}_k \setminus \{0\}$. For graph-theoretic terms that are not explicitly defined in this paper, see [3].

2. Proof of the Main Theorem

We divide the proof into several results.

It is not difficult to see that if G is 1-regular, then $\Sigma_k(G) = \mathbb{Z}_k^*$. For 2-regular graphs, the following remark is a consequence of Theorem 1.7.

Remark 2.1. *Let $k \geq 3$ and G a 2-regular graph. If G has an odd cycle, then*

$$\Sigma_k(G) = \begin{cases} \mathbb{Z}_k^* & \text{if } k \text{ is odd} \\ \{0, 2, \ldots, k-2\} & \text{if } k \text{ is even.} \end{cases}$$

Otherwise, we have $\Sigma_k(G) = \mathbb{Z}_k$.

Clearly, if G is 1-factorable, then G is completely k-magic. The following theorem considers regular graphs that has a factor that is completely k- magic.

Theorem 2.2. *Let $r \geq 2$, $2 \leq h \leq r$, $k \neq 2$, and G an r-regular graph. If G has an h-factor that is completely k-magic, then G is completely k-magic.*

Proof. The case when $h = r$ is trivial, so we assume $h < r$. Let H be an h-factor of G that is completely k-magic. Let $\alpha = c - (r - h) \pmod k$ and f_α be an α-sum k-magic labeling of H for each $c \in \mathbb{Z}_k$.

Define $\ell_c : E(G) \to \mathbb{Z}_k^*$ by

$$\ell_c(e) = \begin{cases} f_\alpha(e) & \text{if } e \in E(H) \\ 1 & \text{if } e \in E(G \setminus H). \end{cases}$$

Observe that ℓ_c is a c-sum k-magic labeling of G for each $c \in \mathbb{Z}_k$. Hence, G is completely k-magic. □

The following construction will be useful.

Remark 2.3. *Let G be an r-regular graph with $E(G) = \{e_1, e_2, e_3, \ldots, e_m\}$, where $r \geq 1$. Then we can construct a graph G' (with parallel edges) such that $V(G') = V(G)$ and $E(G') = E(G) \cup \{e'_1, e'_2, e'_3, \ldots, e'_m\}$, where e'_i is a duplicate edge of e_i in G for each i (that is, edges e_i and e'_i have the same end vertices). By Theorem 1.1, G' has a $2h$-factor H' for each h, $1 \leq h \leq r$. Also, $G' \setminus H'$ is a $(2r-2h)$-factor of G' obtained by removing the edges of H' from G'.*

Theorem 2.4. *Let G be a 5-regular graph. Then $\mathbb{N} \setminus \{2,4\} \subseteq N(G)$.*

Proof. We know from Theorem 1.11 and Theorem 1.6 that $1, 3 \in N(G)$. For $k \geq 5$, we consider two cases.

Case 1. Suppose $k \geq 5$ and $k \neq 8$. Using the construction described in Remark 2.3, let H' and $G' \setminus H'$ be a 2-factor and 8-factor of G', respectively.

Define a zero-sum k-magic labeling ℓ' on G' by

$$\ell'(e) = \begin{cases} k-4 & \text{if } e \in E(H') \\ 1 & \text{if } e \in E(G' \setminus H'). \end{cases}$$

Note that the labeling ℓ on G defined by $\ell(e_i) = \ell'(e_i) + \ell'(e'_i)$ for $e_i \in E(G)$ is a zero-sum k-magic labeling on G.

Case 2. Suppose $k = 8$. Using again the construction in Remark 2.3, let H' and $G' \setminus H'$ be a 4-factor and 6-factor of G', respectively.

Define a zero-sum labeling ℓ' on G' by

$$\ell'(e) = \begin{cases} 2 & \text{if } e \in E(H') \\ 4 & \text{if } e \in E(G' \setminus H'). \end{cases}$$

Observe that the labeling ℓ on G defined by $\ell(e_i) = \frac{1}{2}[\ell'(e_i) + \ell'(e'_i)]$ for $e_i \in E(G)$ is a zero-sum 8-magic labeling on G.

Therefore, $\mathbb{N} \setminus \{2,4\} \subseteq N(G)$. □

Note that an odd-regular graph may not be zero-sum 4-magic. It was remarked in [1, Remark 10] that an odd-regular graph G is not zero-sum 4-magic if G has a vertex such that every edge incident to it is a cut-edge.

Theorem 2.5. *Let G be an r-regular graph, where $r \geq 3$ is odd and $k \geq 5$. Then G is completely k-magic.*

Proof. We know from Theorems 1.5 and 2.4 that $0 \in \Sigma_k(G)$. Let $E(G) = \{e_1, e_2, e_3, \ldots, e_m\}$. As constructed in Remark 2.3, let H' and $G' \setminus H'$ be a 2-factor and $(2r-2)$-factor of G', respectively. We consider two cases.

Case 1. Suppose $r \equiv 1 \pmod{k}$. Then $\gcd(r,k) = 1$. By Theorem 1.9, G is completely k-magic.

Case 2. Suppose $r \not\equiv 1 \pmod{k}$. Assume $\gcd(r,k) = d$ so that $r = ad$ and $k = bd$ for some positive integers a and b. Note that, since r is odd, d is also odd. We consider two sub-cases.

Sub-Case 2.1. Suppose $k \geq 5$ is odd. Then b is odd.

For each $c \in \mathbb{Z}_k^* \setminus \{k-b, k-2b\}$, define $\ell'_c : E(G') \to \mathbb{Z}_k^*$ by

$$\ell'_c(e) = \begin{cases} x & \text{if } e \in E(H') \\ \frac{1}{2}(k+b) & \text{if } e \in E(G' \setminus H'), \end{cases}$$

where $x = \frac{1}{2}(b+c)$ if c is odd, and $x = \frac{1}{2}(b+c+k)$ if c is even. Observe that ℓ'_c is a c-sum k-magic labeling of G' for each $c \neq 0$.

For each $c \notin \{0, k-b, k-2b\}$, define $\ell_c : E(G) \to \mathbb{Z}_k^*$ by $\ell_c(e_i) = \ell'_c(e_i) + \ell'_c(e'_i)$ for $1 \leq i \leq m$. Since ℓ'_c is a c-sum k-magic labeling of G', ℓ_c is a c-sum k-magic labeling of G for each $c \in \mathbb{Z}_k^* \setminus \{k-b, k-2b\}$.

If $k \neq 3b$, then, by Remark 1.3, $k-b, k-2b \in \Sigma_k(G)$. If $k = 3b$, it is enough to show that $k - 2b \in \Sigma_k(G)$. To do that, we provide a different labeling using a different set of factors of G'. Let J' and $G' \setminus J'$ be a 4-factor and $(2r-4)$-factor of G' respectively. In addition, we let $J' = J'_1 \cup J'_2$, where J'_1 and J'_2 are 2-factors of J'.

Define $\ell' : E(G') \to \mathbb{Z}_k^*$ by

$$\ell'(e) = \begin{cases} \frac{1}{2}(b+1) & \text{if } e \in E(J'_1) \\ \frac{1}{2}(b-1) & \text{if } e \in E(J'_2) \\ b & \text{if } e \in E(G' \setminus J'). \end{cases}$$

Since $k = 3b$, $d = 3$ and $r = 3a$. Thus, the magic sum in G' is given by $2[\frac{1}{2}(b+1)] + 2[\frac{1}{2}(b-1)] + b(2r-4) \equiv -2b \pmod{k}$. Define $\ell : E(G) \to \mathbb{Z}_k^*$ by $\ell(e_i) = \ell'(e_i) + \ell'(e'_i)$ for $1 \leq i \leq m$. Note that ℓ is also a $(k-2b)$-sum k-magic labeling of G.

Sub-Case 2.2. Suppose $k \geq 6$ is even. Then b is even.

By labeling all the edges of G with $\frac{1}{2}k$, we see that $\frac{1}{2}k \in \Sigma_k(G)$.

Suppose $r - 1 \equiv \frac{1}{2}k \pmod{k}$. For each $c \in \mathbb{Z}_k^* \setminus \{k-1, \frac{1}{2}k\}$, define $\ell'_c : E(G') \to \mathbb{Z}_k^*$ by

$$\ell'_c(e) = \begin{cases} c & \text{if } e \in E(H') \\ 1 & \text{if } e \in E(G' \setminus H'). \end{cases}$$

Observe that the sum of the labels of the edges incident to each vertex in G' is $2(r-1) + 2c \equiv 2c \pmod{k}$. Using a similar argument as in Sub-Case 2.1, it can be shown that G is also e-sum k-magic for all even $e \neq 0$. Thus, we are left to show that G is c-sum k-magic as well for all odd c.

For each odd $c \neq k-1$, define $\ell_c : E(G) \to \mathbb{Z}_k^*$ by $\ell_c(e_i) = \frac{1}{2}[\ell'_c(e_i) + \ell'_c(e'_i)]$ for each i, $1 \leq i \leq m$. Note that, since ℓ'_c is a $2c$-sum k-magic labeling of G', ℓ_c is a c-sum k-magic labeling of G for each odd $c \neq k-1$. Again, by Remark 1.3, we see that $k - 1 \in \Sigma_k(G)$.

Suppose $r - 1 \equiv r_0 \pmod{k}$, where $r_0 \neq \frac{1}{2}k$. For each $c \in \mathbb{Z}_k^* \setminus \{r_0, r_0 + \frac{1}{2}k, r_0 - 1\}$, define $\ell'_c : E(G') \to \mathbb{Z}_k^*$ by

$$\ell'_c(e) = \begin{cases} c - r_0 & \text{if } e \in E(H') \\ 1 & \text{if } e \in E(G' \setminus H'). \end{cases}$$

Observe that the sum of the labels of the edges incident to each vertex in G' is $2r_0 + 2c - 2r_0 \equiv 2c \pmod{k}$. As in Sub-Case 2.1, it can be shown that G is also even-sum k-magic. So again, we are left to show that G is odd-sum k-magic.

As what we did earlier, for each odd $c \neq r_0 - 1$ (and, possibly, $r_0 + \frac{1}{2}k$), define $\ell_c : E(G) \to \mathbb{Z}_k^*$ by $\ell_c(e_i) = \frac{1}{2}[\ell'_c(e_i) + \ell'_c(e'_i)]$ for all i, $1 \leq i \leq m$. Since ℓ'_c is a $2c$-sum k-magic labeling of G', ℓ_c is a c-sum k-magic labeling of G for each odd $c \neq r_0 - 1$ (and, possibly, $r_0 + \frac{1}{2}k$). If $r_0 - 1$ and $r_0 + \frac{1}{2}k$ are not inverses, then, by Remark 1.3, $\mathbb{Z}_k^* \subset \Sigma_k(G)$.

If $r_0 - 1$ and $r_0 + \frac{1}{2}k$ are inverses, then it is enough to show that $r_0 - 1 \in \Sigma_k(G)$. Define ℓ' on G' by

$$\ell'(e) = \begin{cases} k-1 & \text{if } e \in E(H') \\ 1 & \text{if } e \in E(G' \setminus H'). \end{cases}$$

IOP Conf. Series: Journal of Physics: Conf. Series **893** (2017) 012039 doi:10.1088/1742-6596/893/1/012039

Note that the magic sum using ℓ' is $2r_0-2$. Define ℓ on G by $\ell(e_i) = \frac{1}{2}[\ell'(e_i)+\ell'(e_i')]$ for $e_i \in E(G)$. Clearly, ℓ is an $(r_0 - 1)$-sum k-magic labeling on G. Thus, by Remark 1.3, $r_0 + \frac{1}{2}k \in \Sigma_k(G)$, and so $\mathbb{Z}_k^* \subset \Sigma_k(G)$.

In any case, G is completely k-magic. □

Theorem 2.6. *Let $k \geq 5$ and G a $2r$-regular graph of order $n \geq 3$, where $r \geq 2$.*

(1) *If n is even, then G is completely k-magic.*

(2) *If n is odd, then*

  (i) *G is completely k-magic if k is odd, and*

  (ii) *$\Sigma_k(G) = \{0, 2, 4, \ldots, k-2\}$ if k is even.*

Proof. Let $E(G) = \{e_1, e_2, e_3, \ldots, e_m\}$. By Theorem 1.5, G is zero-sum k-magic.

(1) Suppose $r = 2$. To prove the theorem, we only show that $\mathbb{Z}_k^* \subset \Sigma_k(G)$. We consider two cases.

CASE 1. Suppose k is odd. Then $\gcd(4, k) = 1$. By Theorem 1.9, $\mathbb{Z}_k^* \subseteq \Sigma_k(G)$.

CASE 2. Suppose k is even. It is not difficult to see that, being 4-regular, G is 2-edge connected. By Remark 2.3, we can construct G' so that G' is a 4-edge-connected 8-regular graph. By Theorem 1.2, G' has a 3-factor, say H'. Let $G' \setminus H'$ be the 5-factor of G' obtained by removing the edges of H' from G'.

SUB-CASE 2.1. Let $k = 2d$, d even. For each $c \in \mathbb{Z}_k^* \setminus \{\frac{1}{2}k, \frac{1}{4}k\}$, define $f_c : E(G') \to \mathbb{Z}_k^*$ by

$$f_c(e) = \begin{cases} 2c & \text{if } e \in E(H') \\ k - c & \text{if } e \in E(G' \setminus H'). \end{cases}$$

Observe that the sum of the labels of the edges incident to each of the vertices in G' is equal to $5(k-c) + 3(2c) \equiv c \pmod{k}$. This shows that f_c is a c-sum k-magic labeling of G' for all $c \neq 0, \frac{1}{2}k, \frac{1}{4}k$. By Remark 1.3, $\frac{1}{4}k \in \Sigma_k(G')$.

For each $c \in \mathbb{Z}_k^* \setminus \{\frac{1}{2}k, \frac{1}{4}k\}$, define $\ell_c : E(G) \to \mathbb{Z}_k^*$ by $\ell_c(e_i) = f_c(e_i) + f_c(e_i')$ for all i, $1 \leq i \leq m$. Clearly, ℓ_c is a c-sum k-magic labeling of G for each $c \in \mathbb{Z}_k^* \setminus \{\frac{1}{2}k, \frac{1}{4}k\}$. By Remark 1.3, we see that $\mathbb{Z}_k^* \setminus \{\frac{1}{2}k\} \subset \Sigma_k(G)$.

By Theorem 1.1, G is 2-factorable. Let G_1 and G_2 be the two 2-factors of G. Label the edges in G_1 with d and the edges in G_2 with $\frac{1}{2}(k-d)$. This shows that $d = \frac{1}{2}k \in \Sigma_k(G)$.

SUB-CASE 2.2. Let $k = 2d$, $d \geq 3$ odd. Observe that, for $c \neq 0, \frac{1}{2}k$, the labeling ℓ_c in Sub-Case 2.1 is a c-sum k-magic labeling of G. We are left to show that $\frac{1}{2}k \in \Sigma_k(G)$.

Let $d \neq 3$ and 9. We give a labeling for the factors of G' defined above (namely, H' and $G' \setminus H'$) and the 2-factors of G (namely, G_1 and G_2) to show that G is d-sum k-magic.

Let $f : E(G) \to \mathbb{Z}_k^*$ be defined by

$$f(e) = \begin{cases} d+1 & \text{if } e \in E(G_1) \\ \frac{1}{2}(k-d-1) & \text{if } e \in E(G_2). \end{cases}$$

Clearly, f is $(d+1)$-sum k-magic labeling of G.

Let $g' : E(G') \to \mathbb{Z}_k^*$ be defined by

$$g'(e) = \begin{cases} k-2 & \text{if } e \in E(H') \\ 1 & \text{if } e \in E(G' \setminus H'). \end{cases}$$

Define also $g : E(G) \to \mathbb{Z}_k^*$ by $g(e_i) = g'(e_i) + g'(e_i')$ for all i, $1 \leq i \leq m$. Note that g' is a $(k-1)$-sum k-magic labeling of G', so g is a $(k-1)$-sum k-magic labeling of G.

Finally, define $\ell : E(G) \to \mathbb{Z}_k^*$ by $\ell(e) = f(e) + g(e)$ for all $e \in E(G)$. Since f and g are $(d+1)$-sum and $(k-1)$-sum k-magic labeling of G, respectively, ℓ is a d-sum k-magic labeling of G.

Suppose $d = 3$ or 9. Define $g' : E(G') \to \mathbb{Z}_k^*$ be defined by

$$g'(e) = \begin{cases} 2x & \text{if } e \in E(H') \\ 1 & \text{if } e \in E(G' \setminus H'), \end{cases}$$

where $x = 1$ if $d = 3$, and $x = 3$ if $d = 9$. Note that g' is a 5-sum k-magic labeling of G'. Define a labeling g on G by $g(e_i) = g'(e_i) + g'(e_i') + 1$ for all i, $1 \le i \le m$. Note that g is a d-sum k-magic labeling on G. Thus, $d = \frac{1}{2}k \in \Sigma_k(G)$, and so G is completely k-magic.

Suppose $r \ge 3$ is odd. By Theorem 1.1, G is r-factorable. By Theorem 2.5, the r-factors of G are completely k-magic for all $k \ge 5$. Thus, by Theorem 2.2, G is also completely k-magic.

If $r \ge 4$ is even, then, by Theorem 1.1, G has a 6-factor, say H. Using the case for r is odd, H is completely k-magic. Thus, by Theorem 2.2, G is also completely k-magic.

(2(i)) By Theorem 1.1, G is 2-factorable. Let $G_1, G_2, \ldots, G_r$ be the 2-factors of G. If k is odd, then, by Remark 2.1, $\mathbb{Z}_k^* \subseteq \sum_k(G_i)$ for all i, $1 \le i \le r$. For each i and $c \in \mathbb{Z}_k^*$, let ℓ_c^i be a c-sum k-magic labeling of G_i. We consider two cases.

Case 1. Suppose $r \equiv 1 \pmod{k}$. For each $c \in \mathbb{Z}_k^*$, define $\ell_c : E(G) \to \mathbb{Z}_k^*$ by

$$\ell_c(e) = \begin{cases} \ell_c^1(e) & \text{if } e \in E(G_1) \\ \ell_1^i(e) & \text{if } e \in E(G_i) \text{ for some } i = 2, 3, \ldots, r. \end{cases}$$

Note that ℓ_c is a c-sum k-magic labeling of G for all $c \neq 0$.

Case 2. Suppose $r \not\equiv 1 \pmod{k}$. For each $c \in \mathbb{Z}_k^* \setminus \{r-1 \pmod{k}\}$, define $l_c : E(G) \to \mathbb{Z}_k^*$ by

$$l_c(e) = \begin{cases} l_{c-x}^1(e) & \text{if } e \in E(G_1) \\ l_1^i(e) & \text{if } e \in E(G_i) \text{ for some } i = 2, 3, \ldots, r, \end{cases}$$

where $x \equiv r - 1 \pmod{k}$. The sum of the labels of the edges incident to each vertex is $c \pmod{k}$. Thus, G is c-sum k-magic for each $c \neq x$. By Remark 1.3, G is x-sum k-magic since G is $(k-x)$-sum k-magic. In this case, G is completely k-magic.

(2(ii)) This follows from Remark 2.1, Lemma 1.8, and Theorem 2.2. □

The proof of the following theorems are similar to Theorem 2.5 and Theorem 2.6.

Theorem 2.7. *Let $r \ge 3$, and G a zero-sum 4-magic r-regular graph. Then*

(1) *If the order of G is even, then G is completely 4-magic.*

(2) *If the order of G is odd, then $\Sigma_4(G) = \{0, 2\}$.*

Theorem 2.8. *Let G be an r-regular graph, where $r \ge 3$.*

(1) *If $r \not\equiv 0 \pmod{3}$ or $r \equiv 0 \pmod{6}$, then G is completely 3-magic.*

(2) *If $r \equiv 0 \pmod{3}$ and r odd, then G is completely 3-magic if and only if G has a factor H such that $d_H(v) \equiv 1 \pmod{3}$ for all $v \in V(H)$.*

IOP Conf. Series: Journal of Physics: Conf. Series **893** (2017) 012039 doi:10.1088/1742-6596/893/1/012039

References

[1] Akbari S, Rahmati F, and Zare S 2014 Zero-sum magic labelings and null sets of regular graphs *Electron. J. Combin.* **21**(2) #P2.17
[2] Akiyama J and Kano M 2011 *Factors and Factorizations of Graphs* (Springer-Verlag)
[3] Bondy J A and Murty U S R 2008 *Graph Theory* (Springer)
[4] Dong G and Wang N 2014 A conjecture on zero-sum 3-magic labeling of 5-regular graphs *arXiv* 1406.6870v1
[5] Eniego A A and Garces I J L 2015 Completely k-magic regular graphs *Appl. Math. Sci. (Ruse)* **103** pp 5139–5148
[6] Gallai T 1950 On factorisation of graphs *Acta Math. Hungar.* **1**(1) pp 133–53
[7] Petersen J 1891 Die theorie der regulären graphs *Acta Math.* **15** pp 193–220
[8] Salehi E 2006 Integer-magic spectra of cycle-related graphs *Iran. J. Math. Sci. Inform.* **2** pp 53–63
[9] Sedlacek J 1976 On magic graphs *Math. Slovaca* **26** pp 329–35
[10] Wang T M and Hu S W 2011 Constant sum flows in regular graphs *Frontiers in Algorithmics and Algorithmic Aspects in Information and Management*, ed M Attalah, X Y Li and B Zhu (Berlin Heidelberg: Springer) pp 168–175

IOP Conf. Series: Journal of Physics: Conf. Series **893** (2017) 012040 doi:10.1088/1742-6596/893/1/012040

On the locating-chromatic number for graphs with two homogenous components

Des Welyyanti, Edy Tri Baskoro, Rinovia Simajuntak, and Saladin Uttunggadewa

Combinatorial Mathematics Research Group, Faculty of Mathematics and Natural Sciences Institut Teknologi Bandung, Jl. Ganesa 10 Bandung 40132, Indonesia

E-mail: deswelyyanti@students.itb.ac.id, {ebaskoro, rino, s_uttunggadewa}@math.itb.ac.id

Abstract. The locating-chromatic number of a graph was introduced by Chartrand et al. in 2002. The concept of the locating-chromatic number is a marriage between graph coloring and the notion of graph partition dimension. This concept is only for connected graphs. In [8], we extended this concept also for disconnected graphs. In this paper, we determine the locating-chromatic number of a graph with two components. In particular, we determine such values if the components are homogeneous and each component has locating-chromatic number 3.

1. Introduction

The locating-chromatic numbers was have been determined for some classes of graphs, namely cycles [5], multipartite graphs[5], and some classes of trees. The locating-chromatic number of trees are limited for paths[5], stars[5], double stars[5], amalgamation of stars[1], firecracker graphs[3], homogenous lobsters[6] and complete n-ary graphs[7]. Next, the characterization of all graphs with locating-chromatic number 3 was determined by Asmiati et al.[2] and Baskoro et al.[4].

Until now, the study of the locating-chromatic number of graphs are limited only for connected graphs. Therefore, we introduced the notion of the locating-chromatic numbers for disconnected graphs in [8]. Let c be a k-coloring of a (disconnected) graph $H(V, E)$ and $\Pi = \{C_1, C_2, \cdots, C_k\}$ be the partition of $V(H)$ induced by c, where C_i is the set of all vertices receiving color i. The *color code* $c_\Pi(v)$ of a vertex $v \in H$ is the ordered k-tuple $(d(v, C_1), d(v, C_2), \cdots, d(v, C_k))$, where $d(v, C_i) = min\{d(v, x)|x \in C_i\}$ and $d(v, C_i) < \infty$ for all $i \in [1, k]$. If all vertices of H have distinct color codes, then c is called a *locating-coloring* of H. The *locating-chromatic number* of H, denoted by $\chi'_L(H)$, is the smallest k such that H admits a locating k-coloring, otherwise we say that $\chi'_L(H) = \infty$.

In [8], we determined sufficient conditions so that the locating-chromatic number of a disconnected graph is finite. Futhermore, we also determined the locating-chromatic number of some disconnected graphs. In 2015, we determined on locating-chromatic number for graphs with dominant vertices[9]. Next, we determined the locating-chromatic number of a two non homogenous components graph [10] in 2016.

In this paper, we determine the locating-chromatic number of a graph with two components. We give lower and upper bounds for the locating-chromatic number for that graphs. In particular, we determine such values if the components are homogeneous and each component has locating-chromatic number 3.

2. Result

The locating-chromatic number of trees is one of interesting research in graph theory. In 2013, Baskoro et al.[4] characterized all trees with the locating-chromatic number 3.

Theorem 2.1 *[4] a tree T has locating-chromatic number 3 if only if T is a path P_3 or P_4, a double stars $S_{1,2}$ or $S_{2,2}$, or a subtree of A, B, or C in Figure 1 which contains a path ${}_aP_b$.*

Let c be a locating k-coloring on graph $G(V, E)$. Let $\Pi = \{C_1, C_2, \cdots, C_k\}$ be the partition of $V(G)$ induced by c. A vertex $v \in G$ is called a *dominant vertex* if $d(x, C_i) = 0$ and $d(x, C_j) = 1$ for $i \neq j$. A path connecting two dominant vertices in G is called a *clear path* if all of its internal vertices are not dominant [2].

In next theorem, we determine the locating-chromatic number of disconnected graph $H = 2G$, where G is tree with the locating-chromatic number 3 other than a path and a double star $S_{2,2}$.

Theorem 2.2 *If G is a tree with $\chi_L(G) = 3$ and $|V(G)| \geq 6$ other than a path and a double star $S_{2,2}$, then $\chi'_L(2G) = 4$.*

Proof. Let G be a tree which is not a path and a double star $S_{2,2}$ with $\chi_L(G) = 3$ and $|V(G)| \geq 6$. By Theorem 2.1, G will be isomorphic to a subgraph of graph A, B, or C in Figure 1 which contains a path ${}_aP_b$. Since G is not a path nor a double star $S_{2,2}$, then the locating 3-coloring of G has at least two dominant vertices. So, There is no locating 3-coloring for graph $2G$. Thus, $\chi'_L(2G) \geq 4$.

Next, we will show the upper bound of $\chi_L(2G)$. Let $H = 2G$ be a graph with H_1 and H_2 as its components where $H_1 = H_2 = G$. For every vertex $u \in G$, we define u_j as the corresponding vertex with vertex u in H_j with $j = 1, 2$. Let T be a graph A, B, or C in Figure 1. Let $W = \{a, b, x, y, z\}$ for $T = A$, $W = \{a, b, x, y, v^*, v^{**}, z\}$ for $T = B$, and $W = \{a, b, x, u^*, u^{**}, y, v^*, v^{**}, z\}$ for $T = C$.

For every vertex $u \in W$, we define L_u as some path in T which u is one of the end vertex in L_u and not contain other vertices in W. Define a locating 3-coloring c for T with 3 dominant vertices as follows

For $T = A$,

(i) $c(a) = 3$, $c(x) = 1$, $c(y) = 2$, $c(z) = 3$ and $c(b) = 1$
(ii) Color all vertices in L_a by 3 and 1 alternately
(iii) Color all vertices in L_x by 1 and 2 alternately

IOP Conf. Series: Journal of Physics: Conf. Series **893** (2017) 012040 doi:10.1088/1742-6596/893/1/012040

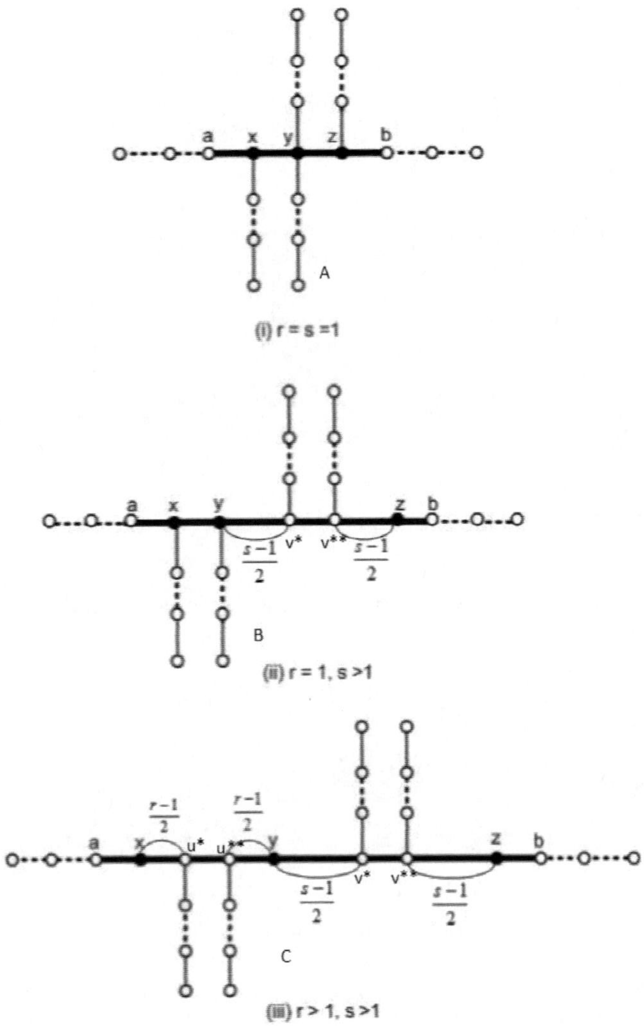

Figure 1. Trees contain ${}_aP_b$ with $\chi_L(T) = 3$.

(iv) Color all vertices in L_y^1 by 2 and 1 alternately
(v) Color all vertices in L_y^2 by 2 and 3 alternately
(vi) Color all vertices in L_z by 3 and 2 alternately
(vii) Color all vertices in L_b by 1 and 3 alternately.

For $T = B$,

(i) $c(a) = 2$, $c(x) = 1$, $c(y) = 2$, $c(z) = 3$ and $c(b) = 1$
(ii) Color all vertices in L_a by 2 and 1 alternately
(iii) Color all vertices in L_x by 1 and 3 alternately
(iv) Color all vertices in L_y by 2 and 1 alternately
(v) Color all internal vertices in ${}_xP_y$ by 3 and 2 alternately
(vi) Color all vertices in L_v^* by 3 and 2 alternately

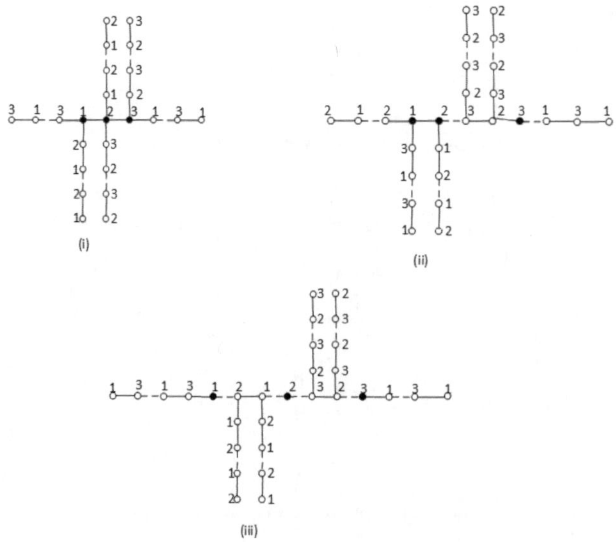

Figure 2. The locating 3-coloring on A, B, and C.

(vii) Color all vertices in L_v^{**} by 2 and 3 alternately
(viii) Color all vertices in L_b by 1 and 3 alternately.

For $T = C$,

(i) $c(a) = 3$, $c(x) = 1$, $c(y) = 2$, $c(z) = 3$ and $c(b) = 1$
(ii) Color all vertices in L_a by 3 and 1 alternately
(iii) Color all vertices in L_u^* by 2 and 1 alternately
(iv) Color all vertices in L_u^{**} by 1 and 2 alternately
(v) Color all internal vertices in ${}_xP_y$ by 2 and 1 alternately
(vi) Color all vertices in L_v^* by 3 and 2 alternately
(vii) Color all vertices in L_v^{**} by 2 and 3 alternately
(viii) Color all internal vertices in ${}_yP_z$ by 2 and 3 alternately
(ix) Color all vertices in L_b by 1 and 3 alternately.

This above 3-coloring of T is illustrated in Figure 2.

Let G be a subtree of T. Define a coloring $c' : V(G) \to \{1, 2, 3\}$ such that $c'(x) = c(x)$, for $x \in V(G)$.

Next, define a coloring $c'' : V(H) \to \{1, 2, 3, 4\}$ such that $c''(x_1) = c''(b_2) = 4$, for $x_1 \in V(H_1)$, $b_2 \in V(H_2)$, and $c''(x) = c(x)$ for $x \in V(H) \setminus \{x_1, b_2\}$.

For $G \subseteq A$. We will show the color codes of all vertices in $2G$ are different. Let $\Pi = \{C_1, C_2, C_3, C_4\}$ be a partition of $V(2G)$ induced by c''. The coloring of $2G$ is illustrated in Figure 3.

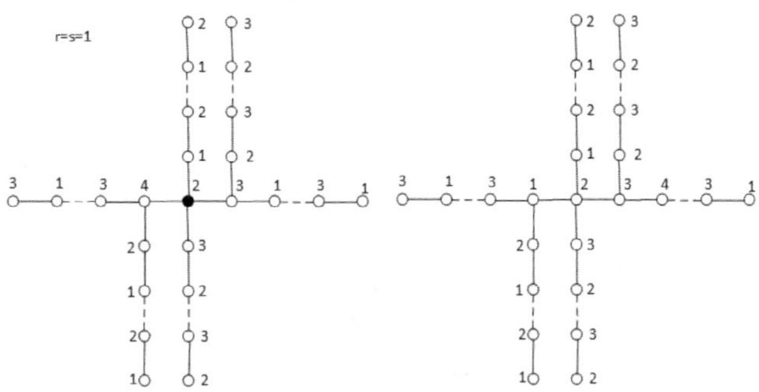

Figure 3. The locating 4-coloring for $2G$ if $G \subseteq A$.

In H_1, we have $c''_{\Pi}(a) = (1,3,0,1)$, $c''_{\Pi}(x) = (2,1,1,0)$, $c''_{\Pi}(y) = (1,0,1,1)$, $c"_{\Pi}(z) = (1,1,0,2)$, and $c''_{\Pi}(b) = (0,2,1,3)$. In H_2, we have $c''_{\Pi}(a) = (1,2,0,4)$, $c''_{\Pi}(x) = (0,1,1,3)$, $c''_{\Pi}(y) = (1,0,1,2)$, $c''_{\Pi}(z) = (2,1,0,1)$, and $c''_{\Pi}(b) = (3,2,1,0)$.

Let u and v be two distinct vertices in $2G$ with $c''(u) = c''(v)$. If $d(u, C_4) \neq d(v, C_4)$, then u and v clearly have different color codes. If $d(u, C_4) = d(v, C_4) = l$, then we have three cases.

Case 1. $c(u) = c(v) = 1$. (i) For $u \in L_y^1$ and $v \in L_x$, we have $c''_{\Pi}(u) = (0,1,k,l)$, for $k \geq 2$, k even and $c''_{\Pi}(v) = (0,1,k,l)$, for $k \geq 1$, k odd. (ii) For $u \in L_a$ and $v \in L_b$, we have $c''_{\Pi}(u) = (0,k,1,l)$, for $k \geq 3$, k odd and $c''_{\Pi}(v) = (0,k,1,l)$, for $k \geq 2$, k even.

Case 2. $c(u) = c(v) = 2$. (i) For $u \in L_x$ and $v \in L_y^1$, we have $c''_{\Pi}(u) = (1,0,k,l)$, for $k \geq 2$, k even and $c''_{\Pi}(v) = (1,0,k,l)$, for $k \geq 1$, k odd. (ii) For $u \in L_y^2$ and $v \in L_z$, we have $c''_{\Pi}(u) = (k,0,1,l)$, for $k \geq 1$, k odd and $c''_{\Pi}(v) = (k,0,1,l)$, for $k \geq 2$, k even.

Case 3. $c(u) = c(v) = 3$. (i) For $u \in L_a$ and $v \in L_b$, we have $c''_{\Pi}(u) = (1,k,0,l)$, for $k \geq 2$, k even and $c''_{\Pi}(v) = (1,k,0,l)$, for $k \geq 3$, k odd. (ii) For $u \in L_y^2$ and $v \in L_z$, we have $c''_{\Pi}(u) = (k,1,0,l)$, for $k \geq 3$, k odd and $c''_{\Pi}(v) = (k,1,0,l)$, for $k \geq 2$, k even.

Consequently, all vertices of $2G$ have different color codes. Thus, $\chi'_L(2G) \leq 4$.

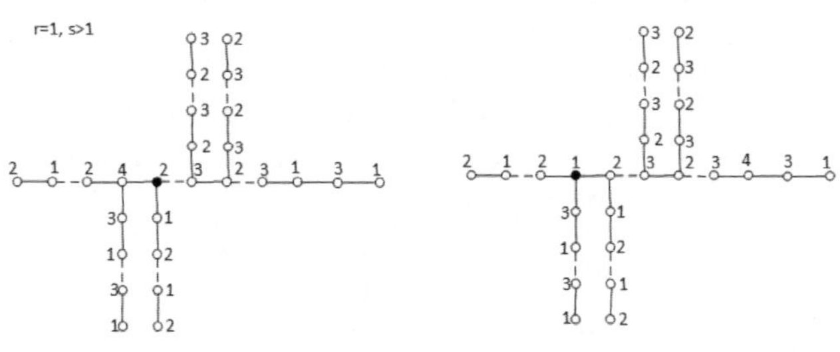

Figure 4. The locating 4-coloring for $2G$ if $G \subseteq B$.

For $G \subseteq B$. G has two clear paths with odd lengths. There clear paths are ${}_xP_y$ and ${}_yP_z$. The length that two path are r and s. Let v^* and v^{**} be two internal vertices in ${}_yP_z$ and $s^* = \frac{s-1}{2}$

be length of ${}_yP_{v^*}$ or ${}_{v^{**}}P_z$.

Let $\Pi = \{C_1, C_2, C_3, C_4\}$ be a partition of $V(2G)$ is induced by c''. Coloring of $2G$ is illustrated in Figure 4.

In H_1, we have $c''_\Pi(a) = (1,0,2,1)$, $c''_\Pi(x) = (2,1,1,0)$, $c''_\Pi(y) = (1,0,1,1)$, $c''_\Pi(z) = (1,1,0,s+1)$, and $c''_\Pi(b) = (0,2,1,s+2)$. In H_2, we have $c''_\Pi(a) = (1,0,2,s+3)$, $c''_\Pi(x) = (0,1,1,s+2)$, $c''_\Pi(y) = (1,0,1,s+1)$, $c''_\Pi(z) = (3,1,0,1)$, and $c''_\Pi(b) = (2,2,1,0)$.

Let u and v be two distinct vertices in $2G$ with $c''(u) = c''(v)$. If $d(u, C_4) \neq d(v, C_4)$, then u and v clearly have different color codes. If $d(u, C_4) = d(v, C_4) = l$, then we have three cases.

Case 1. $c(u) = c(v) = 1$. For $u \in L_a$ and $v \in L_y$, we have (i) $c''_\Pi(u) = (0,1,k,l)$, for $k \geq 3$, k odd and $c''_\Pi(v) = (0,1,k,l)$, for $k \geq 2$, k even. (ii) For $u \in L_x$ and $v \in L_b$, we have $c''_\Pi(u) = (0,k,1,l)$, for $k \geq 2$, k even and $c''_\Pi(v) = (0,k,1,l)$, for $k \geq 3$, k odd.

Case 2. $c(u) = c(v) = 2$. For $u \in L_a$ and $v \in L_y$, we have (i) $c''_\Pi(u) = (1,0,k,l)$, for $k \geq 2$, k even and $c''_\Pi(v) = (1,0,k,l)$, for $k \geq 1$, k odd. (ii) For $u \in L_{u^*}$ and $v \in L_{u^{**}}$, we have $c''_\Pi(u) = (s^* + k, 0, 1, l)$, for $k \geq 3$, k odd and $c''_\Pi(v) = (s^* + k, 0, 1, l)$, for $k \geq 2$, k even.

Case 3. $c(u) = c(v) = 3$. For $u \in L_x$ and $v \in L_b$, we have (i) $c''_\Pi(u) = (1,k,0,l)$, for $k \geq 2$, k even and $c''_\Pi(v) = (1,k,0,l)$, for $k \geq 3$, k odd. (ii) For $u \in L_{u^*}$ and $v \in L_{u^{**}}$, we have $c''_\Pi(u) = (s^* + k, 1, 0, l)$, for $k \geq 1$, k odd and $c''_\Pi(v) = (s^* + k, 1, 0, l)$, for $k \geq 2$, k even.

Consequently, all vertices of $2G$ have different color codes. Thus, $\chi'_L(2G) \leq 4$.

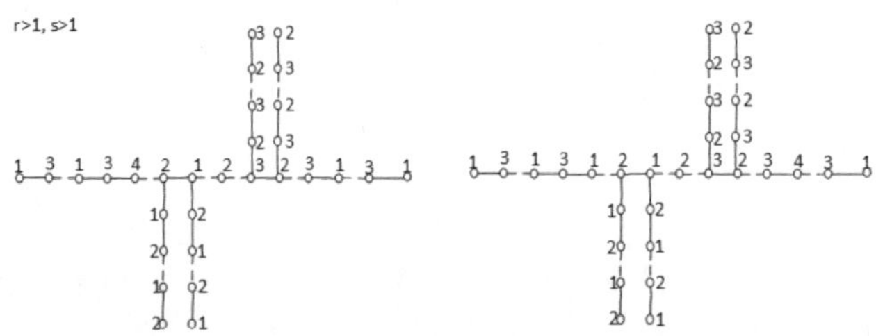

Figure 5. The locating 4-coloring for $2G$ if $G \subseteq C$.

For $G \subseteq C$. G has two clear paths with odd length. There clear paths are ${}_xP_y$ and ${}_yP_z$. The length that two path are r and s. Let u^* and u^{**} be two internal vertices in ${}_xP_z$, $r^* = \frac{r-1}{2}$ be length of ${}_xP_{u^*}$ or ${}_{u^{**}}P_y$, v^* and v^{**} be two internal vertices of ${}_yP_z$, and $s^* = \frac{s-1}{2}$ be length ${}_yP_{v^*}$ or ${}_{v^{**}}P_z$.

Let $\Pi = \{C_1, C_2, C_3, C_4\}$ be a partition of $V(2G)$ is induced by c''. Coloring of $2G$ is illustrated in Figure 5.

In H_1, we have $c''_\Pi(a) = (1, r^*, 0, 1)$, $c''_\Pi(x) = (2,1,1,0)$, $c''_\Pi(y) = (1,0,1,r)$, $c''_\Pi(z) = (1,1,0,r+s)$, and $c''_\Pi(b) = (0,2,1,r+s+1)$. In H_2, we have $c''_\Pi(a) = (1,2,0,r+s+2)$, $c''_\Pi(x) = (0,1,1,r+s+1)$, $c''_\Pi(y) = (1,0,1,s+1)$, $c''_\Pi(z) = (3,1,0,1)$, and $c''_\Pi(b) = (2,2,1,0)$.

Let u and v be two distinct vertices in $2G$ with $c''(u) = c''(v)$. If $d(u, C_4) \neq d(v, C_4)$, then u and v clearly have different color codes. If $d(u, C_4) = d(v, C_4) = l$, then we have three cases.

Case 1. $c(u) = c(v) = 1$. For $u \in L_a$ and $v \in L_b$, we have (i) $c''_{\Pi}(u) = (0, k, 1, l)$, for $k \geq 3$, k odd and $c''_{\Pi}(v) = (0, k, 1, l)$, for $k \geq 4$, k even. (ii) For $u \in L_{u^*}$ and $v \in L_{u^{**}}$, we have $c''_{\Pi}(u) = (0, 1, r^* + k, l)$, for $k \geq 2$, k even and $c''_{\Pi}(v) = (0, 1, r^* + k, l)$, for $k \geq 1$, k odd.

Case 2. $c(u) = c(v) = 2$. For $u \in L_{u^*}$ and $v \in L_{u^{**}}$, we have (i) $c''_{\Pi}(u) = (1, 0, r^* + k, l)$, for $k \geq 2$, k even and $c''_{\Pi}(v) = (1, 0, r^* + k, l)$, for $k \geq 1$, k odd. (ii) For $u \in L_{v^*}$ and $v \in L_{v^{**}}$, we have $c''_{\Pi}(u) = (s^* + k, 0, 1, l)$, for $k \geq 2$, k even and $c''_{\Pi}(v) = (s^* + k, 0, 1, l)$, for $k \geq 1$, k odd.

Case 3. $c(u) = c(v) = 3$. For $u \in L_a$ and $v \in L_b$, we have (i) $c''_{\Pi}(u) = (1, k, 0, l)$, for $k \geq 2$, k even and $c''_{\Pi}(v) = (1, k, 0, l)$, for $k \geq 3$, k odd. (ii) For $u \in L_{v^*}$ and $v \in L_{v^{**}}$, we have $c''_{\Pi}(u) = (s^* + k, 1, 0, l)$, for $k \geq 1$, k odd and $c''_{\Pi}(v) = (s^* + k, 1, 0, l)$, for $k \geq 2$, k even.

Consequently, all vertices of $2G$ have different color codes. Thus, $\chi'_L(2G) \leq 4$. So, $\chi'_L(2G) = 4$.
The characterization of all graphs G other than trees with $\chi_L(G) = 3$ has been also studied in [2]. This graphs is a subgraph of G_a, G_b, G_c, or G_d that contains a odd cycle with x, y, z as dominant vertices (see Figure 6) or a subgraph of G_5 that contains even cycle only (see Figure 7).

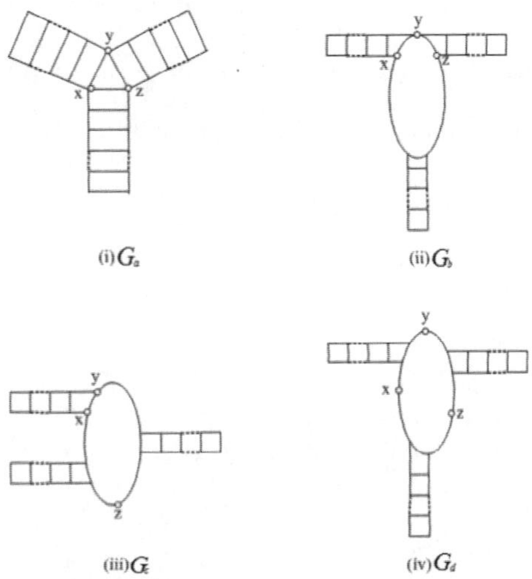

Figure 6. The maximal graphs have odd cycle with $\chi_L(G) = 3$.

Next theorem will discuss about locating-chromatic number of disconnected $H = 2G$ with locating-chromatic number of G is 3 and contains cycle.

Theorem 2.3 *If graph G contains cycle with $|V(G)| \geq 4$ and $\chi_L(G) = 3$, then $\chi'_L(2G) = 4$.*

Proof of Theorem 2.3 is similar with Proof of Theorem 2.2.

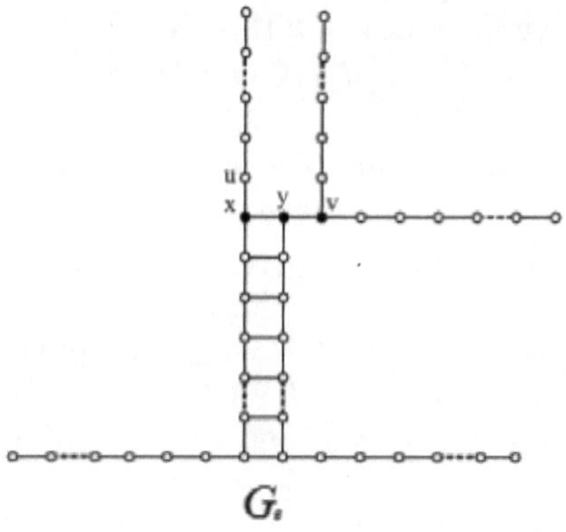

Figure 7. The maximal graph have even cycle with $\chi_L(G) = 3$.

Acknowledgments
This research was supported by Research Grant Program Hibah Riset Unggulan ITB-DIKTI. Ministry of Research, Technology and Higher Education, Indonesia.

References
[1] Asmiati, Assiyatun H, Baskoro ET 2011 Locating-chromatic of amalgamation of stars *ITB J.Sci.* **43A**(1) pp 1–8
[2] Asmiati, Baskoro ET 2012 Characterizing all graphs containing cycles with locating-chromatic number 3 *AIP Conf. Proc.* **1450** pp 351–357
[3] Asmiati, Baskoro ET, Assiyatun H, Suprijanto D, Simanjuntak R, Uttunggadewa S 2012 Locating-chromatic number of firecracker graphs *Far East J. Math. Sci.* **63**(1) pp 11–13
[4] Baskoro ET, Asmiati 2013 Characterizing all trees with locating-chromatic number 3 *Electronic Journal of Graph Theory and Applications* **I**(2) pp 109–117
[5] Chartrand G, Erwin D, Henning MA, Slater PJ, Zhang P 2002 The locating-chromatic number of a graph *Bull. Inst.Combin. Appl.* **36** pp 89–101
[6] Syofyan DK, Baskoro ET, Assiyatun H. On locating-chromatic number of homogeneous lobster 2013 *AKCE Int. J. Graph Comb.* **10**(3) pp 245–252
[7] Welyyanti D, Baskoro ET, Simanjuntak R, Uttunggadewa S 2013 On locating-chromatic number of complete n-ary tree *AKCE Int. J. Graph Comb.* **10**(3) pp 309–315
[8] Welyyanti D, Baskoro ET, Simanjuntak R, Uttunggadewa S 2014 The locating-chromatic number of disconnected graphs *Far East J. Math. Sci.* **94**(2) pp 169–182
[9] Welyyanti D, Baskoro ET, Simanjuntak R, Uttunggadewa S 2015 On Locating-Chromatic Number for Graphs with Dominant Vertices *Procedia Comp. Sci.* **74** pp 89–92
[10] Welyyanti D, Simanjuntak R, Uttunggadewa S, Baskoro ET 2016 Locating-chromatic number for a graph of two component *AIP Conf. Proc.* 1707 020024 10.1063/1.4940825

IOP Conf. Series: Journal of Physics: Conf. Series **893** (2017) 012041 doi:10.1088/1742-6596/893/1/012041

Metric graphic sets

I J L Garces and J B Rosario

Department of Mathematics, Ateneo de Manila University, Quezon City, The Philippines

E-mail: `ijlgarces@ateneo.edu, rosario.trolls@yahoo.com`

Abstract. For an ordered subset $W = \{w_1, w_2, \ldots, w_k\}$ of vertices in a connected graph G and a vertex v of G, the metric representation of v with respect to W is the k-vector $r(v|W) = (d(v, w_1), d(v, w_2), \ldots, d(v, w_k))$, where $d(v, w_i)$ is the distance of the vertices v and w_i in G. The set W is called a resolving set of G if $r(u|W) = r(v|W)$ implies $u = v$. The metric dimension of G, denoted by $\beta(G)$, is the minimum cardinality of a resolving set of G, and a resolving set of G with cardinality equal to its metric dimension is called a metric basis of G. A set T of vectors is called a positive lattice set if all the coordinates in each vector of T are positive integers. A positive lattice set T consisting of n k-vectors is called a metric graphic set if there exists a simple connected graph G of order $n + k$ with $\beta(G) = k$ such that $T = \{r(u_i|S) : u_i \in V(G)\backslash S, 1 \leq i \leq n\}$ for some metric basis $S = \{s_1, s_2, \ldots, s_k\}$ of G. If such G exists, then we say G is a metric graphic realization of T. In this paper, we introduce the concept of metric graphic sets anchored on the concept of metric dimension and provide some characterizations. We also give necessary and sufficient conditions for any positive lattice set consisting of 2 k-vectors to be a metric graphic set. We provide an upper bound for the sum of all the coordinates of any metric graphic set and enumerate some properties of positive lattice sets consisting of n 2-vectors that are not metric graphic sets.

1. Introduction

Let $G = (V(G), E(G))$ be a finite, simple, and connected graph. The *distance* between two vertices u and v of G, denoted by $d(u, v)$, is the length of the shortest u-v path in G. For an ordered subset $W = \{w_1, w_2, \ldots, w_k\}$ of $V(G)$, we refer to the k-vector (ordered k-tuple) $r(v|W) = (d(v, w_1), d(v, w_2), \ldots, d(v, w_k))$ as the *metric representation of vertex v with respect to W*. The set W is called a *resolving set of G* if every two distinct vertices u and v satisfy $r(u|W) \neq r(v|W)$. A *metric basis* of G is a resolving set of G with minimal cardinality, and the *metric dimension of G* refers to this cardinality, denoted by $\beta(G)$.

Remark 1.1. *Although W is treated as an ordered set, when its elements are permuted, the coordinates of $r(v|W)$ will follow correspondingly.*

The metric dimension problem was first introduced by Harary and Melter [5] in 1976 and independently by Slater [7] in 1975. Several authors studied this topic and published numerous results. See, for example, [2] and [4]. Chartrand et. al. [3] proved the following characterizations of connected graphs of order n with metric dimensions 1 and $n - 1$. They also characterized graphs with metric dimension $n - 2$ (see Theorem 4.1), and obtained a formula in computing the metric dimension of a tree that is not a path.

Theorem 1.2 (Chartrand et. al. [3]). *Let G be a connected graph of order $n \geq 2$. Then*

(i) $\beta(G) = 1$ *if and only if* $G = P_n$.

(ii) $\beta(G) = n - 1$ *if and only if* $G = K_n$.

Remark 1.3. *If* G *is a path, then either of the end vertices of* G *constitutes a metric basis. On the other hand, if* G *is a complete graph of order* $n \geq 2$*, then any subset* S *of* $V(G)$ *of order* $n-1$ *constitutes a metric basis.*

The concept of metric dimension is finding a minimal resolving set S of vertices for a given graph G such that the metric representations of the vertices of $V(G)\backslash S$ with respect to S are distinct. The idea of reversing this process suggests the following problem: given any set of vectors whose coordinates are positive integers, is there a graph or graphs with the given set as the metric representations of the vertices with respect to some metric basis of the graph? In other words, can a graph or graphs be realized out of these vectors? This leads us in introducing the concept of metric graphic sets.

In this paper, we introduce the concept of metric graphic sets anchored on the concept of metric dimension and provide some characterizations. We also give necessary and sufficient conditions for any positive lattice set consisting of 2 k-vectors to be a metric graphic set. We provide an upper bound for the sum of all the coordinates of any metric graphic set and enumerate some properties of positive lattice sets consisting of n 2-vectors that are not metric graphic sets.

For graph-theoretic terms that are not explicitly defined in this paper, see [1].

2. Metric Graphic Sets

A set T of vectors is called a *positive lattice set* if all the coordinates in each vector of T are positive integers. That is, for positive integers n and k, a set $T = \{v_1, v_2, \ldots, v_n\}$, where $v_i = (a_{i1}, a_{i2}, \ldots, a_{ik})$ for each i, $1 \leq i \leq n$ is a *positive lattice set* if a_{ij} is a positive integer for each i and j, $1 \leq j \leq k$.

A positive lattice set T consisting of n k-vectors is called a *metric graphic set* if there exists a simple connected graph G of order $n+k$ with $\beta(G) = k$ such that $T = \{r(u_i|S) : u_i \in V(G)\backslash S, 1 \leq i \leq n\}$ for some metric basis S of G. If such G exists, then we say G is a *metric graphic realization of* T.

For example, consider the set $T = \{(1,1)\}$. By Remark 1.3, if S is a metric basis of K_n, then $S = V(K_n)\backslash\{u_i\}$ for some $u_i \in V(K_n)$. Thus, if $V(K_3) = \{u_1, u_2, u_3\}$, then $r(u_i|S) = (1,1)$ where $S = V(K_3)\backslash\{u_i\}$ for some $u_i \in V(K_3)$. Clearly, $T = \{r(u_i|S) : u_i \in V(K_3)\backslash S\}$ for any metric basis S of K_3. Therefore, T is a metric graphic set and K_3 is a metric graphic realization of T. On the other hand, the set $T = \{(1,1),(1,3)\}$ is not a metric graphic set since, for any graph G of order 4 with $\beta(G) = 2$, the diameter of G is 2. That is, the distance between any two distinct vertices in G is at most 2.

3. The Metric Graphic Sets with n 1-vectors and 1 k-vector

Graphs with metric dimension 1 and $n-1$ are characterized in [3] (see also Theorem 1.2). With the aid of these characterizations, we have the following theorems which characterizes metric graphic sets consisting of n 1-vectors and 1 k-vector.

Theorem 3.1. *Let* n *be a positive integer. Then a positive lattice set* T *consisting of* n *1-vectors is a metric graphic set if and only if* $T = \{(1), (2), \ldots, (n)\}$.

Proof. Suppose T is a metric graphic set. Then there exists a simple connected graph G of order $n+1$ with $\beta(G) = 1$ such that $T = \{r(u_i|S) : u_i \in V(G)\backslash S, 1 \leq i \leq n\}$ for some metric basis S of G. Since $\beta(G) = 1$, it follows that G is a path on $n+1$ vertices by Theorem 1.2. Let $u_0, u_1, u_2, \ldots, u_n$ be the vertices of G with end vertices u_0 and u_n. By Remark 1.3, either $S = \{u_0\}$ or $S = \{u_n\}$ is a metric basis of G. It follows that $\{r(u_i|S) : u_i \in V(G)\backslash S\} = \{(1), (2), \ldots, (n)\}$ for any metric basis of G.

Conversely, suppose $T = \{(1), (2), \ldots, (n)\}$. Let P_{n+1} be a path with vertex set $V(P_{n+1}) = \{u_0, u_1, u_2, \ldots, u_n\}$ and edge set $E(P_{n+1}) = \{u_i u_{i+1} : 0 \leq i \leq n-1\}$. Let $S = \{u_0\}$. Then $r(u_i|S) = (i)$ for each i, $1 \leq i \leq n$. Thus, $T = \{r(u_i|S) : u_i \in V(P_{n+1}) \backslash S\}$, and so T is a metric graphic set. □

Corollary 3.2. *Let n be a positive integer. Then the path P_{n+1} is the only metric graphic realization of $T = \{(1), (2), \ldots, (n)\}$.*

Theorem 3.3. *Let k be a positive integer. Then a positive lattice set $T = \{(a_1, a_2, \ldots, a_k)\}$ is a metric graphic set if and only if $a_j = 1$ for each j, $1 \leq j \leq k$.*

Proof. Suppose $T = \{(a_1, a_2, \ldots, a_k)\}$ is a metric graphic set. Then there exists a graph G of order $1 + k$ with $\beta(G) = k$ such that $T = \{r(u_i|S) : u_i \in V(G) \backslash S\}$ for some metric basis S of G. Since $|V(G)| = 1 + k$ and $\beta(G) = k$, it follows that $G = K_{1+k}$ by Theorem 1.2. By Remark 1.3, any metric basis S of G is of order k. Thus, $\{r(u|S)\} = \{(\underbrace{1, 1, \ldots, 1}_{k \text{ times}})\}$, where $\{u\} = V(G) \backslash S$, and so we have $a_j = 1$ for each j.

Conversely, suppose that $a_j = 1$ for each j. Let G be the complete graph of order $1 + k$ with vertices $u_0, u_1, u_2, \ldots, u_k$. Then, by Theorem 1.2, $\beta(G) = k$ and any metric basis S of G is of order k. Since $d(u_i, u_j) = 1$ for every distinct vertices $u_i, u_j \in V(G)$, then $r(u|S) = (\underbrace{1, 1, \ldots, 1}_{k \text{ times}})$ for any metric basis S of G, where $V(G) \backslash S = \{u\}$. Thus, T is a metric graphic set. □

Corollary 3.4. *Let k be a positive integer. Then the complete graph K_{1+k} is the only metric graphic realization of $T = \{(\underbrace{1, 1, \ldots, 1}_{k \text{ times}})\}$.*

4. Characterization of Metric Graphic Sets with 2 k-vectors

In this section, we give necessary and sufficient conditions for any positive lattice set consisting of 2 k-vectors to be a metric graphic set. The next two theorems proved in [3] are important tools in proving our results.

Theorem 4.1 (Chartrand et. al. [3]). *Let G be a connected graph of order $n \geq 4$. Then $\beta(G) = n - 2$ if and only if $G = K_{s,t}$ ($s, t \geq 1$ and $s + t = n$), $G = K_s + \overline{K_t}$ ($s \geq 1, t \geq 2$, and $s + t = n$), or $G = K_s + (K_1 \cup K_t)$ ($s, t \geq 1$ and $s + t + 1 = n$).*

Theorem 4.2 (Chartrand et. al. [3]). *If G is a connected graph with $|V(G)| \geq 2$ and $d = diam(G)$, then $\beta(G) \leq |V(G)| - d$.*

Corollary 4.3. *If G is a connected graph of order $|V(G)| \geq 4$ with $\beta(G) = |V(G)| - 2$, then $diam(G) = 2$.*

Proof. By Theorem 4.2, $\beta(G) \leq |V(G)| - \text{diam}(G)$. It follows that $\text{diam}(G) \leq 2$. If $\text{diam}(G) = 1$, then G is a complete graph, implying that $\beta(G) = |V(G)| - 1$, which is a contradiction. Hence, $\text{diam}(G) = 2$. □

Theorem 4.4. *Let n and k be positive integers. If $T = \{v_1, v_2, \ldots, v_n\}$ is a metric graphic set, where $v_i = (a_{i1}, a_{i2}, \ldots, a_{ik})$ for each i, $1 \leq i \leq n$, then $a_{ij} \leq n$ for each i and j, $1 \leq j \leq k$.*

Proof. Let G be a metric graphic realization of T. By definition, $|V(G)| = n + k$, $\beta(G) = k$, and $T = \{r(u_i|S) : u_i \in V(G) \backslash S, 1 \leq i \leq n\}$ for some metric basis S of G. Since $|V(G)| = n + k \geq 2$ and $\beta(G) = k$, we have $\text{diam}(G) \leq (n + k) - k = n$ by Theorem 4.2. Let $r(u_i|S)$ correspond to v_i so that $d(u_i, s_j)$ corresponds to a_{ij} for each i and j. Since $\text{diam}(G) \leq n$, then $1 \leq d(u_i, s_j) \leq n$ and it follows that $1 \leq a_{ij} \leq n$ for each i and j. □

Corollary 4.5. *Let n and k be positive integers. Then a positive lattice set $T = \{v_1, v_2, \ldots, v_n\}$, where $v_i = (a_{i1}, a_{i2}, \ldots, a_{ik})$ for each i, $1 \le i \le n$, is not a metric graphic set if there exists at least one a_{ij} such that $a_{ij} \ge n+1$ for some i and j, $1 \le j \le k$.*

If $n = 1$, then the converse of Theorem 4.4 is true for all k. If $k = 1$, then the converse of Theorem 4.4 is not true for all $n \ge 2$ since the positive lattice set $T = \{(2), (2), (3), (4), \ldots, (n)\}$ is not a metric graphic set by Theorem 3.1. In fact, the converse of Theorem 4.4 is not true for all $n \ge 2$.

Theorem 4.6. *Let k be a positive integer. Let G be any graph of order $2+k$ with $\beta(G) = k$.*

(i) *Let $S = \{s_1, s_2, \ldots, s_k\}$ be any metric basis of G and $V(G) \backslash S = \{u_1, u_2\}$. Then we have*

$$d(u_1, s_j) + d(u_2, s_j) \le 3 \qquad (\star)$$

for each j, $1 \le j \le k$.

(ii) *For $k \ge 2$ and for some metric basis S of G, equality holds in $(\star)$ if and only if $G = K_{s,t}$ with $s+t \ge 4$ and $s+t = k+2$, or $G = K_1 + (K_1 \cup K_k)$.*

(iii) *Let $S = \{s_1, s_2, \ldots, s_k\}$ be any metric basis of G and $V(G) \backslash S = \{u_1, u_2\}$. Then there exists at least one j such that $d(u_1, s_j) + d(u_2, s_j) = 3$.*

Proof. (i) If $k = 1$, then G is the path P_3 by Theorem 1.2. Thus, if $V(P_3) = \{u_0, u_1, u_2\}$ and $E(P_3) = \{u_0u_1, u_1u_2\}$, then $S = \{u_0\}$ or $S = \{u_2\}$ is a metric basis of P_3 and $d(u_1, u_0) + d(u_2, u_0) = 1 + 2 = 3$ and $d(u_1, u_2) + d(u_0, u_2) = 1 + 2 = 3$.

Let $k \ge 2$. Then $|V(G)| = 2 + k \ge 4$. Since $\beta(G) = k = |V(G)| - 2$, then G is one of the graphs descibed in Theorem 4.1. That is, $G = K_{s,t}$ $(s, t \ge 1)$, $G = K_s + \overline{K_t}$ $(s \ge 1, t \ge 2)$, or $G = K_s + (K_1 \cup K_t)$ $(s, t \ge 1)$. We consider five cases.

CASE 1. $G = K_{s,t}$, where $s + t \ge 4$ and $s = 1$. In this case, $k = s + t - 2 = t - 1$.

Let $X = \{u_1\}$ and $Y = \{v_1, v_2 \ldots, v_t\}$ be the bipartition of $K_{s,t}$. If S is a metric basis of G, then $S = V(G) \backslash \{u_1, v_l\}$ for some l, $1 \le l \le t$. Clearly, $r(u_1|S) = (\underbrace{1, 1, \ldots, 1}_{t-1 \text{ times}})$ and $r(v_l|S) = (\underbrace{2, 2, \ldots, 2}_{t-1 \text{ times}})$.

Therefore, for any metric basis $S = \{s_1, s_2, \ldots, s_k\}$ of $G = K_{s,t}$, where $s + t \ge 4$ and $s = 1$, we have $d(u_1, s_j) + d(v_l, s_j) \le 3$ for each j, $1 \le j \le k$, where $V(G) \backslash S = \{u_1, v_l\}$.

CASE 2. $G = K_{s,t}$, where $s + t \ge 4$ and $s \ge 2$. In this case, $k = s + t - 2$.

Let $X = \{u_1, u_2, \ldots, u_s\}$ and $Y = \{v_1, v_2, \ldots, v_t\}$ be the bipartition of $K_{s,t}$. If S is a metric basis of G, then $S = (X \backslash \{u_i\}) \cup (Y \backslash \{v_l\})$ for some i $(1 \le i \le s)$ and l, $1 \le l \le t$. By Remark 1.1, we can let

$$S = \{u_1, u_2, \ldots, u_{i-1}, u_{i+1}, \ldots, u_s,\\ v_1, v_2, \ldots, v_{j-1}, v_{j+1}, \ldots, v_t\}.$$

Then we have

$$r(u_i|S) = (\underbrace{2, 2, \ldots, 2}_{s-1 \text{ times}}, \underbrace{1, 1, \ldots, 1}_{t-1 \text{ times}})$$

and

$$r(v_l|S) = (\underbrace{1, 1, \ldots, 1}_{s-1 \text{ times}}, \underbrace{2, 2, \ldots, 2}_{t-1 \text{ times}}).$$

Therefore, for any metric basis $S = \{s_1, s_2, \ldots, s_k\}$ of $G = K_{s,t}$, where $s + t \ge 4$ and $s \ge 2$, we have $d(u_i, s_j) + d(v_l, s_j) \le 3$ for each j, $1 \le j \le k$, where $V(G) \backslash S = \{u_i, v_l\}$.

Case 3. $G = K_s + \overline{K_t}$, where $s + t \geq 4$ and $s \geq 2$. In this case, $k = s + t - 2$.

Let $V(K_s) = \{u_1, u_2, \ldots, u_s\}$ and $V(\overline{K_t}) = \{v_1, v_2, \ldots, v_t\}$. If S is a metric basis of G, then $S = (V(K_s)\backslash\{u_i\}) \cup (V(\overline{K_t})\backslash\{v_l\})$ for some i $(1 \leq i \leq s)$ and l $(1 \leq l \leq t)$. Since $d_G(u_i, v_l) = 1$ for each l, then $r(u_i|S) = (\underbrace{1, 1, \ldots, 1}_{s+t-2 \text{ times}})$ and $r(v_l|S)$ consists of $(s-1)$ 1's and $(t-1)$ 2's.

Therefore, for any metric basis $S = \{s_1, s_2, \ldots, s_k\}$ of $G = K_s + \overline{K_t}$, we have $d(u_i, s_j) + d(v_l, s_j) \leq 3$ for each j, $1 \leq j \leq k$, where $V(G)\backslash S = \{u_i, v_l\}$.

Case 4. $G = K_s + (K_1 \cup K_t)$, where $s = 1$ and $t \geq 2$. In this case, $k = s + t - 1 = t$.

Let $V(K_t) = \{u_1, u_2, \ldots, u_t\}$. For convenience, let $V(G) = V(K_t) \cup \{v_1, v_2\}$ and $E(G) = \{v_1v_2, v_1u_i : 1 \leq i \leq t\} \cup E(K_t)$. If S is a metric basis of G, then $S = V(K_t)$ or $S = (V(K_t)\backslash\{u_i\}) \cup \{v_1\}$ or $S = (V(K_t)\backslash\{u_i\}) \cup \{v_2\}$ for some i, $1 \leq i \leq t$.

If $S = V(K_t)$, then $r(v_1|S) = (\underbrace{1, 1, \ldots, 1}_{t \text{ times}})$ and $r(v_2|S) = (\underbrace{2, 2, \ldots, 2}_{t \text{ times}})$. If $S = (V(K_t)\backslash\{u_i\}) \cup \{v_1\}$, then $r(u_i|S) = (\underbrace{1, 1, \ldots, 1}_{t \text{ times}})$ and $r(v_2|S)$ consists of one 1 and $(t-1)$ 2's. If $S = (V(K_t)\backslash\{u_i\}) \cup \{v_2\}$, then $r(v_1|S) = (\underbrace{1, 1, \ldots, 1}_{t \text{ times}})$ and $r(u_i|S)$ consists of one 2 and $(t-1)$ 1's.

Therefore, for any metric basis $S = \{s_1, s_2, \ldots, s_k\}$ of $G = K_s + (K_1 \cup K_t)$, where $s = 1$ and $t \geq 2$, we have $d(v_1, s_j) + d(v_2, s_j) \leq 3$, where $V(G)\backslash S = \{v_1, v_2\}$ or $d(u_i, s_j) + d(v_1, s_j) \leq 3$, where $V(G)\backslash S = \{u_i, v_1\}$ or $d(u_i, s_j) + d(v_2, s_j) \leq 3$, where $V(G)\backslash S = \{u_i, v_2\}$ for each j, $1 \leq j \leq k$.

Case 5. $G = K_s + (K_1 \cup K_t)$, where $s \geq 2$ and $t \geq 1$. In this case, $k = s + t - 1$.

Let $V(K_s) = \{u_1, u_2, \ldots, u_s\}$, $V(K_t) = \{v_1, v_2, \ldots, v_t\}$, and $V(K_1) = \{w_1\}$. Consider first the case when $s = 2$ and $t = 1$. If S is a metric basis of G, then $S = V(G)\backslash\{w_1, u_i\}$ or $S = V(G)\backslash\{v_1, u_i\}$ for some i, $1 \leq i \leq s$. In any case, $\{r(w_1|S), r(u_i|S)\} = \{r(v_1|S), r(u_i|S)\} = \{(1,1), (1,2)\}$.

Now, let $s \geq 2$ and $t \geq 2$. If S is a metric basis of G, then $S = V(G)\backslash\{w_1, u_i\}$ for some i, $1 \leq i \leq s$, $S = V(G)\backslash\{w_1, v_l\}$ for some l, $1 \leq l \leq t$, or $S = V(G)\backslash\{u_i, v_l\}$ for some i and l.

If $S = V(G)\backslash\{w_1, u_i\}$, then $r(u_i|S) = (\underbrace{1, 1, \ldots, 1}_{(s-1)+t \text{ times}})$ and $r(w_1|S)$ consists of $(s-1)$ 1's and t 2's. If $S = V(G)\backslash\{w_1, v_l\}$, then $r(v_l|S) = (\underbrace{1, 1, \ldots, 1}_{s+(t-1) \text{ times}})$ and $r(w_1|S)$ consists of s 1's and $(t-1)$ 2's. If $S = V(G)\backslash\{u_i, v_l\}$, then $r(u_i|S) = (\underbrace{1, 1, \ldots, 1}_{s+t-1 \text{ times}})$ and $r(v_l|S)$ consists of $(s+t-2)$ 1's and one 2.

Therefore, for any metric basis $S = \{s_1, s_2, \ldots, s_k\}$ of $G = K_s + (K_1 \cup K_t)$, where $s \geq 2$ and $t \geq 1$, we have $d(w_1, s_j) + d(u_i, s_j) \leq 3$, where $V(G)\backslash S = \{w_1, u_i\}$, $d(v_l, s_j) + d(u_i, s_j) \leq 3$, where $V(G)\backslash S = \{w_1, v_l\}$, or $d(u_i, s_j) + d(v_l, s_j) \leq 3$, where $V(G)\backslash S = \{u_i, v_l\}$ for each j, $1 \leq j \leq k$.

(ii) From the proof of (i), we see that the graphs $G = K_{s,t}$ with $s + t \geq 4$ and $s + t = k + 2$ or $G = K_1 + (K_1 \cup K_k)$ are the only graphs such that $d(u_1, s_j) + d(u_2, s_j) = 3$ for each j, $1 \leq j \leq k$, for some metric basis $S = \{s_1, s_2, \ldots, s_k\}$ of G, where $V(G)\backslash S = \{u_1, u_2\}$.

(iii) If there exists no j such that $d(u_1, s_j) + d(u_2, s_j) = 3$, then $d(u_1, s_j) + d(u_2, s_j) = 2$ for each j, $1 \leq j \leq k$. This means that $d(u_1, s_j) = d(u_2, s_j) = 1$ for each j, $1 \leq j \leq k$, implying $r(u_1|S) = r(u_2|S) = (\underbrace{1, 1, \ldots, 1}_{k \text{ times}})$, which is a contradiction. □

Consider each graph G in the proof of Theorem 4.6(i). Observe that if there exists a vertex $s_j \in S = \{s_1, s_2, \ldots, s_k\}$ such that $d(u_1, s_j) = d(u_2, s_j) = 1$, where S is a metric basis of G and

$V(G)\backslash S = \{u_1, u_2\}$, then either $r(u_1|S) = (\underbrace{1, 1, \ldots, 1}_{k \text{ times}})$ or $r(u_2|S) = (\underbrace{1, 1, \ldots, 1}_{k \text{ times}})$. Thus, we have the following result.

Corollary 4.7. *Let $k \geq 2$ be a positive integer. Let $T = \{v_1, v_2\}$ be a metric graphic set, where $v_i = (a_{i1}, a_{i2}, \ldots, a_{ik})$ for $i = 1, 2$. If there exists a j such that $a_{1j} + a_{2j} = 2$, then one of the vectors of T is $(\underbrace{1, 1, \ldots, 1}_{k \text{ times}})$.*

The final theorem in this paper is a characterization of metric graphic sets consisting of 2 k-vectors.

Theorem 4.8. *Let k be a positive integer. Let $T = \{v_1, v_2\}$ be a positive lattice set with $v_i = (a_{i1}, a_{i2}, \ldots, a_{ik})$ for $i = 1, 2$. Then T is a metric graphic set if and only if one of the following holds:*

(i) $v_1 = (\underbrace{1, 1, \ldots, 1}_{k \text{ times}})$ *and* $a_{2j} \leq 2$ *for all* j, $1 \leq j \leq k$, *or*

(ii) $a_{1j} + a_{2j} = 3$ *for all* j, $1 \leq j \leq k$.

Proof. Suppose T is a metric graphic set. Let G be a metric graphic realization of T. By definition, $|V(G)| = 2 + k$ and $\beta(G) = k$. Then G is one of the graphs considered in the proof of Theorem 4.6(i). Let $S = \{s_1, s_2, \ldots, s_k\}$ be a metric basis of G, and let $V(G)\backslash S = \{u_1, u_2\}$. Let v_1 correspond to $r(u_1|S)$ and v_2 correspond to $r(u_2|S)$ so that $d(u_1, s_j) = a_{1j}$ and $d(u_2, s_j) = a_{2j}$ for each j, $1 \leq j \leq k$. It then follows that $v_1 = (\underbrace{1, 1, \ldots, 1}_{k \text{ times}})$ and $a_{2j} \leq 2$, $v_2 = (\underbrace{1, 1, \ldots, 1}_{k \text{ times}})$ and $a_{1j} \leq 2$, or $a_{1j} + a_{2j} = 3$ for all j, $1 \leq j \leq k$.

Conversely, suppose $v_1 = (\underbrace{1, 1, \ldots, 1}_{k \text{ times}})$ and $a_{2j} \leq 2$, or $a_{1j} + a_{2j} = 3$ for all j, $1 \leq j \leq k$. It is not difficult to see that one of the graphs considered in the proof of Theorem 4.6(i) is a metric graphic realization of T, implying that T is a metric graphic set. □

5. Further Results on Other Positive Lattice Sets

We enumerate some properties of a positive lattice set consisting of n 2-vectors that are not metric graphic sets. Let $n \geq 3$ be an integer and let T_p contains those vectors of a positive lattice set $T = \{(a_{11}, a_{12}), (a_{21}, a_{22}), \ldots, (a_{n1}, a_{n2})\}$ with first coordinate equal to p, $1 \leq p \leq q$, where $q = \max\{a_{i1} : 1 \leq i \leq n\}$. Then T is not a metric graphic set if T satisfies at least one of the following conditions:

(1) $\sum_{j=1}^{2} \sum_{i=1}^{n} a_{ij} \geq n(n+1)$ except for the case when $(a_{i1}, a_{i2}) = (i, i)$ for each i, $1 \leq i \leq n$.

(2) There exists at least one $a_{ij} \geq n + 1$.

(3) T_1 is empty or $|T_1| \geq 4$.

(4) At least two of $T_1, T_2, \ldots, T_q$ are empty or at least one of $T_{p+1}, T_{p+2}, \ldots, T_q$ is of order greater than 1, whenever T_p is empty for some p, $2 \leq p \leq q - 1$.

(5) $T_{p'} \neq \{p', p' - p\}$ for some p', $p + 1 \leq p' \leq q$, whenever T_p is empty for some p, $2 \leq p \leq q - 1$.

In general, given a positive lattice set T consisting of n k-vectors, it is an open problem to characterize when exactly it is a metric graphic set. However, summing up all the coordinates of each of the vectors in T gives a necessary inequality condition for it to be a metric graphic set. We end this paper by stating the following theorem without proof.

Theorem 5.1. *Let n and k be positive integers. Let $T = \{v_1, v_2, \ldots, v_n\}$ be a metric graphic set with $v_i = (a_{i1}, a_{i2}, \ldots, a_{ik})$ for each $1 \leq i \leq n$.*

(i) *Then*

$$\sum_{i=1}^{n} a_{ij} \leq \frac{n(n+1)}{2} \qquad (\star\star)$$

for each j, $1 \leq j \leq k$.

(ii) *There exists a metric graphic set such that equality in* $(\star\star)$ *holds for each* j, $1 \leq j \leq k$.

References

[1] Bondy J A and Murty U S R 2008 *Graph Theory* (Springer)
[2] Caceres J, Hernando C, Mora M, Pelayo I M, Puertas M L, Seara C and Wood D R 2005 On the metric dimension of some families of graphs *Electron. Notes Discrete Math.* **22** pp 129–133
[3] Chartrand G, Eroh L, Johnson M and Oellerman O 2000 Resolvability in graphs and metric dimension of a graph *Discrete Appl. Math.* **105** pp 99–113
[4] Garces I J L and Rosario J B 2015 Computing the metric dimension of truncated wheels *Appl. Math. Sci. (Ruse)* **9** pp 2761–2767
[5] Harary F and Melter R A 1976 On the metric dimension of a graph *Ars Combin.* **2** pp 191–195
[6] Slater P J 1975 Leaves of trees *Congr. Numer.* **14** pp 549–559
[7] Slater P J 1988 Dominating and reference sets in a graph *J. Math. Phys. Sci.* **22** pp 445–455

A generalized shackle of any graph H admits a super H-antimagic total labeling

Dafik[1,2], Moh Hasan[1,3], Y N Azizah[2], and I H Agustin[3]
[1]CGANT University of Jember, Jember, Indonesia
[2]Mathematics Edu. Depart. University of Jember, Jember, Indonesia
[3]Mathematics Depart. University of Jember, Jember, Indonesia

E-mail: d.dafik@unej.ac.id

Abstract. Let H be a simple and connected graph. A *shackle* of graph H, denoted by $G = shack(H, v, n)$, is a graph G constructed by non-trivial graphs $H_1, H_2, \dots, H_n$ such that, for every $1 \leq s, t \leq n$, H_s and H_t have no a common vertex with $|s - t| \geq 2$ and for every $1 \leq i \leq n-1$, H_i and H_{i+1} share exactly one common vertex v, called *connecting vertex*, and those $k-1$ connecting vertices are all distinct. By a *generalized shackle* of graph, we mean the graph $G = shack(H, v, n)$ by replacing the connecting vertex by any subgraph $K \subset H$ and we denote such a graph as $G = gshack(H, K \subset H, n)$. Graph $G = gshack(H, K \subset H, n)$ admits a H-covering, if every edge in $E(G)$ belongs to at least one subgraph of G isomorphic to a given graph H. A graph G is an (a, d)-H-antimagic total graph if there exists a bijective function $f : V(G) \cup E(G) \to \{1, 2, \dots, |V(G)| + |E(G)|\}$ such that for all subgraphs isomorphic to H, the total H-weights $w(H) = \sum_{v \in V(H)} f(v) + \sum_{e \in E(H)} f(e)$ form an arithmetic sequence $\{a, a+d, a+2d, ..., a+(n-1)d\}$, where a and d are positive integers and n is the number of all subgraphs isomorphic to H. If such a function exists then f is called an (a, d)-H-antimagic total labeling of G. An (a, d)-H-antimagic total labeling f is called super if $f : V(G) \to \{1, 2, \dots, |V(G)|\}$. In this paper, we study a super (a, d)-H antimagic total labeling of $G = gshack(H, e \in H, n)$ by using a partition technique. The result shows that there exist a super(a, d)-H antimagic total labeling for almost feasible difference d up to the determined upper bound.

1. Introduction

For all basic definition of graph we refer to [7]. All graph in this paper are finite, simple and undirected. By a labeling of a graph, we mean any mapping that sends some set of graph elements to a set of positive integers. If the domain is a vertex-set $V(G)$ or an edge-set $E(G)$, the labelings are called respectively vertex labelings or edge labelings. Moreover, if the domain is $V(G) \cup E(G)$ then the labelings are called *total* labelings. Some results regarded to (a, d)-edge antimagic total labeling have been published by Dafik *et. al* in [2, 3, 4]. Inayah *et al.* [8] and [15] introduced the (a, d)-H-antimagic total labeling. A graph G is an (a,d)-H-antimagic total graph if there exists a bijective function $f : V(G) \cup E(G) \to \{1, 2, \dots, |V(G)| + |E(G)|\}$ such that for all subgraphs of G isomorphic to H, the total H-weights $w(H) = \sum_{v \in V(H)} f(v) + \sum_{e \in E(H)} f(e)$ form an arithmetic sequence $\{a, a+d, a+2d, ..., a+(n-1)d\}$, where a and d are positive integers and n is the number of all subgraphs of G isomorphic to H. If such a function exist then f is called an (a,d)-H-antimagic total labeling of G. An (a,d)-H-antimagic total labeling f is called super if $f : V(G) \to \{1, 2, \dots, |V(G)|\}$.

Some relevant results have been published in many journals, some of them can be found in [5, 6, 9, 10, 11, 12, 13, 14]. In 2013, Inayah *et al.* [9] proved that, for H is a non-trivial connected graph and $k \geq 2$ is an integer, $shack(H, v, k)$ which contains exactly k subgraphs isomorphic to H is H-super antimagic. They only covered a shackle of graph when a vertex as a connector, and their paper did not cover all feasible d which are necessary condition for a graph to be a super (a,d)-H-antimagic total graph. Our paper attempt to solve super (a,d)-H-antimagic total graph of *a generalized shackle*, i.e. the shackle of graph in which an edge as a connector, and we have got more coverage for feasible d.

By using a special technique, namely *an integer set partition technique.* It is used in determining the feasible difference d. This concept was introduced in [1, 6]. Let n, c, d and k be positive integers. We consider the partition set $\mathcal{P}^n_{c,d}(k)$ of the set $\{1, 2, \dots, cn\}$ into n columns, $n \geq 2$, c-rows such that the difference between the sum of the numbers in the $(k+1)$th c-rows and the sum of the numbers in the kth c-rows is always equal to the constant d, where $k = 1, 2, \dots, n-1$. Thus these sums form an arithmetic sequence with the difference d. By the symbol $\mathcal{P}^n_{c,d}(k)$ we denote the kth c-rows in the partition with the difference d, where $k = 1, 2, \dots, n$. Let $\sum \mathcal{P}^n_{c,d}(k)$ be the sum of the numbers in $\mathcal{P}^n_{c,d}(k)$, thus $d = \sum \mathcal{P}^n_{c,d}(k+1) - \sum \mathcal{P}^n_{c,d}(k)$. By the notation $\mathcal{P}^n_{c,d}(k) \oplus b$ we mean that we add the constant b to every element in $\mathcal{P}^n_{c,d}(k)$.

2. A Useful Lemma and Corollary

Let G be a *generalized shackle* of graph denoted by $G = gshack(H, e, n)$. Let G and H be a connected graph with $|V(G)| = p_G$, $|E(G)| = q_G$, $|V(H)| = p_H$, and $|E(H)| = q_H$. The vertex set and edge set of the graph $G = gshack(H, e, n)$ can be split into following sets: $V(G) = \bigcup_{k=1}^{n} \{v_{i,k},\ v_{j,k},\ v_{j,k+1},\ 1 \leq i \leq p_H - 4,\ j = 1, 2\}$ and $E(G) = \bigcup_{k=1}^{n} \{e_{i,k},\ e_k,\ e_{k+1},\ 1 \leq i \leq q_H - 2,\}$. Let n, i, j, k be positive integers with $n \geq 2$. Thus $|V(G)| = p_G = n(p_H - 2) + 2$ and $|E(G)| = q_G = n(q_H - 1) + 1$.

A useful lemma proved in [5] gives a necessary condition for feasible d for a graph to be super (a,d)-H- antimagic total labeling.

Lemma 1. *[5] Let G be a simple graph of order p and size q. If G is super (a,d)-H-antimagic total labeling then $d \leq \frac{(p_G - p_H)p_H + (q_G - q_H)q_H}{n-1}$, for $p_G = |V(G)|$, $q_G = |E(G)|$, $p_H = |V(H)|$, $q_H = |E(H)|$, and $n = |H_j|$.*

Thus for $p_G = n(p_H - 2) + 2$ and $q_G = n(q_H - 1) + 1$, we have the following corollary.

Corrollary 1. *For $n \geq 2$, if the graph $G = gshack(H, e, n)$ admits super (a,d)-H-antimagic total labeling then $d \leq (p_H - 2)p_H + (q_H - 1)q_H$*

3. The Results

Prior to show the main theorem of this study, we need to establish the following lemmas. These lemmas are useful to develop the super (a,d)-H antimagic total labeling of $G = gshack(H, e \in H, n)$ by using a partition technique.

Lemma 2. *Let n, c, d and k be positive integers, $\sum_{i=1}^{c} \mathcal{P}^n_{c,d}(i,k) = \{(i-1)n + k,\ 1 \le i \le c\}$ and $\sum_{i=1}^{c} \mathcal{P}^n_{c,d}(i,k) = \{(k-1)c + i;\ 1 \le i \le c\}$ form an aritmatic sequence of difference $d \in \{c, c^2\}$, respectively.*

Proof. By simple calculation, for $1 \le k \le n$, it gives $\sum_{i=1}^{c} \mathcal{P}^n_{c,c}(i,k) = \frac{c^2n-nc}{2} + ck \longleftrightarrow \sum_{i=1}^{c} \mathcal{P}^n_{c,c}(i,k) = \{\frac{c^2n-nc}{2} + c, \frac{c^2n-nc}{2} + 2c, \dots, \frac{c^2n-nc}{2} + (n-1)c, \dots, \frac{c^2n-nc}{2} + nc\}$ and $\sum_{i=1}^{c} \mathcal{P}^n_{c,c^2}(i,k) = \frac{(c-c^2)}{2} + c^2k \longleftrightarrow \sum_{i=1}^{c} \mathcal{P}^n_{c,c^2}(i,k) = \{\frac{(c-c^2)}{2} + c^2, \frac{(c-c^2)}{2} + 2c^2, \dots, \frac{c^2n-nc}{2} + (n-1)c^2, \dots, \frac{c^2n-nc}{2} + nc^2\}$. It concludes the proof. □

Lemma 3. *Let n, c, d and k be positive integers. For $c \equiv 0 \pmod 2$ and $n \equiv 1 \pmod 2$, the following bijection*

$$\sum_{i=1}^{c} \mathcal{P}^n_{c,d}(i,k) = \begin{cases} \{(i-1)n + \frac{k+1}{2}; 1 \le i \le c, i \equiv 1 \pmod 2\} \\ \cup\{\frac{k-n}{2} + in; 1 \le i \le c, i \equiv 0 \pmod 2\}, k \equiv 1 \pmod 2 \\ \\ \{\frac{k+1-n}{2} + in; 1 \le i \le c, i \equiv 1 \pmod 2\} \\ \cup\{(i-1)n + \frac{k}{2}; 1 \le i \le c, i \equiv 0 \pmod 2\}, k \equiv 0 \pmod 2 \end{cases}$$

form an aritmatic sequence of difference $d = \frac{c}{2}$.

Proof. By simple calculation, for $1 \le k \le n$, it gives $\sum_{i=1}^{c} \mathcal{P}^n_{c,\frac{c}{2}}(i,k) = \frac{c}{4}(2cn - n + 1) + \frac{ck}{2} \longleftrightarrow \sum_{i=1}^{c} \mathcal{P}^n_{c,\frac{c}{2}}(i,k) = \{\frac{c}{4}(2cn-n+1) + \frac{c}{2}, \frac{c}{4}(2cn-n+1) + c, \dots, \frac{c}{4}(2cn-n+1) + \frac{c(n-1)}{2}, \dots, \frac{c}{4}(2cn-n+1) + \frac{cn}{2}\}$. □

Lemma 4. *Let $\sum \mathcal{P}^n_{c,d}(k)$ be the sum of partition $\mathcal{P}^n_{c,d}(k)$ for $n, c \ge 2$. Let t be a number of rows of all selected c_s, where s is any natural number. If n is any integer, or $n \equiv 1$ (mod 2) and $c \equiv 0$ (mod 2) then the feasible d are respectively $d \in \pm(c-t)^2 \oplus t \ominus 2 \otimes \{0, 1, \dots, t\}$ where $t = 0, 1, 2, \dots, c$, or $d \in \pm(c-t)^2 \oplus (t-2) \ominus 2 \otimes \{0, 1, \dots, t-2\} \oplus \{\pm 1\}$ where $t = 2, 3, 4, \dots, c$.*

Proof. By simple observation, any partition of $\mathcal{P}^n_{c,d}(k)$ can be formed by the partition of Lemma 2 and Lemma 3.

$$\begin{aligned} \mathcal{P}^n_{c,d}(k) &= \mathcal{P}^n_{c_1,c_1^2}(k) \cup (\mathcal{P}^n_{c_2,-c_2^2}(k) \oplus nc_1) \cup (\mathcal{P}^n_{c_3,c_3}(k) \oplus n(c_1+c_2)) \cup \\ &\quad (\mathcal{P}^n_{c_4,-c_4}(k) \oplus n(c_1+c_2+c_3)) \cup (\mathcal{P}^n_{c_5,\frac{c_5}{2}}(k) \oplus n(c_1+c_2+c_3+c_4)) \cup \\ &\quad (\mathcal{P}^n_{c_6,-\frac{c_6}{2}}(k) \oplus n(c_1+c_2+c_3+c_4+c_5)) \\ \sum \mathcal{P}^n_{c,d}(k) &= \sum \mathcal{P}^n_{c_1,c_1^2}(k) + \sum \mathcal{P}^n_{c_2,-c_2^2}(k) + \sum \mathcal{P}^n_{c_3,c_3}(k) + \sum \mathcal{P}^n_{c_4,-c_4}(k) + \\ &\quad \sum \mathcal{P}^n_{c_5,\frac{c_5}{2}}(k) + \sum \mathcal{P}^n_{c_6,-\frac{c_6}{2}}(k) + n \sum_{e=2}^{6} \sum_{t=1}^{e-1} c_e c_t \end{aligned}$$

where $c = c_1 + c_2 + c_3 + c_4 + c_5 + c_6$ and $d(c) = c_1^2 - c_2^2 + c_3 - c_4 + \frac{c_5}{2} - \frac{c_6}{2}$. Now suppose $c_3 + c_4 + c_5 + c_6 = t \Rightarrow c_1 + c_2 = c - t \Rightarrow d(c-t) = c_1^2 - c_2^2$ and we have two possibilities:

(i) $c_5 = c_6 = 0 \Rightarrow t = c_3 + c_4 \Rightarrow c_3 = t - c_4$. Considering $c_4 \in \{0, 1, \ldots, t\}$ and assumed that either c_1 or c_2 is zero

$$\begin{array}{rcl} d(c) & = & c_1^2 - c_2^2 + c_3 - c_4 + \frac{c_5}{2} - \frac{c_6}{2} \\ & \in & \pm(c-t)^2 \oplus (c_3 - c_4) \\ & \in & \pm(c-t)^2 \oplus (t - 2c_4) \\ & \in & \pm(c-t)^2 \oplus t \ominus 2 \otimes \{0, 1, \ldots, t\} \end{array}$$

It easy to see that the feasible d are alternately even and odd.

(ii) $c_3 + c_4 + c_5 + c_6 = t \Rightarrow$ for $n \equiv 1 \pmod 2$ and suppose $c_5 + c_6 = 2 \Rightarrow t = c_3 + c_4 + 2 \Rightarrow c_3 = t - c_4 - 2$. Considering that $c_4 \in \{0, 1, \ldots, t-2\}$,assumed that either c_1 or c_2 is zero, and assumed that either c_5 or c_6 is zero.

$$\begin{array}{rcl} d(c) & = & c_1^2 - c_2^2 + c_3 - c_4 + \frac{c_5}{2} - \frac{c_6}{2} \\ & \in & \pm(c-t)^2 \oplus (c_3 - c_4) \oplus \frac{c_5}{2} - \frac{c_6}{2} \\ & \in & \pm(c-t)^2 \oplus (t - 2c_4 - 2) \oplus \{\pm 1\} \\ & \in & \pm(c-t)^2 \oplus (t-2) \ominus 2 \otimes \\ & & \{0, 1, \ldots, t-2\} \oplus \{\pm 1\} \end{array}$$

It is clearly to see that if n is any integer, or $n \equiv 1 \pmod 2$ and $c \equiv 0 \pmod 2$ then the feasible d are respectively $d \in \pm(c-t)^2 \oplus t \ominus 2 \otimes \{0, 1, \ldots, t\}$ where $t = 0, 1, 2, \ldots, c$, or $d \in \pm(c-t)^2 \oplus (t-2) \ominus 2 \otimes \{0, 1, \ldots, t-2\} \oplus \{\pm 1\}$ where $t = 2, 3, 4, \ldots, c$. □

Lemma 5. *Let n, c, d and k be positive integers. The sum of $\mathcal{P}^n_{c+2,d}(k) = \mathcal{P}^n_{2,d_1}(k) \cup \mathcal{P}^n_{c,d_2}(k)$ where $\mathcal{P}^n_{2,2(c+1)}(k) = \{1 + (k-1)(c+1),\ \ 1 + k(c+1)\}$ or*

$$\mathcal{P}^n_{2,c+1}(k) = \begin{cases} \{1 + (\frac{k-1}{2} + (i-1)\lceil \frac{n+1}{2} \rceil)(c+1),\ i = 1, 2\}, \\ k \equiv 1 \pmod 2 \\ \{1 + (c+1)(\lceil \frac{n+1}{2} \rceil + \frac{k}{2} - 1 - (i-1)\lfloor \frac{n}{2} \rfloor),\ i = 1, 2\}, k \equiv 0 \pmod 2 \end{cases}$$

$\mathcal{P}^n_{c,(c+1)(c_1-c_2)}(k) = \{i + k(c+1) - c,\ j + (n-k)(c+1) + 1,\ 1 \le i \le c_1, c_1 + 1 \le j \le c,$ $c = c_1 + c_2\}$

form an aritmatic sequence of difference $d = (c+1) \otimes \{0, \pm 1, \pm 2, \ldots, \pm(c+2)\}$.

Proof. By simple calculation, it gives $\sum \mathcal{P}^n_{2,2(c+1)}(k) = 1 - c + 2k(c+1)$, $\sum \mathcal{P}^n_{2,c+1}(k) = 2 + (c+1)(k - 1 + \lceil \frac{n+1}{2} \rceil)$, $\sum \mathcal{P}^n_{c,(c+1)(c_1-c_2)}(k) = \frac{c_1 + c_1^2}{2} - cc_1 + c_2 n(c+1) + \frac{c_1c_2 + 3c_2 + cc_2}{2} + k(c+1)(c_1 - c_2)$, $\sum \mathcal{P}^n_{c+2,d}(k) = \frac{c_1 + c_1^2}{2} - cc_1 + c_2 n(c+1) + \frac{c_1c_2 + 3c_2 + cc_2}{2} + k(c+1)(c_1 - c_2) + \begin{cases} 2 + (c+1)(k - 1 + \lceil \frac{n+1}{2} \rceil), \text{for } d = (c+1)(\pm(c_1 - c_2) \oplus \pm 1) \\ \\ 1 - c + 2k(c+1), \text{for } d = (c+1)(\pm(c_1 - c_2) \oplus \pm 2) \end{cases}$

□

Lemma 6. *Let $\sum \mathcal{P}^n_{c+2,d}(k)$ be the sum of partition $\mathcal{P}^n_{c+2,d}(k)$ for $n, c \ge 2$. Let t be a number of rows of all selected c_s, where s is any natural number. If n is any integers or $n \equiv 1 \pmod 2$ then the feasible d are respectively $d = (c_1 + 1) \otimes \{0, \pm 1, \pm 2, \ldots, \pm(c_1 + 2)\} \oplus (\pm(c_2 - t)^2) \oplus t \ominus 2 \otimes \{0, 1, \ldots, t\}$ where $t = 0, 1, 2, \ldots, c_2$ or $d = (c_1 + 1) \otimes \{0, \pm 1, \pm 2, \ldots, \pm(c_1 + 2)\} \oplus (\pm(c_2 - t)^2) \oplus (t-2) \ominus 2 \otimes \{0, 1, \ldots, t-2\} \oplus \{\pm 1\}$ where $t = 2, 3, \ldots, c_2$, $c = c_1 + c_2$, $0 \le d \le (c+1)(c+2)$.*

Proof. Similarly, any partition of $\mathcal{P}^n_{c,d}(k)$ can be formed by the partition of Lemma 2, Lemma 3, and Lemma 5.

$$\begin{aligned}\mathcal{P}^n_{c+2,d}(k) &= \mathcal{P}^n_{c_1+2,d_1}(k) \cup (\mathcal{P}^n_{c_2,d_2}(k) \oplus n(c_{1_1}+c_{1_2}+1) \oplus 1);\\ \sum \mathcal{P}^n_{c+2,d}(k) &= \sum \mathcal{P}^n_{c_1+2,d_1}(k) + \sum \mathcal{P}^n_{c_2,d_2}(+nc_2(c_{1_1}+c_{1_2}+1)+c_2\end{aligned}$$

where $c+2 = c_1+c_2+2$, $c_1 = c_{1_1}+c_{1_2}$ and $d = d_1+d_2$.

From Lemma 5, we have $d_1 = (c_1+1)(\pm(c_{1_1}-c_{1_2}) \oplus \{\pm1, \pm2\})$. Considering Lemma 4 we have two possibilities:

(i) For any n, $d_2(c_2) = \pm(c_2-t)^2 \oplus t \ominus 2 \otimes \{0,1,\ldots,t\}$ where $t = 0,1,2,\ldots,c_2$.

(ii) For n is odd, $d_2(c_2) = \pm(c_2-t)^2 \oplus (t-2) \ominus 2 \otimes \{0,1,\ldots,t-2\} \oplus \{\pm1\}$ where $t = 0,1,2,\ldots,c_2$.

Combining the desired $d = d_1+d_2$, then $d = (c_1+1)(\pm(c_{1_1}-c_{1_2}) \oplus \{\pm1,\pm2\}) \oplus (\pm(c_2-t)^2) \oplus t \ominus 2 \otimes \{0,1,\ldots,t\}$ for any n, or $d = (c_1+1)(\pm(c_{1_1}-c_{1_2}) \oplus \{\pm1,\pm2\}) \oplus (\pm(c_2-t)^2) \oplus (t-2) \ominus 2 \otimes \{0,1,\ldots,t-2\} \oplus \{\pm1\}$ for odd n. Simply checking, since $c+2 = c_1+c_2+2$ and $c_1 = c_{1_1}+c_{1_2}$, we have $0 \le d \le (c+1)(c+2)$. □

Lemma 7. *Let n, c, x, d and k be positive integers. The sum of $\mathcal{P}^n_{c+4,d}(k) = \mathcal{P}^n_{4,d_1}(k) \cup \mathcal{P}^n_{c,d_2}(k)$ where*

$$\mathcal{P}^n_{4,0}(k) = \{1+(i+k-2)(c+2), 2+(n-k+j-1)(c+2),\ i,j=1,2\}$$

$$\mathcal{P}^n_{4,c+2}(k) = \begin{cases}\{1+(i+k-2)(c+2), i=1,2\}\\ \{2+\lceil\frac{n-k}{2}\rceil(c+2), 2+(\lfloor\frac{n}{2}\rfloor+1+\lfloor\frac{n-k}{2}\rfloor)(c+2)\}, k\equiv 1 \pmod 2\\ \{2+(\lfloor\frac{n}{2}\rfloor+\lceil\frac{n+1-k}{2}\rceil)(c+2), 2+\lfloor\frac{n+1-k}{2}\rfloor(c+2)\}, k\equiv 0 \pmod 2\end{cases}$$

$$\mathcal{P}^n_{4,2(c+2)}(k) = \begin{cases}\{i+\frac{k-1}{2}(c+2), j+(\lfloor\frac{n}{2}\rfloor+\frac{k+1}{2})(c+2), i,j=1,2\}, k\equiv 1 \pmod 2\\ \{i++(\lfloor\frac{n}{2}\rfloor+\frac{k}{2})(c+2), j+\frac{k}{2}(c+2),\ i,j=1,2\}, k\equiv 0 \pmod 2\end{cases}$$

$$\mathcal{P}^n_{4,3(c+2)}(k) = \begin{cases}\{1+(i+k-2)(c+2), i=1,2\}\\ \{2+\frac{k-1}{2}(c+2), 2+(\lfloor\frac{n}{2}\rfloor+\frac{k+1}{2})(c+2)\}, k\equiv 1 \pmod 2\\ \{2+(\lfloor\frac{n}{2}\rfloor+\frac{k}{2})(c+2), 2+\frac{k}{2}(c+2)\}, k\equiv 0 \pmod 2\end{cases}$$

or

$$\begin{aligned}\mathcal{P}^n_{4,4(c+2)}(k) &= \{1+(i+k-2)(c+2), 2+(j+k-2)(c+2), i,j=1,2\}\\ \mathcal{P}^n_{c,(c+2)(c_1-c_2)}(k) &= \{i+k(c+2)-c, j+(n-k)(c+2)+1,\ 1\le i\le c_1,\\ &\quad c_1+1\le j\le c,\ c=c_1+c_2\}\end{aligned}$$

form an arithmetic sequence of difference $d = (c+2) \otimes \{0, \pm1, \pm2, \ldots, \pm(c+4)\}$.

Proof. By simple calculation, it gives $\sum \mathcal{P}^n_{4,0}(k) = 6 + (2n+2)(c+2)$, $\sum \mathcal{P}^n_{4,(c+2)}(k) = 6 + (\lceil \frac{n+1}{2} \rceil + n + 1)(c+2) + k(c+2)$, $\sum \mathcal{P}^n_{4,2(c+2)}(k) = 6 + (2\lceil \frac{n+1}{2} \rceil - 2)(c+2) + 2k(c+2)$, $\sum \mathcal{P}^n_{4,3(c+2)}(k) = 6 + (\lceil \frac{n+1}{2} \rceil - 2)(c+2) + 3k(c+2)$, $\sum \mathcal{P}^n_{4,4(c+2)}(k) = 2 - 2c + 4k(c+2)$, $\sum \mathcal{P}^n_{c,(c+2)(c_1-c_2)}(k) = \frac{1+c_1}{2}c + \frac{cc_2}{2} - cc_1 + c_2 n(c+2) + k(c+2)(c_1 - c_2)$. Thus $\sum \mathcal{P}^n_{c+4,d}(k) = \sum \mathcal{P}^n_{4,d_1}(k) + \sum \mathcal{P}^n_{c,d_2}(k) = \mathcal{S}^n_{4,d_1} + \mathcal{S}^n_{c,d_2} + k(d_1 + d_2)$, where $d_1 = \{0, \pm 1, \pm 2, \pm 3, \pm 4\} \otimes (c+2)$ and $d_2 = (c_1 - c_2)(c+2)$. Since $c = c_1 + c_2$, it gives $d_2 = (c_1 - c_2)(c+2)$. For c is even or odd, it implies respectively $d_2 \in \{0, \pm 2, \ldots, \pm c\} \otimes (c+2)$ or $d_2 \in \{\pm 1, \pm 3, \ldots, \pm c\} \otimes (c+2)$. Thus

$$\begin{aligned} d &= d_1 + d_2, \ d_2 \text{ is even} \\ &= \{0, \pm 1, \pm 2, \pm 3, \pm 4\} \otimes (c+2) \oplus \{0, \pm 2, \ldots, \pm c\} \otimes (c+2) \\ &= \{0, \pm 1, \pm 2, \ldots, \pm (c+4)\} \otimes (c+2) \end{aligned}$$

or

$$\begin{aligned} d &= d_1 + d_2, \ d_2 \text{ is odd} \\ &= \{0, \pm 1, \pm 2, \pm 3, \pm 4\} \otimes (c+2) \oplus \{\pm 1, \pm 3, \ldots, \pm c\} \otimes (c+2) \ . \\ &= \{0, \pm 1, \pm 2, \ldots, \pm (c+4)\} \otimes (c+2) \end{aligned}$$

Thus, $d = (c+2) \otimes \{0, \pm 1, \pm 2, \ldots, \pm (c+4)\}$. □

Remark 1. *For any positive integer x, the partition of $\mathcal{P}^n_{c+2x,d}(k)$ can be formed by the following way:*

$$\begin{aligned} \mathcal{P}^n_{c+2x,d}(k) &= \mathcal{P}^n_{c_1+2x,d_1}(k) \cup (\mathcal{P}^n_{c_2,d_2}(k) \oplus n(c_1 + x) \oplus 1); \text{for } 1 \le k \le n \\ \sum \mathcal{P}^n_{c+2x,d}(k) &= \sum \mathcal{P}^n_{c_1+2x,d_1}(k) + \sum (\mathcal{P}^n_{c_2,d_2}(k) \oplus n(c_1 + x) \oplus 1) \\ &= \sum \mathcal{P}^n_{c_1+2x,d_1}(k) + \sum \mathcal{P}^n_{c_2,d_2}(k) + c_2(n(c_1 + x) + 1); \text{for } 1 \le k \le n, \end{aligned}$$

where $c = c_1 + c_2$ and $d = d_1 + d_2$.

Lemma 8. *Let $\sum \mathcal{P}^n_{c+4,d}(k)$ be the sum of partition $\mathcal{P}^n_{c+4,d}(k)$ for $n, c \ge 2$. Let t be a number of rows of all selected c_s, where s is any natural number. If n is any integers then the feasible d are $d = (c_1+2) \otimes \{0, \pm 1, \pm 2, \ldots, \pm (c_1+4)\} \oplus (\pm (c_2 - t)^2) \oplus t \ominus 2 \otimes \{0, 1, \ldots, t\}$ where $t = 0, 1, 2, \ldots, c_2$, $c = c_1 + c_2$ or $d = (c_1 + 1) \otimes \{0, \pm 1, \pm 2, \ldots, \pm (c_1 + 2)\} \oplus (c_2 + 1) \otimes \{0, \pm 1, \pm 2, \ldots, \pm (c_2 + 2)\} \oplus (\pm (c_3 - t)^2) \oplus t \ominus 2 \otimes \{0, 1, \ldots, t\}$ where $t = 0, 1, 2, \ldots, c_3$, $c = c_1 + c_2 + c_3$. Moreover, if $n \equiv 1 \pmod 2$ then the feasible d are $d = (c_1+2) \otimes \{0, \pm 1, \pm 2, \ldots, \pm (c_1+4)\} \oplus (\pm (c_2 - t)^2) \oplus (t-2) \ominus 2 \otimes \{0, 1, \ldots, t-2\} \oplus \{\pm 1\}$ where $t = 2, 3, \ldots, c_2$, $c = c_1 + c_2$ or $d = (c_1 + 1) \otimes \{0, \pm 1, \pm 2, \ldots, \pm (c_1 + 2)\} \oplus (c_2 + 1) \otimes \{0, \pm 1, \pm 2, \ldots, \pm (c_2 + 2)\} \oplus (\pm (c_3 - t)^2) \oplus (t-2) \ominus 2 \otimes \{0, 1, \ldots, t-2\} \oplus \{\pm 1\}$ where $t = 2, 3, \ldots, c_3$, $c = c_1 + c_2 + c_3$.*

Proof. By using Lemma 2, 3, 5 and 7, also Remark 1, 2 and 3, we have two possibilities:
(i) $\mathcal{P}^n_{c+4,d}(k) = \mathcal{P}^n_{c_1+4,d_1}(k) \cup (\mathcal{P}^n_{c_2,d_2}(k) \oplus n(c_1 + 2) \oplus 1)$
Thus

$$\begin{aligned} \sum \mathcal{P}^n_{c+4,d}(k) &= \sum \mathcal{P}^n_{c_1+4,d_1}(k) + \sum (\mathcal{P}^n_{c_2,d_2}(k) \oplus n(c_1 + 2) \oplus 1) \\ &= \mathcal{S}^n_{c_1+4,d_1} + \mathcal{S}^n_{c_2,d_2} + nc_2(c_1 + 2) + c_2 + kd \\ &= \mathcal{S}^n_{c+4,d} + kd \end{aligned}$$

IOP Conf. Series: Journal of Physics: Conf. Series **893** (2017) 012042 doi:10.1088/1742-6596/893/1/012042

where

$$\begin{aligned}
d_1 &= \{0, \pm1, \pm2, \ldots, \pm(c_1+4)\} \otimes (c_1+2)\\
d_2 &= (\pm(c_2-t)^2) \oplus t \ominus 2 \otimes \{0,1,\ldots,t\}, t = 0,1,2,\ldots,c_2\\
d_2 &= (\pm(c_2-t)^2) \oplus (t-2) \ominus 2 \otimes \{0,1,\ldots,t-2\} \oplus \{\pm1\},\ t = 2,3,\ldots,c_2\\
c &= c_1+c_2,\\
d &= d_1+d_2
\end{aligned}$$

Thus

$$\begin{aligned}
d &= d_1+d_2\\
&= \{0, \pm1, \pm2, \ldots, \pm(c_1+4)\} \otimes (c_1+2) \oplus (\pm(c_2-t)^2) \oplus t \ominus 2 \otimes \{0,1,\ldots,t\},\\
t &= 0,1,2,\ldots,c_2
\end{aligned}$$

or

$$\begin{aligned}
d &= d_1+d_2\\
&= \{0, \pm1, \pm2, \ldots, \pm(c_1+4)\} \otimes (c_1+2) \oplus (\pm(c_2-t)^2)\\
&\quad \oplus(t-2) \ominus 2 \otimes \{0,1,\ldots,t-2\} \oplus \{\pm1\},\\
t &= 2,3,\ldots,c_2
\end{aligned}$$

(ii) $\mathcal{P}^n_{c+4,d}(k) = \mathcal{P}^n_{c_1+2,d_1}(k) \cup (\mathcal{P}^n_{c_2+2,d_2}(k) \oplus n(c_1+1) \oplus 1) \cup (\mathcal{P}^n_{c_3,d_3}(k)$ $\oplus n(c_1+c_2+1) \oplus 1)$

Thus

$$\begin{aligned}
\sum \mathcal{P}^n_{c+4,d}(k) &- \sum \mathcal{P}^n_{c_1+2,d_1}(k) + \sum(\mathcal{P}^n_{c_2+2,d_2}(k) \oplus n(c_1+1) \oplus 1) + \sum(\mathcal{P}^n_{c_3,d_3}(k)\\
&\quad \oplus n(c_1+c_2+1) \oplus 1)\\
&= \mathcal{S}^n_{c_1+2,d_1} + \mathcal{S}^n_{c_2+2,d_2} + \mathcal{S}^n_{c_3,d_3} + nc_3(c_1+c_2+1) + c_3 + kd\\
&= \mathcal{S}^n_{c+4,d} + kd
\end{aligned}$$

where

$$\begin{aligned}
d_1 &= \{0, \pm1, \pm2, \ldots, \pm(c_1+2)\} \otimes (c_1+2)\\
d_2 &= \{0, \pm1, \pm2, \ldots, \pm(c_2+2)\} \otimes (c_2+2)\\
d_3 &= (\pm(c_3-t)^2) \oplus t \ominus 2 \otimes \{0,1,\ldots,t\}, t = 0,1,2,\ldots,c_3\\
d_3 &= (\pm(c_3-t)^2) \oplus (t-2) \ominus 2 \otimes \{0,1,\ldots,t-2\} \oplus \{\pm1\},\ t = 2,3,\ldots,c_3\\
c &= c_1+c_2+c_3\\
d &= d_1+d_2+d_3
\end{aligned}$$

Thus

$$\begin{aligned}
d &= d_1+d_2+d_3\\
&= \{0, \pm1, \pm2, \ldots, \pm(c_1+2)\} \otimes (c_1+2) \oplus \{0, \pm1, \pm2, \ldots, \pm(c_2+2)\}\\
&\quad \otimes(c_2+2) \oplus (\pm(c_3-t)^2) \oplus t \ominus 2 \otimes \{0,1,\ldots,t\},\\
t &= 0,1,2,\ldots,c_3
\end{aligned}$$

or

$$\begin{aligned}
d &= d_1+d_2+d_3\\
&= \{0, \pm1, \pm2, \ldots, \pm(c_1+2)\} \otimes (c_1+2) \oplus \{0, \pm1, \pm2, \ldots, \pm(c_2+2)\}\\
&\quad \otimes(c_2+2) \oplus (\pm(c_3-t)^2) \oplus (t-2) \ominus 2 \otimes \{0,1,\ldots,t-2\} \oplus \{\pm1\},\\
t &= 2,3,\ldots,c_3.
\end{aligned}$$

□

Now we are ready to present the main theorem of our result, i.e. the existence of super $(a,d)-H$ antimagicness of the graph $G = gshack(H,e,n)$, in the following theorem.

Theorem 1. *For $n \geq 2$, the graph $G = gshack(H,e,n)$ admits a super $(a,d)-H$ antimagic total labeling with feasible $0 \leq d \leq (p_H-2)p_H + (q_H-1)q_H$.*

Proof. For $k = 1,2,\ldots,n$, by Lemma 2, 3, 5 and 7 we define the vertex and the edge labels as follows

$$\begin{aligned} g_k(V(G)) &= \mathcal{P}^n_{p_H,d_v}(k) = \mathcal{P}^n_{c_1+4,d_1}(k) \cup (\mathcal{P}^n_{c_2,d_2}(k) \oplus n(c_1+2) \oplus 1) \\ g_k(E(G)) &= \mathcal{P}^n_{q_H,d_e}(k) \oplus p_G = \mathcal{P}^n_{c_3+2,d_3}(k) \oplus p_G \cup (\mathcal{P}^n_{c_4,d_4}(k) \oplus n(c_3+1) \oplus 1 \oplus p_G) \\ W_{g_k} &= \sum \mathcal{P}^n_{p_H,d_v}(k) + \sum \mathcal{P}^n_{q_H,d_e}(k) = \mathcal{S}^n_{p_H,d_v} + \mathcal{S}^n_{q_H,d_e} + q_H p_G + kd \end{aligned}$$

where $\mathcal{P}^n_{c_2,d_2}(k)$ and $\mathcal{P}^n_{c_4,d_4}(k)$ are linear combinations of Lemma 2, 3. It is not difficult to verify by a routine procedure that a system of set W_{g_k} forms arithmetic sequence where $a = \mathcal{S}^n_{p_H,d_v} + \mathcal{S}^n_{q_H,d_e} + q_H p_G$, and d by Lemma 6 and 8 are as follows:

$$\begin{aligned} d_v &= (c_1+2) \otimes \{0,\pm1,\pm2,\ldots,\pm(c_1+4)\} \oplus (\pm(p_H-c_1-4-t)^2) \oplus t \ominus 2 \\ &\quad \otimes\{0,1,\ldots,t\},\ t = 0,1,2,\ldots,p_H-c_1-4 \\ d_v &= (c_{1_1}+1) \otimes \{0,\pm1,\pm2,\ldots,\pm(c_{1_1}+2)\} \oplus (c_{1_2}+1) \otimes \{0,\pm1,\pm2,\ldots,\pm(c_{1_2}+2)\} \\ &\quad \oplus(\pm(p_H-c_1-4-t)^2) \oplus t \ominus 2 \otimes \{0,1,\ldots,t\}, t = 0,1,2,\ldots,p_H-c_1-4, \\ &\quad c_1 = c_{1_1} + c_{1_2} \\ d_v &= (c_1+2) \otimes \{0,\pm1,\pm2,\ldots,\pm(c_1+4)\} \oplus (\pm(p_H-c_1-4-t)^2) \oplus (t-2) \ominus 2 \\ &\quad \otimes\{0,1,\ldots,t-2\} \oplus \{\pm1\}, t = 2,3,\ldots,p_H-c_1-4 \\ d_v &= (c_{1_1}+1) \otimes \{0,\pm1,\pm2,\ldots,\pm(c_{1_1}+2)\} \oplus (c_{1_2}+1) \otimes \{0,\pm1,\pm2,\ldots,\pm(c_{1_2}+2)\} \\ &\quad \oplus(\pm(p_H-c_1-4-t)^2) \oplus (t-2) \ominus 2 \otimes \{0,1,\ldots,t-2\} \oplus \{\pm1\}, \\ &\quad t = 2,3,\ldots,p_H-c_1-4 \\ d_e &= (c_3+1) \otimes \{0,\pm1,\pm2,\ldots,\pm(c_3+2)\} \oplus (\pm(c_4-t)^2) \oplus t \ominus 2 \otimes \{0,1,\ldots,t\}, \\ &\quad t = 0,1,2,\ldots,q_H-c_3-2 \\ d_e &= (c_3+1) \otimes \{0,\pm1,\pm2,\ldots,\pm(c_3+2)\} \oplus (\pm(c_4-t)^2) \oplus (t-2) \ominus 2 \\ &\quad \otimes\{0,1,\ldots,t-2\} \oplus \{\pm1\}, t = 2,3,\ldots,p_H-c_3-2 \end{aligned}$$

where $p_H = c_1+c_2+4$, and $q_H = c_3+c_4+2$. Since $d = d_v + d_e$, there will be eight combinations of d satisfy $0 \leq d \leq (p_H-2)p_H + (q_H-1)q_H$. Thus, for $n \geq 2$, the graph $G = gshack(H,e,n)$ admits a super $(a,d)-H$ antimagic total labeling. □

Concluding Remarks

We have shown the existence of super antimagicness of generalized shackle of any graph in which an edge as a connector. By using a partition technique we can prove that $gshack(H,e,n)$ admits a super(a,d)-H antimagic total labeling for almost feasible difference d up to the determined upper bound. We note that if the shackle of graph contains a subgraph as a connector then finding the labels for feasible d remains widely open. Thus, we propose the following open problem.

Open Problem 1. *Let K be a subgraph of H, and G =gshack$(H,K \subset H,n)$. Does G admit a super $(a,d)-H$ antimagic total labeling for $n \geq 2$ and $d \leq (p_H-p_K)p_H + (q_H-q_K)q_H$?*

Acknowledgments
We gratefully acknowledge the support from DP2M research grant HIKOM-DIKTI and CGANT - University of Jember of year 2017.

References

[1] Bača M, Brankovic L, Lascsáková M, Phanalasy O and Semaničová-Feňovčíková A 2013 On *d*-antimagic labelings of plane graphs *Electron. J. Graph Theory Appl.* **1(1)** pp 28–39
[2] Bača M, Miller M, Ryan J, and Dafik 2009 Antimagic labeling of disjoint union of s-crowns *Util. Math.* **78** pp 193–205
[3] Dafik, Miller M, Ryan J, and Baca M 2011 Super edge-antimagic total labelings of $mK_{n,n}$ *Ars Combin.* **101** pp 97–107
[4] Dafik, Miller M, Ryan J, and Baca M 2008 On antimagic labelings of disjoint union of complete s-partite graphs *J. Combin. Math. Combin. Comput.* **65** pp 41–49
[5] Dafik, Purnapraja A K and Hidayat R 2015 Cycle-Super Antimagicness of Connected and Disconnected Tensor Product of Graphs *Procedia Computer Science* **74** pp 93–99
[6] Dafik, Slamin, Tanna D, Semaničová-Feňovčíková A and Bača M 2017 Constructions of H-antimagic graphs using smaller edge-antimagic graphs *Ars Combin.* **133** pp 233–245
[7] Gross J L, Yellen J and Zhang P 2014 *Handbook of Graph Theory* Second Edition (CRC Press, Taylor and Francis Group)
[8] Inayah N, Salman A N M and Simanjuntak R 2009 On $(a, d) - H$-antimagic coverings of graphs *J. Combin. Math. Combin. Comput.* **71** pp 273–281
[9] Inayah N , Simanjuntak R and Salman A N M 2013 Super $(a, d) - H$-antimagic total labelings for shackles of a connected graph H *The Australasian Journal of Combinatorics* **57** pp 127–138
[10] Jeyanthi P and Selvagopal P 2010 More classes of H-supermagic Graphs *Intern. J. of Algorithms, Computing and Mathematics* **3(1)** pp 93–108.
[11] Lladó A and Moragas J 2007 Cycle-magic graphs *Discrete Math.* **307** pp 2925–2933.
[12] Maryati T K, Salman A N M , Baskoro E T , Ryan J and Miller M 2010 On H- supermagic labelings for certain shackles and amalgamations of a connected graph *Util. Math.* **83** pp 333–342.
[13] Ngurah A A G, Salman A N M and Susilowati L 2010 H-supermagic labeling of graphs *Discrete Math.* **310** pp 1293–1300.
[14] Roswitha M and Baskoro E T 2012 H-magic covering on some classes of graphs *American Institute of Physics Conference Proceedings* **1450** pp 135–138.
[15] Simanjuntak R, Miller M and Bertault F 2000 Two new (a, d)-antimagic graph labelings *Proc. Eleventh Australas. Workshop Combin. Alg. (AWOCA)* **17** pp 179–189.

IOP Conf. Series: Journal of Physics: Conf. Series **893** (2017) 012043 doi:10.1088/1742-6596/893/1/012043

Implementation of Pollard Rho over binary fields using Brent Cycle Detection Algorithm

Intan Muchtadi-Alamsyah, Taufiq Akbari Utomo

Faculty of Mathematics and Natural Sciences, Institut Teknologi Bandung, Jalan Ganesha No. 10 Bandung 40132, Indonesia

E-mail: ntan@math.itb.ac.id

Abstract. The security of Elliptic Curve Cryptography depends on how to solve the Elliptic Curve Cryptography Discrete Logarithm Problem (ECDLP). In this paper we propose the use of modified Pollard Rho Algorithm by using Brent Cycle Detection Algorithm to solve the ECDLP. We give performance comparison on time and the number of iterations between Pollard Rho with Brent Cycle Detection and Pollard Rho with Negation map. In particular, for Koblitz curve, we also give comparison between Pollard Rho with Brent Cycle Detection and Pollard Rho with Negation and Frobenius maps.

1. Introduction

Some public key cryptosystems that is widely used today include RSA and the elliptic curve cryptosystem. Elliptic curve cryptosystem can provide the same level of security with RSA with smaller key sizes. As a result, the elliptic curve cryptosystem consumes less energy, memory, bandwidth, and time to perform encryption and decryption than RSA cryptosystem.

The best way to test the security of a cryptosystem is to position themselves as the attacker. The security of elliptic curve cryptosystem relies on the solution of the elliptic curve discrete logarithm problem. One of the algorithm used to resolve such problems is the Pollard Rho Algorithm.

In previous research we have implemented the Pollard Rho algorithm using the Frobenius and Negation maps [5] and also Basis Conversion [4]. In this research we explore the use of Brent Cycle Detection Algorithm to detect collisions in Pollard Rho Algorithm. We give comparison on time and the number of iterations between Pollard Rho with Brent Cycle Detection and Pollard Rho with Negation map. In particular, for Koblitz curve, we also give comparison between Pollard Rho with Brent Cycle Detection and Pollard Rho with Negation and Frobenius maps.

2. Basic Definition

Let F be binary field. An elliptic curve is an equation of the form

$$y^2 + a_1xy + a_3y = x^3 + a_2x^2 + a_4x + a_6$$

plus certain conditions with $a_1, a_2, a_3, a_4, a_6 \in F$. For binary field $GF(2^n)$ the above equation can be simplified to

$$y^2 + xy = x^3 + ax^2 + b,$$

with $a, b \in GF(2^n)$.

If E is an elliptic curve over $\mathbb{R}$ then the addition operation of two points can be illustrated as follows.

Let $P = (x_1, y_1), Q = (x_2, y_2) \in E$, and $R = P + Q$ with $P \neq Q$. To obtain R, drag a line through P and Q. This line will intersect the curve at three points, two of them are P and Q. Let R' be the third intersection point. Then R is the reflection of R' through the x-axis. Note that if Q is the reflection of P through the x-axis, the line formed with intersect the curve at P, Q and another point called the point at infinity, which we denote by O. Reflecting O through the x-axis gives O itself. So $P+Q = O$ when Q is the reflection of P through the x-axis.

Now let $S = P + P$. To obtain S, draw the tangent line of E at P. This line will intersect the curve at two points, one of them is P. Let R' be the second intersection point. Then S is the reflection of S' through the x-axis.

For E elliptic curve over $GF(2^n)$, the addition rule is as follows.

(i) For all $P \in E$, $P + O = O + P = P$, with O as the point at infinity.

(ii) If $P = (x, y)$, then the negative of P, denotes by $-P$ is $(x, x + y)$.

(iii) If $P = (x_1, y_1)$ and $Q = (x_2, y_2)$ with $Q \neq \pm P$, then $P + Q = (x_3, y_3)$, where $x_3 = \lambda^2 + \lambda + x_1 + x_2 + a$, $y_3 = \lambda(x_1 + x_3) + x_3 + y_1$, and $\lambda = \dfrac{y_1 + y_2}{x_1 + x_2}$.

(iv) If $P = (x_1, y_1)$ then $P + P = (x_4, y_4)$, where $x_4 = \mu^2 + \mu + a = x_1^2 + \dfrac{b}{x_1^2}$, $y_4 = x_1^2 + \mu x_4 + x_4$, and $\mu = x_1 + \dfrac{y_1}{x_1}$.

The points in E together with O form an abelian group with the above addition operation [2].

3. Pollard Rho Algorithm

Let $P \in E$ and k be a positive integer. We define $kP = P + P + \cdots + P$. Moreover if $k = 0$ then $kP = O$, and if k is a negative integer then $kP = (-P) + (-P) + \cdots + (-P)$. We define the order of P as the smallest positive integer m such that $mP = O$.

Sending a message from Alice to Bob using Elliptic Curve Cryptography is done in the following way:

(i) Alice and Bob agree on the elliptic curve E and a point P of order m.

(ii) Bob chooses a positive integer $k \in \{1, 2, \cdots, m-1\}$, which is confidential. After that, Bob computes $Q = kP$ and announces Q.

(iii) Let u be the message that Alice will send. Alice converts u into a point M on the curve.

(iv) Alice chooses a positive integer $j \in \{1, 2, \cdots, m-1\}$, which is confidential. After that, Alice computes $C_1 = jP$ and $C_2 = M + jQ$.

(v) Alice sends (C_1, C_2) to Bob.

(vi) Bob can get M from (C_1, C_2) by computing $M = C_2 - jQ = C_2 - j(kP) = C_2 - k(jP) = C_2 - kC_1$.

(vii) Bob converts M back into the message u.

Eve as a third party only knows E, P, Q, n, C_1 and C_2. If she wants to know M, then she must know j or k. This is the discrete logarithm problem.

More formally, given an elliptic curve E and a point $P \in E$ of order m, the Elliptic Curve Discrete Logarithm Problem (ECDLP) is to find $k \in \{1, \cdots, m-1\}$, if $Q = kP$ is known. The Pollard Rho Algorithm is one of the algorithms for solving the discrete logarithm problem. The main idea of this algorithm is to generate a sequence $\{X_i\}_{i=1}^{\infty}$, with $X_i \in E$, until the collision occurred, i.e. when $X_i = X_j$ for some $i \neq j$. The computation of each term in the sequence is deterministic so that we can obtain information on X_i such that ECDLP can be solved. However, the sequence $\{X_i\}_{i=0}^{\infty}$ is generated such that the sequence behaves like a random sequence. Thus, based on the birthday paradox, expectations of the number of terms prior to the collision is $\sqrt{\frac{\pi m}{2}}$, with m is the order of P [3].

Let E be an elliptic curve over $GF(2^n)$, $P \in E$ of order m, and $Q = kP$, with k is unknown. The Pollard Rho Algorithm can be used to find k and is given as follows :

(i) Divide E into three sets with almost equal cardinalities, namely S_1, S_2, and S_3.

(ii) Set $X_0 = P$, and define the iteration function f:

$$X_{i+1} = f(X_i) = \begin{cases} X_i + P, & X_i \in S_1 \\ 2X_i, & X_i \in S_2 \\ X_i + Q, & X_i \in S_3 \end{cases}$$

(iii) Write $X_i = s_iP + t_iQ$ for all i.

(iv) We get $s_0 = 1, t_0 = 0$ and

$$s_{i+1} = \begin{cases} s_i + 1, & X_i \in S_1 \\ 2s_i, & X_i \in S_2 \\ s_i, & X_i \in S_3 \end{cases} \quad \text{and}$$

$$t_{i+1} = \begin{cases} t_i, & X_i \in S_1 \\ 2t_i, & X_i \in S_2 \\ t_i + 1, & X_i \in S_3. \end{cases}$$

(v) Continue the iterations until we obtain j, l with $j \neq l$ but $X_j = X_l$.

(vi) For j and l as in 5, we get

$$s_jP + t_jQ = s_lP + t_lQ.$$

Since $Q = kP$ and the order of P is m, we get $s_j + t_jk \equiv s_l + t_lk \mod m$. If $gcd(t_j - t_l, m) = 1$, then

$$k = \frac{s_l - s_j}{t_j - t_l} \mod m.$$

Notice that if $gcd(t_j - t_l, m) \neq 1$, then $t_j - t_l$ has no inverse over multiplication modulo m. However in this research we choose P such that m is a prime number, so that k can always be found except if $t_j = t_l$.

4. Modified Pollard Rho

In Pollard Rho Algorithm, the detection is done by checking whether the point generated on the latest iteration is the same as one of the points generated in the previous iteration. In other words, we run iterations until we obtain j, l with $j \neq l$ but $X_j = X_l$. After that, we obtain $X_j = s_jP + t_jQ$ and $X_l = s_lP + t_lP$. Since $Q = kP$ and the order of P is m, we get

$$s_j + t_jk \equiv s_l + t_lk \mod m.$$

IOP Conf. Series: Journal of Physics: Conf. Series **893** (2017) 012043 doi:10.1088/1742-6596/893/1/012043

Then we solve that equation to get k.

If this is done, then we need to store every X_i generated through iterations. In addition, we also need to store s_i and t_i on each iteration. Thus, although Pollard Rho Algorithm can solve ECDLP within $O(\sqrt{m})$ time, we need to store $O(\sqrt{m})$ points.

However, storing one point X_j requires minimal $4n$ bits, i.e. for each n bit we store the abscissa of X_j, the ordinate of X_j, s_j and t_j. For a large n, for example $n > 60$, keeping all points will require big memory. To decrease memory usage, cycle detection can be done by using the Brent Cycle Detection Algorithm. In detecting the iteration cycle performed in Pollard Rho Algorithm, the Brent Cycle Detection Algorithm goes as follows:

(i) Suppose the sequence that we want to generate is $\{X_i\}_{i=0}^{\infty}$ with initial value X_0 and iteration function f such that $X_{i+1} = f(X_i)$. Set $j = 0, k = 0$, and $l = 1$.

(ii) Replace k with $k + 1$, then check whether $X_j = X_k$. If $X_j = X_k$, then we get a collision.

(iii) If the collision has not occurred, repeat step 2 until a collision occurs or $k = 2l - 1$.

(iv) If $k = 2l - 1$, replace j with k and replace l with $2l$.

(v) Repeat steps 2-4 until a collision occurred.

The expectation of the number of iterations performed is $1.9828\sqrt{m}$, where m is the order of P [1]. It is more than $\sqrt{\frac{\pi m}{2}} \cong 1.2533\sqrt{m}$, i.e. the expectation of the number of iterations in Pollard Rho Algorithm unmodified. However, we only need to store two points, instead of $O(\sqrt{m})$ points.

5. Experimental Results

In previous research we gave a comparison between standard Pollard Rho and Pollard Rho Algorithm with Negation and Frobenius maps for Koblitz curves (see [4] [5]).

In this section we will give a comparison between three variants Pollard Rho Algorithm on elliptic curves over binary field : the standard Pollard Rho Algorithm, Pollard Rho Algorithm with Negation map, and Pollard Rho using Brent Cycle Detection Algorithm. In addition, the possibility of using Frobenius on Koblitz curves gives more comparison between variants of Pollard Rho Algorithm.

First, the following is the list of curves used in this research.

Table 1 The list of curves used in this research

n	a	b	P_x	P_y	m	k
7	1	1	125	17	71	10
11	1	1	337	1892	991	62
13	0	1	3164	3281	2003	612
17	1	1	18909	112804	65587	13690
17	90907	8300	130691	68402	65449	43818
19	1	1	82773	3149	262543	228885
19	34563	310103	233684	277725	261631	74551
23	1	1	715839	1674296	4196903	852098
23	3437961	8316424	1609557	8085378	4196333	448137
29	516948584	317015777	253272920	62475630	268419847	81495421
31	945199535	1617508530	1680131681	910251609	1073784797	952674630
37	94314897882	71408299382	123023382464	21593014607	68719385843	207711372
41	0	1	1535619221598	212045604889	549756390943	457396657162
41	212802682593	828991065662	1595161278299	85985792012	1099510488257	875102879016

In Table 2 we give comparison between standard Pollard Rho and Pollard Rho with Brent Cycle Detection Algorithm. In Table 3 we give the same comparison for Koblitz curves without Frobenius map, meanwhile the same comparison for Koblitz curves with Frobenius map is given in Table 4.

Table 2 Comparison between standard Pollard Rho and Pollard Rho with Brent Cycle Detection Alg

Bit	Standard		Brent	
	Iteration	Time (second)	Iteration	Time (second)
17	352	0.156248	473	0.203124
19	895	0.453124	1571	0.765614
23	6098	3.765658	12555	7.656267
29	15689	12.687533	30471	23.515687
31	58678	51.593854	99505	83.578315
37	251950	290.188130	351431	364.672700
41	2310322	4993.765471	3869603	4943.581061

Table 3 Comparison between std Pollard Rho and Pollard Rho with Brent Alg(Koblitz)

Bit	Standard		Brent	
	Iteration	Time (second)	Iteration	Time (second)
7	16	0.015624	30	0.015626
11	43	0.015637	51	0.015624
13	130	0.062496	168	0.062498
17	89	0.046876	97	0.046874
19	1819	0.906245	3479	1.703126
23	398	0.249999	636	0.390642
41	1125258	2098.714221	3165644	4122.895303

Table 4 Comparison between standard Pollard Rho and Pollard Rho with Brent Cycle Detection Algorithm for Koblitz curves, both use Frobenius map

Bit	Standard		Brent	
	Iteration	Time (second)	Iteration	Time (second)
7	3	0.015641	6	0.015625
11	21	0.046877	25	0.031263
13	12	0.031254	24	0.031230
17	131	0.218761	162	0.234359
19	123	0.265640	207	0.375001
23	861	2.125028	1065	2.359400
41	87869	648.544495	90326	558.360148

From the experiment, we see that the use of Frobenius map generally reduces the number of iterations required. However, the time required is not always shorter. This is due to the additional time required to generate equivalence class of each iteration.

In Table 5 we give comparison between Pollard Rho with Frobenius map and Pollard Rho without Frobenius map for Koblitz curves, both use Brent Cycle Detection Algorithm. In Table 6 we give the same comparison but without Brent Cycle Detection Algorithm and without Negation map. Meanwhile the same comparison without Brent Cycle Detection Algorithm but with Negation map is given in Table 7.

Table 5 Comparison between Pollard Rho with Frobenius map and Pollard Rho without Frobenius map for Koblitz curves, both use Brent Cycle Detection Algorithm

Bit	With Frobenius		Without Frobenius	
	Iteration	Time (second)	Iteration	Time (second)
7	30	0.015626	6	0.015625
11	51	0.015624	25	0.031263
13	168	0.062498	24	0.031230
17	97	0.046874	162	0.234359
19	3479	1.703126	207	0.375001
23	636	0.390642	1065	2.359400
41	3165644	4122.895303	90326	558.360148

Table 6 Comparison between Pollard Rho with Frobenius map and Pollard Rho without Frobenius map for Koblitz curves, both without Brent Cycle Detection Algorithm and without Negation map

Bit	With Frobenius		Without Frobenius	
	Iteration	Time (second)	Iteration	Time (second)
7	16	0.015624	3	0.015641
11	43	0.015637	21	0.046877
13	130	0.062496	12	0.031254
17	89	0.046876	131	0.218761
19	1819	0.906245	123	0.265640
23	398	0.249999	861	2.125028
41	1125258	2098.714221	87869	648.544495

Table 7 Comparison between Pollard Rho with Frobenius map and Pollard Rho without Frobenius map for Koblitz curves, both without Brent Cycle Detection Algorithm but with Negation map

Bit	With Frobenius		Without Frobenius	
	Iteration	Time (second)	Iteration	Time (second)
7	11	0.015616	2	0.031252
11	25	0.015626	13	0.046874
13	113	0.062489	17	0.046894
17	411	0.218751	63	0.171890
19	495	0.296888	102	0.328139
23	3959	2.906526	696	2.609399
41	1142980	2449.223652	66879	666.827368

From the experiment, we see that the use of Negation map generally reduce the number of iterations required. However, if Negation map is used without Frobenius map, almost 10 percent (276,344 out of 2,773,726) iterations are repeated due to fruitless cycles. As a result, the acceleration factor of $\sqrt{2}$ which was originally predicted is not achieved.

6. Conclusion and Further Research

The use of Brent Cycle Detection Algorithm to detect collisions in Pollard Rho Algorithm needs more iterations and generally takes longer than storing every point and check it out.

Nevertheless, we can not store all points for Pollard Rho Algorithm with large binary field, for example $GF(2^n)$ with $n > 60$.

The use of Frobenius map generally reduces the number of iterations required. However, the time required is not always shorter. This is due to the additional time required to generate equivalence class of each iteration. The use of Negation map generally reduce the number of iterations required. However, if Negation map is used without Frobenius map, almost 10 percent (276,344 out of 2,773,726) iterations are repeated due to fruitless cycles. As a result, the acceleration factor of $\sqrt{2}$ which was originally predicted is not achieved.

For further research, one can investigate Nivash Cycle Detection algorithm. Theoretically, Nivasch Cycle Detection algorithm need $\frac{5}{2}\sqrt{\frac{\pi m}{8}}$ iterations, less than Brent Cycle Detection Algorithm in detecting collisions. However, expectations for the number of points that need to be stored is $\ln h + O(1)$, with h is the number of iterations [1].

Acknowledgment
This research is supported by Asahi Glass Foundation research grant 2014.

Reference

[1] Cohen H and Frey G 2006 *Handbook of Elliptic and Hyperelliptic Curve Cryptography* (Chapman and Hall/CRC)
[2] Fulton, W 1969 *Algebraic Curves: An Introduction to Algebraic Geometry* (W. A. Benjamin)
[3] Hankerson D, Menezes A, and Vanstone S 2004 *Guide to Elliptic Curve Cryptography* (Springer)
[4] Muchtadi-Alamsyah I, Ardiansyah T, Carita S S 2014 *Adv. Sc. Let.* **20(1)** 340-343.
[5] Muchtadi-Alamsyah I, Ardiansyah T, Carita S S 2013 *Far East J. Math Sc.* **Special Volume, No 4** 385-402.

IOP Conf. Series: Journal of Physics: Conf. Series **893** (2017) 012044 doi:10.1088/1742-6596/893/1/012044

Magnetohydrodynamics of unsteady viscous fluid on boundary layer past a sliced sphere

Rahmat Nursalim, Basuki Widodo, and Chairul Imron

Department of Mathematics, Institut Teknologi Sepuluh Nopember, Surabaya, Indonesia

E-mail: rahmatnursalim@gmail.com

Abstract. Magnetohydrodynamics (MHD) is important study in engineering and industrial fields. By study on MHD, we can reach the fluid flow characteristics that can be used to minimize its negative effect to an object. In decades, MHD has been widely studied in various geometry forms and fluid types. The sliced sphere is a geometry form that has not been investigated. In this paper we study magnetohydrodynamics of unsteady viscous fluid on boundary layer past a sliced sphere. Assumed that the fluid is incompressible, there is no magnetic field, there is no electrical voltage, the sliced sphere is fix and there is no barrier around the object. In this paper we focus on velocity profile at stagnation point ($x = 0^o$). Mathematical model is governed by continuity and momentum equation. It is converted to non-dimensional, stream function, and similarity equation. Solution of the mathematical model is obtained by using Keller-Box numerical method. By giving various of slicing angle and various of magnetic parameter we get the simulation results. The simulation results show that increasing the slicing angle causes the velocity profile be steeper. Also, increasing the value of magnetic parameter causes the velocity profile be steeper. On the large slicing angle there is no significant effect of magnetic parameter to velocity profile, and on the high the value of magnetic parameter there is no significant effect of slicing angle to velocity profile.

1. Introduction

In decades, MHD has been widely studied in many variations of geometry forms and fluid types. [1] studied MHD viscoelastic fluid with cylinder geometry form. [2] studied MHD fluid with sphere geometry form. [3] studied Viscoelastic Fluid Flow Pass A Porous Circular Cylinder When Magnetic Field Included. On the Mohammads (et al) research about MHD viscous fluid with sphere geometry form, there are some interesting conclusion. One of the conclusion is if the magnetic parameter value M increases then flow velocity significantly increases and boundary layer thickness is $\eta = 2$. Sliced sphere is a different geometry form with a sphere, hence it is very interesting to study MHD with sliced sphere geometry form. By giving a sliced sphere geometry form, it will be appear interesting questions, is there any influence of the sliced angle to velocity profile and boundary layer thickness, and how is the influence of the slicing angle to velocity profile, if magnetic parameter variation is given.

A sliced sphere discussed is a sphere which is sliced by pair of symmetric angle (θ_s) where $0^o \leq \theta_s < 90^o$ and sliced surface faces the fluid flow. The study focus on stagnation point ($x = 0$).

IOP Conf. Series: Journal of Physics: Conf. Series **893** (2017) 012044 doi:10.1088/1742-6596/893/1/012044

2. Research Methods

Steps on this research is:

(i) Modeling the problem
Governing equation is constructed by physical model and then it is modified to non dimensional model, stream function and similarity equations.

(ii) Solve the Model
The model is solved by using Keller-Box numerical method. Steps on the method are modify the similarity equations to first order equation, convert to descret form, linearize, and elimination block matrix.

(iii) Simulate the Model
The model is simulated by giving various of slicing angle and various of magnetic parameter.

3. Mathematical Model

While construct the mathematical model, we use coordinate system. The coordinate system is shown on fig 1.

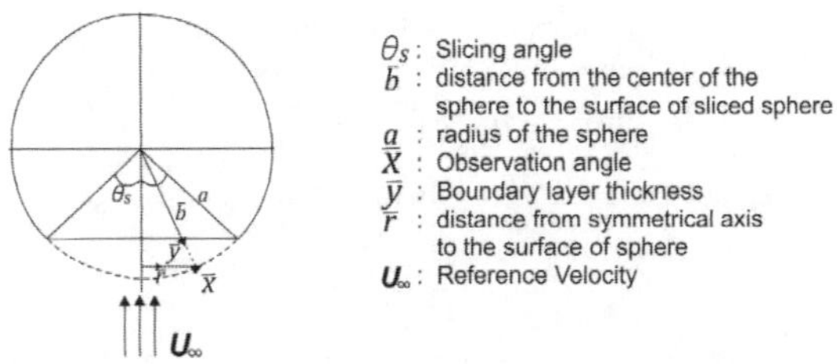

Figure 1. Coordinate system of the problem.

According to the coordinate system we can construct a governing equation which contains continuity equation and momentum equation, and also initial and boundary condition.

3.1. Dimensional Model

Continuity equation

$$\frac{\partial \bar{r}\bar{u}}{\partial \bar{x}} + \frac{\partial \bar{r}\bar{v}}{\partial \bar{y}} = 0$$

x Momentum equation

$$\frac{\partial \bar{u}}{\partial \bar{t}} + \bar{u}\frac{\partial \bar{u}}{\partial \bar{x}} + \bar{v}\frac{\partial \bar{u}}{\partial \bar{y}} = -\frac{1}{\rho}\frac{\partial \bar{p}}{\partial \bar{x}} + \frac{\mu}{\rho}(\frac{\partial^2 \bar{u}}{\partial \bar{x}^2} + \frac{\partial^2 \bar{u}}{\partial \bar{y}^2}) - \frac{\sigma}{\rho}B_0^2\bar{u}$$

y Momentum equation

$$\frac{\partial \bar{v}}{\partial \bar{t}} + \bar{u}\frac{\partial \bar{v}}{\partial \bar{x}} + \bar{v}\frac{\partial \bar{v}}{\partial \bar{y}} = -\frac{1}{\rho}\frac{\partial \bar{p}}{\partial \bar{y}} + \frac{\mu}{\rho}(\frac{\partial^2 \bar{v}}{\partial \bar{x}^2} + \frac{\partial^2 \bar{v}}{\partial \bar{y}^2}) - \frac{\sigma}{\rho}B_0^2\bar{v}$$

initial and boundary condition

$$\bar{t} < 0 : \bar{u} = \bar{v} = 0 \text{ for any } \bar{x}, \bar{y}$$
$$\bar{t} \geq 0 : \bar{u} = \bar{v} = 0 \text{ at } \bar{y} = 0$$
$$\bar{u} = u_e(x) \text{ at } \bar{y} \to \infty$$
$$\text{where } u_e(x) = \tfrac{3}{2}U_\infty sin(\tfrac{\bar{x}}{b})$$

3.2. Non-Dimensional Model

By substituting the non-dimensional variable
$x = \frac{\bar{x}}{a}$, $y = Re^{1/2}\frac{\bar{y}}{a}$, $r(x) = \frac{r(\bar{x})}{a}$, $u = \frac{\bar{u}}{u_\infty}$, $M = \frac{\sigma}{\rho}B_0^2$
$v = Re^{1/2}\frac{\bar{v}}{v_\infty}$, $u_e = \frac{u_e(\bar{x})}{U_\infty}$, $t = \frac{U_\infty \bar{t}}{a}$, $u = \frac{\bar{u}}{U_\infty}$, $b = \frac{\bar{b}}{a}$

we obtain the non-dimensional model

$$\frac{\partial ru}{\partial x} + \frac{\partial rv}{\partial y} = 0$$

$$\frac{\partial u}{\partial t} + u\frac{\partial u}{\partial x} + v\frac{\partial u}{\partial y} = -\frac{\partial p}{\partial x} + \frac{1}{Re}\frac{\partial^2 u}{\partial x^2} + \frac{\partial^2 u}{\partial y^2} - Mu$$

$$\frac{1}{Re}(\frac{\partial v}{\partial t} + u\frac{\partial v}{\partial x} + v\frac{\partial v}{\partial y}) = -\frac{\partial p}{\partial y} + \frac{1}{Re^2}\frac{\partial^2 v}{\partial x^2} + \frac{1}{Re}\frac{\partial^2 v}{\partial y^2} - \frac{1}{Re}Mv$$

initial and boundary condition

$$t < 0 : u = v = 0 \text{ for any } x, y$$
$$t \geq 0 : u = v = 0 \text{ at } y = 0$$
$$u = u_e(x) \text{ at } y \to \infty$$
$$\text{where } u_e(x) = \frac{3}{2}sin(\frac{x}{b})$$

By boundary layer approach, when the Reynold number is large such that $\frac{1}{Re} \to 0$ we obtain

$$-\frac{\partial p}{\partial y} = 0$$

$$-\frac{\partial p}{\partial x} = u_e\frac{\partial u_e}{\partial x} + Mu_e$$

Then Non-dimensional model is simplified to

$$\frac{\partial ru}{\partial x} + \frac{\partial rv}{\partial y} = 0$$

$$\frac{\partial u}{\partial t} + u\frac{\partial u}{\partial x} + v\frac{\partial u}{\partial y} = u_e\frac{\partial u_e}{\partial x} + \frac{\partial^2 u}{\partial y^2} + M(u - u_e)$$

3.3. Stream Function

The Non-dimensional model is converted to stream function by using this equation: $u = \frac{1}{r}\frac{\partial \psi}{\partial y}$, and $v = -\frac{1}{r}\frac{\partial \psi}{\partial x}$.
By substituting the equation to the non-dimensional model we get stream function:

$$\frac{\partial^2 \psi}{\partial x \partial y} = \frac{\partial^2 \psi}{\partial y \partial x}$$

$$\frac{1}{r}\frac{\partial^2 \psi}{\partial y \partial t} + \frac{1}{r^2}\frac{\partial \psi}{\partial y}\frac{\partial^2 \psi}{\partial y \partial x} - \frac{1}{r^3}\frac{dr}{dx}(\frac{\partial \psi}{\partial y})^2 - \frac{1}{r^2}\frac{\partial \psi}{\partial x}\frac{\partial^2 \psi}{\partial y^2}$$
$$= u_e\frac{\partial u_e}{\partial x} + \frac{1}{r^3}\frac{\partial^3 \psi}{\partial y^3} + M(u_e - \frac{1}{r}\frac{\partial \psi}{\partial y})$$

$$t < 0 : \psi = \frac{\partial \psi}{\partial y} = 0 \text{ for any } x, y$$
$$t \geq 0 : \psi = \frac{\partial \psi}{\partial y} = 0 \text{ at } y = 0$$
$$\frac{\partial \psi}{\partial y} = u_e(x)r(x) \text{ at } y \to \infty$$
$$\text{where } u_e(x) = \frac{3}{2}sin(\frac{x}{b})$$

3.4. Similarity Equation

In unsteady condition, the function depends on time. It is important to distinguish between small time ($t \leq t^*$) and large time ($t > t^*$) where t^* can be any value, and in this research we define $t^* = 1$. The governing equations are transformed by using the following similarity variables

for small time
$\psi = t^{1/2}u_e(x)r(x)f(x,\eta,t)$, and $\eta = y/t^{1/2}$

for large time
$\psi = u_e(x)r(x)F(x,Y,t)$, and $\eta = Y$

By substituting similarity variable to the stream function we obtain similarity equation. The similarity equation contains two equations that is for small time and for large time.
Small Time

$$\frac{\partial^3 f}{\partial\eta^3} + \frac{\eta}{2}\frac{\partial^2 f}{\partial\eta^2} + t\frac{\partial u_e}{\partial x}[1 - (\frac{\partial f}{\partial\eta})^2 + f\frac{\partial^2 f}{\partial\eta^2}] + Mt(1 - \frac{\partial f}{\partial\eta})$$
$$= t\frac{\partial^2 f}{\partial\eta\partial t} + tu_e[\frac{\partial f}{\partial\eta}\frac{\partial^2 f}{\partial\eta\partial x} - \frac{\partial f}{\partial x}\frac{\partial^2 f}{\partial\eta^2} - \frac{1}{r}\frac{dr}{dx}f\frac{\partial^2 f}{\partial\eta^2}]$$

$t < 0 : f = f' = 0$, for any x dan η
$t \geq 0 : f = f' = 0$, at $\eta = 0$
$f' = 1$, at $\eta \to \infty$

Large Time

$$\frac{\partial^3 F}{\partial Y^3} + \frac{\partial u_e}{\partial x}[1 - (\frac{\partial F}{\partial Y})^2 + F\frac{\partial^2 F}{\partial Y^2}] + Mt(1 - \frac{\partial F}{\partial Y})$$
$$= \frac{\partial^2 F}{\partial Y\partial t} + u_e[\frac{\partial F}{\partial Y}\frac{\partial^2 F}{\partial Y\partial x} - \frac{\partial F}{\partial x}\frac{\partial^2 F}{\partial Y^2} - \frac{1}{r}\frac{dr}{dx}f\frac{\partial^2 F}{\partial Y^2}]$$

$F = F' = 0$, at $Y = 0$
$F' = 1$, at $Y \to \infty$

3.5. Initial Condition

We have initial condition for the function when $t = 0$. By substituting $t = 0$ to the similarity equation we get

$$\frac{\partial^3 f}{\partial\eta^3} + \frac{\eta}{2}\frac{\partial^2 f}{\partial\eta^2} = 0$$

Then, we substitute the boundary condition to the equation, we get:

$$f = \eta erf(\frac{\eta}{2}) + \frac{2}{\sqrt{\pi}}[e^{-\frac{\eta^2}{4}} - 1]$$

$$f' = erf(\frac{\eta}{2})$$

$$f'' = \frac{1}{\sqrt{\pi}} e^{-\frac{\eta^2}{4}}$$

4. Results and Discussion

By using Keller-Box numerical method we solve the similarity equations. Simulation results are shown at figure (2), figure (3), figure (4), and figure (5).

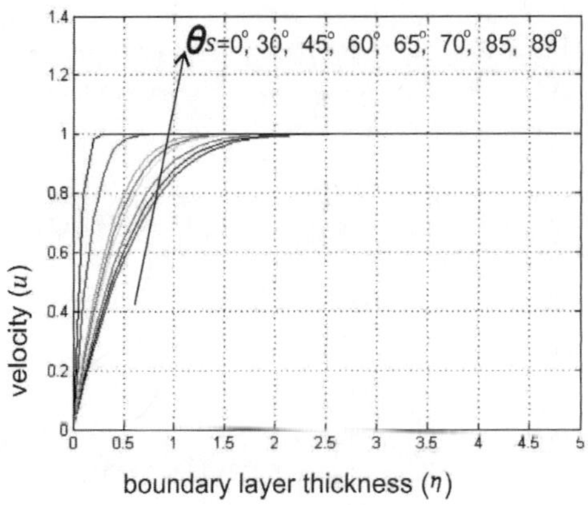

Figure 2. Velocity Profile on $M = 0$ and various slicing angle (θ_s).

From figure (2) we can conclude that "the larger the slicing angle, the steeper the velocity profile".

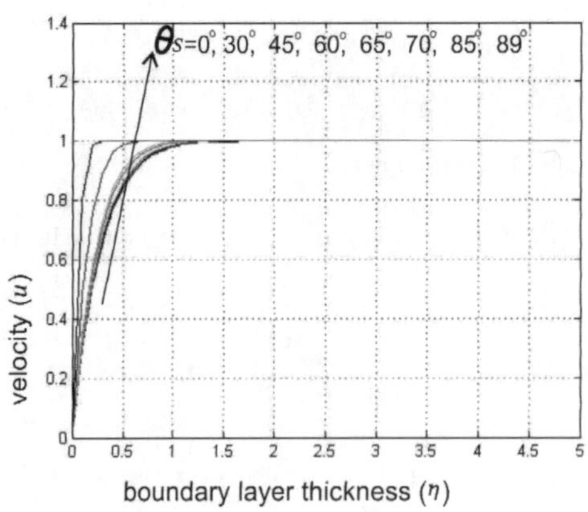

Figure 3. Velocity Profile on $M = 10$ and various slicing angle (θ_s).

From figure (3) we can conclude that "on the high magnetic parameter there is no significant effect of slicing angle on velocity profile".

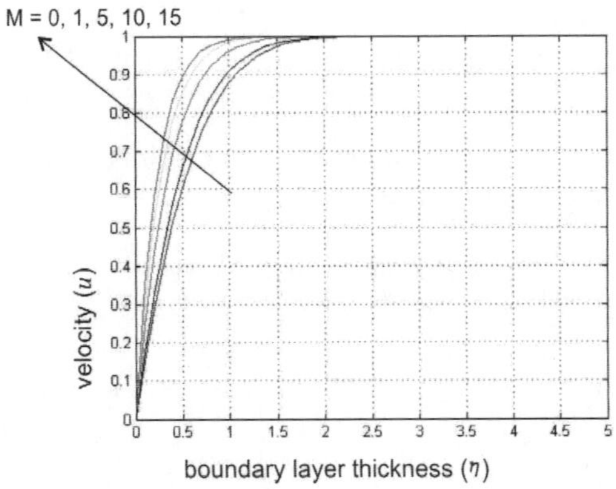

Figure 4. Velocity Profile on slicing angle ($\theta_s = 30^o$) and various magnetic parameter (M) .

From figure (4) we can conclude that "the higher the magnetic parameter, the steeper the velocity profile".

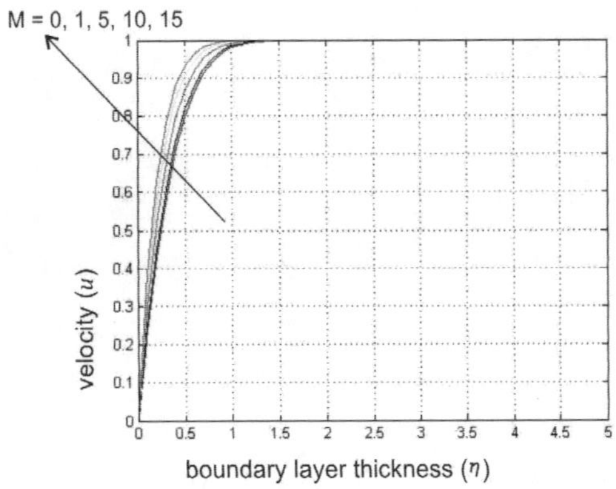

Figure 5. Velocity Profile on slicing angle ($\theta_s = 70^o$) and various magnetic parameter (M).

From figure (5) we can conclude that "on the large Slicing angle there is no significant effect of magnetic parameter on velocity profile".

5. Conclusion

Both increasing slicing angle and increasing magnetic parameter cause the velocity profile be steeper and boundary layer thickness be thinner. On the large slicing angle, there is no significant effect of magnetic parameter to velocity profile. And on the high magnetic parameter, there is no significant effect of slicing angle to velocity profile.

Acknowledgment

The authors would like to thank to the LPPM-ITS Surabaya Indonesia for encouraging us to present this paper on the International conference through the funding Agreement Latter with number of 01695/IT2.11/PN.08/20.

IOP Conf. Series: Journal of Physics: Conf. Series **893** (2017) 012044 doi:10.1088/1742-6596/893/1/012044

References

[1] Rahman M, Alim M, and Chowdhury M 2009 Magnetohydrodynamics Mixed Convection Around a Heat Conducting Horizontal Circular Cylinder in a Rectanguler Lid-driven Cavity with Joule Heating *Journal of Sciencetific Research* **1**(3) pp 461–472

[2] Mohammad N F, Mohd Kasim A R, Ali A, and Shafie S 2013 Effect of MHD on Unsteady boundary layer flow past a sphere *In Proceedings of The 3rd Annual International Conference Syiah Kuala University (AIC Unsyiah)*

[3] Widodo B, Imron C, Aisyah N G, Siswono O, Rahayuningsih T, and Purbandini 2016 Viscoelastic Fluid Flow Pass A Porous Circular Cylinder When Magnetic Field Included *Far East Journal of Mathematical Sciences (FJMS)* **99**(2) pp 173–186

IOP Conf. Series: Journal of Physics: Conf. Series **893** (2017) 012045 doi:10.1088/1742-6596/893/1/012045

Short segment search method for phylogenetic analysis using nested sliding windows

A A Iskandar[1], A Bustamam[1], and H Trimarsanto[2]

[1] Department of Mathematics, Universitas Indonesia, Depok, Indonesia

[2] Bioinformatics Laboratory, Eijkman Institute for Molecular Biology, Indonesia

E-mail: {afifai,alhadi}@sci.ui.ac.id, anto@eijkman.go.id

Abstract. To analyze phylogenetics in Bioinformatics, coding DNA sequences (CDS) segment is needed for maximal accuracy. However, analysis by CDS cost a lot of time and money, so a short representative segment by CDS, which is envelope protein segment or non-structural 3 (NS3) segment is necessary. After sliding window is implemented, a better short segment than envelope protein segment and NS3 is found.This paper will discuss a mathematical method to analyze sequences using nested sliding window to find a short segment which is representative for the whole genome. The result shows that our method can find a short segment which more representative about 6.57% in topological view to CDS segment than an Envelope segment or NS3 segment.

1. Introduction

Dengue fever is a disease caused by dengue virus and transmitted by Aedes aegepti [1]. This infection is characterized by high fever, pain in the joints and red spots on the skin. This disease is common in tropical climates such as Asia, Africa, the Middle East, the Pacific, and Central and South America [1]. Based on the clinical level, dengue disease is categorized into ordinary dengue fever, dengue hemorrhagic fever (DHF) and dengue shock syndrome (DSS). DHF and DSS is a disease that can be life-threatening. Dengue virus (DENV) included in the group B.arthropod-borne viruses (Arboviruses) in the flavivirus genus (*family flaviviridae*) [2]. Dengue virus consists of single-stranded RNA enveloped by the nucleocapsid which diameter is 30 nm, nucleocapsid enveloped by a fatty sheath with thickness is 10 nm so that the overall diameter is 50 nm virion [2].

Research for dengue virus at the level of molecular biology will involve analysis of the gene sequences [1] [2]. Analysis of the gene sequences is used for molecular epidemiology studies, identification of strains, as well as evolution and phylogenetic studies [3]. Analysis using the entire gene sequences of dengue virus genome, particularly from all segments of the coding DNA sequence (CDS), can improve the accuracy and quality of analysis and the research itself. Since costs of the experimental sorting line to get the base sequence of the entire genome are not cheap and need a long time, many researchers only analyzed the regional envelope, assuming that the envelope protein is a protein that is the most varied and must often adapt to the immune system of a patient. However, some recent studies indicate that some of the analysis, especially phylogenetic analysis, which is done by simply using the envelope area can give different results with the overall segment analysis using ORF. This indicates that the envelope area does not represent the entire genome of dengue virus. Therefore, if there are other areas of the dengue virus genome that can represent the whole genome of the virus to the analysis of the gene sequences, especially for phylogenetic analysis, it will increase the accuracy and quality of research at the molecular level without having to sacrifice time and lower costs.

To reduce the cost and time to perform the phylogenetic analysis, we need a short segment of the genome that can be used as representative of the whole genome sequences. The purpose of this research is to propose a method of finding a short segment of DNA sequences for phylogenetic analysis, the study of evolution and classification and identification of virus strains.

2. Material and Methods

2.1. Data Acquisition

This research uses DNA sequence data from virus DENV 4 which can be downloaded at http://www.ncbi.nlm.nih.gov. This study used 152 samples virus DENV 4 with the following criteria:

- Sequences only contain characters 'A','G','T' and 'C'
- One sample of DNA contains $8000 \leq x \leq 10000$ bp
- Data was aligned using multiple sequences alignment

2.2. Evolutionary Model

The evolutionary model is a mathematical model to describe the nucleotide mutation that took place in the gene sequences[4]. Nucleotide Mutation is defined as the process of changing a nucleotide to another nucleotide[6]. This substitution is divided into two, namely the transition and transversion[4]. The transition is a substitution that took place among purine or pyrimidine, whereas transversion was the substitution of one purine with a pyrimidine or otherwise[6]. Each nucleotide position assumed to be independent when mutations in the gene sequences occur. Therefore, mutations in the sequence can be described by a mutation which took place on only a single site.

The objective of evolutionary models is to determine the evolutionary distance between rows, the shorter the distance of its evolution, the closer kinship of the species (or sequences) are concerned. One of the simplest evolutionary models is genetic proportional distance. model. This method is said to be simple because to calculate the distance between two sequences counted for the number of base pairs (bp) are different which are further divided by the length of the sequence is assuming the sequence is already aligned. To write the distances between pairs of sequences in the database, used a matrix which entry of the matrix represents the distance between each pair of species.

To build a genetic proportional distance, we use the following algorithm:

2.3. Neis Minimum Genetic Distance

Nei's minimum genetic distance is a one of popular model for calculate the genetic distance [5] [6]. In this study, the method used to calculate how informative segments are taken for performing phylogenetic analysis.

Consider X and Y are populations, we can compute their genetic distance as :

$$D_m = \frac{J_x + J_y}{2} - J_{xy}$$

In this research, we use this model to compute a distance between a CDS and a short segment.

Algorithm 1 Build a Distance Matrix

```
procedure P_DISTANCE(mseq)
    matrix ← zero_matrix(length(mseq))
    i ← 1
    while i ≤ length(mseq) − 1 do
        j ← i + 1
        while j ≤ length(mseq) do
            distance ← 0
            k ← 1
            while k ≤ length(mseq[0]) do
                if mseq[i] ≠ mseq[j] then
                    distance ← distance + 1
                end if
                k ← k + 1
            end while
            distance ← distance / mseq[0]
            matrix[i][j] ← distance
            matrix[j][i] ← distance
            j ← j + 1
        end while
        i ← i + 1
    end while
    return matrix
end procedure
```

2.4. General Time Reversible Evolutionary Model

The general time reversible (GTR) is the evolutionary model that used for analysis of DENV sequences. Suppose Q is a matrix form of GTR model, so we can write:

$$Q = \begin{bmatrix} -\mu x & a\mu\pi_C & b\mu\pi_G & c\mu\pi_T \\ g\mu\pi_A & -\mu y & d\mu\pi_G & e\mu\pi_T \\ h\mu\pi_A & i\mu\pi_C & -\mu z & f\mu\pi_T \\ j\mu\pi_A & k\mu\pi_C & l\mu\pi_G & -\mu w \end{bmatrix}$$

Where:
$x = a\mu\pi_C + b\mu\pi_G + c\mu\pi_T$
$y = g\mu\pi_A + d\mu\pi_G + e\mu\pi_T$
$z = h\mu\pi_A + i\mu\pi_C + f\mu\pi_T$
$w = j\mu\pi_A + k\mu\pi_C + l\mu\pi_G$

GTR model is a complex models because this model is a generalization of the evolutionary model based on Markov Chain [4].

2.5. Parsimony Informative Site

Parsimony method of tree construction work by finding all the possible topology of the tree and construct the line of ancestors that requires a minimum number of evolutionary ranks obtained today [7].

To save computing time, only a small number of sites with the highest phylogenetic information that is used to construct the tree with this method. This site is called parsimony informative sites [6].

To find a parsimony informative site, we built an algorithm as follows:

Algorithm 2 Parsimony Informative Site

procedure PIS(*multisequence*)
  $PIS\_location \leftarrow \emptyset$
  $j \leftarrow 1$
  while $j \leq length(multisequence[0])$ **do**
    if contains at least two types of nucleotides **then**
      $PIS\_location \leftarrow \{j\}$
    end if
    $j \leftarrow j + 1$
  end while
  return PIS_location
end procedure

2.6. Sliding Window Analysis

The segment of DNA sequence is depends on the position, we can not take a random segment of a sequence to find a short representative segment. So, we need a sliding window analysis to find a segment without changing a position of that segment on the whole sequence. Sliding window analysis is a method to analyze the data by analyzing a subset of the data using the technique of 'shift' until all subsets of the data are analyzed.

2.7. Short Segment Search using Nested Sliding Window

Unlike the sliding window analysis that we discussed earlier, before sliding window which is more focused on the formation of scatter map that will be analyzed, nested sliding window is a sliding window that is focused on short segments of the CDS. However, to determine that the segment is candidates for the segment, nested sliding window is very dependent on the sliding window analysis that previously discussed. Illustration of nested sliding window can be seen from the following figure:

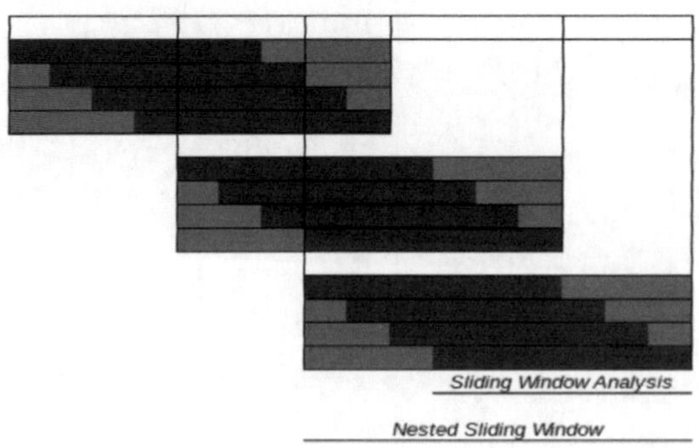

Figure 1: Nested sliding window illustration

2.8. Tree Comparison using Disagree Method

The disagree method is a method for comparing two trees using the novel algorithm and return taxa that make the disagreement between those trees [8]. To perform this process, we use TopD software. If this algorithm returns small number of taxa branch, we can say that those trees are similar.

3. Experiment
This study uses the following framework:

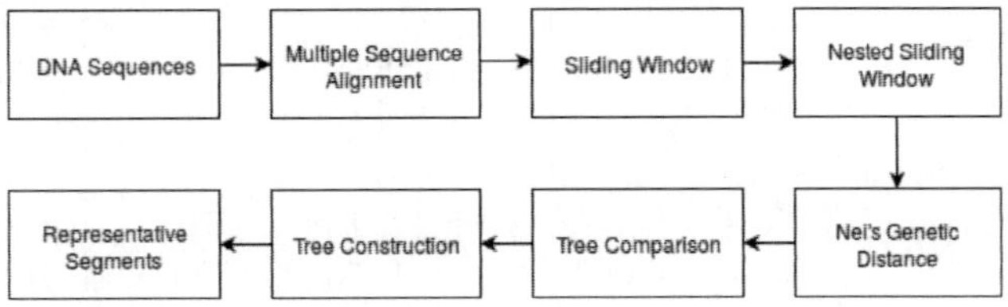

Figure 2: Short segment search method flow

In this research, we found 30 segments which is more representative than envelope segment and NS3 segment. The results for each step explained as follow:

3.1. Sliding Window Plot
A sliding window computation for CDS is used as a population which needed for computing a distance between a short segment and CDS. A plot of CDS can be seen in figure below

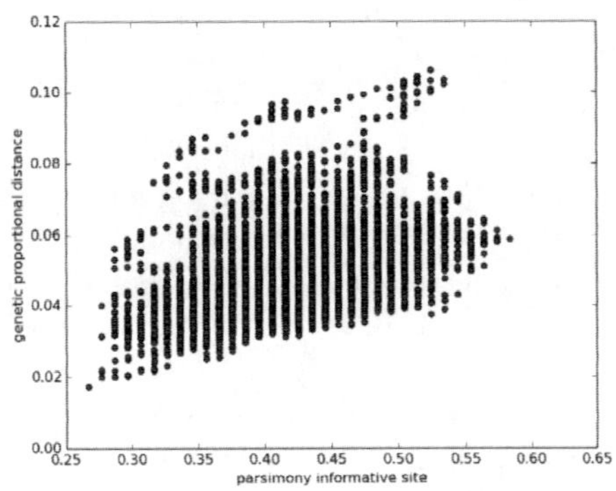

Figure 3: Sliding window result on CDS

Fig.3 show a variance point of parsimony informative site and proportional genetic distance from CDS. We can see that many points accumulates in one coordinate. The aim of this analysis method is to minimize accumulate points without change phylogenetic information, so we need a short segment that represents that points. We did the same thing for an envelope segment and NS3 segment, so we have other pairs of parsimony informative sites and proportional genetic distance points.

TABLE 1 shows the results of minimum Nei's genetic distance using some weighting function.

After the computation using sliding window, we choose two function $f(x) = log(5x)$ and $f(x) = log(x)$ for short segment filtration using nested sliding window.

3.2. Finding short segment
At this stage, we found a representative segment using nested sliding window with a large window size is 1000 bp with three codons shifts step (9bp), and a small window size is 100 bp with 1bp shift step. A length of aligned sequences of CDS are about 10164 bp, with sliding window analysis that we defined before, we can find about 1019 short segments. However, from our methods, it reduced to 200 segments.

Table 1: Sliding window computation result

Functions	Protein	Mean	Standard Deviation
Without Weighted Function	Envelope	0.172274829359	0.117744229463
	NS3	0.188885006042	0.133148237803
$f(x) = log(x)$	Envelope	0.172274829359	0.117744229463
	NS3	0.188885006042	0.133148237803
$f(x) = log(5x)$	Envelope	-0.145138999776	0.106608939153
	NS3	-0.1470009567	0.102611490732
$f(x) = log(5y)$	Envelope	0.802650015027	0.0556421676208
	NS3	0.0891129912456	0.0645141471741

Table 2: Nested sliding window result

Functions	Total Segments
log x	44 segments
log 5x	198 segments

Table 3: Tree distance based on disagreement

Segment	Tree Distance
Envelope	0,361842105
NS3	0,361842105
Best Segment 1	0,296052632
Best Segment 2	0,296052632
Best Segment 3	0,309210526

Next, we constructed a phylogenetic tree from all segments that we have got from sliding window analysis with PhyML software, and then we compare those trees with CDS tree.

From TABLE III, we can see that our method can find a better tree distance with disagreement method, there are 30 segments which have a smaller disagreement value than NS3 or envelope. Tree that have a small disagreement is considered as a tree that has a close resemblance to the CDS, that are considered to represent the entire genome.

4. Conclusion and Future Works

Phylogenetic analysis using CDS (Coding DNA Sequence) or a whole genome are not cheap and take a long time for the computational side. In this research, we proposed a method for searching a short segment for phylogenetic analysis which better than the envelope segment and the NS3 segment on Dengue Virus sample. So, the result from this methods can be used for further phylogenetic analysis more efficient than using a whole genome. To make computational process faster we plan to apply parallel computing for sliding window process [9].

Acknowledgement

This research was supported by research grant HIBAH PUPT 2016 and Bioinformatics laboratory at Eijkman Institute for Molecular Biology

References

[1] Lozano-Fuentes S, Hayden M H, Welsh-Rodriguez C et al. 2012 The Dengue Virus Mosquito Vector Aedes aegypti at High Elevation in Mxico. In: *he American Journal of Tropical Medicine and Hygiene* **7**(5) pp

902–909 http://doi.org/10.4269/ajtmh.2012.12-0244

[2] Zhang H et al. 2014 Spatiotemporal characterization of dengue virus in mainland China : insight into the whole genome from 1978 to 2011. In: *PLoS ONE* **9** pp 1–10

[3] Hochreiter S 2008 Bioinformatics I: Sequence Analysis and Phylogenetics *http://www.bioinf.jku.at[On-line]*

[4] Isaev A 2004 *Introduction to mathematical method in bioinformatics* Berlin: Springer

[5] Nei M 1987 *Molecular genetics* Columbia University New York Press

[6] Lemey P, Salemi M, Vandamme A 2010 *The phylogenetic handbook* New York: Cambridge

[7] Xiong J 2006 *Essential bioinformatics* New York: Cambridge

[8] Puigbo et al. 2007 TOPD/FMTS: a new software to compare phylogenetic trees human diseases: principles and evolution approaches In: *Oxford Journal* **23**(12) pp 1556–1558

[9] Bustamam A, Ardaneswari G, Lestari D 2014 Performance Evaluation of Fast Smith-Waterman Algorithm for Sequence Database Searches using CUDA GPU-Based Parallel Computing In: *Journal of Next Generation Information Technology (JNIT)* **5**(2) pp 38–46

Implementation of plaid model biclustering method on microarray of carcinoma and adenoma tumor gene expression data

Gianinna Ardaneswari, Alhadi Bustamam, and Devvi Sarwinda

Faculty of Mathematics and Natural Sciences, Universitas Indonesia, Depok, Indonesia 16424

E-mail: {ninna, alhadi, devvi}@sci.ui.ac.id

Abstract. A Tumor is an abnormal growth of cells that serves no purpose. Carcinoma is a tumor that grows from the top of the cell membrane and the organ adenoma is a benign tumor of the gland-like cells or epithelial tissue. In the field of molecular biology, the development of microarray technology is used in the data store of disease genetic expression. For each of microarray gene, an amount of information is stored for each trait or condition. In gene expression data clustering can be done with a bicluster algorithm, thats clustering method which not only the objects to be clustered, but also the properties or condition of the object. This research proposed Plaid Model Biclustering as one of biclustering method. In this study, we discuss the implementation of Plaid Model Biclustering Method on microarray of Carcinoma and Adenoma tumor gene expression data. From the experimental results, we found three biclusters are formed by Carcinoma gene expression data and four biclusters are formed by Adenoma gene expression data.

1. Introduction

Bioinformatics is a branch of science that studies the application of computational techniques in managing and analyzing biological data. Molecular Biology research evolves through the development of the technologies used for carrying them out. It is not possible to research on a large number of genes using traditional methods. DNA Microarray is one such technology which enables the researchers to investigate and address issues which were once thought to be non traceable. One can analyze the expression of many genes in a single reaction quickly and in an efficient manner. DNA Microarray technology has empowered the scientific community to understand the fundamental aspects underlining the growth and development of life as well as to explore the genetic causes of anomalies occurring in the functioning of the human body. Scientists use DNA microarrays to measure the expression levels of large numbers of genes simultaneously or to genotype multiple regions of a genome.

DNA Microarray or gene-chip was developed in the 1990s to determine the transcriptional thousands of genes in one (or several) experiments on a small membrane or glass slide. There are a number of ways to make a DNA microarray. One method starts with PCR amplification of gene sequences that are known from an organism.

The development of microarray technology has provided an opportunity for researchers to explore the availability of genes an organism associated with genes we are studying. Microarray data provided are processed with the help of computerized measurement expression profiling, or by looking at the proximity of a pair of genes through stringensi degrees. From here generated a database that can predict the linkage of a gene with other genes. These results are usually displayed in the form of clusters which is the result of grouping genes that have similar expression patterns.

One of the most topics that used in bioinformatics is about clustering and also biclustering for two-dimensional data. In 2014, Das et al. proposed two-phase method for finding a bicluster. In the first phase a modified k-means algorithm is applied. In the second phase the residue score of Cheng and Church is applied to each of the columns of the clusters [1]. Biclustering is a very useful data mining technique which identifies coherent patterns from microarray gene expression data. A bicluster of a gene expression dataset is a subset of genes which exhibit similar expression patterns along a subset of conditions. Biclustering is a powerful analytical tool for the biologist and has generated considerable interest over the past few decades.

The aim of biclustering is to identify subset pairs (each pair consisting of a subset of genes and a subset of conditions) by clustering both the rows and columns of an expression matrix. Hence, biclustering algorithms must guarantee that the output biclusters are meaningful. This is usually done by accompanying statistical model or a heuristic scoring method that defines which of the many possible sub matrices represent a significant biological behavior. The biclustering problem is to find a set of significant biclusters in a matrix. Usually, gene expression data is arranged in a data matrix, where each gene corresponds to one row and each condition to one column [1].

Clustering is a search method for hidden patterns that may exist in datasets. It is a process of grouping data objects into disjointed clusters so that the data in each cluster are similar, yet different to the other clusters. K-means is one of the most famous and typical clustering algorithms and applied in many application areas. K-means has the advantages of fast convergence and ease of implementation, but it has poor performance in some applications with large dataset.

Poor performance as the computational problem can be solved by parallel computing simultaneously. Parallel processing is the processing of program instructions by dividing them among multiple processors with the objective of running a program in less time. Parallel computing is using multiple independent processors simultaneously or utilize GPU computing to solve a problem. In this issue typically used for processing very large data and high computational demands.

Therefore, by the development of bioinformatics analysis, related to increasing the number of data that to be analyzed, encourage researchers to propose methods for better results and efficient. In this study, we proposed a two-phase method. Two-phase in this method consists of two phases, the first phase is the implementation of parallel algorithms k-means on gene expression data to generate k clusters. The second phase is the implementation of bicluster algorithm, we applied plaid model biclustering algorithm. The application of the two-phase method on these gene expression data aims to find good-quality bicluster effectively.

2. Material and Methods

2.1. K-Means Algorithm

K-means is one of the simplest unsupervised learning algorithms that solve the well known clustering problem. The procedure follows a simple and easy way to classify a given data set through a certain number of clusters (assume k clusters) fixed a priori. The main idea is to define k centroids, one for each cluster. These centroids shoud be placed in a cunning way because of different location causes different result. So, the better choice is to place them as much as possible far away from each other. The next step is to take each point belonging to a given data set and associate it to the nearest centroid. When no point is pending, the first step is completed and an early groupage is done. At this point we need to re-calculate k new centroids as barycenters of the clusters resulting from the previous step. After we have these k new centroids, a new binding has to be done between the same data set points and the nearest new centroid. A loop has been generated. As a result of this loop we may notice that the k centroids change their location step by step until no more changes are done. In other words centroids do not move any more [7].

2.2. Parallel K-Means Algorithm

K-means algorithm is one of the most famous unsupervised clustering algorithms. K-means has the advantages of fast convergence and ease of implementation, but it has poor performance in some applications with large dataset.

Meanwhile, poor performance as the computational problem can be solved by parallel computing simultaneously. Parallel computing is a type of computational method that the execution of processes are carried out simultaneously. Large problems can often be divided into smaller ones, which can be solved at the same time. So, we can apply parallel computing in this k-means algorithm.

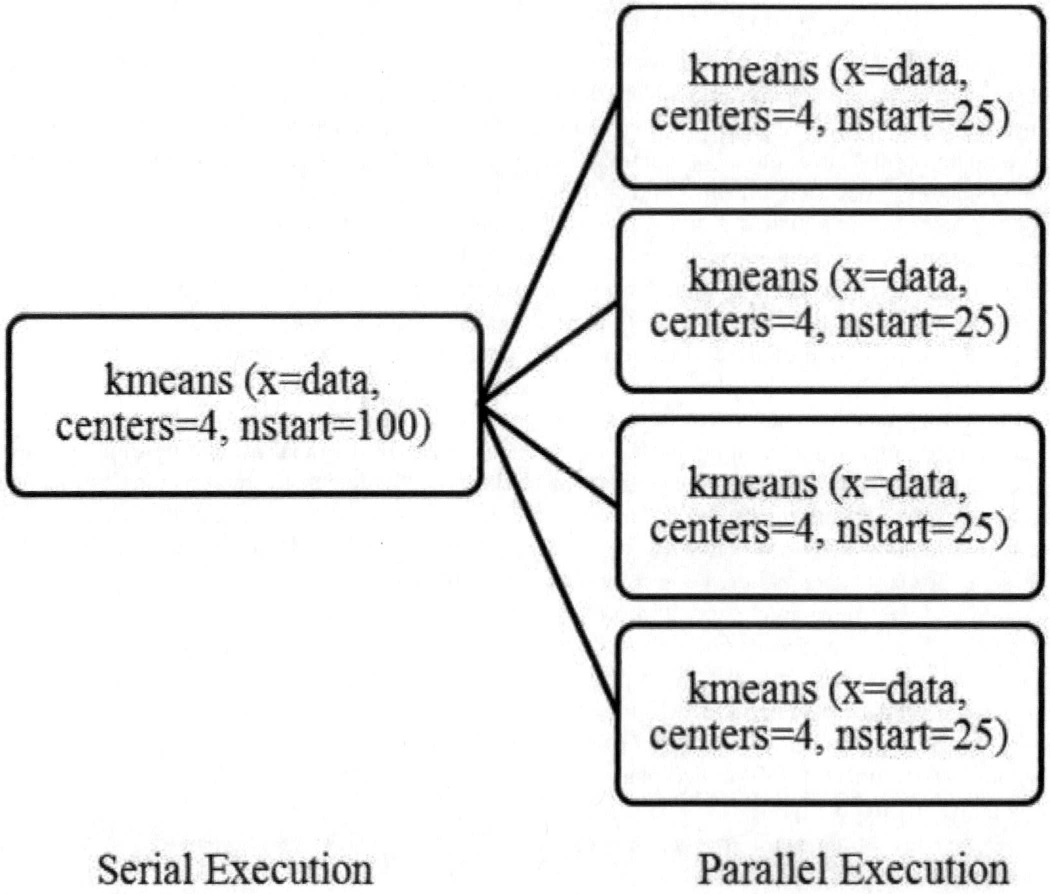

Figure 1. Parallel K-Means

2.3. Plaid Model Biclustering

In the field of molecular biology, the development of microarray technology is used in the data store of disease genetic expression. Biclustering has become a popular technique for the study of gene expression data. For each of microarray gene, an amount of information is stored for each trait or condition. In gene expression data clustering can be done with a bicluster algorithm, thats clustering method which not only the objects to be clustered, but also the properties or condition of the object.

Lazzeroni and Owen proposed plaid models. In this approach, the genes-condition matrix is represented as a superposition layers, corresponding to bicluster [2].

$$Y_{ij} = \sum_{k=0}^{k} \theta_{ijk} \rho_{ik} \kappa_{jk} \tag{1}$$

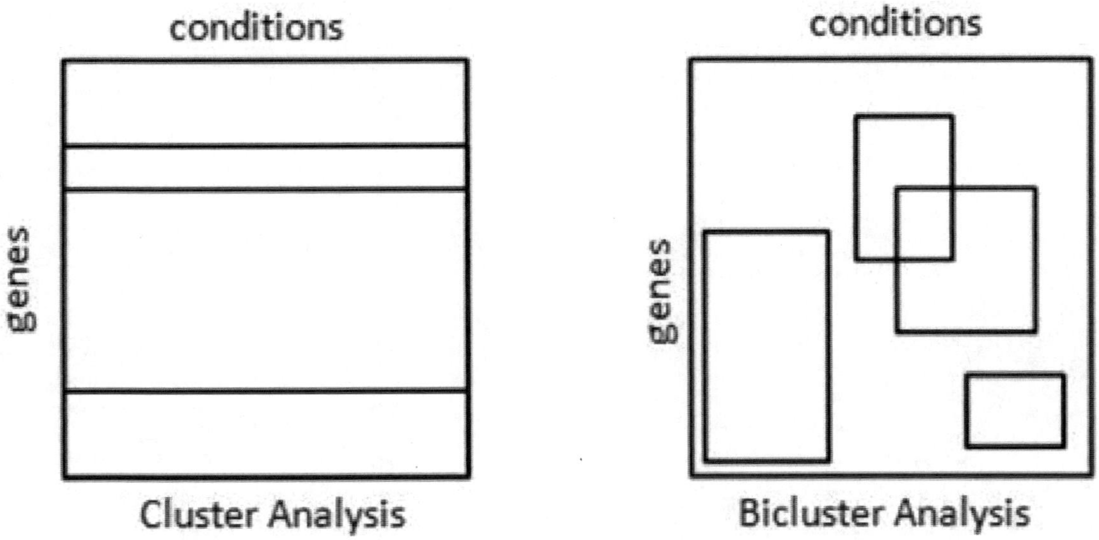

Figure 2. Clustering & Biclustering Analysis

Where Y_{ij} refers to the expression level of gene i under sample j in the input matrix, k is the number of biclusters, θ_{ij0} describes the background layer and θ_{ijk} represents four different types of models, depending on the types of biclusters. Each $\rho_{ik} \in \{0,1\}$ is 1 if gene i is in the kth bicluster, zero otherwise. Similarly, each $\kappa_{jk} \in \{0,1\}$ is 1 if sample j is in the kth bicluster, zero otherwise. Using this equation, a bicluster is assumed to be the sum of a bicluster background level plus row-specific and column-specific constants.

In order to find k biclusters in the data, Lazzeroni and Owen proposed a greedy algorithm that adds one layer at a time. The process seeks for a plaid model minimizing the sum of squared errors when approximating the data matrix to the model [4]. In the eq. (1) allows a gene to be more than one bicluster or in none at all.

In order to find k biclusters in the data, Lazzeroni and Owen proposed a greedy algorithm that adds one layer at a time. The process seeks for a plaid model minimizing the sum of squared errors when approximating the data matrix to the model. For this purpose, an iterative approach is adopted with each cycle updating h values, q values and j values in turns. Assuming that residual data becomes more and more unstructured noise as each layer is removed from the data, authors propose a simple rule for stopping the process, in which only a small number of extra layers can be extracted once the data have been reduced to noise.

2.4. Proposed Method

Biclustering is one of data mining techniques are very useful to identify a coherent pattern of microarray gene expression data. A bicluster of gene expression dataset is a subset of genes that showed similar expression patterns along the subset of the conditions.

By the development of bioinformatics analysis, related to increasing the number of data that to be analyzed, encourage researchers to propose methods for better results and efficient. In this study, we proposed a two-phase method. Two-phase in this method consists of two phases, the first phase is the implementation of parallel algorithms k-means on gene expression data to generate k clusters. The second phase is the implementation of bicluster algorithm, we applied plaid model biclustering algorithm. The application of the two-phase method on these gene expression data aims to find good-quality bicluster effectively.

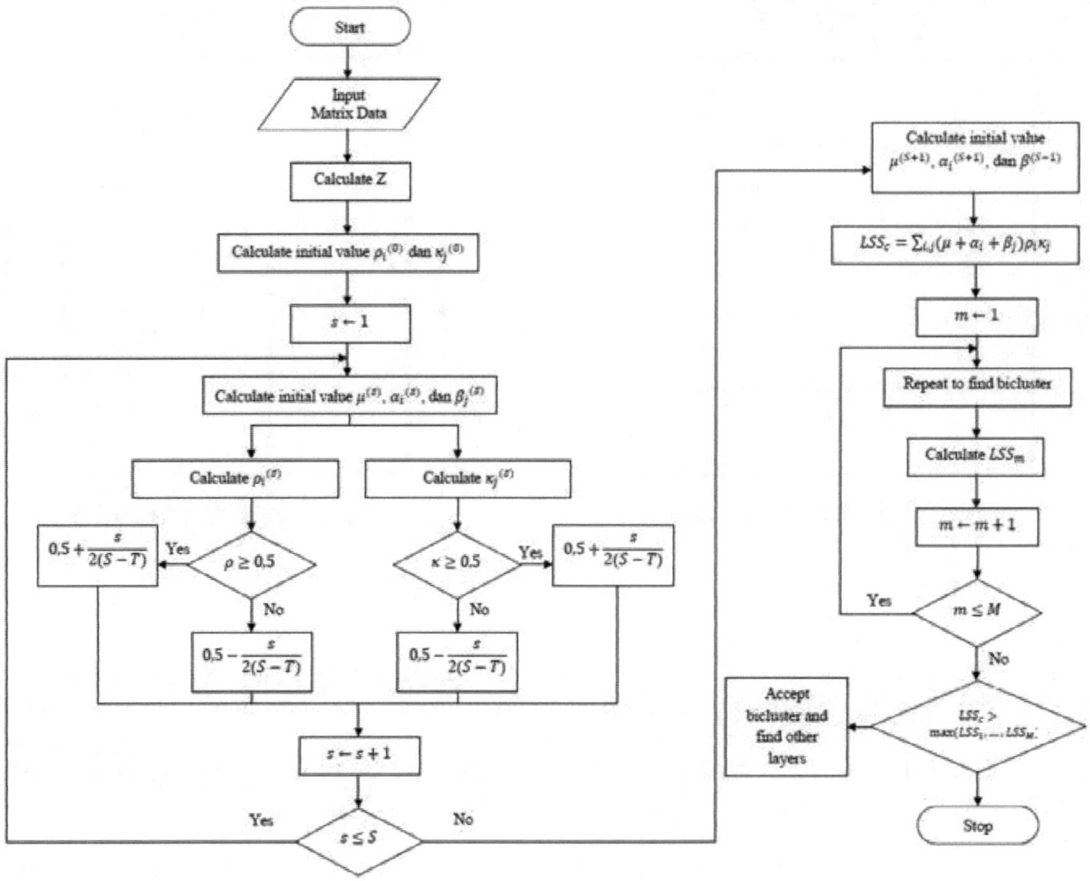

Figure 3. Plaid Model Biclustering Algorithm

3. Results

In this study, experiments are conducted on the Carcinoma and Adenoma microarray database that can be accessed via http://genomics-pubs.princeton.edu/.

- Carcinoma Data: There are 7457 genes and 36 samples.
- Adenoma Data: There are 7086 genes and 8 samples.

From the experimental result, we found three biclusters are formed by Carcinoma gene expression data and four biclusters are formed by Adenoma gene expression data. We got sum of squares (SS) of the clusters is 66.9% for Carcinoma gene expression data and 71.5% for Adenoma gene expression data.

In Carcinoma gene expression data, the three biclusters shows that they have matrices size 11×15, 16×2, and 3×12, respectively.

In Adenoma gene expression data, the four biclusters shows that they have matrices size 4×3, 4×3, 4×5, and 25×4, respectively.

Furthermore, we can identify what genes that affect the most, according to samples or conditions. So, from the experimental results it can be used for further identification or diagnosis.

4. Conclusion

In this paper, the proposed method is introduced which is divided in two phase. To do bicluster microarray of carcinoma and adenoma gene expression data, we can use plaid model biclustering

IOP Conf. Series: Journal of Physics: Conf. Series **893** (2017) 012046 doi:10.1088/1742-6596/893/1/012046

Algorithm 1 Two-Phase Algorithm

Input: a gene expression matrix
Output: bicluster

```
procedure TWO–PHASE (gene_expression_matrix)
    Phase 1 :
    Generate clusters using parallel k-means
    Initialize the centroid k = 1 : K
    for i = 1 to K do
        for j = 1 to J do
            read C_ij
        end for
    end for
    Create k clusters
    Recompute the centroid for each cluster until there is
    no change in the centroids
    Phase 2 :
    for each cluster do
        Apply plaid model biclustering algorithm
    end for
end procedure
```

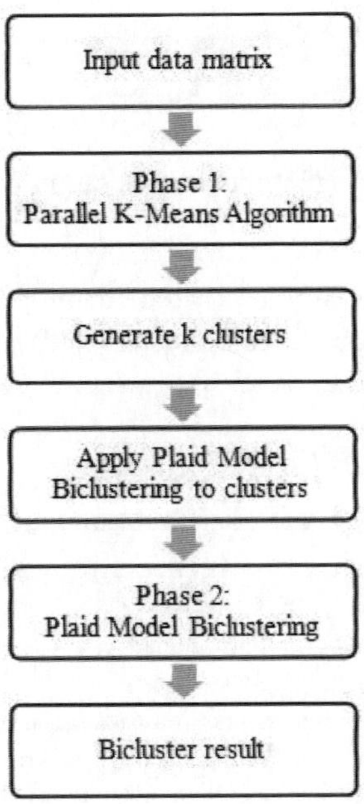

Figure 4. Resume of Two-Phase Method

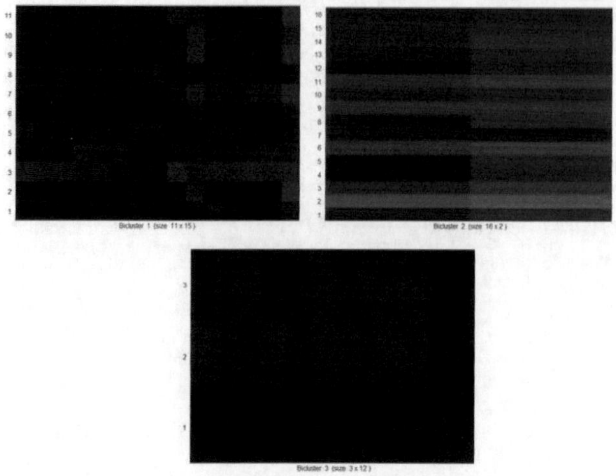

Figure 5. Heatmap of Carcinoma Biclusters

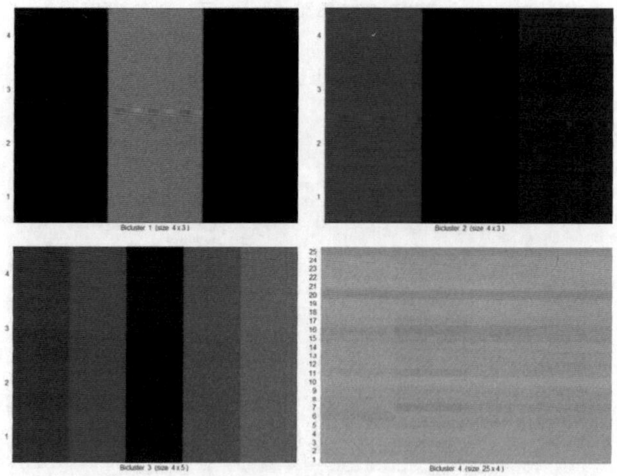

Figure 6. Heatmap of Adenoma Biclusters

and K-means or parallel K-means algorithm as a tool to preprocessing data. Our results show that Carcinoma gene expression data produces three biclusters and Adenoma gene expression data produces four biclusters.

Acknowledgment
This research is supported by Hibah PITTA UI 2016 research grant.

References
[1] Das M, Borah B 2014 Biclustering of Gene Expression Data Using Two-Phase Method *International Journal of Computer Applications* **103**(13) November 2014 pp 0975–8887
[2] Lazzeroni L, Owen A 2002 Plaid models for gene expression data *Statistica Sinica* **12** pp 61–86
[3] Notterman D A, Alon U, Sierk A J, Levine A J 2001 Transcriptional Gene Expression Profiles of Colorectal Adenoma, Adenocarcinoma, and Normal Tissue Examined by Oligunucleotide Arrays. *Cancer Research* **61** pp 3124–3130
[4] Pontes B, Giraldes R, Aguilar-Ruiz S 2015 Bioclustering on expression data: A review *Elseiver Journal of Biomedical Informatics* **57** pp 163–180

[5] Turner H, Bailey T, Krzanowski W 2005 Improved biclustering of microarray data demonstrated through systematic performance tests *Elseiver Computational Statistics & Data Analysis* **48** pp 235–254
[6] Cheng Y and Church G 2000 Biclustering of Expression Data *International Conference on Intelligent Systems for Molecular Biology* pp 93–103
[7] MacQueen J B 1967 Some Methods for classification and Analysis of Multivariate Observations *Proceedings of 5-th Berkeley Symposium on Mathematical Statistics and Probability Berkeley, University of California Press* **1** pp 281–297
[8] Pollard K S, Van der Laan M J 2002 Statistical inference for simultaneous clustering of gene expression data *Math. Biosci.* **176** pp 99–121
[9] Kanungo T, Mount D, Netanyahu N, *et al.* 2002 An efficient k-means clustering algorithm: analysis and implementation *IEEE Transactions on Pattern Analysis and Machine Intelligence* **24**(7) pp 881–892
[10] Madeira S C and Oliveira A L 2004 Biclustering Algorithms for Biological Data Analysis: A Survey *IEEE/ACM Transactions on Computational Biology and Bioinformatics* **1** pp 24–45
[11] Bustamam A, Ardaneswari G, Lestari, Dian, Tasman H 2014 Performance Evaluation of Fast Smith-Waterman Algorithm for Sequence Database Searches using CUDA GPU-Based Parallel Computing *Journal of Next Generation Information Technology* **5**(2) pp 38–46
[12] Permata T S, Bustamam A 2015 Clustering protein-protein interaction network of TP53 tumor suppressor protein using Markov clustering algorithm *International Conference on Advanced Computer Science and Information Systems (ICACSIS) IEEE Proceeding* pp 221–226

IOP Conf. Series: Journal of Physics: Conf. Series **893** (2017) 012047 doi:10.1088/1742-6596/893/1/012047

Collision attack against Tav-128 hash function

Fajar Hariyanto and Bety Hayat Susanti
Sekolah Tinggi Sandi Negara
E-mail: harifajar1994@gmail.com, bety.hayat@stsn-nci.ac.id

Abstract. Tav-128 is a hash function which is designed for Radio Frequency Identification (RFID) authentication protocol. Tav-128 is expected to be a cryptographically secure hash function which meets collision resistance properties. In this research, a collision attack is done to prove whether Tav-128 is a collision resistant hash function. The results show that collisions can be obtained in Tav-128 hash function which means in other word, Tav-128 is not a collision resistant hash function.

1. Introduction

Hash function is a cryptographic technique that can be used to ensure the integrity and authenticity of data [8]. It is divided into two types, i.e. Message Authentication Code (MACs) or keyed hash functions and Modification Detection Codes (MDCs) or unkeyed hash function. As the input, MACs requires a secret key and a message, while MDCs just requires a message [6].

In 2008, Tav-128 hash function is designed for an authentication protocol Radio Frequency Identification (RFID) [10]. Tav-128 is a hash function based on Merkle-Damgard construction [11]. It is expected to be a hash function that fulfill the cryptographic security service [4].

A good hash function must satisfy three properties, i.e. preimage resistance, second preimage resistance, and collision resistance. Collision resistance means that it is computationally difficult to obtain two different inputs that have the same hash value [2]. Based on that, collision attack against Tav-128 was performed in 2010 by Kumar and Sanadhya [3]. The results show that collisions found in Tav-128.

In this research, collision attack will be applied on Tav-128 hash function with the different method from [3]. The goal of this research is to know the resistance of Tav-128 against collision attack.

2. Theoretical Background

2.1. Hash Function

Hash function maps an arbitrary length input to the fixed length output [5]. Based on [9], hash function which is denoted by $h(m)$ must provide the following properties:

(i) Compression: for any input m, output $y = h(m)$ is small. In practice, the hash function produces output with a fixed size, regardless of the length of input, with the length of a typical output in the range of 128 to 512 bits.

(ii) Efficiency: it must be efficient to calculate $h(m)$ of any input m. Computational effort depends on the length of m.

(iii) One-way: it is difficult to invert the hash, i.e. if y is given, it is difficult to find the value of m where $h(m) = y$.

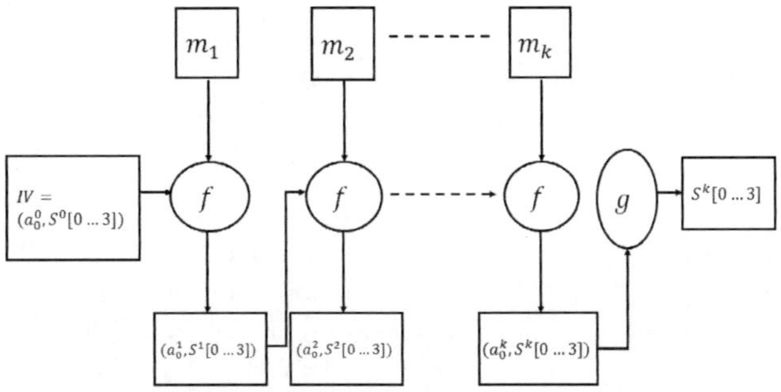

Figure 1. Structure of Tav-128 hash function.

(iv) Weak collision resistance: given m and $h(m)$, it is difficult to find w, with $w \neq m$, where $h(w) = h(m)$.

(v) Strong collision resistance: it is hard to find a pair of m and w, with $m \neq w$, where $h(m) = h(w)$.

2.2. Merkle-Damgard construction

Merkle-Damgard construction is the construction of a hash function $h : \{0,1\}^* \rightarrow \{0,1\}^n$ constructed by iterating the compression function $f : \{0,1\}^m \times \{0,1\}^n \rightarrow \{0,1\}^n$. The process is carried out as follows [7]:

(i) Pad and split the message M into k blocks $m_1, ..., m_k$ with each block measuring t bits.

(ii) Set the value of h_0 as initialization value (IV).

(iii) Calculate $h_k = f(h_{k-1}, m_k)$ for each message block.

(iv) Obtained output $h^f(M) = h_k$.

Padding is usually done by adding a bit of "1" followed by bits "0" as much as needed to complete the k block. See [7] for more details discussion.

2.3. Tav-128

Tav-128 is a hash function that is designed for Radio Frequency Identification (RFID) authentication protocol. It has 160-bits internal state and generate 128-bits output [4].

The hash value of the input m of length $k \times 32$ bits obtained by calculating the value of output of the compression function f that is iterated k times by Merkle-Damgard construction and transformation g. The compression function f constructed by four functions, i.e. A,B,C, and D functions. Each of A,B,C, and D functions process 64-bits input and produce 32-bits output. The output of the compression function f on the last block becomes the input to the transformation g, so that the resulting output hash value that is 4×32 last bits of input values at the transformation g [4]. The structure of Tav-128 hash function shown in Figure 1. See [4] for more details.

2.4. Collision Attack On Tav-128

In [4], Kumar *et al.* stated that the collision resistance of the hash function which has Merkle-Damgard construction affected by the compression function. Therefore, collision attack on Tav-128 is done by utilizing the compression function of it. Collision attack on Tav-128 can be applied by following steps [4]:

(i) Make a correspondence table with the size of 2^{32} for all messages m consisting of (h_0, h_1, m). h_0 and h_1 are output values after one application of $C\&D$ functions.

(ii) Sort this table.

(iii) Find a pair in a row where h_0 and h_1 have the same value.

3. Research Method

In general, the collision attack in this research is based on Kumar *et al.* [4]. But the difference from Kumar *et al.*, there is a grouping of $C\&D$ functions output into classes based on a range of values. There are 64 classes where the range between the classes is 400000000000000 (hexadecimal). The first class that is $0000000000000000 - 03ffffffffffffff$ and so on until the 64th class is $fc00000000000000 - ffffffffffffffff$. The grouping is useful to find collisions at the output of $C\&D$ functions efficienctly.

After grouping is done, then sort the output values in each class. The sorting technique used in this research is merge sort [1]. The use of sorting is aimed to make the finding of collisions in each class becomes easier and faster. In details, to do collision attack against Tav-128, we use the following steps:

(i) Calculate the output from one application of $C\&D$ functions using 2^{32} inputs.

(ii) Group all of the output values into classes according to a range of values.

(iii) Sort the output in each class with the order of the smallest value to the largest value.

(iv) After all the class is sorted, find output pairs which have the same value in each class.

(v) Keep the same output values into an output column $(h_0||h_1)$ in correspondence table.

(vi) To determine the input that has the same output based on the correspondence table, generate all pairs of input and output values only to the class that contains the same output values.

(vii) Find the input values that produce output based on the correspondence table.

(viii) Finally, the values of the corresponding input has been found and accommodated into inputs (m) and outputs $(h_0||h_1)$ column in the correspondence table.

This research was performed by implementing Tav-128 using C programming language and MinGW Dev C++ Compiler. The number of inputs used in this attack are 2^{32} which have 32 bits length of each input without padding process.

4. Results and Analysis

After the application of the collision attack using 2^{32} inputs against Tav-128, we obtain the collision result. Table 1 shows collision results of 2^{32} inputs. Class column shows list of classes that are used in grouping the output values that obtained from one application of $C\&D$ functions. Output range column indicates the range of output values in each class. Collision column shows the number of collisions that obtained in each class.

From Table 1, it can be seen that the collision can be obtained at class 43 and 47. At class 43, we obtain one collision which means that there is one different inputs pair which have the same output value. In class 47 was obtained one collision. So that the total collision obtained are two collisions as shown in Table 2.

A collision can be obtained easily because output values from one application of $C\&D$ functions, which are grouped into each class has been sorted. Hence, the process used to find collisions in each class becomes 2^{26}(because it is already divided into 64 classes).

With the discovery of a collision in the compression function f, then it will be shown how to find a collision in Tav-128 with a starting collision on the compression function f. From collision attack on $C\&D$ functions, we have obtained two inputs that have the same output value in one application of them. Figure 2 shows that m_1^1 is a notation for input 1 and m_1^2 is a notation for input 2 where $m_1^1 \neq m_1^2$.

Table 1. Collision Result of 2^{32} Input On First Iteration Of $C\&D$ Function.

Class	Output Range	Collisions
1	$0000000000000000 - 03ffffffffffffff$	0
2	$0400000000000000 - 07ffffffffffffff$	0
3	$0800000000000000 - 0bffffffffffffff$	0
4	$0c00000000000000 - 0fffffffffffffff$	0
5	$1000000000000000 - 13ffffffffffffff$	0
...	...	...
...	...	...
...	...	...
40	$9c00000000000000 - 9fffffffffffffff$	0
41	$a000000000000000 - a3ffffffffffffff$	0
42	$a400000000000000 - a7ffffffffffffff$	0
43	$a800000000000000 - abffffffffffffff$	1
44	$ac00000000000000 - afffffffffffffff$	0
45	$b000000000000000 - b3ffffffffffffff$	0
46	$b400000000000000 - b7ffffffffffffff$	0
47	$b800000000000000 - bbffffffffffffff$	1
48	$bc00000000000000 - bfffffffffffffff$	0
49	$c000000000000000 - c3ffffffffffffff$	0
50	$c400000000000000 - c7ffffffffffffff$	0
...	...	...
...	...	...
...	...	...
60	$ec00000000000000 - efffffffffffffff$	0
61	$f000000000000000 - f3ffffffffffffff$	0
62	$f400000000000000 - f7ffffffffffffff$	0
63	$f800000000000000 - fbffffffffffffff$	0
64	$fc00000000000000 - ffffffffffffffff$	0

Table 2. Correspondence Table Of Input Pairs That Has The Same Output Value.

No.	Inputs (m)	Outputs $(h_0 \| h_1)$ from one application of $C\&D$ functions
1	$8fa00d2b$	$abdf190b20c8fdac$
	$fb86cc90$	$abdf190b20c8fdac$
2	$49390f23$	$bb8e44aa1472b576$
	$8099ba12$	$bb8e44aa1472b576$

(a)

(b)

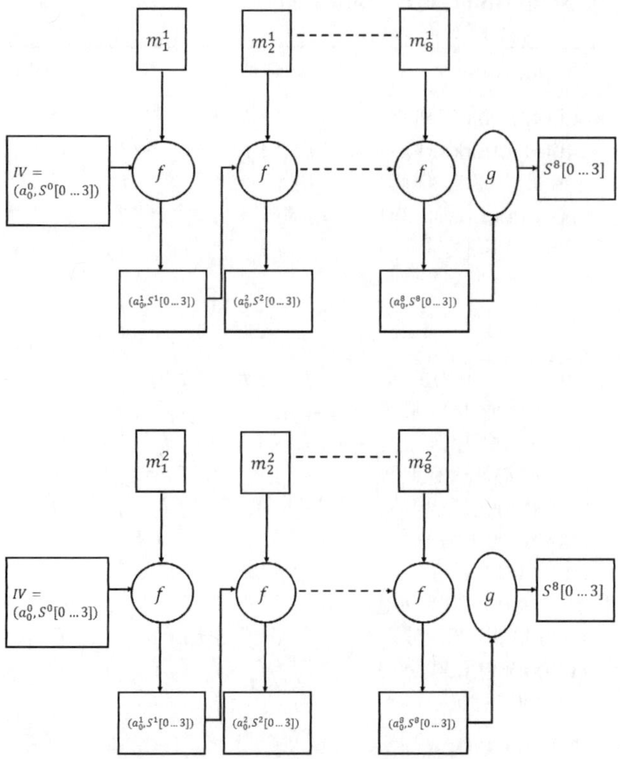

Figure 2. Collision Attack on $C\&D$ functions with input 1 (a) and input 2 (b).

In this stage, we want to prove that the compression function is not collision resistance, resulting that Tav-128 is also not collision resistance. We use the method by adding some value behind each inputs (m_1^1 and m_1^2) so that each input size of 256 bits. The use of 256-bits input only to indicate the property of compression on Tav-128. Two inputs after additional bits are:
Input 1 $= m_1^1 \| m_2^1 \| m_3^1 \| m_4^1 \| m_5^1 \| m_6^1 \| m_7^1 \| m_8^1$, and
Input 2 $= m_1^2 \| m_2^2 \| m_3^2 \| m_4^2 \| m_5^2 \| m_6^2 \| m_7^2 \| m_8^2$.

By processing the compression function f each of 32 bits, then input 1 and input 2 are processed within 8 blocks.

- First block:
  Assuming that m_1^1 and m_1^2 are two inputs which have the same output value of $C\&D$ functions so that:
  $f(m_1^1, a_0^0) = f(m_1^2, a_0^0) = (a_0^1, S^1[0 \dots 3])$

- Second block:
  $f(m_2^1, a_0^1) = (a_0^2, S^2[0 \dots 3])$
  $f(m_2^2, a_0^1) = (a_0^2, S^2[0 \dots 3])$
  So that $f(m_2^1, a_0^1) = f(m_2^2, a_0^1) = (a_0^2, S^2[0 \dots 3])$ because $m_2^1 = m_2^2$.

- Third block:
  $f(m_3^1, a_0^2) = (a_0^3, S^3[0 \dots 3])$

$f(m_3^2, a_0^2) = (a_0^3, S^3[0 \dots 3])$
So that $f(m_3^1, a_0^2) = f(m_3^2, a_0^2) = (a_0^3, S^3[0 \dots 3])$ because $m_3^1 = m_3^2$.

- Fourth block:
  $f(m_4^1, a_0^3) = (a_0^4, S^4[0 \dots 3])$
  $f(m_4^2, a_0^3) = (a_0^4, S^4[0 \dots 3])$
  So that $f(m_4^1, a_0^3) = f(m_4^2, a_0^3) = (a_0^4, S^4[0 \dots 3])$ because $m_4^1 = m_4^2$.

- Fifth block:
  $f(m_5^1, a_0^4) = (a_0^5, S^5[0 \dots 3])$
  $f(m_5^2, a_0^4) = (a_0^5, S^5[0 \dots 3])$
  So that $f(m_5^1, a_0^4) = f(m_5^2, a_0^4) = (a_0^5, S^5[0 \dots 3])$ because $m_5^1 = m_5^2$.

- Sixth block:
  $f(m_6^1, a_0^5) = (a_0^6, S^6[0 \dots 3])$
  $f(m_6^2, a_0^5) = (a_0^6, S^6[0 \dots 3])$
  So that $f(m_6^1, a_0^5) = f(m_6^2, a_0^5) = (a_0^6, S^6[0 \dots 3])$ because $m_6^1 = m_6^2$.

- Seventh block:
  $f(m_7^1, a_0^6) = (a_0^7, S^7[0 \dots 3])$
  $f(m_7^2, a_0^6) = (a_0^7, S^7[0 \dots 3])$
  So that $f(m_7^1, a_0^6) = f(m_7^2, a_0^6) = (a_0^7, S^7[0 \dots 3])$ because $m_7^1 = m_7^2$.

- Eighth block:
  $f(m_8^1, a_0^7) = (a_0^8, S^8[0 \dots 3])$
  $f(m_8^2, a_0^7) = (a_0^8, S^8[0 \dots 3])$
  So that $f(m_8^1, a_0^7) = f(m_8^2, a_0^7) = (a_0^8, S^8[0 \dots 3])$ because $m_8^1 = m_8^2$.

Furthermore, the output of the compression function f of eighth block becomes input of the transformation g. The output of the transformation g is a hash value of each input on Tav-128, so that:
Hash value for input 1 is

$$g(a_0^8, S^8[0 \dots 3]) = S^8[0 \dots 3] \tag{1}$$

Hash value for input 2 is

$$g(a_0^8, S^8[0 \dots 3]) = S^8[0 \dots 3] \tag{2}$$

Based on (1) and (2) we can see:

$$\begin{aligned} &Tav - 128(m_1^1 \| m_2^1 \| m_3^1 \| m_4^1 \| m_5^1 \| m_6^1 \| m_7^1 \| m_8^1) \\ = &Tav - 128(m_1^2 \| m_2^2 \| m_3^2 \| m_4^2 \| m_5^2 \| m_6^2 \| m_7^2 \| m_8^2) \\ &= S^8[0 \dots 3] \end{aligned} \tag{3}$$

Finally, it is proven that input 1 and input 2 have the same hash value to Tav-128 hash function. This indicates that the compression function on Tav-128 is not a collision resistance. Thus, we can conclude that Tav-128 is also not a collision resistance.

Table 3 shows examples of two different 256-bits inputs which have the same output value of one application of $C\&D$ functions. Then, if the two inputs are added by any bits at the end, it will still have the same hash value.

Table 3. Collisions For 256 Inputs Of Tav-128 Hash Function.

No.	Inputs (m) 256 bits	Outputs from one application of $C\&D$ functions	Hash value $H(M)$ 128 bits
1	$8fa00d2b$ $c5a8e24f$ $d66f56ab$ $4185389e$ $afde5459$ $abffeca0$ $01762bbd$ $88e0cbda$	$abdf190b$ $20c8fdac$	$44d71501$ 69689192 $fca69199$ $e6c8b361$
	$fb86cc90$ $c5a8e24f$ $d66f56ab$ $4185389e$ $afde5459$ $abffeca0$ $01762bbd$ $88e0cbda$	$abdf190b$ $20c8fdac$	$44d71501$ 69689192 $fca69199$ $e6c8b361$
2	$49390f23$ 12121212 $aaaaaaaa$ $ffffffff$ 30303030 $abababab$ $1f1f1f1f$ $dcdcdcda$	$bb8e44aa$ $1472b576$	$6e6bb26b$ $945ea6bd$ $9308ee17$ $f8f0155d$
	$8099ba12$ 12121212 $aaaaaaaa$ $ffffffff$ 30303030 $abababab$ $1f1f1f1f$ $dcdcdcdc$	$bb8e44aa$ $1472b576$	$6e6bb26b$ $945ea6bd$ $9308ee17$ $f8f0155d$

5. Conclusion

In this paper, collision attack on Tav-128 was performed. The results show that it is obtained two collisions from the total of 2^{32} inputs. We use different methods to do collision attack on Tav-128. Those methods are output grouping and merge sorting. Collisions can be obtained with 2^{26} computing. This proves that this method is faster and easier compared with the actual computing which is 2^{128}. Finally, it can be concluded that the Tav-128 is not a collision resistant hash function.

Acknowledgment

This research was supported by Sekolah Tinggi Sandi Negara, Bogor, West Java. We would like to thank to the reviewers for their valuable comments.

References

[1] Dalal A C 2004 *Searching and Sorting Algorithms* Supplementary Lecture Notes

[2] Menezes A J, van Oorschot P C and Vanstone S A 1997 *Handbook of Applied Cryptography* (Boca Raton : CRC Press LLC)

[3] Kumar A and Sanadhya S K 2010 *Attacking the Tav-128 Hash Function* IIIT-Delhi Technical Report

[4] Kumar A, Sanadhya S K, Gauravaram P, Safkhani M and Naderi M 2010 *Cryptanalysis of Tav-128 Hash Function* 11th Int. Conf. Crypto. INDOCRYPT 2010 LNCS 6212

[5] Ege B 2010 *Structural Testing Of Hash Function* Applied Mathematics Of Middle East Technical University

[6] Preenel B 2003 *Analysis and Design of Cryptographic Hash Functions* Thesis Katholieke Universiteit Leuven

[7] Andreeva E, Bouillaguet C, Fouque P, Hoch J J, Kelsey J, Shamir A and Zimmer 2008 *Second Preimage Attack on Dithered Hash Functions* 27th Annual Int. Conf. on the Theory and Applications of Cryptographic Techniques EUROCRYPT 2008 LNCS 4965

[8] Saarinen M O 2009 *Cryptanalysis of Dedicated Cryptographic Hash Functions* Departement of Mathematics, University of London

[9] Stamp M and Richard L M 2007 *Applied Cryptanalysis Breaking Ciphers in the Real World* (John Wiley & Sons : San Jose State University)

[10] Lopez P P, Castro J C C, Tapiador J M E and Ribagorda A 2008 *An Efficient Authentication Protocol for RFID Systems Resistant to Active Attacks* Computer Science Department, Carlos III University of Madrid

[11] Lopez P P 2008 *Lightweight Cryptography in Radio Frequency Identification (RFID) Systems* Computer Science Department, Universidad Carlos III De Madrid

IOP Conf. Series: Journal of Physics: Conf. Series **893** (2017) 012048 doi:10.1088/1742-6596/893/1/012048

Multicollision attack on CBC-MAC, EMAC, and XCBC-MAC of AES-128 algorithm

Alfonso Brolin Sihite and Bety Hayat Susanti

Sekolah Tinggi Sandi Negara

E-mail: alfonso.brolin.s.17@gmail.com, bety.hayat@stsn-nci.ac.id

Abstract. A Message Authentication Codes (MAC) can be constructed based on a block cipher algorithm. CBC-MAC, EMAC, and XCBC-MAC constructions are some of MAC schemes that used in the hash function. In this paper, we do multicollision attack on CBC-MAC, EMAC, and XCBC-MAC construction which uses AES-128 block cipher algorithm as basic construction. The method of multicollision attack utilizes the concept of existential forgery on CBC-MAC. The results show that the multicollision can be obtained easily in CBC-MAC, EMAC, and XCBC-MAC construction.

1. Introduction

Hash function is the function that process and maps an arbitrary length string to a fixed output length as a hash value [1]. During a transmission, the error of the message can be detected by the hash function by attaching the hash value with the message. If the hash value of the received message was not the same as the hash value that attached, then there was an error in the message [2].

Hash functions are categorized into unkeyed hash function (Modification Detection Codes) and keyed hash function (Message Authentication Codes). In addition to integrity, Message Authentication Codes (MAC) also guarantees authentication of data origin because it uses keys in the process [3]. MAC can be designed by defining a new mode of operation for existing primitive [4]. Some of them are the Cipher Block Chaining Message Authentication Code (CBC-MAC), Encrypted Message Authentication Code (EMAC), and XCBC-MAC. In this research, we use CBC-MAC, EMAC, and XCBC-MAC construction for constructing the MAC.

In 2001, National Institute of Standards and Technology (NIST) specified Advanced Encryption Standard (AES) as a block cipher standard algorithm [5]. AES has been proved to have good security and easy to implement as a block cipher algorithm [4]. AES process 128 bit data block using varied secret key length, i.e. 128-bit, 192-bit, and 256-bit. AES with 128-bit key called AES-128 [5]. In this research, we implement AES-128 in CBC-MAC, EMAC, and XCBC-MAC construction.

A hash function must satisfy three main requirements, i.e. preimage resistance, second preimage resistance, and collision resistance to be a good hash function [6]. Collision resistance is a condition where it is computationally difficult to find two messages x and x' with $x{\neq}x'$ which satisfy $h(x) = h(x')$ [3]. In this research, a multicollision attack will be applied on CBC-MAC, EMAC, and XCBC-MAC of AES-128 algorithm. The method of multicollision attack will utilize the concept of existential forgery on CBC-MAC. The goal of this research is to find the multicollision on CBC-MAC, EMAC, and XCBC-MAC construction based on a block cipher algorithm.

2. Theoretical Background

2.1. Hash Function

A hash function takes an arbitrary length string as an input then maps it to a fixed output length called as a hash value [1]. A hash value can be used as an image representation of an input string if it can be uniquely identified with an input string. That is the basic idea of a cryptographic hash function [3].

A hash function must satisfy the following three properties, i.e: preimage resistance (given a hash value y, then it is computationally difficult to find m which satisfy $F(m) = y$), second preimage resistance (given $(m, F(m))$, it is computationally difficult to find m, with $m{\neq}m$, which satisfy $F(m) = F(m)$), and collision resistance (it is computationally difficult to find two messages m and m, with $m{\neq}m'$, which satisfy $F(m) = F(m)$) [3]. There are three ways to construct a MAC algorithm, i.e. based on a block cipher algorithm (OMAC, CBC-MAC and PMAC), based on a cryptographic hash function (HMAC), and based on hash in general (universal hashing[7].

2.2. Cipher Block Chaining Message Authentication Code (CBC-MAC)

CBC-MAC is a Message Authentication Code (MAC) based on block cipher with Cipher Block Chaining (CBC) mode. CBC-MAC performs compression to messages M (fixed length mn) with a key K, where n is the length of a message block and m is the number of message blocks [7]. For details, we refer to [7].

2.3. Encrypted Message Authentication Code (EMAC)

EMAC is a Message Authentication Code (MAC) construction which encrypt the result of CBC-MAC with the second key K_2 [7]. EMAC performs compression to messages M that have fixed length with two keys K_1 and K_2, where n is the length of a block message and m is the number of message blocks. We refer to [7] for details.

2.4. XCBC-MAC

XCBC-MAC is an improvement of MAC construction of CBC-MAC that uses 3 keys i.e. K_1, K_2, and K_3. See [4] for more detail discussion.

2.5. Advanced Encryption Standard (AES)

Advanced Encryption Standard (AES) is a symmetric block cipher algorithm that performs encryption (encipher) and decryption (decipher) of information. It processes 128 bit data block using varied secret key length, i.e. 128 bit, 192 bit, and 256 bit, and has 10, 12, 14 rounds which called as AES-128, AES-192, and AES-256 respectively. Either AES-128, AES-192, and AES-256 processes 128 bits input and yield 128 bit output. The AES algorithm performs key expansion session before encryption or decryption process. For details, we refer to [5].

2.6. Existential Forgery on CBC-MAC [3]

Existential Forgery on a CBC-MAC consists of 2 stages. Fig 1 is the first stage and Fig 2 is the second stage.

Fig 1 shows the first stage of existential forgery on a CBC-MAC. The input of the CBC-MAC is $x = x_1||x_2$. E is a block cipher encryption function and its key is K. The encryption results are $y_1$and y_2. The stage begin by calculating z that satisfy $z = x_2{\oplus}y_1$. Then, calculate MAC value of x yield y_2. After that, it is begin to the second stage that is explained in Fig 2.

We can see that the input of the second stage is $x' = x'_1||x'_2$. x'_1 can be randomly chosen but satisfy $x'_1{\neq}x_1$. x'_2 is determined by $x'_2 = z{\oplus}y'_1$ where z is obtained in the first stage. After that, calculate the MAC value of x' yield y_2. It is equals to the MAC of x. Both x and x' will generate $E_K(z) = y_2$ as long as the values of z and K are the same. It can be proven by:

$$x'_2 \oplus y'_1 = z \oplus y'_1 \oplus y'_1 = z \tag{1}$$

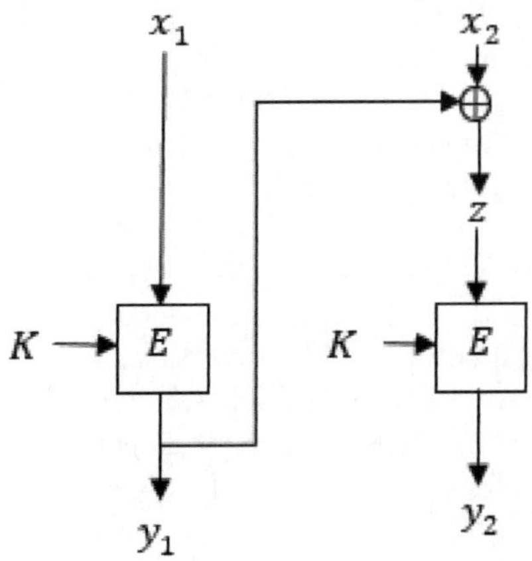

Figure 1. The first stage existential forgery on a CBC-MAC.

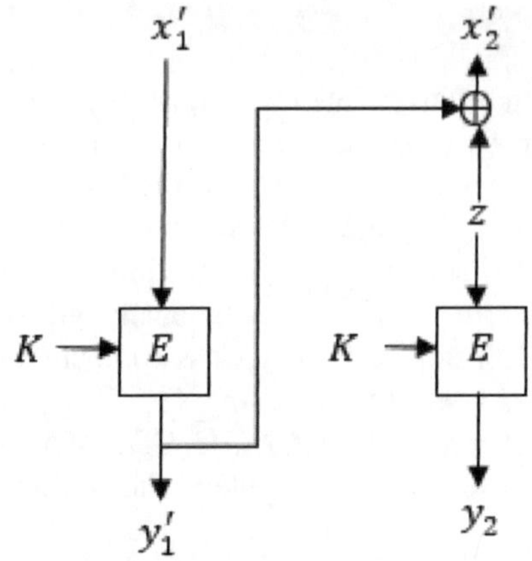

Figure 2. The second stage existential forgery on a CBC-MAC.

2.7. Multicollision Attack

Multicollision attack is finding different inputs $\{M_1, M_2, ..., M_s\} \subseteq M$ that satisfy $f(M_1) = f(M_2) = ... = f(M_s)$ [8]. In finding collisions, the padding process can be ignored if the messages that are used have the same length even though the messages are different because the padding process to both messages will yield the same pad. Moreover, if the intermediate hash chaining values collide at some point in the hash computation of two different messages and the next message blocks are same, at the end of the hash computation will yield the same hash value (collision). Thus, for the messages that have the same length, collisions that happened on the unpadded messages clearly lead to collisions on the padded messages [9].

3. Research Method

The multicollision attack method to CBC-MAC, EMAC, and XCBC-MAC utilize the concept of existential forgery on CBC-MAC. The multicollision in this research is four different messages that collide (4-collision). Assumed the adversary has the oracle $C = MAC()$ and has generated a message $M = x_1||x_2||x_3||...||x_m$. To get four messages that have same MAC, the adversary can apply the multicollision attack method, i.e.:

(i) Calculate the encryption result of the first block of message $M = x_1||x_2||x_3||x_4||x_5||...||x_m$ yield $y_1 = E_K(x_1)$. (2)

(ii) Calculate z that satisfy $z = x_2 \oplus y_1$. (3)

(iii) Formed message $M' = x_1^i||x_2||x_3||x_4||x_5||...||x_m$ where $x_1^i \neq x_1$.

(iv) Calculate the encryption result of the first block of message $M' = x_1^i||x_2||x_3||x_4||x_5||...||x_m$ yield $y_1^i = E_K(x_1^i)$. (4)

(v) Determine x_2^j that satisfy $x_2^j = z \oplus y_1^i$. (5)

(vi) Replace the second block message x_2 of message M' with x_2^j so that we can obtain message $M' = x_1^i||x_2^j||x_3||x_4||x_5||...||x_m$.

(vii) Then, on message M, calculate the chaining process until the third block yield $y_3 = E_K(E_K(E_K(x_1) \oplus x_2) \oplus x_3)$. (6)

(viii) Calculate z' that satisfy $z' = x_4 \oplus y_3$. (7)

(ix) Formed message $M'' = x_1||x_2||x_3^k||x_4||x_5||...||x_m$ where $x_3^k \neq x_3$.

(x) On message M'', calculate the chaining process until the third block yield $y_3^k = E_K(E_K(E_K(x_1) \oplus x_2) \oplus x_3^k)$. (8)

(xi) Determine x_4^l that satisfy $x_4^l = z' \oplus y_3^k$. (9)

(xii) Replace the fourth block message x_4 of message M'' with x_4^l so that we can obtain message $M' = x_1||x_2||x_3^k||x_4^l||x_5||...||x_m$.

(xiii) From the messages M, M', and M'', four messages can be constructed that will collide (4-collision). The four messages are:

$$M_1 = x_1||x_2||x_3||x_4||x_5||...||x_m;$$
$$M_2 = x_1||x_2||x_3^k||x_4^l||x_5||...||x_m;$$
$$M_3 = x_1^i||x_2^j||x_3||x_4||x_5||...||x_m;$$
$$M_4 = x_1^i||x_2^j||x_3^k||x_4^l||x_5||...||x_m.$$

The four message M_1, M_2, M_3, and M_4 will satisfy: $MAC(M_1) = MAC(M_2) = MAC(M_3) = MAC(M_4)$.

This experiment performed by implementing AES-128 block cipher algorithm into CBC-MAC, EMAC, and XCBC-MAC construction using C programming language and MinGW Dev C++ Compiler. Afterthat, we apply the multicollision attack to that constructions. The number of samples that is used are 5 extreme and pseudorandom inputs which has 640 bits length of each sample. In addition, 640 bits can be operated without padding process.

For each sample, we do 2^{20} minor modifications to the original input. For the first block of the sample, we do a modification by increment. The second block is determined appropriated to the multicollision attack method. For the third block, minor modification performed by changing 20 least significant bits as many as 1048576 possibilities. Every modification to third block followed by the modification to fourth block appropriated to the multicollision attack method.

4. Results and Analysis

After the application of the multicollision attack method for CBC-MAC, EMAC, and XCBC-MAC constructions, we can obtain the multicollision results to all inputs, both 5 extreme and pseudorandom inputs. Table I shows multicollision results of 1 extreme and 1 pseudorandom input for each construction of CBC-MAC, EMAC, and XCBC-MAC. The table shows the 4 messages which are multicollision (4-collision) that obtained as many as 2^{20} for each sample. Change column shows the order of the changes of a sample. Block 1st, Block 2nd, Block 3rd, and Block 4th column shows the first, second, third, and fourth block of input, respectively. MAC Value column shows the MAC value of the input. All values in Table I are represented in hexadecimal. MAC values of each construction show the multicollision (4-collision) that obtained.

From Table 1, we can see that the multicollision (4-collision) can be obtained in every modification to the third block followed by the modification to fourth block appropriated to the multicollison attack method for each sample of each MAC construction. As many as 2^{20} modifications for each sample, we can obtain the multicollision (4-collision) as many as 2^{20} too. But, we can see that the first and second block is the same for every modification. It is because we just make a modification to the first and the second block.

CBC-MAC, EMAC, and XCBC-MAC constructions have been analyzed to find the reason why the multicollision (4-collision) can be obtained easily. It can be proven mathematically and to facilitate it, we can see Fig 3.

As shown in Fig 3, there are two different messages. The first message in the left i.e. $x_1||x_2||x_3||x_4||x_5||...$ and the second message in the right i.e. $x_1^i||x_2^j||x_3^k||x_4^l||x_5||...$ with block messages after the fourth block on these two messages are the same. We can see that collision occurs in point z and z' (the collision is indicated by dotted horizontal lines). The proof of collision that occurred in point z can be explained as follows. Based on the equation (3) and the equation (5), we can obtain

$$y_1^i \oplus x_2^j = y_1^i \oplus (z \oplus y_1^i) = z. \quad (10)$$

Table 1. Multicollision (4-Collision) Result of Extreme Input and Pseudorandom Input on CBC-MAC, EMAC, and XCBC-MAC Construction

Sample	Change	Message				MAC Value		
		Block 1st	*Block 2nd*	*Block 3rd*	*Block 4th*	*CBC-MAC*	*EMAC*	*XCBC-MAC*
1	1	000000000000 000000000000 00000000	000000000000 000000000000 00000000	000000000000 000000000000 00000000	000000000000 000000000000 00000000	28a32a13e693 d5bea7155f9ee bb09050	783cf0d17365 9485b73a9d79 c273d882	daf3286c0185 d9b1022158ec 7de44569
		000000000000 000000000000 00000000	000000000000 000000000000 00000000	000000000000 000000000000 00000001	88ebd585b78d 72b38bd3f85c 1ea79c06	28a32a13e693 d5bea7155f9ee bb09050	783cf0d17365 9485b73a9d79 c273d882	daf3286c0185 d9b1022158ec 7de44569
		000000000000 000000000000 00000001	2ae5164c2e09 270c90b696fe 7e693ba9	000000000000 000000000000 00000000	000000000000 000000000000 00000000	28a32a13e693 d5bea7155f9ee bb09050	783cf0d17365 9485b73a9d79 c273d882	daf3286c0185 d9b1022158ec 7de44569
		000000000000 000000000000 00000001	2ae5164c2e09 270c90b696fe 7e693ba9	000000000000 000000000000 00000001	88ebd585b78d 72b38bd3f85c 1ea79c06	28a32a13e693 d5bea7155f9ee bb09050	783cf0d17365 9485b73a9d79 c273d882	daf3286c0185 d9b1022158ec 7de44569
	⋮	⋮	⋮	⋮	⋮	⋮	⋮	⋮
	2^{20}	000000000000 000000000000 00000000	000000000000 000000000000 00000000	000000000000 000000000000 00000000	000000000000 000000000000 00000000	28a32a13e693 d5bea7155f9ee bb09050	783cf0d17365 9485b73a9d79 c273d882	daf3286c0185 d9b1022158ec 7de44569
		000000000000 000000000000 00000000	000000000000 000000000000 00000000	000000000000 000000000000 00100000	95094619c6c4 3b698f48e212 aa6173af	28a32a13e693 d5bea7155f9ee bb09050	783cf0d17365 9485b73a9d79 c273d882	daf3286c0185 d9b1022158ec 7de44569
		000000000000 000000000000 00000001	2ae5164c2e09 270c90b696fe 7e693ba9	000000000000 000000000000 00000000	000000000000 000000000000 00000000	28a32a13e693 d5bea7155f9ee bb09050	783cf0d17365 9485b73a9d79 c273d882	daf3286c0185 d9b1022158ec 7de44569
		000000000000 000000000000 00000001	2ae5164c2e09 270c90b696fe 7e693ba9	000000000000 000000000000 00100000	95094619c6c4 3b698f48e212 aa6173af	28a32a13e693 d5bea7155f9ee bb09050	783cf0d17365 9485b73a9d79 c273d882	daf3286c0185 d9b1022158ec 7de44569
2	1	c153c96f9d87c 6701c5f3153ac 4529e4	fe0ad1a2c88ec 11454f9b6010f c5ae96	6bb1ddf47ea0 dbef30d85410 11a36794	b0e147db0564 6c5a2873a3e9 5868ea95	9a292a3998be aad3c898f2b5e 05a78f9	d1019f69c023 92fe82321100 6c90fea3	f7858151e897 b9a3cba8dc24 11f9150f
		c153c96f9d87c 6701c5f3153ac 4529e4	fe0ad1a2c88ec 11454f9b6010f c5ae96	6bb1ddf47ea0 dbef30d85410 11a36795	fc5b1d049c1de 40a19d01a63b e63fae6	9a292a3998be aad3c898f2b5e 05a78f9	d1019f69c023 92fe82321100 6c90fea3	f7858151e897 b9a3cba8dc24 11f9150f
		c153c96f9d87c 6701c5f3153ac 4529e5	eb6789c5f15fa 8e8bfb964c6ea 1c010a	6bb1ddf47ea0 dbef30d85410 11a36794	b0e147db0564 6c5a2873a3e9 5868ea95	9a292a3998be aad3c898f2b5e 05a78f9	d1019f69c023 92fe82321100 6c90fea3	f7858151e897 b9a3cba8dc24 11f9150f
		c153c96f9d87c 6701c5f3153ac 4529e5	eb6789c5f15fa 8e8bfb964c6ea 1c010a	6bb1ddf47ea0 dbef30d85410 11a36795	fc5b1d049c1de 40a19d01a63b e63fae6	9a292a3998be aad3c898f2b5e 05a78f9	d1019f69c023 92fe82321100 6c90fea3	f7858151e897 b9a3cba8dc24 11f9150f
	⋮	⋮	⋮	⋮	⋮	⋮	⋮	⋮
	2^{20}	c153c96f9d87c 6701c5f3153ac 4529e4	fe0ad1a2c88ec 11454f9b6010f c5ae96	6bb1ddf47ea0 dbef30d85410 11a36794	b0e147db0564 6c5a2873a3e9 5868ea95	9a292a3998be aad3c898f2b5e 05a78f9	d1019f69c023 92fe82321100 6c90fea3	f7858151e897 b9a3cba8dc24 11f9150f
		c153c96f9d87c 6701c5f3153ac 4529e4	fe0ad1a2c88ec 11454f9b6010f c5ae96	6bb1ddf47ea0 dbef30d85410 11b36794	c2d4c76da17e d3f0b7556f201 d8d22b2	9a292a3998be aad3c898f2b5e 05a78f9	d1019f69c023 92fe82321100 6c90fea3	f7858151e897 b9a3cba8dc24 11f9150f
		c153c96f9d87c 6701c5f3153ac 4529e5	eb6789c5f15fa 8e8bfb964c6ea 1c010a	6bb1ddf47ea0 dbef30d85410 11a36794	b0e147db0564 6c5a2873a3e9 5868ea95	9a292a3998be aad3c898f2b5e 05a78f9	d1019f69c023 92fe82321100 6c90fea3	f7858151e897 b9a3cba8dc24 11f9150f
		c153c96f9d87c 6701c5f3153ac 4529e5	eb6789c5f15fa 8e8bfb964c6ea 1c010a	6bb1ddf47ea0 dbef30d85410 11b36794	c2d4c76da17e d3f0b7556f201 d8d22b2	9a292a3998be aad3c898f2b5e 05a78f9	d1019f69c023 92fe82321100 6c90fea3	f7858151e897 b9a3cba8dc24 11f9150f

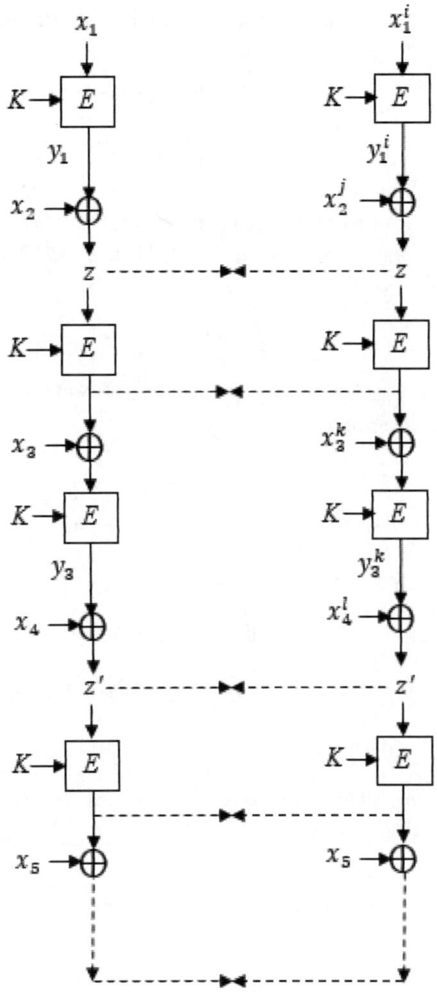

Figure 3. Collisions in multicollision attack method to CBC-MAC, EMAC, and XCBC-MAC constructions.

So it can be proven that the collision occurs at the point z, i.e.

$$y_1 \oplus x_2 = y_1^i \oplus x_2^j = z. \quad (11)$$

For the proof of collision that occurred in point z' can be explained as follows. Based on the equation (7) and the equation (9), we can obtain

$$y_3^k \oplus x_4^l = y_3^k \oplus (z' \oplus y_3^k) = z'. \quad (12)$$

So it can be proven that the collision occurs at the point z', i.e.

$$y_3 \oplus x_4 = y_3^k \oplus x_4^l = z'. \quad (13)$$

If the collisions occur in point z and z' while the block messages after the fourth block are the same, so until the end of the process, both messages will collide. From these two messages, we can formed four messages that will collide at the end of the process which will become the multicollision (4-collision), i.e.:

$$M_1 = x_1||x_2||x_3||x_4||...||x_m;$$
$$M_2 = x_1||x_2||x_3^k||x_4^l||...||x_m;$$
$$M_3 = x_1^i||x_2^j||x_3||x_4||...||x_m;$$
$$M_4 = x_1^i||x_2^j||x_3^k||x_4^l||...||x_m.$$

Thus, the multicollision (4-collision) can be obtained easily in CBC-MAC, EMAC, and XCBC-MAC constructions by modifications to four blocks of the message.

5. Conclusion

In this paper, we do multicollision attack on CBC-MAC, EMAC, and XCBC-MAC which use AES-128 as the block cipher. The results show that by the modifications as many as 2^{20} for each sample of each MAC construction, the multicollisions (4-collision) can be obtained as many as 2^{20} too. To obtain at least a multicollision (4-collision), it is easily by modifications to four blocks of the message in CBC-MAC, EMAC, and XCBC-MAC constructions.

Acknowledgements

This research was supported by Sekolah Tinggi Sandi Negara, Bogor, West Java. We would like to thank to the reviewers for their valuable comments.

References

[1] Bartkewitz T 2009 *Building Hash Function from Block Cipher, Their Security, and Implementation Properties* Ruhr-University Bochum Germany

[2] Bakhtiari S, Safavi-Naini R and Pieprzyk J 1995 *Cryptographic Hash Function : A Survey* Technical Report 95-09 Department of Computer Science University of Wollongong

[3] Menezes A J, Van Oorschot P C and Vanstone S A 1997 *Handbook of Applied Cryptography* (Boca Raton: CRC Press LLC)

[4] Black J and Rogaway P 2000 *CBC MACs for Arbitrary-Length Messages: The Three-Key Constructions* 20th Int. Crypto. Conf. CRYPTO 2000 LNCS 1880 p 197-215

[5] NIST 2001*Federal Information Processing Standards Publication (FIPS) 197* Springfield : National Institute of Standards and Technology (NIST)

[6] Rjasko M 2008 *Properties of Cryptographic Hash Functions* Department of Computer Science Bratislava

[7] Jia K, Wang X, Yuan Z and Xu G 2009 *Distinguishing Attack and Second-Preimage Attack on the CBC-like MACs* 8th Int. Conf. CANS

[8] Upadhyay J K 2010 *Generic Attack on Hash Function* Thesis of University of Waterloo

[9] Joux A 2004 *Multicollisions In Iterated Hash Function Application To Cascaded Construction* 24th Int. Crypto. Conf. CRYPTO 2004 LNCS 3152 p 306-316

IOP Conf. Series: Journal of Physics: Conf. Series **893** (2017) 012049 doi:10.1088/1742-6596/893/1/012049

Isometries and binary images of linear block codes over $\mathbb{Z}_4 + u\mathbb{Z}_4$ and $\mathbb{Z}_8 + u\mathbb{Z}_8$

Virgilio Sison and Monica Remillion

Institute of Mathematical Sciences and Physics, University of the Philippines Los Baños, College, Laguna 4031, Philippines

E-mail: {vpsison,mnremillion}@up.edu.ph

Abstract. Let $\mathbb{F}_2$ be the binary field and $\mathbb{Z}_{2^r}$ the residue class ring of integers modulo 2^r, where r is a positive integer. For the finite 16-element commutative local Frobenius non-chain ring $\mathbb{Z}_4 + u\mathbb{Z}_4$, where u is nilpotent of index 2, two weight functions are considered, namely the Lee weight and the homogeneous weight. With the appropriate application of these weights, isometric maps from $\mathbb{Z}_4 + u\mathbb{Z}_4$ to the binary spaces $\mathbb{F}_2^4$ and $\mathbb{F}_2^8$, respectively, are established via the composition of other weight-based isometries. The classical Hamming weight is used on the binary space. The resulting isometries are then applied to linear block codes over $\mathbb{Z}_4+u\mathbb{Z}_4$ whose images are binary codes of predicted length, which may or may not be linear. Certain lower and upper bounds on the minimum distances of the binary images are also derived in terms of the parameters of the $\mathbb{Z}_4 + u\mathbb{Z}_4$ codes. Several new codes and their images are constructed as illustrative examples. An analogous procedure is performed successfully on the ring $\mathbb{Z}_8 + u\mathbb{Z}_8$, where $u^2 = 0$, which is a commutative local Frobenius non-chain ring of order 64. It turns out that the method is possible in general for the class of rings $\mathbb{Z}_{2^r} + u\mathbb{Z}_{2^r}$, where $u^2 = 0$, for any positive integer r, using the generalized Gray map from $\mathbb{Z}_{2^r}$ to $\mathbb{F}_2^{2^{r-1}}$.

1. Introduction

In this present work, we give much consideration to the fact that the finite commutative ring $\mathbb{Z}_{2^r} + u\mathbb{Z}_{2^r}, u^2 = 0$ is a Frobenius ring. The reason is that Frobenius rings, which include the residue class rings $\mathbb{Z}_M$ and the Galois fields $\mathbb{F}_q$, constitute the most general class of rings that admit a homogeneous weight, a far reaching generalization of the Lee metric over $\mathbb{Z}_4$. Moreover, since two of the classical theorems in coding theory, namely the Extension Theorem and the MacWilliams identities, generalize to the case of finite Frobenious rings, these rings are viewed as the most appropriate alphabet for code-theoretic purposes. Moreover, a homogeneous weight on $\mathbb{Z}_{2^r} + u\mathbb{Z}_{2^r}$ induces an isometry into its binary image. An isometry is basically a weight-preserving map.

The material is organized as follows. Section 2 introduces the basic definitions and the relevant concepts such as Frobenius rings, homogeneous weights, linear block codes, and the structural properties of the alphabet ring. Section 3 gives the main results of the paper, while Section 4 summarizes the method used for a possible generalization. We refer the reader to [1] for a characterization of Frobenius rings. Greferath and Schmidt [2] obtained an existence and characterization theorem for (left) homogeneous weights on arbitrary finite rings, and Honold [3] showed that a homogeneous weight on a finite Frobenius ring can be expressed in terms of its generating character [please see (2) below].

2. Preliminaries and definitions

2.1. Frobenius rings and homogeneous weight

Let R be a finite ring with identity $1 \neq 0$, and $\mathbb{T}$ be the multiplicative group of unit complex numbers. The group $\mathbb{T}$ is a one-dimensional torus. A *character* of R (considered as an additive abelian group) is a group homomorphism $\chi : R \to \mathbb{T}$. The set of all characters $\widehat{R}$ (called the *character module of* R) is a right (resp. left) R-module whose group operation is pointwise multiplication of characters while scalar multiplication is given by $\chi^r(x) = \chi(rx)$ (resp. ${}^r\chi(x) = \chi(xr)$). A character χ of R is called a *right (resp. left) generating character* if the mapping $\phi : R \to \widehat{R}$ given by $\phi(r) = \chi^r$ (resp. $\phi(r) = {}^r\chi$) is an isomorphism of right (resp. left) R-modules. It is known that χ is a right (resp. left) generating character if and only if $\ker \chi$ contains no nonzero right (resp. left) ideals, and that for finite rings, χ is a right generating character if and only if it is a left generating character. We now call R as *Frobenius* if R admits a right (resp. left) generating character.

Alternatively we say that R is Frobenius if $Soc(R) \cong R/J(R)$ as right (resp. left) R-modules. Here, $Soc(R)$ denotes the socle of R, the sum of all minimal (or simple) right (resp. left) ideals of R, while $J(R)$, the Jacobson radical of R, is the intersection of all maximal right (resp. left) ideals of R. In a commutative local ring the Jacobson radical is necessarily the unique maximal ideal.

A homogeneous weight on an arbitrary finite ring R with identity is defined in the sense of [2]. Let $\mathbb{R}$ be the set of real numbers and Rx be the principal (left) ideal generated by $x \in R$. A weight function $w : R \to \mathbb{R}$ is called (left) homogeneous if the following conditions are satisfied.

(i) If $Rx = Ry$, then $w(x) = w(y)$ for all $x, y \in R$.

(ii) There exists a real number $\Gamma \geq 0$ such that

$$\sum_{y \in Rx} w(y) = \Gamma \cdot |Rx|, \text{ for all } x \in R \setminus \{0\}. \tag{1}$$

Right homogeneous weights are defined accordingly. If a weight is both left homogeneous and right homogeneous, we call it simply as a homogeneous weight. The constant Γ in (1) is called the *average value* of w. The homogeneous weight is said to be *normalized* if $\Gamma = 1$. The weight w is extended naturally to R^n, the free module of rank n consisting of n-tuples of elements from R, via $w(z) = \sum_{i=0}^{n-1} w(z_i)$ for $z = (z_0, z_1, \ldots, z_{n-1}) \in R^n$. The homogeneous distance metric $\delta : R^n \times R^n \longrightarrow \mathbb{R}$ is defined by $\delta(x, y) = w(x - y)$ for $x, y \in R^n$.

It was proved in [3] that, if R is Frobenius with generating character χ, then every homogeneous weight w on R can be expressed in terms of χ as follows.

$$w(x) = \Gamma \left[1 - \frac{1}{|R^\times|} \sum_{v \in R^\times} \chi(xv) \right] \tag{2}$$

where $R^\times$ is the group of units of R.

2.2. Linear block codes

A linear block code C of length n over a finite ring R is a nonempty subset of R^n. The code C is called *right (resp. left) R-linear* if C is a right (resp. left) R-submodule of R^n. If C is both left R-linear and right R-linear, especially in the case when R is commutative, we simply call C a linear block code over R. A matrix $G \in R^{k \times n}$ is a *generator matrix* of a rate-k/n linear block code C if the k rows of G span C and no proper subset of the rows of G generates C, in which case, $C = \{v \in R^n \mid v = uG, u \in R^k\}$. If C is a free module, then C is called a *free code.*

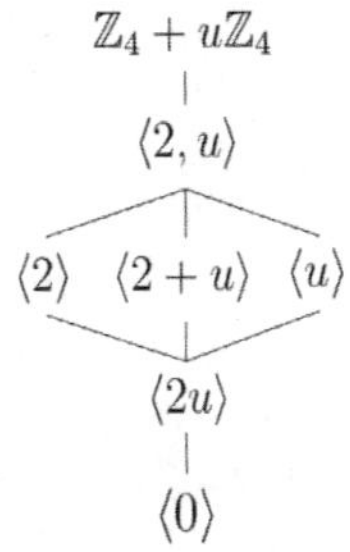

Figure 1. Lattice of Ideals of $\mathbb{Z}_4 + u\mathbb{Z}_4$

The minimum homogeneous distance $\delta_{\mathtt{min}}$ of C is defined to be $\delta_{\mathtt{min}} = \min\{\delta(x,y) \mid x,y \in C, x \neq y\}$, while the minimum homogeneous weight $w_{\mathtt{min}}$ of C is defined to be $w_{\mathtt{min}} = \min\{w(x) \mid x \in C, x \neq 0\}$. For linear codes, the minimum distance is equal to the minimum weight.

We begin with the algebraic structure of the code alphabet $R = \mathbb{Z}_4 + u\mathbb{Z}_4 = \{a + ub \mid a, b \in \mathbb{Z}_4, u^2 = 0\}$. Martinez-Moro and Szabo [4] gave the 21 commutative local rings with identity of order 16. Of these, 15 are Frobenius, where 8 are chain rings and 7 are non-chain rings, as given in [5]. The ring R belongs to the class of finite 16-element Frobenius commutative rings with identity which are local, non-chain, and non-principal ideal rings. Its characteristic is 4 and its additive group is $\mathbb{Z}_4 \times \mathbb{Z}_4$. The Jacobson radical is the unique maximal non-principal ideal $\langle 2, u\rangle = \{0, 2, u, 2u, 3u, 2+u, 2+2u, 2+3u\}$ which consists of the zero element and all the zero divisors of R. The remaining 8 elements in $R \setminus \langle 2, u\rangle$ form the multiplicative group of units of R which can be described as $a + ub, a \in \{1, 3\}$, the group of units of $\mathbb{Z}_4$. The lattice in Fig. 1 also shows the unique minimal principal ideal $\langle 2u\rangle = \{0, 2u\}$ which is significant in deriving the homogeneous weight on R. This minimal ideal is the socle. Thus clearly, $R/J(R) \cong Soc(R) \cong \mathbb{Z}_2$ as R-modules. The generating character of R is given by $\chi(a + ub) = e^{\frac{\pi i}{2}(a+b)} = i^{a+b}$. It can be shown that R is isomorphic to the quotient ring $\mathbb{Z}_4[x]/\langle x^2\rangle$.

Yildiz and Karadeniz in [6] emulated the Gray map $\mathbb{Z}_4 \to \mathbb{F}_2^2$ to define the map $\phi_3 : R \to \mathbb{Z}_4^2$ by $\phi_3(a + ub) = (b, a + b)$, and then extended the classical Lee weight on $\mathbb{Z}_4$ to R by $w_L(a + ub) = w_L(b, a + b) = w_L(b) + w_L(a + b)$. However, this weight is not homogeneous.

3. Results and discussion

3.1. Linear block codes over $\mathbb{Z}_4 + u\mathbb{Z}_4$

With the structure of R and its generating character known, from (2) we derive directly the explicit form of the homogeneous weight w_{hom} on R.

Corollary 3.1 *The homogeneous weight on $\mathbb{Z}_4 + u\mathbb{Z}_4$ is given by*

$$w_{hom}(x) = \begin{cases} 0 & \text{if } x = 0 \\ 2\Gamma & \text{if } x = 2u \\ \Gamma & \text{otherwise} \end{cases} \tag{3}$$

Observe that the highest possible homogeneous weight is given by the nonzero element of the minimal ideal, while all the other nonzero elements of R take the average value Γ as their common weight. Later we will choose $\Gamma = 4$.

To obtain certain binary images, in this paper we apply the mapping ϕ_2 from $\mathbb{Z}_4^2$ onto $\mathbb{F}_2^4$ as a natural extension of the Gray map on $\mathbb{Z}_4$. The map ϕ_2 is given by

$$\phi_2(x, y) = (x_2, x_1 + x_2, y_2, y_1 + y_2) \tag{4}$$

where $x_1 + 2x_2$ and $y_1 + 2y_2$ are the binary expansions of x and y respectively. We remark that ϕ_2 is a non-linear bijection but is an isometry from ($\mathbb{Z}_4^2$, Lee weight) onto $\mathbb{F}_2^4$, Hamming weight). Then we take the composition $\phi_2\phi_3$ from $\mathbb{Z}_4 + u\mathbb{Z}_4$ onto $\mathbb{F}_2^4$ to get

$$\phi_2\phi_3(a + ub) = (b_2, b_1 + b_2, c_2, c_1 + c_2) \tag{5}$$

where $b = b_1 + 2b_2$, $c = c_1 + 2c_2$ are the 2-adic representations of b and $c = a + b$ respectively. The following results are determined quite straightforwardly.

Theorem 3.2 *The map $\phi_2\phi_3$ is a weight preserving map from ($\mathbb{Z}_4 + u\mathbb{Z}_4$, Lee weight) to ($\mathbb{F}_2^4$, Hamming weight).*

Theorem 3.3 *Let B be a linear block code over $\mathbb{Z}_4 + u\mathbb{Z}_4$ of length n, then the associated image $\phi_2\phi_3(B)$ is a (possibly nonlinear) binary block code of length $4n$. If B is free of rank k, then necessarily the binary image has rank $4k$.*

We now proceed with determining an isometry, in terms of the homogeneous weight, from R to the binary space $\mathbb{F}_2^8$, which has twice the length of the binary image under $\phi_2\phi_3$. Define the map ϕ_4 from R to $\mathbb{F}_2^8$ by $\phi_4(a + ub) = (b_2, b_2 + b_1, b_2 + a_2, b_2 + a_2 + b_1, b_2 + a_1, b_2 + a_1 + b_1, b_2 + a_1 + a_2, b_2 + a_1 + a_2 + b_1)$ where $a = a_1 + 2a_2$ and $b = b_1 + 2b_2$.

Note that ϕ_4 is a nonlinear non-surjective map. This seemingly convoluted map is in fact a composition of various isometries $R \to \mathbb{F}_2^4 \to \mathbb{Z}_{16} \overset{\gamma_2}{\to} \mathbb{F}_2^8$, where the last component γ_2 is the generalized Gray map given by Carlet [7] as follows: let $x \in \mathbb{Z}_{16}$ where $x = x_1 + 2x_2 + 4x_3 + 8x_4$ is its binary expansion. The image of x under γ_2 is the following Boolean function f_x on $\mathbb{F}_2^3$ given by

$$\begin{array}{cccc}
f_x: & \mathbb{F}_2^3 & \to & \mathbb{F}_2 \\
 & (y_1, y_2, y_3) & \mapsto & x_4 + x_1y_1 + x_2y_2 + x_3y_3 \\
 & (0,0,0) & \mapsto & x_4 \\
 & (0,0,1) & \mapsto & x_4 + x_3 \\
 & (0,1,0) & \mapsto & x_4 + x_2 \\
 & (0,1,1) & \mapsto & x_4 + x_2 + x_3 \\
 & (1,0,0) & \mapsto & x_4 + x_1 \\
 & (1,0,1) & \mapsto & x_4 + x_1 + x_3 \\
 & (1,1,0) & \mapsto & x_4 + x_1 + x_2 \\
 & (1,1,1) & \mapsto & x_4 + x_1 + x_2 + x_3
\end{array}$$

The main idea for this composition is to preserve in a systematic manner the isometry from R to $\mathbb{F}_2^8$ as seen in the next theorems.

Theorem 3.4 *With $\Gamma = 4$, the map ϕ_4 is a weight preserving map from ($\mathbb{Z}_4 + u\mathbb{Z}_4$, Homogeneous weight) to ($\mathbb{F}_2^8$, Hamming weight).*

Theorem 3.5 *Let B be a linear block code over $\mathbb{Z}_4 + u\mathbb{Z}_4$ of length n, then the associated image $\phi_4(B)$ is a (possibly nonlinear) binary block code of length $8n$. If B is free of rank k, then necessarily the binary image has rank $8k$.*

3.2. Linear block codes over $\mathbb{Z}_8 + u\mathbb{Z}_8$

The ring $\mathbb{Z}_8 + u\mathbb{Z}_8 = \{a + ub \mid a, b \in \mathbb{Z}_8, u^2 = 0\}$ is a finite 64-element commutative ring with identity of characteristic 8 which is a local, non-chain, and non-principal ideal ring. It behaves quite similarly with $\mathbb{Z}_4 + u\mathbb{Z}_4$. The Jacobson radical is the unique maximal non-principal ideal

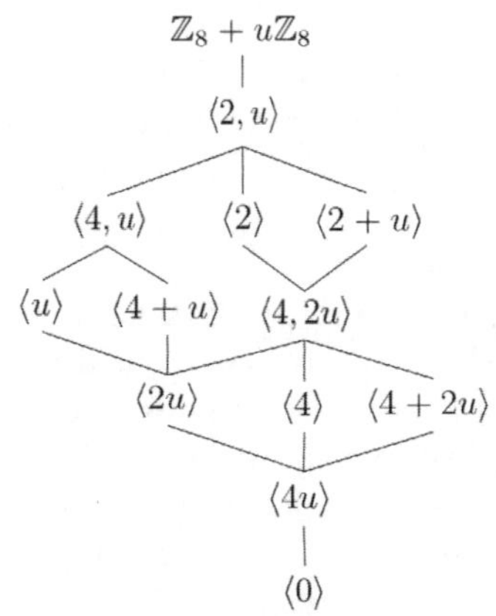

Figure 2. Lattice of Ideals of $\mathbb{Z}_8 + u\mathbb{Z}_8$

$\langle 2, u\rangle$ consisting of 32 elements of the form $a + ub$ where $a \in \{0, 2, 4, 6\}$, the unique maximal ideal of the chain ring $\mathbb{Z}_8$. The group of units of $\mathbb{Z}_8 + u\mathbb{Z}_8$ consists of 32 elements of the form $a + ub$ where $a \in \{1, 3, 5, 7\}$, the group of units of $\mathbb{Z}_8$. The lattice in Fig. 2 shows the unique minimal principal ideal $\langle 4u\rangle = \{0, 4u\}$ which is the socle. Consequently this ring is Frobenius and its generating character is given by $\chi(a+ub) = e^{\frac{\pi i}{4}(a+b)}$. It can also be shown that $\mathbb{Z}_8 + u\mathbb{Z}_8$ is isomorphic to the quotient ring $\mathbb{Z}_8[x]/\langle x^2\rangle$.

We extend the Lee weight w_L on $\mathbb{Z}_8$ given by $w_L(x) = \min\{x, 8 - x\}$ to $\mathbb{Z}_8 + u\mathbb{Z}_8$ by

$$w_L(a + ub) = w_L(b) + w_L(a + b).$$

As usual, this weight is not homogeneous. The homogeneous weight is derived analogously.

Corollary 3.6 *The homogeneous weight on $\mathbb{Z}_8 + u\mathbb{Z}_8$ is given by*

$$w_{hom}(x) = \begin{cases} 0 & \text{if } x = 0 \\ 2\Gamma & \text{if } x = 4u \\ \Gamma & \text{otherwise} \end{cases} \tag{6}$$

Again observe that the highest possible homogeneous weight is given by the nonzero element of the minimal ideal, while all the other nonzero elements take the average value Γ as their common weight. We obtain the following similar results.

Theorem 3.7 *There is a weight preserving linear map from $(\mathbb{Z}_8 + u\mathbb{Z}_8,$ Lee weight$)$ to $(\mathbb{Z}_8^2,$ Lee weight$)$.*

Theorem 3.8 *There is a weight preserving non-linear map ϕ_5 from $(\mathbb{Z}_8+u\mathbb{Z}_8,$ Homogeneous weight$)$ to $(\mathbb{F}_2^{32},$ Hamming weight$)$.*

The map ϕ_5 utilizes the generalized Gray isometry γ_3 from $\mathbb{Z}_{64}$ to $\mathbb{F}_2^{16}$ whose image is the Boolean function on $\mathbb{F}_2^5$ defined by $(y_1, y_2, y_3, y_4, y_5) \mapsto x_4 + x_1y_1 + x_2y_2 + x_3y_3 + x_4y_4 + x_5y_5$.

Theorem 3.9 *Let B be a linear block code over $\mathbb{Z}_8+u\mathbb{Z}_8$ of length n, then the associated image $\phi_5(B)$ is a (possibly nonlinear) binary block code of length $32n$.*

4. Summary and conclusion

The method is evidently possible for the class of finite commutative rings $\mathbb{Z}_{2^r} + u\mathbb{Z}_{2^r} = \{a + ub \mid a, b \in \mathbb{Z}_{2^r}, u^2 = 0\}$ with 4^r elements and characteristic 2^r. This ring is local, non-principal and a non-chain ring. The Jacobson radical is the unique maximal ideal $\langle 2, u\rangle$ consisting of 2^{2r-1} elements of the form $a + ub$ where a is in the unique maximal ideal $\langle 2\rangle$ with 2^{r-1} elements of the chain ring $\mathbb{Z}_{2^r}$. Thus $|\mathbb{Z}_{2^r} + u\mathbb{Z}_{2^r}/\langle 2, u\rangle| = 2$. The group of units of $\mathbb{Z}_{2^r} + u\mathbb{Z}_{2^r}$ is the difference $\mathbb{Z}_{2^r} + u\mathbb{Z}_{2^r} \setminus \langle 2, u\rangle$ which has cardinality 2^{2r-1} and takes the form $a + ub$ where a is a unit in $\mathbb{Z}_{2^r}$. The socle is the unique minimal ideal $\langle 2^{r-1}u\rangle$ which clearly has only two elements $\{0, 2^{r-1}u\}$. Hence $\mathbb{Z}_{2^r} + u\mathbb{Z}_{2^r}$ is Frobenius and it induces a homogeneous weight similar to (3) and (6). To ensure that weights or distances are preserved, we compose certain isometries that capitalize on the generalized Gray map from $\mathbb{Z}_{2^r}$ to $\mathbb{F}_2^{2^{r-1}}$ of Carlet [7] which he defined as follows: let $x \in \mathbb{Z}_{2^r}$ and $x = \sum_{i=1}^{r} 2^{i-1}x_i$ as its binary expansion, $x_i \in \mathbb{F}_2$. The image of x by the generalized Gray map is the Boolean function G on $\mathbb{F}_2^{r-1}$ defined by $G(x) : (y_1, y_2, \cdots, y_{r-1}) \rightarrow x_r + \sum_{i=1}^{r-1} x_i y_i$. It will also be interesting to consider the linearity conditions on the binary image code.

Acknowledgment

The first author acknowledges travel support from the UP System Research Dissemination Grant and the UPLB Academic Development Fund.

References

[1] Wood J 1999 Duality for modules over finite rings and applications to coding theory *Amer. J. Math.* **121**(3) pp 555–575
[2] Greferath M and Schmidt S E 2000 Finite-ring combinatorics and MacWilliams equivalence theorem *J. Combin. Theory Ser. A* **92** pp 17–28
[3] Honold H 2001 A characterization of finite frobenius rings *Arch. Math. (Basel)* **76** pp 406–415
[4] Martinez-Moro E and Szabo S 2015 On codes over local frobenius non-chain rings of order 16 *Contemp. Math.* **634** pp 227–241
[5] Dougherty S, Salturk E and Szabo S On codes over local frobenius rings: generator matrices, generating characters and MacWillliams identities *Preprint (version i)* gr-qc/0401010
[6] Yildiz and Karadeniz S 2014 Linear codes over $\mathbb{Z}_4+u\mathbb{Z}_4$: MacWillimans identities, projections and formally self-dual codes *Finite Fields Appl.* **27** pp 24–40
[7] Carlet C 1998 $\mathbb{Z}_{2^k}$-linear codes *IEEE Trans. Inform. Theory* **44**(4) pp 1543–1547

New modified map for digital image encryption and its performance

Suryadi MT, Maria Yus Trinity Irsan, and Yudi Satria

Department of Mathematics, Faculty of Mathematics and Natural Sciences
Universitas Indonesia, Depok 16424

E-mail: yadi.mt@sci.ui.ac.id, maria.irsan@sci.ui.ac.id, ysatria@sci.ui.ac.id

Abstract. Protection to classified digital data becomes so important in avoiding data manipulation and alteration. The focus of this paper is in data and information protection of digital images form. Protection is provided in the form of encrypted digital image. The encryption process uses a new map, $x_{n+1} = \frac{r\lambda x_n}{1+\lambda(1-x_n)^2} \pmod 1$, which is called MS map. This paper will show: the results of digital image encryption using MS map and how the performance is regarding the average time needed for encryption/decryption process; randomness of key stream sequence with NIST test, histogram analysis and goodness of fit test, quality of the decrypted image by PSNR, initial value sensitivity level, and key space. The results show that the average time of the encryption process is relatively same as the decryption process and it depends to types and sizes of the image. Cipherimage (encrypted image) is uniformly distributed since: it passes the goodness of fit test and also the histogram of the cipherimage is flat; key stream, that are generated by MS map, passes frequency (monobit) test, and runs test, which means the key stream is a random sequence; the decrypted image has same quality as the original image; and initial value sensitivity reaches 10^-17, and key space reaches 3.24×10^{634}. So, that encryption algorithm generated by MS map is more resistant to brute-force attack and known plaintext attack.

1. Introduction

Encryption algorithm of digital image based on chaotic system has been widely introduced, as in [1–3]. The study of chaotic systems and their possible applications to cryptography has received considerable attention during the last years in part of the scientific community due to the properties of chaotic system [4]. The properties of chaotic system was described in [4], that is ergodicity, sensitivity of initial condition/ control parameter, mixing property, deterministic dynamic, and structure complexity. [1] has introduce an encryption algorithm of digital image and its performance by using logistic map as a chaotic map/ key stream generator. In this paper we use MS map [5] as our key stream generator. MS map shown at equation (1)

$$x_{n+1} = \frac{r\lambda x_n}{1+\lambda(1-x_n)^2} \pmod 1 \tag{1}$$

where $r, \lambda \in \mathbb{R}$ and $x_n \in (0, 1)$.

The value of all parameters chosen by looking into the bifurcation diagram of MS map. Figure 1 shows the bifurcation diagram for $r = 3.5$. [6]

First 100 iteration from MS map shown on Figure 2 with $r = 3.5; \lambda = 2.1; x_0 = 0.3$. [6]

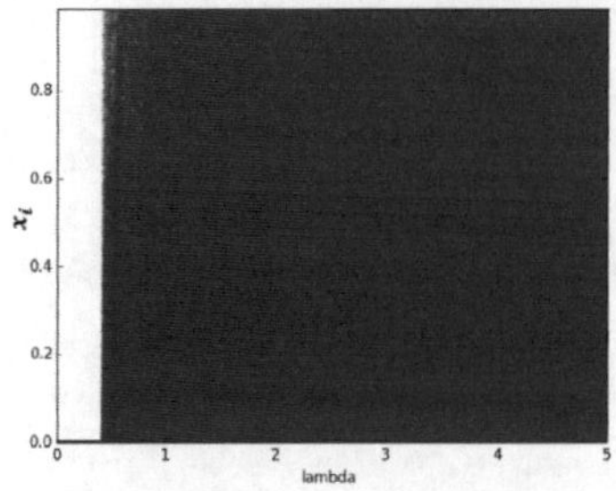

Figure 1: Bifurcation diagram for $r = 3.5$

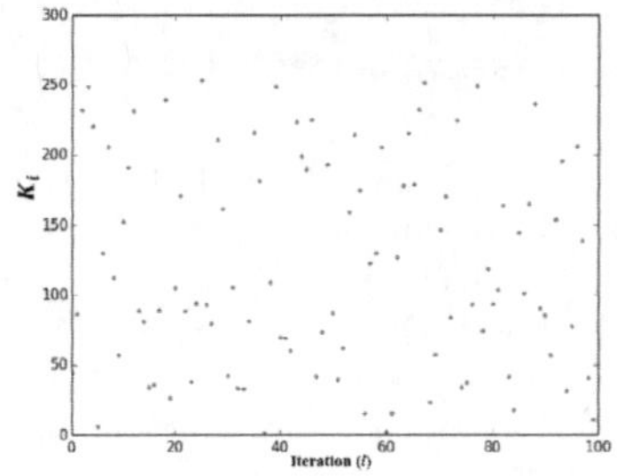

Figure 2: First 100 iteration of MS map with $r = 3.5$; $\lambda = 2.1$; $x_0 = 0.3$

2. MAIN RESULT

2.1. Encryption and Decryption Algorithm

The encryption process shown on Figure 3, and the decryption process shown on Figure 4. [5]

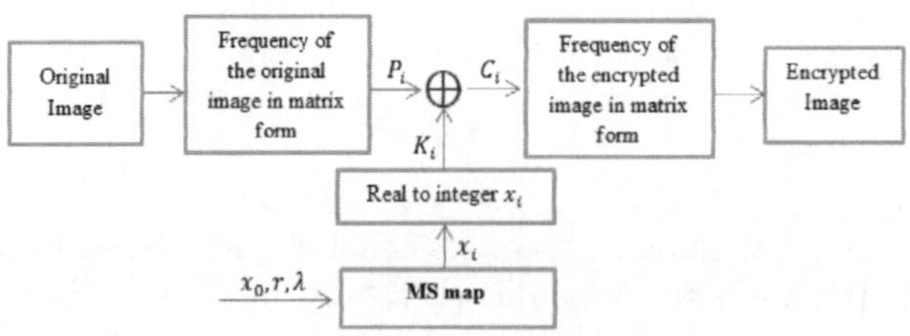

Figure 3: Encryption Process

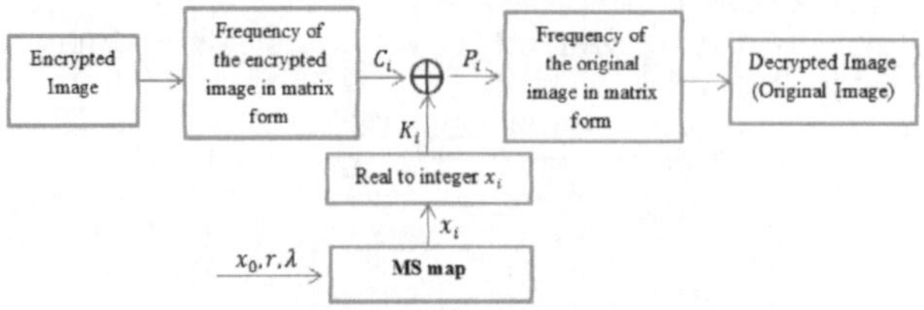

Figure 4: Decryption Process

Let the original image has pixel size $p \times q$. The key stream (K_i) obtained by:

$$K_i = (1000 \times x_i) \pmod{256} \tag{2}$$

IOP Conf. Series: Journal of Physics: Conf. Series **893** (2017) 012050 doi:10.1088/1742-6596/893/1/012050

where x_i generated by MS map and $i = 0, 2, , 3pqd - 1$, d is the image dimension. $d = 1$ for greyscale image and $d = 3$ for color image.

2.2. Testing Data and Encryption Result

We use baboon.jpg in two version (greyscale and color) as our testing data with 5 different pixel size for each version [6].The data testing shown below :

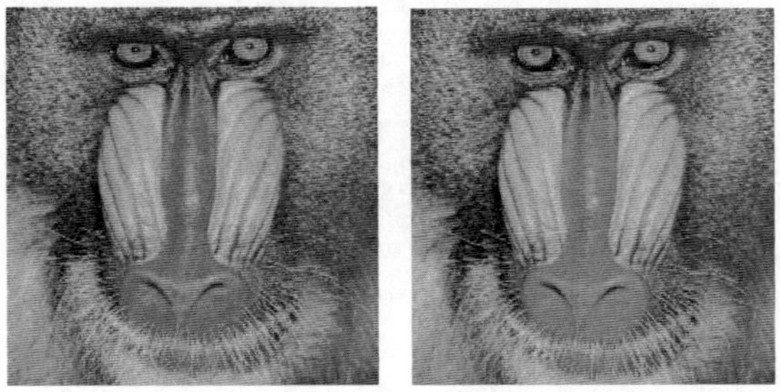

Figure 5: Display Data

Table 1: Data Testing

Image Version	Data	Size (Pixel)
Greyscale	1	77×77
Greyscale	2	212×212
Greyscale	3	512×512
Greyscale	4	1590×1590
Greyscale	5	2783×2783
Color	6	77×77
Color	7	212×212
Color	8	512×512
Color	9	1590×1590
Color	10	2783×2783

Example of encryption process result shown on Figure 6, and the average time of encryption and decryption process of each data shown on Figure 7.

Figure 6: Encrypted image of data 3 (left side) and data 8 (right side)

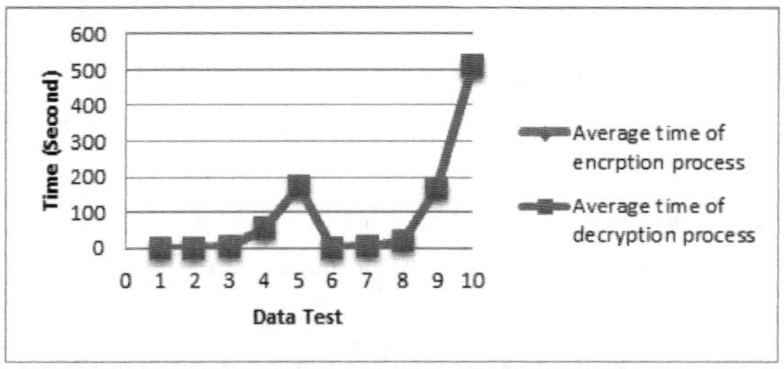

Figure 7: Average time of Encryption and decryption process

2.3. Algorithm's Performance

2.3.1. Randomness Key Stream Analysis The procedure of this test was described on [7]. For $r = 3.5; \lambda = 2.1; x_0 = 0.3$, and generated by 100 terms of key stream, we obtained:

(a) Frequency (monobit) test
  (1) $n = 100 \times 8 = 800$ bit
  (2) $|S_{800}| = 12$
  (3) $s_{obs} = \frac{|S_{800}|}{\sqrt{800}} = \frac{12}{28.28427} = 0.4242640687119285$
  (4) $P_{value} = erfc(\frac{s_{obs}}{\sqrt{2}}) = erfc(\frac{0.494974746 83}{\sqrt{2}} = 0.6713732405408726$

  Since $P_{value} \geq 0.01$, then we accept the key stream sequence as random.

(b) The Run Test
  (1) $n = 100 \times 8 = 800$ bit
  (2) $\tau = \frac{2}{\sqrt{800}} = 0.07071067811865475$
  (3) $\pi = 0.4925$
  (4) $V_{800}(obs) = 419$
  (5) $P_{value} = erfc(\frac{|V_{800}(obs) - 2n\pi(1-\pi)|}{(2\sqrt{2n\pi}(1-\pi))}) = 0.34337010342129703$

  Since $P_{value} \geq 0.01$, then we accept the key stream sequence as random.

2.3.2. Histogram Analysis We will test the distribution of cipherimage pixel's values using Goodness of fit test [8]. The test result shown on Table 2 for grayscale image and Table 3 for color image.

Table 2: Test Statistic Value of Grayscale Image

Data	**Test Statistic Value**
1	267.892561983
2	260.573869705
3	270.802734375
4	239.502114632
5	276.859198777

With degrees of freedom $256 - 1 = 255$ and 1% significance level, then from Chi-Squared table we obtained the critical value is 310.4573882199. Since all the test statistic values are less than the critical value, then we conclude that the distribution of cipherimage pixel's values is uniform. Figure 8 shown the histogram of cipherimage of Data 3 and Figure 9 shown the histogram of cipherimage of Data 8.

Table 3: Test Statistic Value of Color Image

Data	Test Statistic Value		
	Red	Green	Blue
6	232.832349469	234.041322314	240.604317760
7	238.530437878	248.714845141	287.606977572
8	262.916015625	228.000000000	220.248046875
9	222.893353902	246.747802698	275.719416162
10	249.128751264	271.990065318	259.869918473

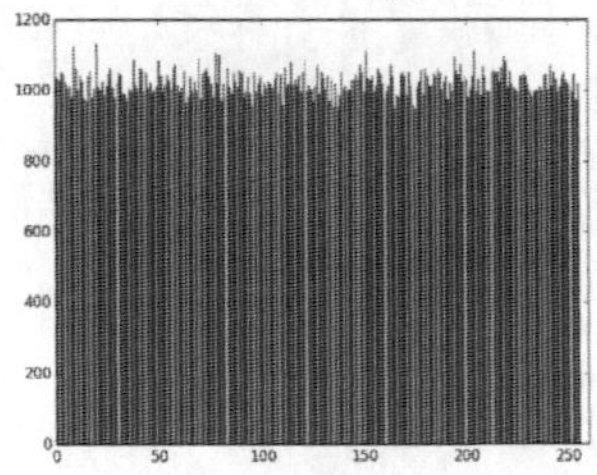

Figure 8: Histogram of cipherimage of Data 3

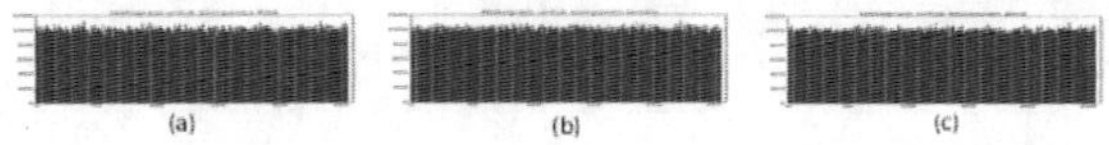

Figure 9: Histogram of cipherimage of Data 8

2.3.3. Quality of Image Tests We will compare the quality of original image and decrypted image, by Peak Signal to Noise Ratio (PSNR) which is described in [9] with formula $PSNR = 10\log_{10}\frac{L^2}{MSE}$, where $MeanSquaredError(MSE) = \frac{1}{N}\Sigma_{i=0}^{p-1}\Sigma_{i=0}^{q-1}(x_{i,j} - {y_{i,j}}^2)$, $N = pq$, $x_{i,j}$ and $y_{i,j}$ are the (i,j) entry of original image matrix and decrypted image matrix, respectively. (assume that the image size is $(p \times q)$ pixel). If $PSNR = \infty$ then we conclude that the original image and the decrypted image has same quality. Using python we obtained :

Table 4: PSNR Result

Data	MSE	PSNR
1	0	∞
2	0	∞
3	0	∞
4	0	∞
5	0	∞
6	0	∞
7	0	∞
8	0	∞
9	0	∞
10	0	∞

From Table 4 we conclude that the quality between the original image and the decrypted image is same.

2.3.4. Key Sensitivity Analysis We will decrypt Data 8 with different x_0. In encryption process we used $x_0 = 0.3$, in this part we will decrypt the cipherimage with three different x_0 as shown on Table 5.

Table 5: Key Sensitive Test Result

$x_0 = 0.3 \times 10^{-5}$	$x_0 = 0.3 \times 10^{-15}$	$x_0 = 0.3 \times 10^{-17}$
		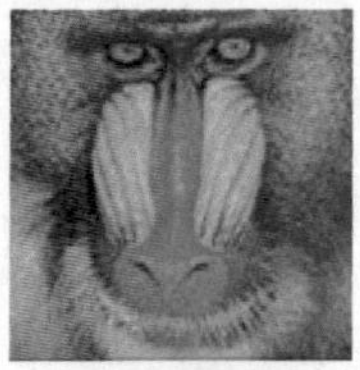
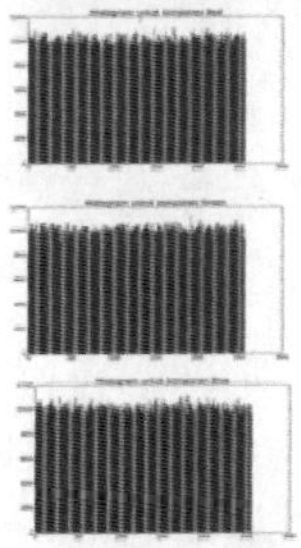	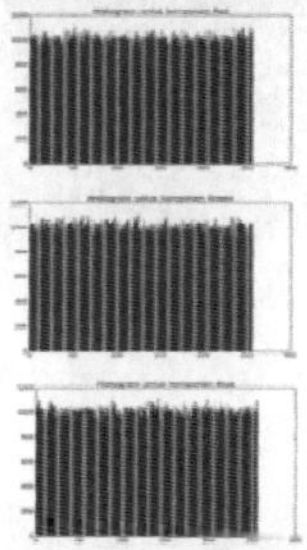	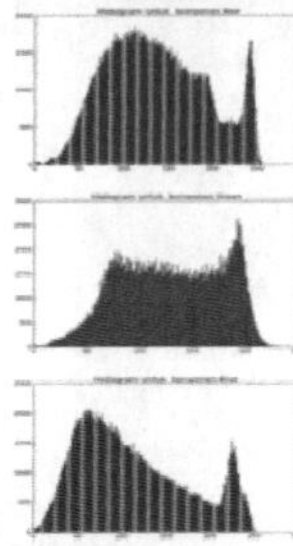

From Table 5 we seen the original image will be produced when $x_0 = 0.1 + 10^{-17}$, means that the sensitivity of initial value of this algorithm is reaches 10^{-17}.

2.3.5. Key Space Analysis In encryption and decryption algorithm using MS map, there are three parameters, x_0,r,λ, whrere $x_0 \in (0, 1)$ and $r\lambda \in$. In Python, the maximum value of float is $1.7976931348623157e + 308 \approx 1.8 \times 10^{308}$. And for real number in interval $(0, 1)$, the precision level reaches 10^{-19}. So the key space of our algorithm is,

$$1.8 \times 10^{308} \times 1.8 \times 10^{308} \times 10^{18} = 3.24 \times 10634$$

Probabilistically, the probability to obtain a right key pair is

$$\frac{1}{1.8 \times 10^{308n}} \times \frac{1}{1.8 \times 10^{308n}} \times \frac{1}{10^{18}} = \frac{1}{3.24 \times 10634} \approx 0.31 \times 10^{-634}$$

3. Conclusion

From the result above, we conclude that:

(i) MS map can be used on encryption digital image

(ii) Performance of algorithm are:

  (a) The encryption algorithm is very difficult to be cracked by known plaintext attack, due to the distribution of cipherimage pixel's values is uniform (by goodness of fit test and also by histogram of cipherimage). And so the key streams were generated, proved random by the frequency (monobit) test with $P_{value} = 0.6713732405408726 \geq 0.01$; and by the run test with $P_{value} = 0.34337010342129703 \geq 0.01$.

  (b) Key sensitivity of the algorithm reaches 10^{-17} and key space reaches 3.24×10^{634}. So that, the encryption algorithm which developed by MS map is resistant to *brute-force attack*.

  (c) The original image and the decrypted image have same quality by PSNR.

  (d) Encryptions time and decryptions time is relative same

IOP Conf. Series: Journal of Physics: Conf. Series **893** (2017) 012050 doi:10.1088/1742-6596/893/1/012050

Acknowledgment
The authors would like to thank Kemenristekdikti and DRPM UI for the grant PUPT Universitas Indonesia 2016 who support us in this research

References
[1] Suryadi M T, Nurpeti E, and Widya 2014 Performance of Chaos-Based Encryption Algorithm for Digital Image *TELKOMNIKA Telecommunication, computing, electronics, and control* **12**(3)
[2] Sukirman E, Suryadi M T, and Mubarak M A 2014 The Implementation of Henon Map Algorithm for Digital Image Encryption *TELKOMNIKA Telecommunication, computing, electronics, and control* **12**(3)
[3] Nurpeti E and Suryadi M T 2013 Chaos-Based Encryption Algorithm for Digital Image *Proceedings IndoMS International Conference on Mathematics and Its Application 2013, Indonesia Mathematical Society*
[4] Kocarev L and Lian S Chaos-Based Cryptography: theory, algorithms, and applications *Springer Science & Business Media* **354**
[5] Suryadi M T, Maria Y T I, and Yudi S 2015 Encryption algorithm using new modified map for digital image *IndoMS International Conference of Mathematics and Application (IICMA)*
[6] Maria Y T I 2016 Analisis Kinerja New Modified Map pada Enkripsi Digital *Thesis of Mathematics FMIPA Universitas Indonesia*
[7] Rukthin A, Soto J, Nechvatal J, Barker E, Leigh S, Levenson M, Banks D, Heckert d A, Dray J, and Vo S 2010 Statistical test suite for random and pseudorandom number generators for cryptographic applications *NIST special publication*
[8] Walpole R E, Myers R H, Myers S L and Ye K 2012 Probability and Statistics for Engineers & Scientists (9th edition) *Boston:Prentice Hall*
[9] Verma A and Kaur I 2013 Techniques and Algorithm Design for the Detection of PSNR in Digital Image *International Journal of Computer Applications (0975-8887): National Conference on Structuring Innovation Through Quality SITQ-2013*

IOP Conf. Series: Journal of Physics: Conf. Series **893** (2017) 012051 doi:10.1088/1742-6596/893/1/012051

Naive Bayes as opinion classifier to evaluate students satisfaction based on student sentiment in Twitter Social Media

Fahmi Candra Permana[1], Yusep Rosmansyah[2], and Atje Setiawan Abdullah[1]

[1] Department of Computer Science, Faculty of Mathematics and Natural Sciences Universitas Padjadjaran, Indonesia
[2] School of Electrical Engineering and Informatics, Institut Teknologi Bandung, Indonesia

E-mail: fahmicandrap@gmail.com

Abstract. Students activity on social media can provide implicit knowledge and new perspectives for an educational system. Sentiment analysis is a part of text mining that can help to analyze and classify the opinion data. This research uses text mining and naive Bayes method as opinion classifier, to be used as an alternative methods in the process of evaluating studentss satisfaction for educational institution. Based on test results, this system can determine the opinion classification in Bahasa Indonesia using naive Bayes as opinion classifier with accuracy level of 84% correct, and the comparison between the existing system and the proposed system to evaluate students satisfaction in learning process, there is only a difference of 16.49%.

1. Introduction

Social media sites like Twitter and Facebook provide great venues for students to share joy and struggle, vent emotion and stress, and seek social support[1]. On social media sites, students discuss and share their everyday encounters in an informal and casual manner. Students activity on social media creates large amounts of implicit knowledge and a whole new perspective for teacher and educational researchers to evaluate and monitor students experiences outside the controlled classroom environment. Therefore, students satisfaction is one key to the success of teachers and academic staff in developing the existing education system. by analyzing a range of opinions given by the students, A teacher can know the response is positive, negative and questions from students the teaching materials or systems that have been done so that it can be a matter of evaluation of teachers and academic staff.

Opinion classification system development can be develop using text mining in sentiment analysis [3]. Process classification of textual documents with sentiment analysis is done with dividing the document types into three categories, namely category sentiment positive, neutral, and negative [3]. Many approaches are being made to perform sentiment analysis, of which the most famous is the Machine Learning Based (training-based machines), Naive Bayes as a method, Support Vector Machine and Maximum Entrophy[4].

Twitter social media activity have huge development in the world, one of them is in Indonesia. Indonesia is a country with a Twitter user fifth largest in the world with 50 million registered accounts [2], even when compared with the number of Internet users in a country, in Asia Pacific Indonesia ranks first [3].

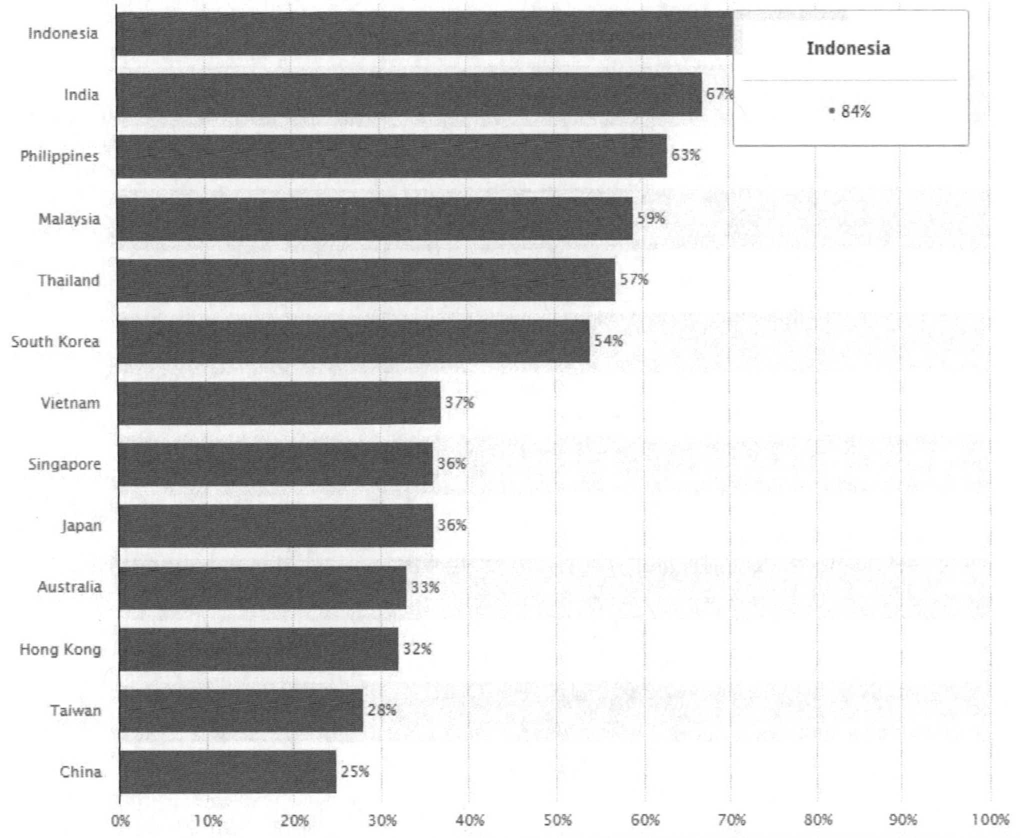

Figure 1. Twitter users in Asia Pacific[7]

With the number of users in Indonesia and the characteristics of Twitter itself is only composed of a maximum of 140 letters in once post, Twitter can produce information that is simple and make sense, the use of Twitter at this time may be authorized to make an evaluation in various fields, including educational system. Therefore, at this time the need for a new system that could facilitate an educational institution in classifying the opinion of all the students who are in the process of education they do into the kind of positive opinion, negative or neutral that can be a useful tools for performance evaluation of educational institution. The purpose of this research is to create and review the results of the application of the system that can monitor and evaluate satisfaction level of student learning by students sentiment on social media, so it can be used as an alternative method for evaluating the performance of an educational institution.

2. Theoretical Background

To do research in Sentiment analysis and monitoring student satisfaction we doing literature reviews in related topics. Literature map is a graphic that helps to visualize the relationship between some of the papers that have between one mapping helps to identify issues such as connection and closeness of ideas and findings [2]. Literature map is useful because it helps divide the paper into sections or chapters.

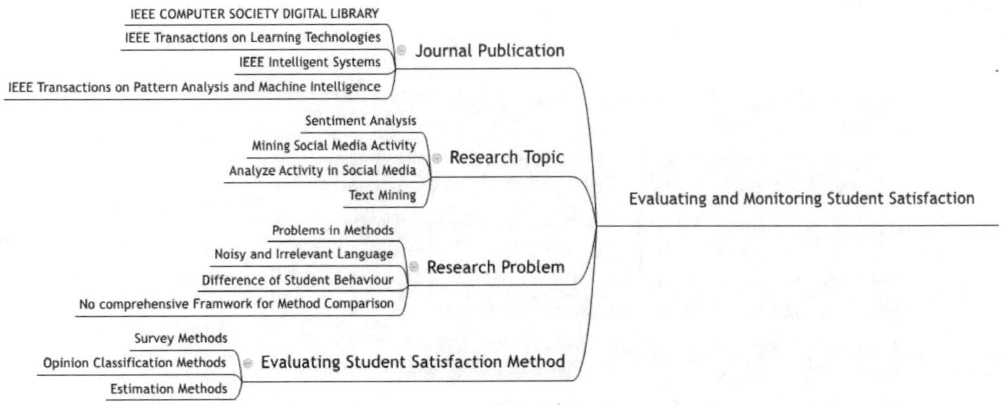

Figure 2. Literature Map.

Figure 2 shows the research literature on the map Sentiment Analysis on social media and monitoring student satisfaction. Realizing the importance and issues of sentiment analysis can make better evaluate to teacher and academic staff who will guide and monitoring student satisfaction in social media.

2.1. Mining Twitter Data

Researchers from diverse fields have analyzed Twitter content to generate specific knowledge for their respective subject domains. For example, someone analyzes tweets with mentioned @edulabindonesia using histograms, user networks, and frequencies of top keywords to quantify online activism. In this research, we built a classification model based on inductive content analysis. This model was applied and validated on a brand new dataset. Therefore, we emphasize not only the insights gained from one dataset, but also the application of the classification algorithm to other datasets for detecting student problems. The human effort Is thus augmented With large-scale Data analysis.

2.2. Machine Learning Technique

Machine Learning techniques use a training set and a test set for classication. Training set contains input feature vectors and their corresponding class labels. Using this training set, a classication model is developed which tries to classify the input feature vectors into corresponding class labels. Then a test set is used to validate the model by predicting the class labels of unseen feature vectors. A number of machine learning techniques like Naive Bayes (NB), Maximum Entropy (ME), and Support Vector Machines (SVM) are used to classify reviews [3]

2.3. Classification Method with Naive Bayes

Opinion classification using Naive Bayes method proved to have a fairly good accuracy, especially in the English-speaking opinion. Based on research on some English-speaking opinion data obtained through the website of a company, the level of accuracy of the method Naive Bayes classification opinion in doing quite well with the percentage of accuracy 81.43322% [4]. The existence of many differences in the structure of language, the term non-standard, and the expression in English and Indonesian prompted the authors to conduct research building classification system in Indonesian language opinion obtained via the social media Twitter by using Naive Bayes.

2.4. Purposive Sampling
Purposive sampling (e.g. judgment and quota) is an approach where members conform to certain criteria for selection. In judgment sampling a researcher may only want to survey those who meet a certain criteria. Judgment sampling is also appropriate in early stages of research where selection is made based on screening criteria[8]. Recruiting survey respondents through Twitter was highly effective for the research conducted but has several limitations. The survey was conducted with purposive sampling which is a non-probabilty sampling technique.

3. Methodology
This research separated in three step, the first step is mine the student sentiment in Twitter with text mining, the next steps is classify and analyze the data from twitter data to evaluate and monitor student satisfaction every mid semester, and the last step is compare the result from this system with existing system at academic consultant.

3.1. Text Mining
There are two basic methodologies to detect sentiments from text. They are Symbolic techniques and Machine Learning techniques [3]. This Research used the Machine Learning techniques with supervised learning as a output the technique because Machine Learning techniques are simpler and efcient than Symbolic techniques and these techniques can be applied for twitter sentiment analysis[3] and Naive Bayes as the classifier method, due to the accuracy of the most well among other classification process[3].

3.2. Sentiment Analysis
In Sentiment Analysis an overview of steps and techniques commonly used in sentiment classification approaches, as shown in Figure 3. Part of speech model in which a document is represented as a vector, whose entries correspond to individual terms of a vocabulary. Part-of-speech information is supposed to be a significant indicator of sentiment expression. The work on subjectivity detection [4] reveals a high correlation between the presence of adjectives and sentence subjectivity.

After get the database from twitter with text mining using the NodeXL Excel Template. In the rst phase preprocessing is done. Then a feature vector is created using relevant features. Finally using Naive Bayes as classiers, tweets are classied into positive, negative or neutrals classes. Based on the number of tweets in each class, the nal sentiment is derived. And this research develop a system to evaluate and monitoring student satisfaction based on sentiment analysis classification.

3.3. Compare This Proposed System with Existing System
After we get the data from twitter, uses sentiment analysis and have a data who was classified we develop a system that will be used every mid semester to evaluate and monitor student satisfaction and compared with manual survey to student which common use in the consultant academic.

4. Implementation
This section describe about the result of ourresearch and discuss it to make sure this research can well implemented, this is part of our plan result after mine the twitter data, using Naive Bayes as Classifier. And can be compare with existing system at consultant academic.

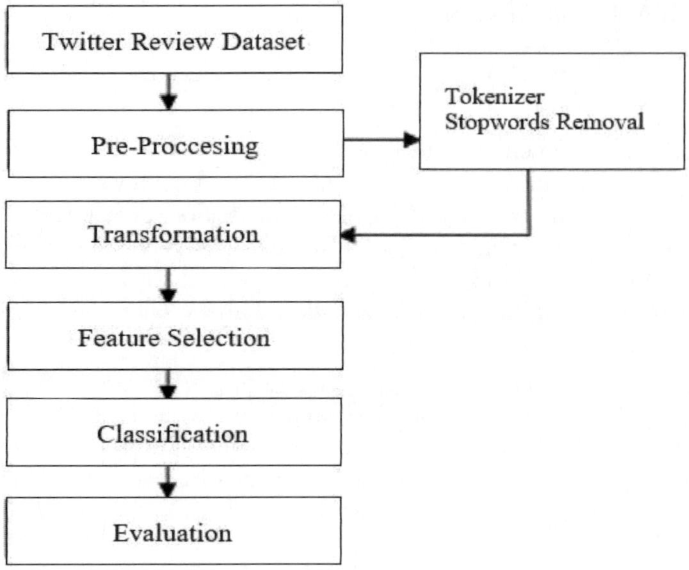

Figure 3. Steps and techniques used in sentiment classification.

4.1. Opinion Chronology Classifier System
Sentiment analysis system developed consists of a system capable of classifying opinions into positive and negative types using Naive Bayes. In this system, the raw data obtained in advance is separated into two types of data that training data and test data.

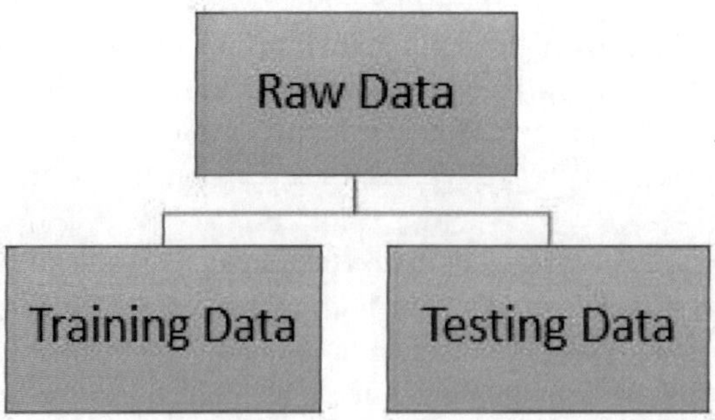

Figure 4. Raw data division

4.2. Dataset
The document level sentiment classification dataset in sample courses at this academic consultant, processed with this system and get the data, this research focus only on positive, negative and neutrals opinion, and after that this result will be use every mid semester to evaluate student satisfaction and will be compare with manual surveys of students which is commons used in the consultant academic.

4.3. Implementation

The system of evaluating students level of satisfaction that we develop and designed to classify students opinion on social media in several categories: positive, negative and neutral, then the system can provide the value of the students level of satisfaction. In the implementation of the system is very dependent on the database, because the database connection is used to store and add training data that serves as the core classification system that opinion. This database connection serves to connect the program or application that the we made to the database "twittersentimentanalyzer" in which there is a table "table-tweettraining"

4.4. Validation Test System

After getting the results of the system that the we made, the next step is to test the system validation to determine the level of accuracy and performance of the system that has the we made, before they were used and compared with the already running previously. The system that the we created consisting of 300 data option is divided into two pieces of data, training data and test data, with the ratio of training data: test data = 70: 30, consisting 90 of 210 training data and test data. validation of test results and test the accuracy of the classification system based on opinion can be seen in Figure 5.

SENTIMENT DOMINANT = 77.0% POSITIVE		
TINGKAT KEPUASAN SISWA EDULAB = 3.85		
TWEET POSITIVE	POSITIVE	
[1]cape matematika @EdulabIndonesia	F	0
[2]@EdulabIndonesia tapi sekarang dating sama kang Haroki yeyeye	T	1
[3]@edulabindonesia edulab enak belajarnya, feel like home :bd	T	1
[4]TO edulab tadi siang susah tapi semoga hasilnya bagus, semangat to2 berikutnyaa @EdulabIndonesia	T	1
[5]Ga kerasa to @EdulabIndonesia tinggal 2x lagi..semangat buat to selanjutnya :) :)	T	1
[6]Daripada mamingan gada kerjaan mending belajar @EdulabIndonesia , semangat SBMPTN, hehe demi masa depan :) :)	T	1
[7]@EdulabIndonesia kesan aku belajar di edu itu pokoknya ga bosenin, seneng boleh tambahan kpn aja, bisa dating~ asik laah.	T	1
[8]@EdulabIndonesia mungki tempat belajar yang bisa ngajain saya malem hanya guru2 di edulab , Thanks :-)	T	1
[9]cuma di @EdulabIndonesia yang gurunya bisa diajak belajar tambahan sampai jam segini, always feel like home :)	T	1
[10]Gasabar pengen cepet les di @EdulabIndonesia , susah kalau sendiri	T	1
[11]Suasana belajar di @EdulabIndonesia enak banget, bikin nyaman belajar	T	1
[12]RT @HasnaSittiH: Gasabar pengen cepet les di @EdulabIndonesia , susah kalau sendiri	T	1
[13]RT @HasnaSittiH: Suasana belajar di @EdulabIndonesia enak banget, bikin nyaman belajar	T	1

Figure 5. Data validation test results.

The test results in Figure 5 can be used to calculate the level of accuracy of the system.

$$\begin{aligned} Level\ of\ accuracy &= \frac{the\ number\ of\ tests\ is\ true}{amount\ of\ data\ validation} \times 100\% \\ &= \frac{75}{90} \times 100\% = 84\% \end{aligned}$$

Based on the calculation of the percentage of accuracy, the accuracy of the values obtained are quite high, with an accuracy of 84%. Implementation of the proposed system in an environment there will always be comparisons with the old system and has been existence, therefore we compare the test results of satisfaction level assessment of student learning in the Edulab using the old system can be seen in Figure 6.

KUESIONER PELAYANAN KONSUMEN

Edulab sebagai konsultan pendidikan senantiasa terus memperbaiki kualitas pelayanan. Demi mencapai tujuan tersebut, Edulab mengharapkan para siswa mengisi kuesioner dengan sebenar-benarnya.

1 Pilih dan lingkari jawaban yang paling sesuai dengan pilihan anda.
2 Jawaban diisi pada LJK yang sudah disediakan
3 Jika pertanyaan berupa kondisi, maka jawaban terdiri atas A (Sangat Baik), B (Baik), C (Cukup), D (Kurang), E (Sangat Kurang)
4 Jika pertanyaan berupa pernyataan, maka jawaban terdiri atas A (Sangat Setuju), B (Setuju), C (Cukup Setuju), D (Kurang Setuju), E (Sangat Kurang Setuju)
5 Soal tidak boleh dicorat-coret
6 Jika ada saran, mohon ditulis dibelakang LJK

PEMBOBOTAN

35%

40 %

RUANG KELAS						
1	Kondisi AC	A	B	C	D	E
2	Kursi meja	A	B	C	D	E
3	Karpet	A	B	C	D	E
4	Pewangi ruangan	A	B	C	D	E
5	Kondisi penerangan	A	B	C	D	E
6	Kondisi dan kebersihan ruangan	A	B	C	D	E

Figure 6. Student satisfaction questionennaire.

In the Figure 6 there are 54 aspects of assessment, and conducted a survey to all students in the branch Edulab K7, with a total of 326 student respondents and divided into two categories namely CSI "Customer Satisfied Indicator" and CEI "Costumer Enggagment Indicator", while the results of the survey can be seen in Figure 7.

From Figure 7 can be seen that the category CSI "Customer Satisfied Indicator" get results with a 3.40 rating, while the category of CEI "Costumer Enggagment Indicator" get assessment results 3.25, so the combined Edulab student learning satisfaction level can be averaged averaged as follows:

$$Edulab\ student\ satisfaction = \frac{CSI + CEI}{2} = \frac{3{,}40 + 3{,}21}{2} = 3{,}305$$

42	*Customer Service* Edulab selalu memberikan informasi kepada saya dengan tepat dan akurat	3,69	3,59
43	Edulab memberikan pelayanan penanganan keluhan saya dengan cepat dan responsif	3,23	3,10
	CSI	3,40	3,37

No	Cabang	K7	
	Program	D	S
CUSTOMER ENGAGEMENT			
44	Secara umum, seberapa puaskah anda dengan Edulab?	3,48	3,46
45	Seberapa besar kemungkinan Anda untuk tetap memilih Edulab?	3,65	3,62
46	Seberapa besar kemungkinan Anda untuk merekomendasikan Edulab pada kawan/kerabat Anda?	3,52	3,48
47	Edulab adalah lembaga yang saya percaya	3,46	3,52
48	Edulab selalu memenuhi janji kepada siswanya	2,83	2,80
49	Edulab selalu memperlakukan saya dengan adil	3,00	3,00
50	Jika saya mengalamai masalah, saya dapat mengandalkan Edulab untuk memberikan solusi yang tepat	3,09	2,93
51	Saya bangga menjadi siswa Edulab	3,41	3,38
52	Edulab selalu memperlakukan saya dengan baik	3,48	3,43
53	Edulab merupakan konsultan pendidikan yang sangat sesuai bagi saya	3,30	3,28
54	Saya tidak dapat membayangkan dunia tanpa konsultan pendidikan seperti Edulab	2,48	2,43
	CEI	3,25	3,21

Figure 7. Result of level student satisfaction in edulab.

The Systems that we develope is a system that has the same goal with a system that is already running in Edulab, that the system evaluator satisfaction levels of student learning based on student sentiment on social media, the system uses the Naive Bayess method as a classifier student opinion in social media. Sentiment analysis system that has been built based on the data that is included @edulabindonesia tweet on his tweet, in the period October 2014 to May 2015, and after the process of training and testing data in the system result in that student satisfaction rates reached 3.85, from the results obtained both types of measurement methods of evaluation and monitoring of student satisfaction levels can be calculated accuracy edulab both types of measurement methods in the following :

$$
\begin{aligned}
Improvement &= \frac{Proposed\ Method - Old\ Method}{Old\ Method} \times 100\% \\
&= \frac{3{,}85 - 3{,}305}{3{,}305} \times 100\% = 16,49\%
\end{aligned}
$$

5. Conclusion

Based on test results, this system can determine the opinion classification in Bahasa Indonesia using naive Bayes as opinion classifier with accuracy level of 84% correct, and the comparison between the existing system and the proposed system to evaluate students satisfaction in learning process, there is only a difference of 16.49%. So the proposed system can be used as an alternative system to evaluate student's satisfaction on educational system.

Acknowledgments
The work reported in this paper is supported by the Academic Leadership Grant Unpad 2016, so the authors thank to Rector Universitas Padjadjaran and Directorate Research and Community Services Unpad for supporting funding of research and paper dissemination.

References
[1] Konopnicki D, Shmueli-Scheuer M, Cohen D, Sznajder B, Herzig J, Raviv A, Zwerling N, Roitman H, Mass Y 2013 A statistical approach to mining customers' conversational data from social media *IBM Journal of Research and Development* **57**(3/4) pp 14:1 –14:13

[2] Chen X, Vorvoreanu M, Madhavan K Mining Social Media Data for Understanding Students' Learning Experiences *Learning Technologies, IEEE Transactions on* **PP**(99) p 1

[3] Pang B, Lee L, Vithyanathan S 2002 Thumbs Up ? SentimentClassification Using Machine Learning Techniques *Proceedings of TheACL-02 conference on Empirical methods in natural language processing(pp. 79-86). Stroudsburg: Association for Computational Linguistic*

[4] Neethu M S, Rajasree R 2013 Sentiment analysis in twitter using machine learning techniques *Computing, Communications and Networking Technologies (ICCCNT), 2013 Fourth International Conference* p 1,5

[5] Bahrainian, Seyed-Ali, Dengel, Andreas 2013 Sentiment Analysis and Summarization of Twitter Data *Computational Science and Engineering (CSE), 2013 IEEE 16th International Conference* pp 227 –234

[6] Omar Y, Ashaari N S 2010 Futuristic model for school's content management systems: A beginning *Information Technology (ITSim), 2010 International Symposium* **3** pp 1387 –1392

[7] http://www.statista.com/statistics/254803/twitter-penetration-in-selected-apac-countries/ (accesed at 25 May 2015)

[8] Sibona, Christopher, Walczak, Steve 2012 Purposive Sampling on Twitter: A Case Study *45th Hawaii International Conference on System Sciences*

Correcting circular non-uniform illumination of Sundanese lontar images

S Hadi, M Suryani, E Paulus, and A K Supriatna

Department of Computer Science, University of Padjadjaran
Jalan Raya Bandung Sumedang KM21 Jatinangor 45363, INDONESIA

E-mail: {setiawanahdi, mira.suryani, erick.paulus, a.k.supriatna}@unpad.ac.id

Abstract. This paper presents the method for enhancing quality of degraded images which is essential for information automatic extraction from the digitized ancient manuscript known as lontar. The degraded lontar images have circular non-uniform illumination characteristics that are represented by more brightness on inner center of images and darker in the outer ellipse-like area of the images. The purpose of enhanced image processing is to leveling the distribution of pixels so the images have similar characteristics. The proposed method is developed based on DEWA Framework, a novel and optimized method for object detection, combined with background subtraction technique, which is implemented in several color spaces. Background images are estimated using morphological mathematics with specific type and size of structuring element. This estimation process has to be done due to the unavailability of exact background image captured during image acquisition step. Theoretical foundations of the proposed method along with experimental results are reported in this paper. The proposed method has been tested on 12 Sundanese lontar images which shows this approach is different and unique from most of the existing methods in this field of study. Experiment demonstrated the successful results of the method which is represented by displaying profile characteristics of each images on row 128.

1. Introduction

Ancient manuscripts can be considered as a very important and valuable ethno-based documents, not only in the sense of cultural heritage, but also as part of ethnocultural documents and country intellectual properties. On one hand, some documents are treated and maintained properly. On the other hand, they are badly stored in a place that are rarely attended and on poor condition. In facts, ancient documents, which are written on palm leaves called lontar, can consist of important data or information that might be related to our modern culture and daily life. It is not impossible that some documents can explain mathematical theories or computing concepts that have not been solved today. Efforts have been performed to preserve those documents by digitizing them into digital images, however there are problems found in the acquisition process [1]. This problem becomes a challenge for researcher in this field [2].

In digital image acquisition process, some images might not be taken properly which can cause problems for the next step of processing [3]. These images, for some reasons, are pictured using camera with bad light condition, are taken using flashlight incorrectly, and are captured in uncontrolled environment. The effects of this process shows that intensity distribution of images are not uniform circularly which is represented by circular ellipse-like shape pixels distribution.

The images show more brightness at the center, but darker in the surrounded outer region. In this paper we proposed a novel combination of image processing method to normalize and leveling the intensity distribution of degraded lontar digital images by implementing two background subtraction methods and three color space models. Background images are generated by the estimation using the concept of morphological mathematics.

The experiment has been conducted by involving 12 given Sundanese lontar images, written in the period of 15th century. The quality of proposed method are visualized using histogram profile representation on row 128 for each color models and for each subtraction methods. It is showed that our proposed algorithm can solve the problem of circular non-uniform illumination problems of lontar images. The result is expected to be used for supporting the next step further processing of lontar images such as automatic pattern recognition and indexation [4]

2. Methods

The proposed method is derived partially from DeWa framework concept [5]. In this novel and optimized object detection framework, the space of input images are transformed into 3 color spaces: normalized RGB (nRGB), Hue-Saturation-Value (HSV), and YCbCr [6]. The next step is background image generation. This is performed by estimating each background in each color space using morphological mathematics concept. As we have background images, we can do the next step by subtracting each input image with background image for each color model. Two background subtraction techniques are applied, those are Direct Subtraction (DS) method, and Retrospective Correction (RC) method [8].

2.1. Color Space Transformation

The color transformation is required to obtain simpler yet accurate color separation information of the object. As previously mentioned, the color spaces used in this research are RGB, normalized RGB, HSV, and YCbCr. These color space models make the processing is focused not on the luminance but on the hue of the image. The conversion formula can be described as follows:

Conversion from RGB to Normalize RGB

$$r = \frac{R}{R+G+B} \qquad g = \frac{G}{R+G+B} \qquad b = \frac{B}{R+G+B}. \tag{1}$$

Conversion from RGB to HSV

$$\begin{aligned} H &= \arccos \frac{0.5((R-G)+(R-B))}{\sqrt{((R-G)^2+(R-B)(G-B))}} \\ S &= 1 - 3\frac{\min(R,G,B)}{R+G+B} \\ V &= \tfrac{1}{3}(R+G+B). \end{aligned} \tag{2}$$

Conversion from RGB to YCbCr

$$\begin{aligned} Y &= 0.299R + 0.587G + 0.114B \\ C_r &= R - Y \\ C_b &= B - Y. \end{aligned} \tag{3}$$

where R, G, B are the pixel intensity values on channel Red, Green, and Blue.

2.2. Background Image Estimation

The background image generation is estimated by applying opening morphological mathematics technique which is consist of erosion followed by dilation with structuring element parameter 'disk' with size of 15. The theoretical concept of this method can be defined as follows:

Definition 1 *Erosion can be formulated as* $A - B = \cup_{b \notin B} A_b = \{p | B_p \cap A\}$.

Definition 2 *Dilation can be formulated as* $A \oplus B = \cup_{b \notin B} A_b = \{p | B_p \cap A \neq \emptyset\}$.

This mathematics definition of erosion can be simplified descriptively as follows:

- Let A is set of Euclidean coordinate which correspondence with binary image as data input
- Let B is structuring element that consist of small elements that are become a determination elements for erosion process
- B_x is transformation process of B for all image region
- Erosion A by B simply can be defined as a set of all points in x which has criteria that B_x is a subset of A.

Similar with above, dilation operation can be described as:

- Let A is set of Euclidean coordinate which correspondence with binary image as data input
- Let B is structuring element that consist of small elements that are become a determination elements for dilation process
- B_x is transformation process of for all image region
- So the dilation A by B can be defined as a set of all points x that follows the rule so the intersection between B_x with A is not empty.

The combination of erosion followed by dilation is known as opening operator. The dilation operation followed by erosion operation is known as closing operator.

2.3. Background Subtraction Techniques

Two techniques are applied in this part, those are Direct Subtraction (DS) technique, and Retrospective Correction(RC) technique. DS technique is the simplest method which is merely do subtract operation between input image and background image.

The RC technique is more complex step that involve statistical calculation. All methods are applied in grayscale image. The formula for RC techniques can be stated as follows:

$$g(x, y) = \frac{f(x, y)}{b(x, y)} \cdot C$$

where

$$C = \overline{f(x, y)} \cdot \frac{1}{\overline{\left(\frac{f(x,y)}{b(x,y)}\right)}}.$$

$f(x, y)$ is the original image, $b(x, y)$ is the estimated background image, and C is a normalization constant that is used to recover the initial intensities.

2.4. Measurement

The result of the proposed method is presented by histogram profiling [7], snapshot on row 128 in 2D. This is visualization of pixel distribution characteristics on specified row the intensity value on each pixel on a row.

The other measurement method is to calculate the simple statistics of each image which is represented by number of pixels mean, standard deviation, minimum value, and maximum value. This measurement show how the intensity distribution of each image.

IOP Conf. Series: Journal of Physics: Conf. Series **893** (2017) 012052 doi:10.1088/1742-6596/893/1/012052

3. Implementation

Lontar images used in this experiment are obtained from personal repository which cannot be considered as good quality images. The images are not only taken using average-quality camera but also difficult to be re-taken due to complex procedures and cultural constraints. However, these images are challenging for us to develop special image processing algorithms for correcting the intensity distribution. Figure 1 shows the 3 of 12 selected-cropped lontars.

Figure 1. Pictures of 3 of 12 selected-cropped Lontar used in this research

We can see that the image intensities are unevenly distributed with more brightness on the center circularly. This is usually caused by the bad quality of flashlight. The intensities on the outer circle are darker especially on the most left and right side.

Getting internal characteristics of the images is performed by histogram profiling. Theoretically this can be done vertically and/or vertically. In this case we took rows 128 horizontally as displayed on the Figure 2.

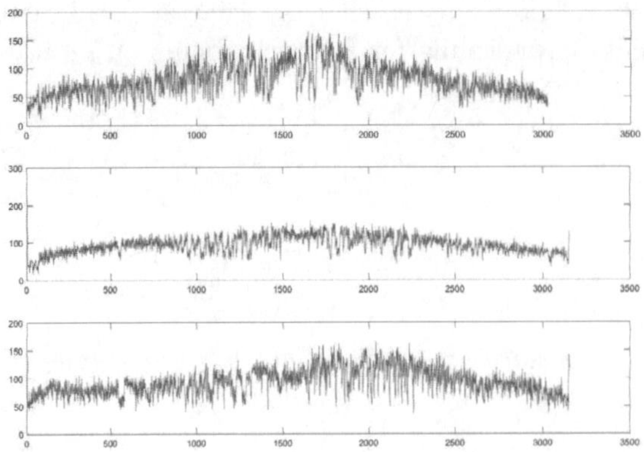

Figure 2. Profile Histogram of Lontar on Row 128

We can see the internal characteristics of those three images by evaluating each histogram profile. The lower histogram is shown on the left and right side. We can infer that the bar is getting higher to the right and after reaching the center then getting lower. It is looked like sinusoidal wave. The profile informs us that the intensities on the center is brighter than on the sides. What we are going to do is to normalize the sinusoidal-like form of profile into straight-like line.

The algorithm that is applied to those input images to normalize the images can be described as follows. It is noted that we are working on grayscale images, so we have to preprocess the images by transforming from color images into grayscale images.

(i) Convert preprocessed input image to selected color model
(ii) Generate estimated background images using morphological opening using 'disk' structuring element with size 15
(iii) Perform background subtraction using
  (a) Direct subscription
  (b) Retrospective Correction
(iv) Perform histogram profiling for the result images

Figure 3 and Figure 4 show the histogram profile of corrected lontar 3306 using DS and RC background subtraction techniques respectively.

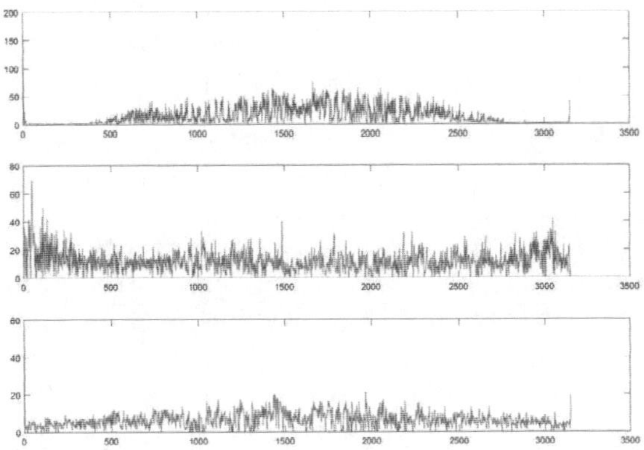

Figure 3. Profile of Corrected Lontar 3308 using DS Technique

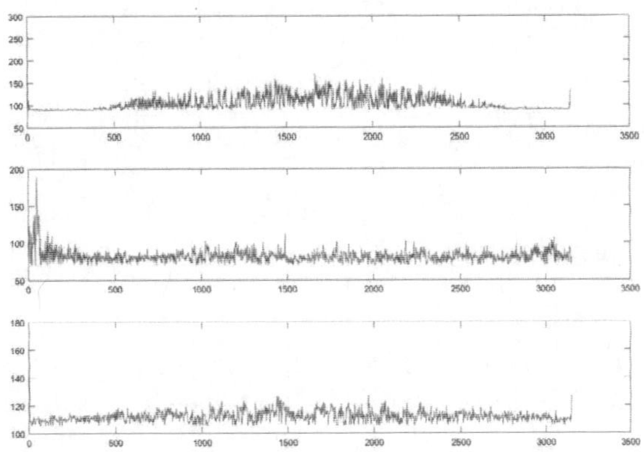

Figure 4. Profile of Corrected Lontar 3308 using RC Technique

To view the numerical value of the intensity distribution, simple image statistics has been applied. The result of lontar 3308 image statistics and its pixel-range normalized are presented on the Table 1 and Table 2. These tables consist of columns that represent: type of color space (Color Space), method of correction used (Corr.), number of pixels (Sum), pixel intensities average (Mean), standard deviation (Std), minimum and maximum intensity value of pixels (Min, Max).

Table 1. Image Statistics of Corrected Lontar 3308

Color Space	Corr.	Sum	Mean	Std	Min	Max
RGB	None	104727728	96.65	40.08	2	255
nRGB	None	115281829	106.39	33.13	38	255
HSV	None	91906324	84.82	13.37	0	154
YCbCr	None	121735669	112.34	11.07	90	176
RGB	RC	103359441	95.39	30.70	60	255
nGB	RC	114555097	105.72	26.47	89	255
HSV	RC	91658512	84.59	19.68	0	255
YCbCr	RC	121893531	112.49	7.74	106	184
RGB	DS	36262725	33.47	29.54	0	232
nGB	DS	18165601	16.76	27.61	0	208
HSV	DS	13491113	12.45	9.35	0	122
YCbCr	DS	7151192	6.60	7.74	0	74

Table 2. Image Statistics of Corrected Lontar 3308 with Pixel-range Adjusted

Color Space	Corr.	Sum	Mean	Std	Min	Max
RGB	None	74732957	68.97	46.67	0	255
nRGB	None	35070265	32.36	49.53	0	255
HSV	None	195067216	180.02	42.49	0	255
YCbCr	None	53127569	49.03	47.30	0	255
RGB	RC	51272922	47.32	39.21	0	255
nRGB	RC	27828681	25.68	40.41	0	255
HSV	RC	35085530	32.38	36.60	0	255
YCbCr	RC	38067007	35.13	35.22	0	255
RGB	DS	46347949	42.77	36.53	0	255
nRGB	DS	26565527	24.52	39.81	0	255
HSV	DS	65732134	60.66	43.29	0	255
YCbCr	DS	34155388	31.52	36.03	0	255

As previously mentioned, we used 12 lontar images for the experiment, so we do averaging for all image intensity adjusted results. The result is presented on the Table 3. To get more impressive result, we presented the standard deviation values for all correction methods on Figure 5.

The implementation of proposed methods on Lontar images can be considered as a novelty of this paper. The numerical result presented on the tables followed the resulting images of the proposed method, so the numerical result can be inferred as a novelty as well.

4. Conclusion and Future Works

We have developed an algorithm for correcting circular illumination problem in Lontar images. The method is based on background subtraction technique which is implemented in two approaches, the first is using direct subtraction approach, and the second is using retrospective

Table 3. Image Statistics of Corrected All Images (12 Lontars)

Color Space	Corr.	Sum	Mean	Std	Min	Max
RGB	None	69216487	65.98	45.92	0	255
nRGB	None	34150915	32.58	47.49	0	255
HSV	None	182561967	173.80	43.70	0	255
YCbCr	None	48864597	46.62	45.51	0	255
RGB	RC	48415162	46.08	37.63	0	255
nRGB	RC	27466880	26.16	39.99	0	255
HSV	RC	28704876	27.37	35.27	0	255
YCbCr	RC	34251970	32.60	34.43	0	255
RGB	DS	43302293	41.22	35.25	0	255
nRGB	DS	24751186	23.57	38.40	0	255
HSV	DS	61088057	58.25	43.49	0	255
YCbCr	DS	31335609	29.86	34.83	0	255

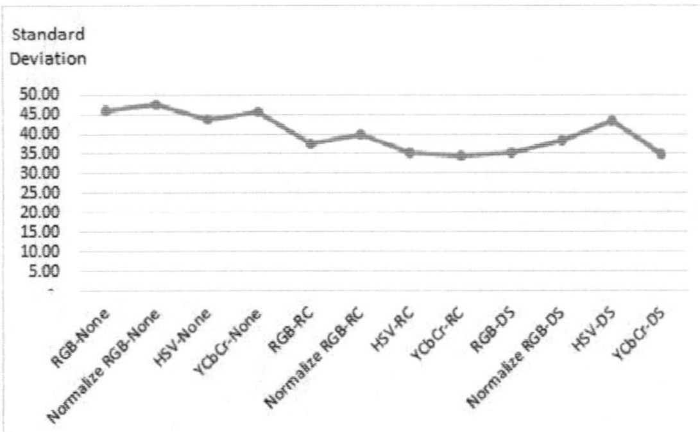

Figure 5. Standard Deviation Visualization for All Correction Methods

correction approach. Both methods are applied in 3 different color space models: normalized RGB, HSV, and YCbCr. The background images are estimated using mathematical morphology idea using disk structuring element with size of 15. The results are evaluated by computing the simple statistics for each results and displaying using profile characteristic histogram taken on row 128.

Based on conducted experiment that involved 12 lontar images, we can see that the result is promising. The histogram profile shows the changing from sinusoidal-like form into straight-line form. It means that the pixel intensity is uniformly distributed. The image statistics present the decreasing value of standard deviation which means the distribution value is not too sparse. However, even we have better intensity distribution, the resulting image is visually not better than the original image. This can be other challenge to make the output image is clearer especially in the area where the written text is located.

Acknowledgments

The authors would like to thank the Rector of Unpad, the Dean of Faculty of Math. and Nat. Sciences Unpad, and Prof. Jean-Christophe Burie from the University of La Rochelle, for supporting of this research. Great appreciation is pointed to the AMADI team Unpad for the hard work and contribution.

References

[1] Yahya S R, Abdullah S N H S, Omar K, Zakaria M S and Liong C Y 2010 Review on Image Enhancement Methods of Old Manuscript with Damaged Background, *Int. Jour. on EE and Informatics* **2** p 1

[2] Kesiman M W A et al. 2017 Southeast Asian Palm Leaf Manuscript Images: A New Challenge for Handwritten Text Line Segmentation Method, *Int. Jour. of Electronic Imaging* **26** p 1

[3] Paulus E, Suryani M and Hadi S 2016 Preliminary Study On Binarization Methods Of Ancient Sundanese Manuscript Images *Asian Mathematical Conference*

[4] Suryani M and Hadi S 2016 Sundanese Ancient Manuscripts Search Engine Using Probabilistic Approach *Asian Mathematical Conference*

[5] Hadi S, Suwardi I S and Ahmad A S 2007 DeWa : A multi aspect approach for multiple face detection in complex scene digital image *ITB Journal of ICT* **1C** p 1

[6] Dubois E 2007 *The Structure and Properties of Color Spaces and the Representation of Color Images*, (Morgan & Claypool Publishers)

[7] C Lundström 2006 Technical report: Measuring digital image quality *diva-portal.org*

[8] Gonzalez R C, Woods R E and Eddins S L 2009 Digital Image Processing Using MATLAB 2nd edition (Gatesmark Publishing)

Geographic Information Systems for Mapping the National Exam Result of Junior High School in 2014 at West Java Province

Atje Setiawan Abdullah[1], Budi Nurani Ruchjana[2], Juli Rejito[1], Rudi Rosadi[1], and Fahmi Candra Permana[1]

[1] Department of Computer Science, Faculty of Mathematics and Natural Sciences Universitas Padjadjaran, Indonesia
[2] Department of Mathematics, Faculty of Mathematics and Natural Sciences Universitas Padjadjaran, Indonesia

E-mail: atjesetiawan@gmail.com

Abstract. National Exam level of schooling is implemented by the Ministry of Education and Culture for the development of education in Indonesia. The national examinations are centrally evaluated by the National Education Standards Agency, and the expected implementation of the national exams can describe the successful implementation of education at the district, regency/city, provincial, or national level. In this study, we evaluate, analyze, and explore the implementation of the national exam database of the results of the Junior High School in 2014, with the Junior High School (SMP/MTs) as the smallest unit of analysis at the district level. The method used in this study is a data mining approach using the methodology of Knowledge Discovery in Databases (KDD) using descriptive analysis and spatial mapping of national examinations. The results of the classification of the data mining process to national exams of Junior High School in 2014 using data 6,878 SMP/MTs in West Java showed that 81.01 % were at moderate or medium levels. While the results of the spatial mapping for SMP/MTs in West Java can be explained 36,99 % at the unfavourable or bad level. The visualisation of evaluation results in graphic is done using ArcGIS to provide position information quality of education in regency/city, provincial or national level. The results of this study can be used by education management to make decision to improve educational services based on the national exam database in West Java.

Keywords: KDD, spatial mapping, national exam, junior high school

1. Introduction

Law of the Republic of Indonesia No. 20/2003 on the National Education [7], states that in order to control the quality of education nation in evaluation as a form of accountability of education providers to the parties concerned. National Exam (UN) is a standard evaluation system of primary and secondary education nationally and equal quality education levels among regions conducted by the Center for Educational Assessment. The evaluation was done by an independent agency on a regular basis, through, transparent and systematic, in order to assess the achievement of national standards for the evaluation and monitoring process to be done on an ongoing basis. UN benefit is to measure the level of success of teaching and learning in every area, to map the competencies of students, schools and local governments, as well as for evaluating and mapping the development of national education quality. Based on the above, the purpose of this study are to evaluate the results of UN 2014 by mapping the value of the results of the 2014 UN SMP / MTs in West Java province, using spatial data mining. This is according to research conducted by [1] with a team of researchers Academic Leadership Grant Universitas Padjadjaran with the title "Development of Space Time Models and Its Application in Industry, Organization, and the Environment, to support the Common Goals West Java Year 2015-2019", in particular the application of spatial data mining for mapping the quality of education in the province of West Java.

2. Research Method

The first step in this research is a build of the study design, utilising the science of the latter being developed, known as data mining, which is defined to extract knowledge automatically from large databases, to obtain the patterns interesting, thus forming a knowledge [4]. Data mining process includes data preprocessing, data mining, and postprocessing. Data preprocessing steps including a data cleaning, data transformation, data integration and data selection, the process of data mining including the use of models to process data, while the postprocessing including processing visualisation and interpretation of results, and conclusion about the result to becomes a knowledge. The data used in this study is taken from the database of the Research Department of Education Balitbang Diknas [2], covers the results of the 2014 UN arranged in a hierarchy at West Java province, 27 districts/cities, 617 districts, and 6,878 Junior High School (SPM/MTs). While the external data source are the coordinates of cities and districts were taken from Information Agency Geographical (BIG), contour map is processed using software applications ArcGIS [3], while the development of application programs for searching the data of schools by district/city and sub-district, using the Java programming and Database in Access [5]. In this study, the authors describe the category of results of quality examination using a model calculation of average and standard deviation of the results of UN 2014, to divide into three groups the UN, the green group or good category (A), yellow or medium (B), and red or bad (C), provided the group A¿ = mean + 1/2 STDEV, group B between the average-half STDEV and average + 1/2 STDEV, while group C ¡mean-half STDEV [1]. Stages of research methodology follows the data mining process using Knowledge Discovery in Databases (KDD) [4], is described in Figure 1 below.

Figure 1 illustrates the stages of the research methodology used, the preprocessing stage begins with an exploration database of the UN in 2014, and to transform the data into groups, provinces, counties, districts and schools, as well as the process of merging the data with external data such as coordinates of the district / city and districts. Stages data mining, is done by calculating the mean and standard deviation for the district and sub-district level, and then perform data classification based on three groups of green or good (A), yellow or medium (B), and red or bad (C). While on stage postprocessing, visualize data in the form of tables, images and maps to illustrate each district/city and district. As well as the stage of making conclusions to get a knowledge.

In this study, we built Java-based application program, to do the search results of the average value of the UN, at the level of provincial, district/municipal, district and school as described in Figure 2. In Figure 2, we explain about sample search application schools in the district/city, to find a name and position of school SMP/MTs, for example the application give the names of the schools in Garut Regency, complete with a school identity, average and percentage results of the UN for the learning Indonesian, English, Mathematics, Science, and the total value of the UN.

Figure 3 explains a sample search application schools in some districts, in order to find a school SMP/MTs that exist, for example in the application seen the names of existing schools in

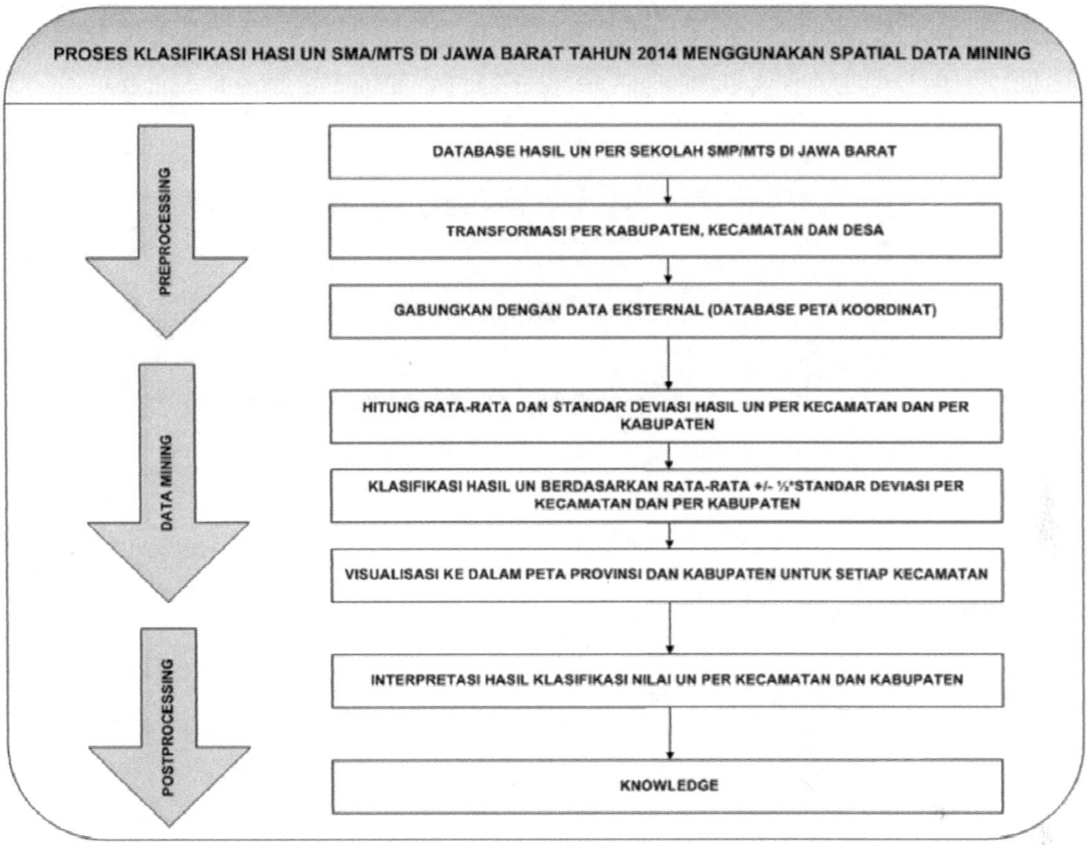

Figure 1. Stages of Research for UN SMP using Spatial Data Mining

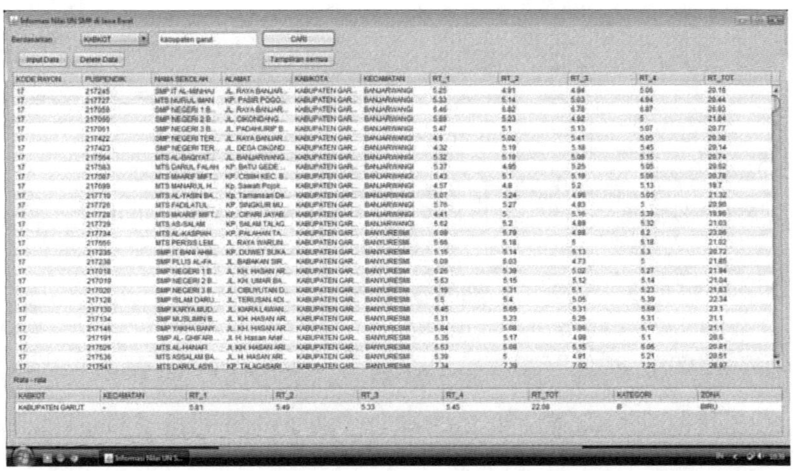

Figure 2. Application Search the Average Value of the UN School in the Regency/City

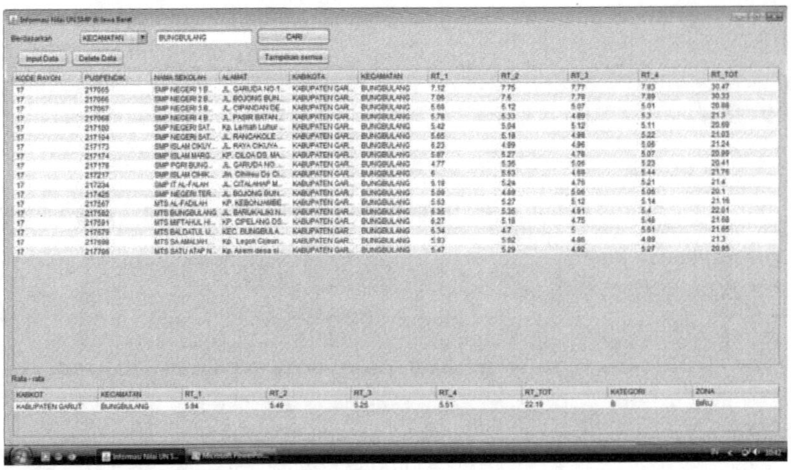

Figure 3. Example Sample Search of District

the district Pameungpeuk Garut, complete with a school identity, the average and the percentage results of the UN school subjects for Indonesian, English, Mathematics, Science, and the total number of UN.

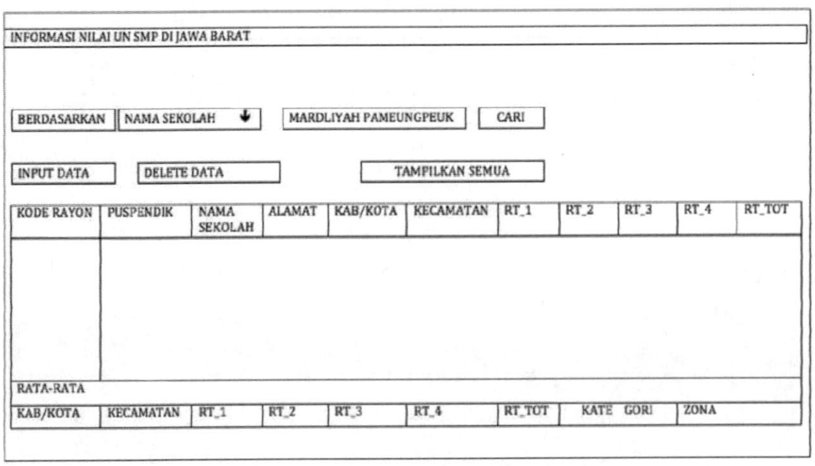

Figure 4. Example Sample Search of Regency/City

Figure 4 illustrates, sample search application names of schools in the province of West Java, to find a school SMP/MTs that exist, for example in applications we have the school name , the following statement names the district and sub-district, complete with a school identity, average and the percentage of school examination results for subjects Indonesian (RT1), English (RT2), Mathematics (RT3), Science (RT4), and the total number of UN(RTTOT). Furthermore, in the end of table we have the average for Regency/City, category and Zona.

3. Main Result

We built a prototype of software application based on Java, and following the KDD method , and we have some important result of our research as described in Figure 5.

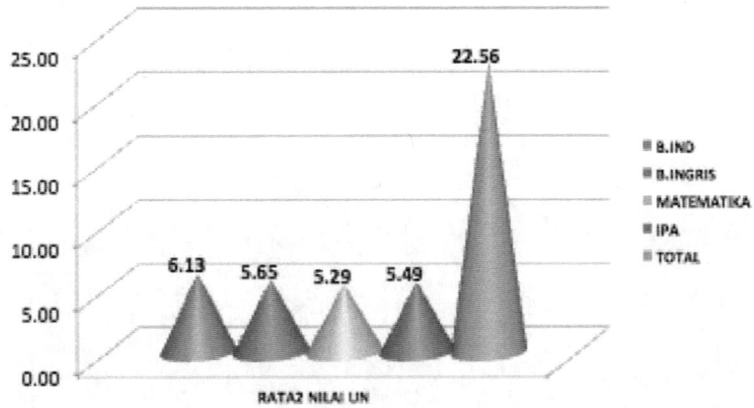

Figure 5. Example of Pie Graphics of UN SMP

Figure 5 illustrates the average results of the UN SMP/MTs in Regency/City at the province of West Java. The average is still relatively low, for the subjects English, Mathematics and Science, it is still less than value 6, while the value of Indonesia Language has average value of 6.13. Furthermore, the classification of the value of UN at West Java Province can be described at Figure 6.

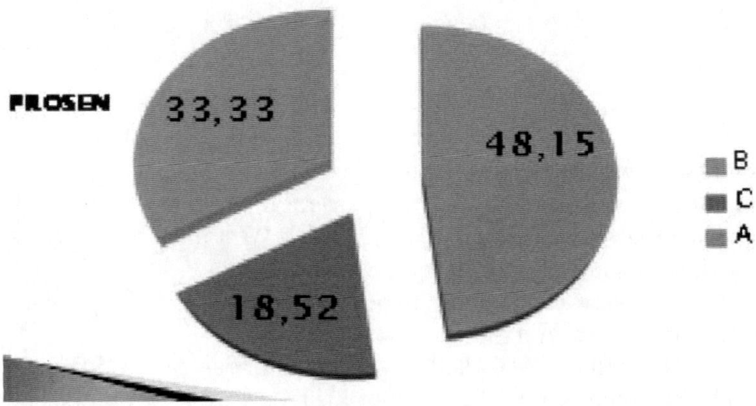

Figure 6. Classification of Averages the UN in West Java

Figure 6 describes the percentage of average achieved results of the UN SMP / MTs at Regency/City in the province of West Java, classified into three categories: orange (A), blue (B) and red (C), it can be seen that are categorized (A) is 33.33%, the medium category (B) 48.15% and red (C) is 18.52%. This shows still needs improvement, especially on the improvement of the red group, as well as an increase in the blue and orange groups. This result can be describe by the map as Figure 7.

Figure 7 describes the classification results the average value of UN Regency/City in West Java, it appears that there are five City and two Regencies in the category of green or good (A), in the Northeast and South of West Java province. In the south has the quality of education less than North area, in the category of red or bad (C), while the rest is categorized as yellow

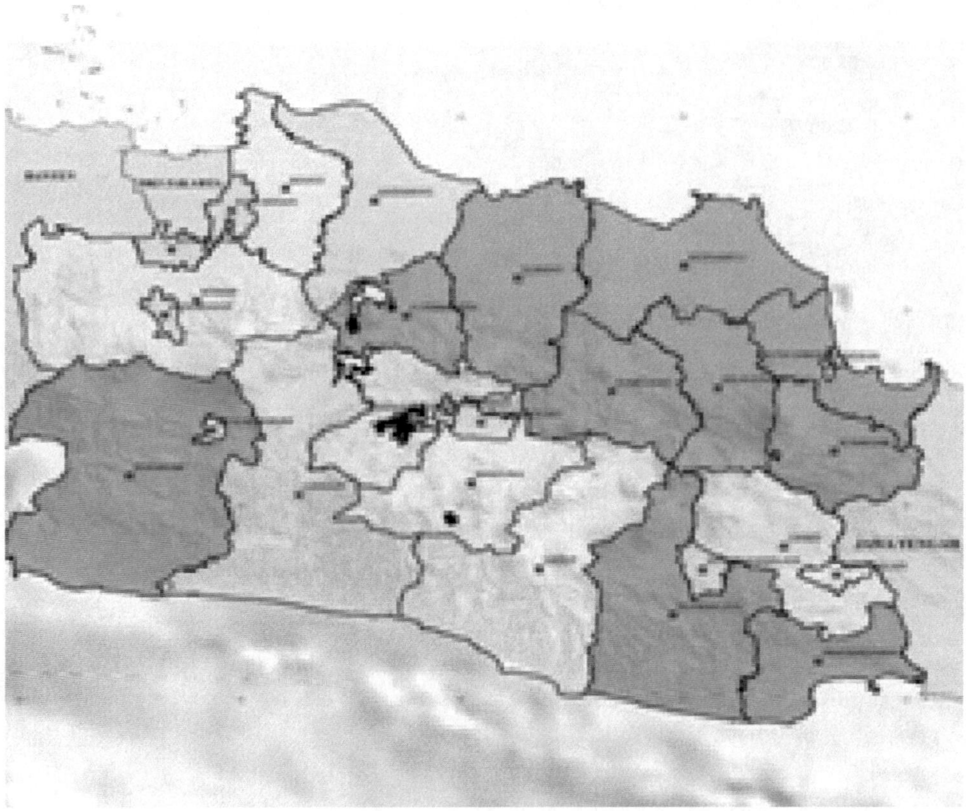

Figure 7. Classification Results of the UN Regency / City in West Java

or medium (B). The Government need to do revise in academic including add of equipment in the red group, and should be increase the group of yellow and also to maintain a green area.

Figure 8 describes the classification results of the average value of the UN sub-districts in West Java, it appears that the sub-districts in part northern at West Java generally has the category of red or bad (C), while in southern and central of West Java generally has the category green or good (A) and category yellow or medium (B). The Government need to repair a group of red and yellow and maintain a green group.

Figure 9 describes the percentage of the acquisition of the average value of the total results of the UN SMP/MTs in Tasikmalaya, classified into three categories: orange or good (A), blue or medium (B) and red or bad (C), it can be seen that are categorized as good (A) is 36.11%, the medium category (B) is 38.89% and red or bad (C) is 25.00%. This result shows that there is still need to improve the quality of UN in Tasikmalaya, especially in has a red category (C). Furthermore, we can build a map of UN classification based on district at Tasikmalaya city as Figure 10.

Figure 10 describes the classification results of the average value of the UN based on districts in the city of Tasikmalaya, the map shows that there are seven districts in the category of green or good (A), and one district in the category of yellow or medium (B), and no districts in the city of Tasikmalaya in the category of red or bad (C). This shows that the quality of education in the city of Tasikmalaya relatively good. The quality of Indonesian subjects and Mathematics entered in good category, while subjects in English and science in one district has the bad category. Finally we can make comparison about the result of average UN at West Java province as Figure 11.

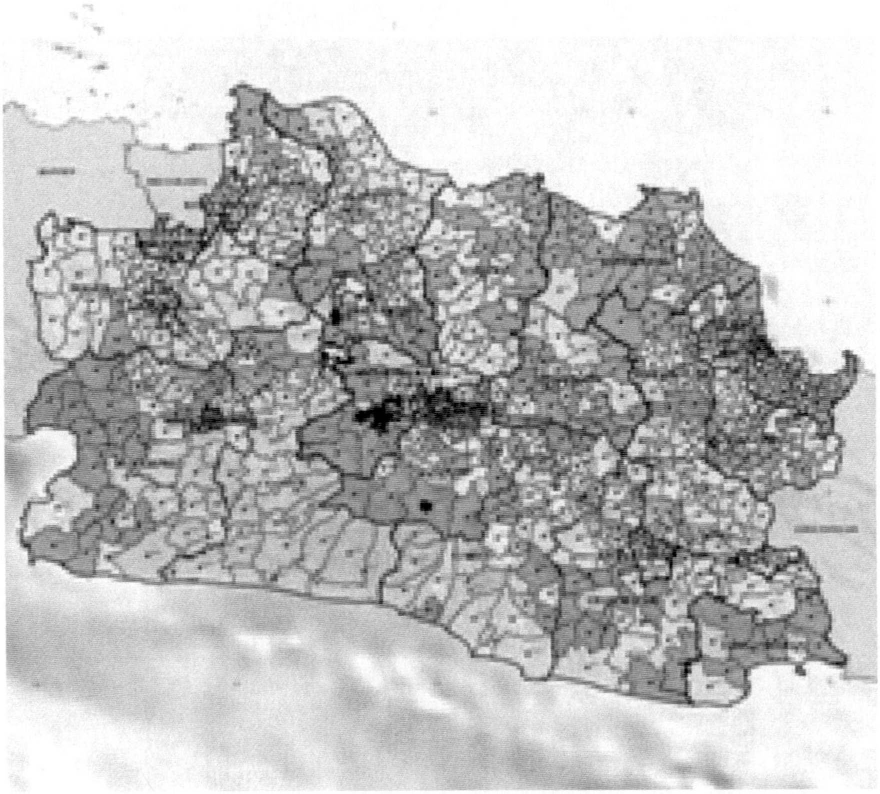

Figure 8. Classification Results of the UN at Districts in West Java

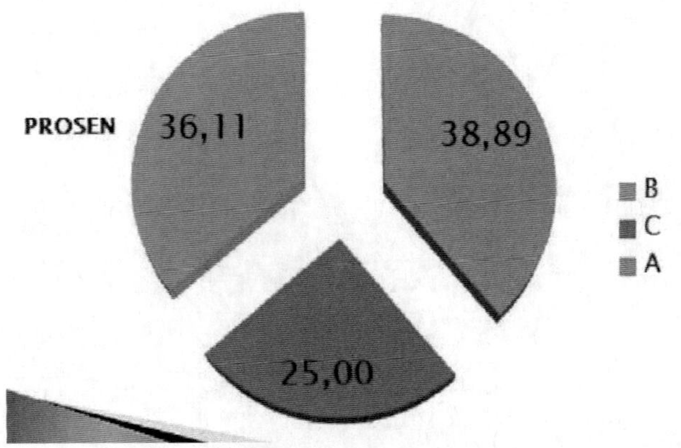

Figure 9. Pie Diagram of the UN Regency / City in West Java

Figure 11 explains that in general the results of examination results of the UN in the Regency in West Java is still relatively medium, for the sub-districts of Cianjur relatively good , sub-districts in the Regency of Garut, Cianjur, and Bekasi relatively balanced between category good, medium and bad, while Sumedang Regency has a bad category.

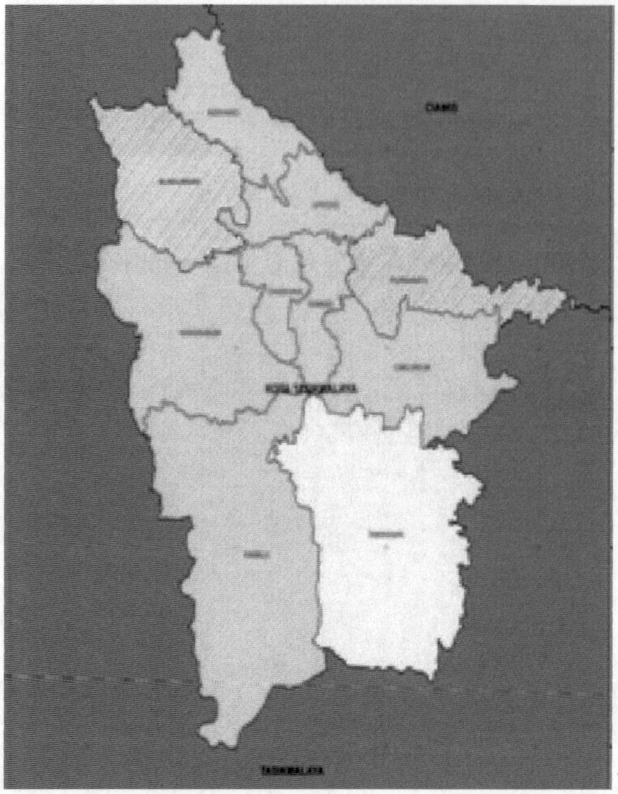

Figure 10. Classification results of the UN regency / cities in West Java

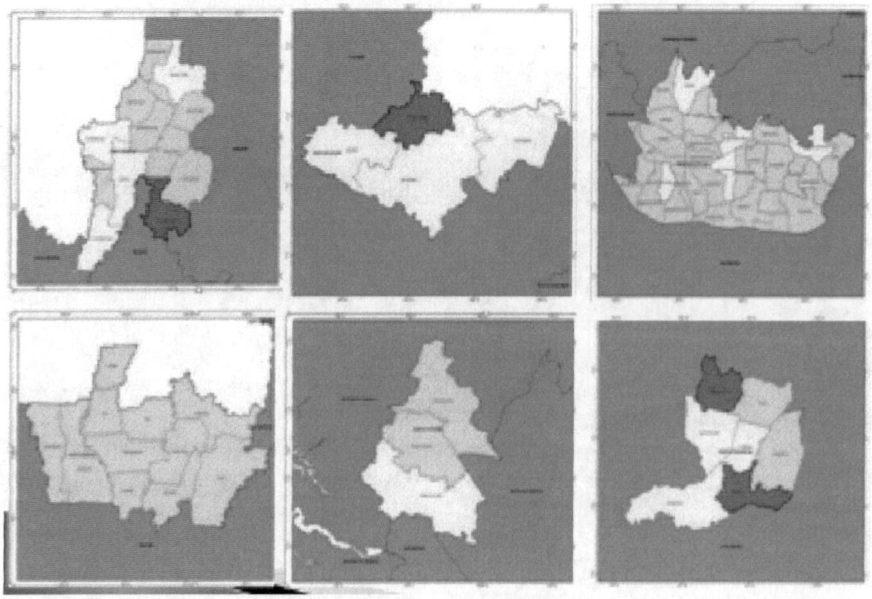

Figure 11. Classification results of the UN regency/ cities in West Java

4. Conclusion

The average results of the UN SMP/MTs Regency/City in West Java province is relatively bad, for the subjects of English, Mathematics and Science, the average is less than the value 6, while the value Bahasa Indonesia has the average value at 6.13.

Using the software we can describe the category of UN at regency/City or sub-disctrics at West Java province at level A (good or green), B (medium or yellow) and C (bad or red). From this classification, the Government can take a decision to improve a quality of education at West Java province.

Acknowledgments

The work reported in this paper is supported by the Academic Leadership Grant Unpad 2016, so the authors thank to Rector Universitas Padjadjaran and Directorate Research and Community Services Unpad for supporting funding of research and paper dissemination.

References

[1] Abdullah A S 2009 Implementation of Spatial Data Mining using SAR-Kriging Model to Predict a Quality of Education in Indonesia. Unpublished. Dissertation. Yogyakarta: Universitas Gadjah Mada
[2] Balitbang Diknas 2015 Rekapitulasi Data Hasil UN SMP/MTs
[3] http://learn.arcgis.com
[4] Jiawei H, Kamber M, and Jian P 2012 Data Mining Concepts and Technique, Morgan Kaufmann Publishers USA
[5] Patrick N 1997 Java Handbook, Konsep Dasar Pemrograman Java McGraw_Hill Inc.
[6] Silberschatz, Korth, and Sudarshan 2002 Database System Concepts, Mc Graw-Hill, North America
[7] UU 20/2003 tentang Sisdiknas

Finite element model of size, shape and blood pressure on rupture of intracranial saccular aneurysms

Jennica Rica Nabong and Guido David

Institute of Mathematics, University of the Philippines, Diliman, Quezon City

E-mail: jennyrica@gmail.com, gdavid@science.upd.edu.ph

Abstract. Rupture of intracranial saccular aneurysms is a primary concern for neurologists and patients because it leads to stroke and permanent disability. This paper examines the role of blood pressure, in connection with size of and wall thickness, in the rupture of saccular aneurysms. A bulb-shaped geometry of a saccular aneurysm is obtained from angiographic images of a patient and modeled using Finite Elements based on the principle of virtual work under the Fung stress-strain relationship. The numerical model is subjected to varying levels of systolic blood pressure. Rupture is assumed to occur when the wall stress exceeded its mechanical strength. The results show which sizes of this class of aneurysms are at high risk of rupture for varying levels of blood pressure.

1. Introduction

Intracranial or cerebral aneurysms are thin-walled, balloon-like dilatations of the arterial wall that occur in an artery located in the front part of the brain that supplies oxygen-rich blood to the brain tissue. Aneurysms occur in one out of twenty people. Cerebral aneurysms, at some stage, may suddenly rupture, causing bleeding within the brain. The incidence of reported aneurysm rupture is about 10 in every 100,000 persons per year in the USA. Aneurysm rupture is one of the primary causes of stroke, and in most cases leads to disability or death. Currently, the two endovascular methods of choice are the deployment of small coils within aneurysms having well-defined necks or the placement of porous intravascular stent for sidewall lesions. The main goal of these treatments is to decrease the risk of subarachnoid hemorrhage, either initially or from a repeated episode of bleeding. However, these treatments are highly dangerous and costly [2],[12],[19]. Depending on the patient's situation, the doctor will make recommendations for intervention that is appropriate. To help the doctors determine if an aneurysm must undergo treatment, mathematical models of aneurysms are used to predict aneurysm rupture.

In a previous study, an ideal class of spherical aneurysms were modeled using principles of biomechanics applied to biomembranes [3],[15]. The assumption was that rupture occurred when the stresses in the aneurysm wall exceeded the limits of its mechanical strength [1],[2]. The results of the study showed that spherical aneurysms with a diameter of 7 to 8 mm were at risk of rupture when the patient had high blood pressure in the range of 140 to 180 mmHg. Spherical aneurysms with diameter of 9 mm were at risk of rupture for patients with pre-hypertension or blood pressure between 120 to 140 mmHg. Spherical aneurysms with diameter of 10 mm and above were at risk of rupture for patients with normal blood pressure (120 mmHg and below) [5]. The findings provided general recommendations on which size of aneurysms must be treated under any circumstance, and which aneurysms may be controlled by monitoring the patient's blood pressure level. The results were in accordance with general recommendations from neurologists.

The present study aims to extend the analytical results of the previous results by considering non-spherical geometry of a saccular aneurysm. The study uses Finite Element methods [6],[4],[17] applied to a model aneurysm obtained from angiographic images, in order to determine the size of aneurysms of this particular shape that are at risk of rupture for various levels of blood pressure.

2. Methods

Let $\mathbf{X}$ denote the initial position of a material point in the initial domain Ω_0 and let $\mathbf{x}$ denote its position in the deformed configuration Ω. We partition each domain into quadrilateral elements. Let $\mathbf{X}^e$ be the vector of the four elemental nodes in the initial domain, and let $\mathbf{x}^e$ be the deformed elemental nodes. Then we can write each material point within each element in the initial and deformed configurations as

$$\mathbf{X} = \mathbf{N}(\eta, \zeta)\mathbf{X}^e \tag{1}$$

and

$$\mathbf{x} = \mathbf{N}(\eta, \zeta)\mathbf{x}^e \tag{2}$$

where $\mathbf{N}$ is the quadrilateral shape function in terms of local coordinates η, ζ [11],[8]. The local metric tensors in each element are thus given by:

$$\mathbf{G_1} = \frac{\partial \mathbf{N}}{\partial \eta}\mathbf{X}^e \tag{3}$$

$$\mathbf{G_2} = \frac{\partial \mathbf{N}}{\partial \zeta}\mathbf{X}^e. \tag{4}$$

The vectors $\mathbf{G_1}$ and $\mathbf{G_2}$ are not orthogonal in general. A local orthonormal basis is computed by:

$$\mathbf{t_1} = \frac{\mathbf{G_1}}{|\mathbf{G_1}|} \tag{5}$$

and

$$\mathbf{t_2} = \frac{\mathbf{G_1} \times \mathbf{G_2}}{\|\mathbf{G_1} \times \mathbf{G_2}\|} \times \mathbf{t_1}. \tag{6}$$

The local orthonormal basis $\{\mathbf{t_1}, \mathbf{t_2}\}$ implicitly define a coordinate system s_1, s_2. The Cauchy-Green strain tensor can be then calculated by:

$$\mathbf{C} = \begin{pmatrix} \frac{\partial \mathbf{x}}{\partial s_1} \cdot \frac{\partial \mathbf{x}}{\partial s_1} & \frac{\partial \mathbf{x}}{\partial s_1} \cdot \frac{\partial \mathbf{x}}{\partial s_2} \\ \frac{\partial \mathbf{x}}{\partial s_2} \cdot \frac{\partial \mathbf{x}}{\partial s_1} & \frac{\partial \mathbf{x}}{\partial s_2} \cdot \frac{\partial \mathbf{x}}{\partial s_2} \end{pmatrix} \tag{7}$$

The study used the finite element formulation based on [8]. Consider a constant pressure $p\mathbf{n}$ per unit deformed area acting in the direction of the normal vector $\mathbf{n}$ acting on the deformed membrane Ω. The principle of virtual work for a finite element analysis of an inflated membrane is given by

$$\int_{\Omega_0} \delta w dA = \int_{\Omega} p\mathbf{n} \cdot \delta \mathbf{x} da \tag{8}$$

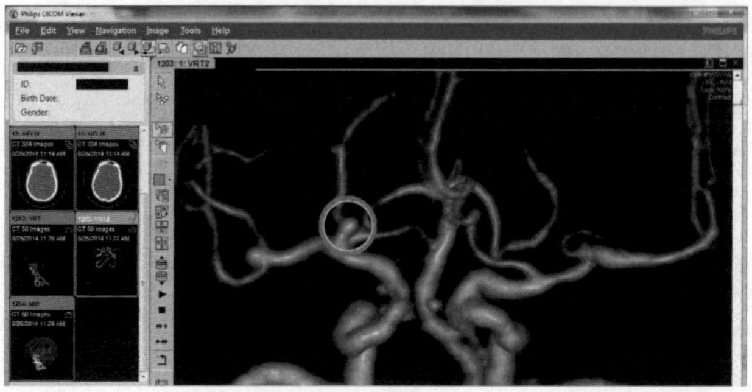

Figure 1. Angiographic image of a bulb-shaped aneurysm.

where w is the strain energy function per deformed area, and $\delta\mathbf{x}$ is the virtual displacement. The strain energy function w is based on the Fung model, which has been shown to be a reasonable model of the mechanical behaviour of soft tissue [7],[9], is given by

$$w = c(e^Q - 1) \tag{9}$$

where

$$Q = c_1 E_{11}^2 + c_2 E_{22}^2 + 2c_3 E_{11} E_{22}. \tag{10}$$

Here, c is a material parameter having units of force per length, c_i's are non-dimensional material parameters and E_{ij} are the components of the Green Strain Tensor $\mathbf{E} = 0.5(\mathbf{C} - \mathbf{I})$. After a change of variables and substituting the values of $\mathbf{n}$, (8) can be written as

$$g(\mathbf{x}^e) = \iint ||\mathbf{G_1} \times \mathbf{G_2}||\delta w - p\mathbf{N}\left(\frac{\partial \mathbf{N}}{\partial \eta}\mathbf{x}^e \times \frac{\partial \mathbf{N}}{\partial \zeta}\mathbf{x}^e\right) d\eta d\zeta. \tag{11}$$

The function $g(\mathbf{x}^e) = 0$ was then solved numerically by assembling the elemental nodes as a global vector and iterating the Newton-Raphson method using the stiffness matrix $K(\mathbf{x}^e) = \Delta g(\mathbf{x}^e)$ from an initial value of $\mathbf{x}^e$. The integrals in $g(\mathbf{x}^e)$ and $K(\mathbf{x}^e)$ were evaluated at the Gaussian quadrature points $\eta, \zeta = \pm 1/\sqrt{3}$.

3. Results and Discussion

Images of intracranial saccular aneurysms of patients from the Philippine General Hospital (PGH) were gathered and was viewed using Philips DICOM Viewer® c/o Dr. Carmela Sales. Figure 1 presents the angiographic image showing the aneurysm. A numerical model was constructed based on the geometry of the aneurysm under the assumption that the aneurysm was axially symmetric with respect to the $z-$ axis. For the stability of the numerical model, we considered the quarter surface in the first octant, subject to rolling boundary conditions along the planes that bound the first octant. The numerical model consisted of 433 nodes and 380 quadrilateral elements.

The maximum wall strength, or the stresses on the aneurysm wall when rupture occurs, was obtained from studies that tested the mechanical limits of the arterial or aneurysm wall. This value was reported to be in the range of 0.5 to 2.0 MPa [10],[16]. In this paper, the assumed critical wall stress was 1.5 MPa.

The Fung parameter c as measured in force per length was obtained using $c = c_0 H$ where H is the reference thickness of the aneurysm and c_0 is the material parameter in force per length squared. Based on previous results, $c_0 = 0.28$ MPa, thus $c = 0.0092$ N/mm [13]. For the other

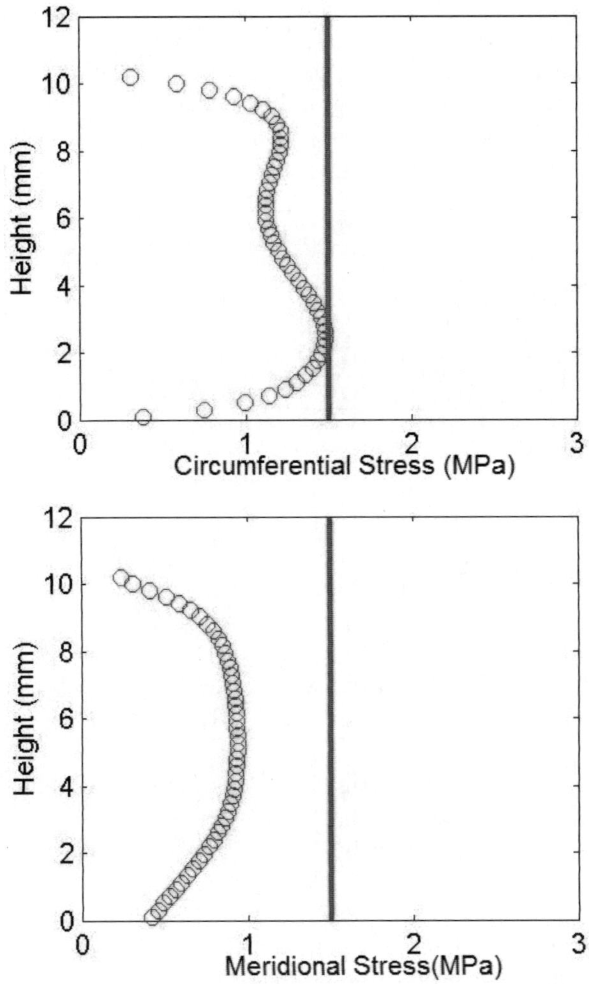

Figure 2. Height vs. Stress Plot of Aneurysm with Uniform Thickness of Height 10 mm with Blood Pressure from 120 mmHg to 180 mmHg.

material parameters, best fit values were based on [13],[14],[18], giving us $c_1 = 17.58$, $c_2 = 12.19$, $c_3 = 7.57$.

Intracranial aneurysms are assumed to have a reference thickness $H = 0.033$ mm [1]. Three thickness profiles were assumed: constant thickness (H_0), linearly varying with thickest at neck (H_1) and linearly varying with thickest at the fundus (H_2). The minimum and maximum thickness were assumed to be 0.0165 mm and 0.0495 mm, respectively.

To determine the effects of blood pressure on the size and geometry of the aneurysm, different values of height from 6 to 12 mm were used. Stresses from the circumferential and meridional directions were determined. Plots of wall stresses vs. height were then obtained using MATLAB®. These plots were used to determine the point at which the aneurysms were at risk of rupture. In the graphs, the maximum wall strength of 1.5 MPa was marked with a solid red line to indicate which aneurysms were at risk of rupture. Figure 2 shows the circumferential and meridional stresses for an aneurysm with height 10 mm at blood pressure of 140 mmHg (prehypertension).

IOP Conf. Series: Journal of Physics: Conf. Series **893** (2017) 012054 doi:10.1088/1742-6596/893/1/012054

Table 1. Summary of numerical pressure test on model aneurysm with height of 6, 8, 10 and 12 mm. An 'x' indicates that the aneurysm was at risk of rupture due to stresses exceeding the maximum wall stress of 1.5 MPa in some region in the aneurysm. H_0 assumed a constant wall thickness of 0.033 mm, H_1 assumed a linearly decreasing thickness with maximum thickness at the neck, and H_2 assumed a linearly decreasing thickness with maximum thickness at the fundus.

Height 6 mm	H_0	H_1	H_2
120 mmHg	-	-	-
140 mmHg	-	-	-
160 mmHg	-	-	x
180 mmHg	-	-	x
Height 8 mm	H_0	H_1	H_2
120 mmHg	-	-	x
140 mmHg	-	x	x
160 mmHg	-	x	x
180 mmHg	x	x	x
Height 10 mm	H_0	H_1	H_2
120 mmHg	-	x	x
140 mmHg	x	x	x
160 mmHg	x	x	x
180 mmHg	x	x	x
Height 12 mm	H_0	H_1	H_2
120 mmHg	x	x	x
140 mmHg	x	x	x
160 mmHg	x	x	x
180 mmHg	x	x	x

Internal blood pressure of 120 mmHg, 140 mmHg, 160 mmHg and 180 mmHg, with a cerebrospinal fluid pressure of 3 mmHg on the external surface. Table 1 shows the effects of blood pressure on the model aneurysm, for heights of 6, 8, 10 and 12 mm and the three thickness assumptions. The x's denote risk of rupture, either in the circumferential or meridional direction.

Table 1 shows that a small aneurysm with heigh 6 mm was not at risk of rupture under the constant thickness assumption H_0 and assumption of linearly decreasing thickness with maximum thickness at the neck H_1. However the aneurysm was at risk of rupture at hypertensive levels of greater than 160 mmHg blood pressure under the assumption of linearly decreasing thickness with maximum thickness at the fundus H_2. For an aneurysm with height 8 mm, the aneurysm was at risk of rupture at greater than 180 mmHg under the constant thickness assumption H_0, at prehypertensive levels greater than 140 mmHg under the H_1 assumption, and at blood pressure levels greater than 120 mmHg under the H_2 assumption. An aneurysm with height 10 mm was not at risk of rupture only at normal blood pressure (less than 120 mmHg) only for the constant thickness assumption H_0. Finally, an aneurysm with height 12 mm or more was at risk of rupture at any blood pressure level above 120 mmHg.

4. Conclusions

An Finite Element model of a bulb-shaped aneurysm obtained from angiographic images was constructed in order to determine what height and blood pressure level increased its risk of rupture. For this study, three types of thickness profiles were assumed: constant thickness, thickest at the neck and decreasing linearly, or thickest at the fundus and decreasing linearly. The

Finite Element membrane model was inflated with pressure using the virtual work principle and assumed the Fung pseudo-strain energy function for biomaterials. The aneurysm was assumed to be at high risk of rupture when its wall stress exceeded the mechanical strength of the aneurysm wall, assumed to be 1.5 MPa. Internal blood pressure from 120 mmHg to 180 mmHg, less the cerebro spinal fluid pressure of 3 mmHg, was numerically applied and corresponding wall stresses of the aneurysm were computed. The results showed that an aneurysm of this geometry was not at risk of rupture when the height was 6 mm, was at risk of rupture at high blood pressure level of greater than 140 mmHg when the height was 10 mm, and was at risk of rupture even at normal blood pressure when the height exceeded 12 mm. This suggested that for this type of geometry and constant thickness assumption, aneurysms below 10 mm in height could safely be controlled by monitoring blood pressure, but an aneurysm with height at least 12 mm should be treated immediately. The study also showed that changing the thickness assumption increased the risk of rupture for the same aneurysm geometry: an aneurysm thickest at the neck with height 10 mm or higher should be treated immediately, while an aneurysm thickest at the fundus with height 8 mm or more should be treated immediately.

The results of the study may provide a guide for neurologists in order to make recommendations seeking treatment of aneurysms based on the height of the aneurysm for this particular geometry type. Further data would be needed in order to validate the findings of the study. Moreover, the study can be extended further by considering other geometries of aneurysms based on additional angiographic images.

Acknowledgments

The authors would like to thank Dr. Carmela Sales of the Philippine General Hospital for providing images for the aneurysm. This project was supported by Department of Science and Technology.

References

[1] Chaudry H, Lott D, Prestigiacomo C and Findley T 2006 Mathematical model for the rupture of cerebral saccular aneurysms through three dimensional stress distribution in the aneurysm wall *Center for Applied Mathematics and Statistics Report*
[2] Costalat C, Sanchez M, Ambard D, Thines L and Lonjon N 2011 Biomechanical wall properties of human intracranial aneurysms resected following surgical clipping *Journal of Biomechanics*
[3] David G and Humphrey J 2003 Further evidence for the dynamic stability of intracranial saccular aneurysms *Journal of Biomechanics* **36**
[4] David G and Humphrey J 2007 Finite element model based of stresses in the anterior lens capsule of the eye *Computer Methods in Biomechanics and Biomedical Engineering* **10**
[5] David G and Nabong J 2014 Rupture model of intracranial saccular aneurysms due to hypertension*Journal of Mechanics in Medicine and Biology* **15**
[6] David G, Pedrigi RM, R.Heistand M and Humphrey J 2007 Regional multiaxial mechanical properties of the porcine anterior lens capsule *Journal of Biomedical Engineering* **129**
[7] Fung Y 1993 Biomechanics: Mechanical Properties of Living Tissues *Springer-Verlag, New York*
[8] Gruttman F and Taylor RL 1992 Theory and Finite element formulation of rubberlike membrane shells using principal stretches *International Journal for Numerical Methods in Engineering* **35**
[9] Humphrey J 2002 Cardiovascular Solid Mechanics *Springer, New York*
[10] Kroon M, and Holzapfel G 2008 Modeling of saccular aneurysm growth in a human middle cerebral artery *Journal of Biomedical Engineering* **130**
[11] Kyriacou S and Humphrey J 1996 Influence of size, shape and properties on the mechanics of axisymmetric saccular aneurysms *Journal of Biomechancs* **29**
[12] Parlea L, Fahrig R, Holdsworth D, and Lownie S 1999 An analysis of the geometry of saccular intracranial aneurysms *American Journal of Neuroradiology* **20**
[13] Seshaiyer P, Hsu F, Shah A, Kyriacou S and Humphrey J 2001 Multi-meridional mechanical behavior of human saccular aneurysms *Computer Methods in Biomechanics and Biomedical Engineering* **4**
[14] Seshaiyer P and Humphrey J 2001 On the potentially protective role of contact constraints on saccular aneurysms *Journal of Biomechanics* **34**

IOP Conf. Series: Journal of Physics: Conf. Series **893** (2017) 012054 doi:10.1088/1742-6596/893/1/012054

[15] Shah A and Humphrey J, 1999 Finite strain elastodynamics of intracranial saccular aneurysms *Journal of Biomechanics* **32**
[16] Sommer G, Mechanical properties of healthy and diseased human arteries *Monographic Series TU Graz*
[17] Toth B, Raffai G, and Bojtar I 2005 Analysis of the mechanical parameters of human brain aneurysm *Acta of Bioengineering and Biomechanics* **5**
[18] Valencia A, Amd PT, Rivera R, Galvez M and Bravo E 2009 A mechanical study of patient-specific cerebral aneurysm models: The correlations between stress and displacement with geometrical indices *Mechanics Research Communications* **36**
[19] Wolfe S, Baskaya M, Heros R, and Tummala R 2006 Cerebral aneurysms: learning from the past and looking toward the future *Clinical Neurosurgery* **53**

Image Indexing using Color Histogram and k-means Clustering for Optimization CBIR in Image Database

Juli Rejito[1], Atje Setiawan Abdullah[1], Akmal[1], Deni Setiana[1], and Budi Nurani Ruchjana[2]

[1] Department of Computer Science, Faculty of Mathematics and Natural Sciences Universitas Padjadjaran, Indonesia
[2] Department of Mathematics, Faculty of Mathematics and Natural Sciences Universitas Padjadjaran, Indonesia

E-mail: jrejito@gmail.com

Abstract. Retrieving visually similar images from image database needs high speed and accuracy. Various text and content based image retrieval techniques are being investigated by the researchers in order to exactly match the image features. In this paper, a content-based image retrieval system (CBIR), which computes color similarity among images, is presented. CBIR is a set of techniques for retrieving semantically relevant images from an image database based on automatically derived image features. Color is one important visual features of an image. This document gives a brief description of a system developed for retrieving images similar to a query image from a large set of distinct images with histogram color feature based on image index. Result from the histogram color feature extraction, then using k-means clustering to produce the image index. Image index used to compare to the histogram color feature of query image and thus, the image database is sorted in decreasing order of similarity. The results obtained by the proposed system obviously confirm that partitioning of image objects helps in optimization retrieving of similar images from the database. The proposed CBIR method is compared with our previously existed methodologies and found better in the retrieval accuracy. The retrieval accuracy are comparatively good than previous works proposed in CBIR system.

1. Introduction

In CBIR, feature vector are extracted from images. Query feature vector are matches with the stored feature vector on one to one basis. This results in slow down the processing time. To improve the speed of executing and better result many researchers are paying attention to uses the clustering algorithm for CBIR. Clustering is a process of separating a dataset into groups in such a way that the object in one group is more similar to those objects in the other group [1].

Chin-Chin Lai et.al. [2] have proposed an interactive genetic algorithm (IGA) to reduce the gap between the retrieval results and the users expectation called semantic gap. They have used HSV color space that corresponds to human way of perceiving the colors and separate the luminance component from chrominance ones. They have also used texture features like the entropy based on the grey level co-occurrence matrix and the edge histogram. They compared this method with others approaches and achieved better results.

Kannan A, et al [3] have proposed Clustering and Image Mining Technique for fast retrieval of Images. The main objective of the image mining is to remove the data loss and extracting the meaningful information to the human expected needs. The images are clustered based on RGB Components, Texture values and Fuzzy C mean algorithm.

Chakravarti and Meng [4] have published a paper on Color Histogram Based Image Retrieval. They have used color histogram technique to retrieve the images. This method allows retrieval of images that have been transformed in terms of their size as well as translated through rotations and flips.

Kumar, et.al [5] have published on Content Based Image Retrieval using Color Histogram. They have used Color Histogram technique to retrieve the similar images. To speed up the retrieval, they have used the proposed grid-based indexing to obtain the nearest neigh bours of the query image and exact images are retrieved. Indexing can be performed in vector space to improve retrieval speed. Mainly, they have implemented CBIR using color histogram technique and is refined with help of grid technique to improve the image retrieval performance.

CBIR with clustering algorithm is an alternative that is expected to improve the performance and accuracy of searches in the query image. k-means is the basic clustering algorithm but in case k-means is very sensitive for initial clustering and sensitive to outliers and noise [1,6]. Also to form the clusters, k-mean depends on initial condition, which causes the algorithm to give suboptimal solution. As compare to k-means ant colony algorithm are more prominent for initializing the cluster. Due to the global optima nature of particle swarm optimization algorithm give the optimum solution for clustering. These most prominent features of ant colony algorithm and particle swarm optimization are used for clustering.

The implementation of a query optimization proposed in this paper was related to CBIR in image databases aimed at obtaining image in database with a high image content similarity level during an image searching process. In this proposal, proposed by carrying out a color extraction process of each image database, a process of clustering with k-means algorithm, and results of clustering of each image database were used to a filtering process based on query images.

2. Literature Review

2.1. CBIR

CBIR is a method that is used to look at image features like (color, shape, texture) to find a query image from database. The difficulties of CBIR lie in reducing the differences of contents based feature and the semantic based features. This problem in giving effective retrieval images and channelize the researchers to use (CBIR) system, to take global color and texture features to achieve, the good retrieval, where others used local color and texture features [6]. The method in [7] presented the holistic representation of spatial envelop with a very low dimensionality for making the incident image.

The method in [8] proposed a modern approach for image classification with the open field design and the concept of over-completeness methodology to achieve a preferable result. As reported in [8], this method achieved the best classification performance with much lower feature spatiality compared to that of the former schemes in image classification task. For similarity search [9] the user need to enter keywords along with the query image that might appear in the text of patents. Higher average precision and recall rates compared to the traditional Dominant Color method were obtained successfully [10]. In another system the image is represented by a Fuzzy Attributed Relational Graph (FARG) that describes each object in the image, its attributes and spatial relation. The texture and color attributes are computed in a way that model the Human Vision System (HSV) [11].

2.2. Color Histogram Feature Extraction

Obtained by extraction of the color histogram feature extraction image pixels for each color component R, G, and B then calculated the frequency of each color index from 0 to 255 and raised in the form of histogram value for each color component, and is written as a vector shown in the equation as follows:

$$H = H[0], H[1], H[2], H[3], , H[i], , H[n] \tag{1}$$

Where i is the color in the color histogram storage and $H\ [i]$ indicates the number of pixels of color i in the image, and n is the number of colors used in the storage of the color histogram.

Results histogram value for each color component (x, y, z) of the image is sought (*Hq*) and record images (*Hi*) then calculate resemblance to calculate the distance to the known color histogram. Histogram Intersection Technique (HIT) [12], using the formula in equation as follows:

$$S(H_q, H_i) = \frac{\sum x \in X, y \in Y, z \in Y min(H_q(x,y,z), H_i(x,y,z))}{\sum x \in X, y \in Y, z \in Y H_q(x,y,z)} \tag{2}$$

Possess the formula distance values tend to have small differences so that the formula developed into the equation as follows:

$$S(H_q, H_i) = \frac{\sum x \in X, y \in Y, z \in Y min(H_q(x,y,z), H_i(x,y,z))}{min[\sum x \in X, y \in Y, z \in Y H_q(x,y,z), \sum x \in X, y \in Y, z \in Y H_i(x,y,z)]} \tag{3}$$

2.3. k-means Algorithm

The *k*-means algorithm is effective in producing good clustering results for many applications [13]. The reasons for the popularity of *k*-means are easy and simplicity of implementation, scalability, speed of convergence and adaptability to sparse data [14]. *k*-means clustering is a partitioning clustering technique in which clusters are formed with the help of centroids. On the basis of these centroids, clusters can vary from one another in different iterations. Also, data elements can vary from one cluster to another, as clusters are based on the random numbers known as centroids [15]. The *k*-means algorithm is the most extensively studied clustering algorithm and is generally effective in producing good results. The *k*-means algorithm is computationally expensive and requires time proportional to the product of the number of data items, number of clusters and the number of iterations. *k*-means is formally described by Algorithm 1.

Algorithm 1 : Basic *k*-means algorithm.

1: Select k points as initial centroids.

2: repeat

3: Form k clusters by assigning each point to its closest centroid.

4: Recompute the centroid of each cluster.

5: until centroids do not change.

2.4. Measurement Image Quality

The measurement of image quality could use a conventional measurement system that applies a technique of computing SNR (Signal to Noise Ration), MSE (Mean Square Error) and PSNR (Peak Signal to Noise Ration) values. Meanwhile, PSNR was the value of a comparison between the maximum values of images as a reconstructed result with Mean Square Error (MSE) values. For a 8-bit image, the maximum pixel value was 255. The higher the result of PSNR, the better the quality of picture. PSNR was expressed in a decibel 9dB unit. The minimum value of PSNR at interpolation that fell into a category of good was 30dB. Mathematically, it was formulated as follows:

$$PSNR = 10 log_{10} \left[\frac{253^2}{MSE} \right] dB \tag{4}$$

3. Proposed Approach

To solve the above problem, the necessary stages of completion as shown in Figure 1.

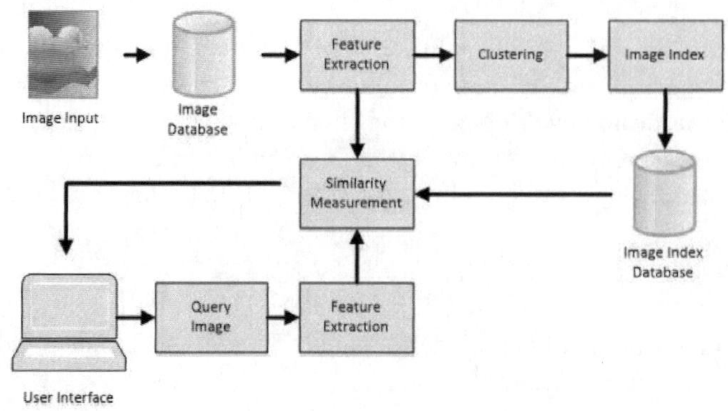

Fig. 1 Architecture Systems

From the figure 1, it can be explained the sequence of the process is following:

(i) Put the whole picture of wang database to a database of images. Using equation 1, the color feature extraction process to obtain the value vector R, G, B based on a base image and calculate the difference in distance using the formula of HIT in equation 2 and 3.

(ii) Result values $S(H_q, H_i)$ was used to classify the image by using the k-means algorithm 6. Save the cluster for each image as the image index.

(iii) To perform image searches done by inserting the image of the user interface applications that have been created in the form of a query image. By using the image index were then determined the position of the cluster image and serve as a filter record in accordance with his cluster group.

(iv) Using equation 4 calculating the similarity between the query image and the image record.

4. Results and Discussion

The database image used in proposed method is WANG database, and the delphi program has been implemented. The WANG database contains 1,000 images in JPEG format. The size of each image is 256x386 and 384x256. It consists of 10 classes such as (Africa, beach, monuments, buses, dinosaurs, elephants, flowers, horses, mountains and food) each class contains 100 images. Figure 2 shows sample of WANG database.

Fig. 2 Example Image from each of the 10 Classes of WANG Database

IOP Conf. Series: Journal of Physics: Conf. Series **893** (2017) 012055 doi:10.1088/1742-6596/893/1/012055

The retrieval effectiveness of the proposed method is measured by using 30 different query images from each class. It is tested for 300 query images. In the beginning, the image size is changed to [128, 128] to get a similar size for each image. The color features are extracted according to HSV histogram values. Euclidean distance is used to measure similarity between query image and database images. Top 100 images are retrieved depending on the minimum distance. Next stage, the shape features are extracted from 100 images that resulted from the first stage and similarity measurements is performed between shape features of query image and the 100 images. Top 50 images are retrieved depending on the minimum distance. The final stage in the proposed method is extracted the first order features from 50 images and compared with query image. The nearest 10 images of the query image are retrieved from the image database.

4.1. The Result of k-means Clustering in WANG Database

The implementation of the clustering was applied in several cluster groups, namely, 2 clusters, 4 clusters, 8 clusters, and the amounts of iteration of each cluster and the values of minimum PSNR and maximum PSNR for each cluster were shown in Table 1 and Figure 3.

Table 1. Clustering of 1,000 Wang Database Records in 2, 4 and 8 clusters

Cluster		Centroid PSNR (db)		Record
Cluster	N	Minimum	Maximum	Count
2	1	7.238144	12.298757	838
2	2	3.217335	7.589639	162
4	1	6.318496	11.190640	529
4	2	6.894847	14.625369	155
4	3	2.638533	6.818268	119
4	4	9.449756	12.881852	197
8	1	10.119414	13.778683	110
8	2	8.000599	11.858669	209
8	3	5.688476	11.963560	188
8	4	6.923792	10.427970	147
8	5	7.305264	17.609415	24
8	6	4.749914	9.994102	95
8	7	6.782889	13.985441	121
8	8	2.518710	6.560355	106

Table 1 shows that the record is processed in WANG database of 1,000 records with the results of the process for the 2 clusters show that the first cluster has a value of 7.238144 db PSNRMin centroid and centroid PSNRMax of 12.298757 db with a record number of 838, while the second cluster has a value of 3.217335 db PSNRMin centroid and centroid PSNRMax amounted to 7.589639 db with a record number of 162. Likewise for the formation of 4 and 8 clusters, as shown in Table 1.

Figure 3 shows the graph plots each record based on the minimum and maximum PSNR value on the formation of 2, 4, and 8 clusters using the k-means algorithm database. Each cluster is plotted in a different color, for example in the formation of 2 clusters is shown in black for the first cluster, while the second cluster is ploted in red color. It is also shown for 4 and 8 clusters with variety of color.

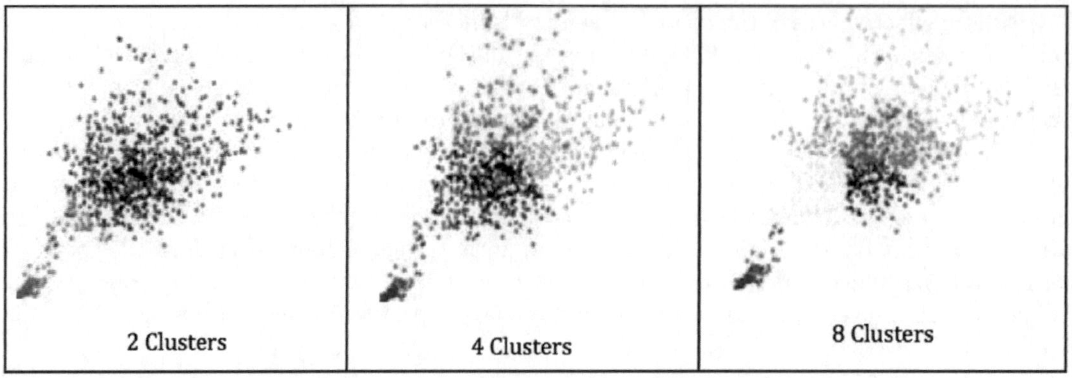

Fig. 3 Plot for k-means Clustering in 2,4, and 8 clusters for WANG Database.

4.2. The Results of Precision

The precision and recall are used to measure performance of retrieval. Recall is used to measure the system's ability to retrieve all the images that are relevant, while precision is used to measure the system's ability to retrieve only the images that are relevant. The equation of the recall and precision are illustrated in the following :

$$precision = \frac{Number\ of\ relevant\ images\ retrivied}{Total\ number\ of\ images\ retrieved} = \frac{N}{N+B} \tag{5}$$

$$recall = \frac{Number\ of\ relevant\ images\ retrivied}{Total\ number\ of\ relevant\ retrieved} = \frac{N}{N+C} \tag{6}$$

Where the number of retrieved images is represented with N and the irrelevant images are represented with B, while the C represents the number of relevant images not retrieved. The similarity measurement used in this work is Euclidean distance. Table 2 shows the results of precision when using color, shape and texture features. The best results are obtained in using cascade color, shape and texture features.

Table 2 shows the results of precision value group by categories obtained by our proposed method (histogram color feature with 5 clusters) with other retrieval systems : Histogram color, GLCM (Gray-level co-occurrence matrix) texture, histogram color + GLCM texture, and histogram color + GLCM texture with sub-block. The best results are obtained in using histogram color feature with 5 clusters.

In Table 2 shows that the average value of precision in a row are 0.33, 0.33, 0.35, 0.45 and 0.68. Value 0.68 is the greatest value of the proposed proposal.

Figure 4 shows the precision obtained by the different methods. It shows that the histogram color feature with 5 clusters gave the best result in image retrieval.

Table 2. The Result of Precision Value Group By Categories Obtained By Proposed Method with Other Retrieval Systems

No	Categories	Histogram Color	GLCM Texture	Hist.Clr+ GLCM Texture	Hist.Clr+ GLCM Texture +sub-block	Histogram Color with 5 Clusters
1	Africa	0.36	0.21	0.34	0.41	0.92
2	Beaches	0.27	0.35	0.21	0.32	0.42
3	Building	0.38	0.50	0.24	0.37	0.38
4	Bus	0.45	0.22	0.51	0.66	0.52
5	Dinosaur	0.26	0.29	0.39	0.43	1.00
6	Elephant	0.30	0.24	0.26	0.39	0.68
7	Flower	0.65	0.73	0.81	0.87	0.92
8	Horses	0.19	0.25	0.28	0.35	0.97
9	Mountain	0.15	0.18	0.20	0.34	0.25
10	Food	0.24	0.29	0.25	0.45	0.68
	Average	**0.33**	**0.33**	**0.35**	**0.45**	**0.68**

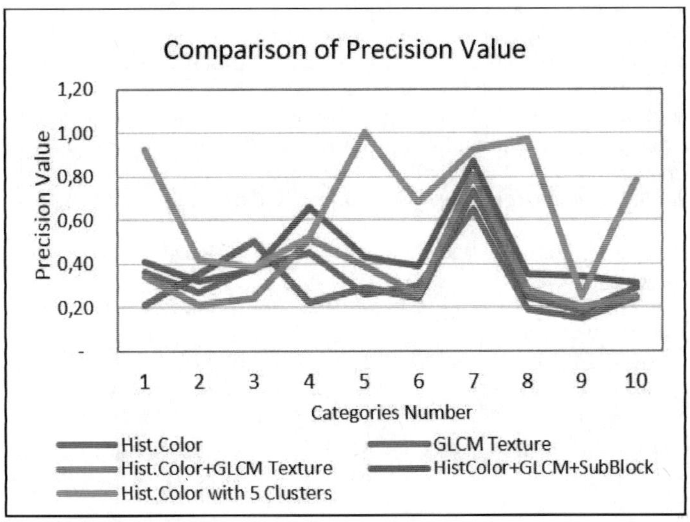

Fig. 4 The Result of Precision Value Group by Categories Obtained by Proposed Method with Other Retrieval Systems

5. Conclusion

Clustering that began with a cluster initialization process would be a main key in an information retrieval process of a query using PSNR min and PSNR Maximum. This research found a new algorithm for cluster initialization by computing in advance the square root of the sum of minimum the square of PSNR and that of maximum PSNR for each record in WANG database used as a base of record order, and then the distance between clusters was determined by computing total record divided by the number of clusters and ended by determining each cluster that was taken from ordered records based on changes in their distances. The results of this CBIR query in WANG database by using color histogram that was taken randomly by an amount of 1,000 showed that highest level of accuracy achieved in 5 clusters shown on the precision value of 0.68.

Acknowledgment
The authors thank to Rector Universitas Padjadjaran who gave funding for this research through Academic Leadership Grant (ALG) Universitas Padjadjaran Indonesia 2016, under the research project entitled on the Development of Space-time Model and Its Application to Industry, Organization and Environment to Support the West Java Common Goals for the Year of 2015-2019.

References
[1] Rejito J, Wardoyo R, Hartati S and Harjoko A 2012 Optimization CBIR using K-Means Clustering for image database *International Journal of Computer Science and Information Technologies* **3** (4) pp 4789–4793
[2] Chin-Chin Lai and Ying - Chuan Chen A 2011 User-oriented image retrieval system based on interactive genetic algorithm *IEEE transactions on instrumentation and measurement* **60** (10)
[3] Kannan A, Mohan V and Anbazhagan N 2010 Image clustering and retrieval using image mining techniques *IEEE Conference*
[4] Chakravarti R and Meng X 2009 A study Of color histogram based image retrieval *Sixth International Conference on Informational Technology*
[5] Kumar A R and Saravanan D 2013 Content based image retrieval using color histogram *International Journal of Computer Science and Information Technologies* **4** (2) pp 242–245.
[6] Datta R, Joshi D, Li J and Wang J Z 2008 Image retrieval *ACM Comput. Surv.* **40** (2) pp 1-60
[7] Oliva A and Torralba A 2001 Modeling the shape of the scene: A holistic representation of the spatial envelope *Int. J. Comput. Vis.* **42** (3) pp 145 –175
[8] Jia Y, Huang C and Darrell T 2012 Beyond spatial pyramids: Receptive field learning for pooled image features *Proc. IEEE Conf. Comput. Vis.Pattern Recognit. (CVPR)* pp 3370–3377
[9] Tiwari A and Bansal V 2004 PATSEEK: Content based image retrieval system for patent database *Proceedings of international conference on electronic business* pp 1167–1171
[10] Krishnan N, Banu M S and Callins C 2007 Content based image retrieval using dominant color identification based on foreground objects *International Conference on Computational Intelligence and Multimedia Applications* **3** pp 190–194
[11] Ahmed H A, Gayar N E and Onsi H 2008 A new approach in content-based image retrieval using fuzzy logic *INFOS*
[12] Smith J R 1997 Integrated spatial and feature image systems: retrieval, analysis and compression *PhD thesis* (New York: Columbia University)
[13] Nikman T A and Bakbak A 2009 An efficient hybrid approach based on PSO, ACO and K-Means for Cluster Analysis *Elsevier*
[14] Ghosh A, Parikh J, Sengar V and Haritsa J 2002 Query clustering for plan selection, tech report *DSL/SERC* (India: Indian Institute of Science)
[15] Kumar M V, M V Chaitanya and Madhavan M 2012 Segmenting the banking market strategy by clustering *International Journal of Computer Applications* **45**

IOP Conf. Series: Journal of Physics: Conf. Series **893** (2017) 012056 doi:10.1088/1742-6596/893/1/012056

Junction trees constructions in Bayesian networks

Linda Smail

Department of Mathematics and Statistics, *Journal of Physics: Conference Series*, IOP Publishing, Zayed University, P.O.Box 19282 Dubai, UAE

E-mail: `linda.smail@zu.ac.ae`

Abstract. unction trees are used as graphical structures over which propagation will be carried out through a very important property called the ruining intersection property. This paper examines an alternative method for constructing junction trees that are essential for the efficient computations of probabilities in Bayesian networks. The new proposed method converts a sequence of subsets of a Bayesian network into a junction tree, in other words, into a set of cliques that has the running intersection property. The obtained set of cliques and separators coincide with the junction trees obtained by the moralization and triangulation process, but it has the advantage of adapting to any computational task by adding links to the Bayesian network graph.

1. Introduction

Inference is one of the important tasks in Bayesian networks (BNs). It consists of computing marginal probabilities P_A, the probability of $X_A = (X_i)_{i \in A}$ where A is a subset of the BN variables), and conditional probabilities $P_{A|B}$ (probability of X_A conditioned on X_B, where A and B are two disjoint subsets of the BN variables). It is essentially an optimization calculation problem, as it becomes increasingly heavy with the complexity of the graph relative to the number of variables and their values. The correspondence between the graphical structure and the associated probabilistic structure bring many of the problems with inference to graphs problems. However these problems are relatively complex and give rise to a lot of research: [2], [7], [12]. Therefore, inference in BNs is a complex and a difficult task. [4] showed that this problem is NP-hard, although more research has been undertaken to develop efficient algorithms for such problems ([3], [10]).

There are several other algorithms for exact inference in BNs. They are often based on methods called clustering; for example, we cite the Shenoy-Shafer algorithm ([11]), Hugin ([6]), and the lazy propagation algorithm ([9], [17]) (See also [15], [14], [13], etc). All of these inference algorithms share an exponential complexity relative to the size of the associated junction tree's largest clique ([8], [16]).

There are two primary methods in literature (and in software) for constructing junction trees. The most commonly used method is by moralization and triangulation, see [5], [6], [8] for details. The other method recursively constructs a proper sequence of cliques that satisfies the running intersection property (see definition in section 3 below), introduced in [1] and [5].

For our part, after presenting the concept of the junction tree (JT), or equivalently, a proper sequence satisfying the running intersection property (RIP), we propose an algorithm that allows

IOP Conf. Series: Journal of Physics: Conf. Series **893** (2017) 012056 doi:10.1088/1742-6596/893/1/012056

constructing a sequence of subsets of the BN that has the RIP, without restrictive conditions on the set of subsets, as is the case with the junction tree algorithm.

Section 2 provides the details on the above-mentioned inference problem and the motivation behind our work. Section 3 offers complementary information about junctions trees and proper covering sequences that satisfy the running intersection property. Section 4 presents our new technique for constructing a JT on a given family of subsets in the BN.

2. Bayesian Networks and Junction Trees

Consider $X_I = (X_i)_{i\in I}$ a family of discrete random variables. For each i, X_i has values in the finite domain Ω_i. For each $J \subseteq I$, we denote with X_J, the subset $(X_i)_{i\in J}$, which has values in the finite domain $\Omega_J = \prod_{i\in J}\Omega_i$ and P_J the joint probability of X_J.

Given $\mathcal{C}$, a collection of subsets that covers I, let us recall that $\mathcal{C}$ and the probability P_I of the family $(X_i)_{i\in I}$ are compatible if P_I is written as follows:

$$\forall\ x_I \in \Omega_I \quad P_I(x_I) \ = \ \frac{1}{\alpha}\prod_{C\in\mathcal{C}} f_C(x_C), \tag{1}$$

where for each C, f_C is a known function from Ω_C in $\mathbb{R}_+$ and usually called potential of C; we can write (1) as: $P_I = \frac{1}{\alpha}\prod_{C\in\mathcal{C}} f_C$, where α is the normalization constant.

Given a BN on I, $\mathcal{C}$, in this case, is the set of subsets $\{i \cup pa(i)\}$, where $pa(i)$ is the set of parents of i and $f_{\{i\cup pa(i)\}}$ is the probability of X_i conditioned on $X_{pa(i)} = (X_j)_{j\in pa(i)}$, α is known and is 1.

The computation algorithms seek to only sum over subsets Ω_C where a smaller C is desired than I and well chosen in function of the probability and conditional probability of the subsets of X_I that we would like to compute.

However, if P_I is compatible with $\mathcal{C}$, a fundamental algorithm proposed by Lauritzen [8], the JT algorithm, provides a computing method for all probabilities P_C of subsets $X_C(= (X_i)_{i\in C})$ under the condition that $\mathcal{C}$ is a JT.

Given a set $\mathcal{K}$ of subsets of I, to compute the probability P_K for each $K \in \mathcal{K}$.

We notice that if $\mathcal{C}$ is a set of subsets of I such that $\forall\ K \in \mathcal{K}\ \ \exists\ C \in \mathcal{C}\ s.t.\ K \subseteq C$, and we are able to obtain the probabilities P_C, we can deduce each P_K by choosing a C such as $K \subseteq C$ and by eliminating the random variables in $C - K$; that is, if we decompose x_C to the pair (x_K, x_{C-K}), we have

$$P_K(x_K) = \sum_{x_{C-K}\in\Omega_{C-K}} P_C(x_C). \tag{2}$$

Given P_I is compatible with the covering $\mathcal{J}$, a method of computing the probability of all targets P_K $(K \in \mathcal{K})$ is given as follows:

A. Construct a collection of cliques $\mathcal{C}$ of $\mathcal{C} \cup \mathcal{K}$, which is a JT.

B. Use the compatibility of P_I with $\mathcal{C}$ to compute P_C.

C. Compute P_K from P_C.

Step A, which is purely from the graph theory, consists of constructing a JT that is a covering of $\mathcal{C} \cup \mathcal{K}$. This step can be heavy. We propose in what follows a new method for constructing a proper covering sequence that has the RIP, We think our method will significantly improve the original method and it can be adapted to any computation task.

3. Junction trees and proper covering Sequence that has the RIP

3.1. Junction Tree - Separators

Definition 1 Given a set I, we call a proper covering sequence of I any sequence $(C_1, \ldots, C_k)$

IOP Conf. Series: Journal of Physics: Conf. Series **893** (2017) 012056 doi:10.1088/1742-6596/893/1/012056

of subsets of I such that

$$\left| \begin{array}{l} \cup_{i=1}^{k} C_i = I \\ \text{for every pair } (j, j') \text{ of distinct indices, } C_j - C_{j'} \neq \emptyset. \end{array} \right.$$

In other words, a covering is proper if there is no inclusion among its elements.

Clearly, each non-proper covering can be associated with a proper covering by simply removing the parts that are subsets of others, and keeping just one part in case they are equal parts.

Definition 2 Given I a finite set, we call a JT on I every tree $(\mathcal{C}, \mathcal{H})$, where $\mathcal{C}$ is a proper covering of I satisfying the following property: for each pair (C, C') of non-adjacent elements of $\mathcal{C}$, each intermediate vertex C'' on the path in $\mathcal{H}$ joining C and C' satisfies $C \cap C' \subset C''$.

Remark 1 If card $(\mathcal{C}) \leq 2$, the property interfering in the definition of junctions trees is empty. They are trivially JTs when:

-for $k = 1$, $\mathcal{C} = \{I\}$ (a tree with only one element, so zero arcs),

-for $k = 2$, any pair $\{C, C'\}$ such that $C \cup C' = I$, $C - C' \neq \emptyset$, $C' - C \neq \emptyset$.

Definition 3 Given $(\mathcal{C}, \mathcal{H})$ a JT on a set I, the separators associated with this tree are the intersections $C \cap C'$, where C and C' are two elements of $\mathcal{C}$ that are adjacent for the structure of the tree $\mathcal{H}$.

We notice that if the cardinality of $\mathcal{C}$ is k, there a $k-1$ separators (number of arcs in a tree with k vertices).

An equivalent presentation of JTs is given by the following theorem.

Theorem 1 Consider $\mathcal{C}$ a proper covering of I and $\mathcal{H}$ a tree structure on $\mathcal{C}$.
$(\mathcal{C}, \mathcal{H})$ is a JT if and only if, for each $i \in I$, the set of subsets of $\mathcal{C}$ to which i belongs, noted as $\mathcal{C}_i$, constitutes a subtree of $(\mathcal{C}, \mathcal{H})$

3.2. Proper Covering Sequences that have the RIP

Definition 4 A sequence of subsets of a set I, $(C_1, \ldots, C_k)$, is said to have the RIP if

$$\forall\ j \in \{2, \ldots, k\}\ \ \exists\ \ell \in \{1, \ldots, j-1\}\ s.t.\ C_j \cap (\cup_{h=1}^{j-1} C_h) \subseteq C_\ell.$$

The RIP can be also written as follows:

$$\forall\ j \in \{2, \ldots, k\}\ \ \exists\ \ell \in \{1, \ldots, j-1\}\ s.t.\ \forall\ \ h < j\ \ C_j \cap C_h \subseteq C_\ell.$$

We notice that this property is always satisfied for $j = 2$, with $\ell = 1$.

We are particularly interested in proper covering sequences that have the RIP. We notice the following two such sequences:

- The sequence has one element (I).
- Any proper covering sequence with 2 elements: (C_1, C_2) such that $C_1 \cup C_2 = I$, $C_1 - C_2 \neq \emptyset$, $C_2 - C_1 \neq \emptyset$.

3.3. Link between JTs and RIP

The most important property of junctions trees, for our needs, is given by the following theorem.

Theorem 2 Any JT on I gives induces to as many proper covering sequences with the RIP as there are topological orderings on this tree. Conversely, any proper covering sequence that satisfies the RIP can be associated with the structure of a JT.

We omit the proof of the theorem due to space limitation.

4. An Alternative Method

4.1. Context

Given a finite covering $\mathcal{F}$ of I, our goal is to construct step by step a proper covering sequence, $(C_1, \ldots, C_p)$, that has the RIP and such that every $F \in \mathcal{F}$ is included in at least one of the C_j $(1 \leq j \leq p)$.

The algorithm requires that there be an order on I denoted $I = \{1, \ldots, n\}$; we consider the smallest element of each F $(F \in \mathcal{F})$ for this ordering, and we take the union of subsets F that have same smallest element. Ordering these unions by their smallest elements, we obtain a covering sequence $(Q_1, \ldots, Q_r)$ (where $r \leq n$) such that, if we denote α_k (where $1 \leq k \leq r$) the smallest element of Q_k, the sequence $(\alpha_1, \ldots, \alpha_r)$ is strictly increasing.

The imposed condition by our problem–that each F is included in one of the C_j– is obviously equivalent to the inclusion of each Q_k in one of the C_j.

The sequence $(Q_1, \ldots, Q_r)$ is not necessarily proper, which will not change anything to the problem if we clean it first; in other words, removing each subset that is already strictly included in another one. We notice in this regard, by construction, that the inclusion of $Q_j \subseteq Q_{j'}$ cannot hold with $j < j'$, because we will then have $\alpha_j \in Q_j$ but $\alpha_j \notin Q_{j'}$ because all of the elements of $Q_{j'}$ are greater or equal to $\alpha_{j'}$, which is strictly greater than α_j.

The cleaning consists of eliminating all Q_j satisfying $\exists\, j' < j$ *s.t.* $Q_j \subset Q_{j'}$, but the algorithm we are proposing does not require that the sequence $(Q_1, \ldots, Q_r)$ be proper.

The problem that the algorithm should solve is as follows:

Given a covering sequence $(Q_j)_{1 \leq j < r}$ of non-empty subsets of $I = \{1, \ldots, n\}$ such that, if we note $\alpha_j = \inf(Q_j)$, the sequence $(\alpha_1, \ldots, \alpha_r)$ is strictly increasing. We have to construct a proper covering sequence $(C_1, \ldots, C_p)$ that satisfies the RIP such that for each j $(1 \leq j \leq r)$, exists k $(1 \leq k \leq p)$ satisfying $Q_j \subseteq C_k$.

We notice that, necessarily, $\alpha_1 = 1$.

In the case of BNs, where $\mathcal{F}$ contains $\{\{i \cup pa(i)\}/i \in I\}$, we achieve the construction of $(Q_1, \ldots, Q_r)$ as follows:

- For I, we adopt an anti-hierarchical order relative to the structure of the directed acyclic graph: for each i, its parents are superior to it (in other words, given I is $\{1, \ldots, n\}$, the order $(n, n-1, \ldots, 1)$ is hierarchical).
- For each i we define $\tilde{Q}_i$ as the union of $\{i\}$, $pa(i)$ and, if they exist, subsets belonging to $\mathcal{F}$, not of the form $\{i \cup pa(i)\}$, where i is the smallest element.
- Optionally, we clean the sequence $(\tilde{Q}_1, \ldots, \tilde{Q}_n)$.

The alternative method will construct a sequence $(C_1^\star, \ldots, C_r^\star)$ of non-empty subsets of I of a length equal to the given sequence $(Q_1, \ldots, Q_r)$, that has the RIP such that for each j, k exists satisfying $Q_j \subseteq C_k^\star$. The constructed sequence is not necessarily proper but has no inclusions $C_{j'}^\star \subseteq C_j^\star$ only if $j' < j$. As a second step, we remove all elements included in others so we obtain the sequence $(C_1, \ldots, C_p)$ that satisfies $\forall\, j \;\; \exists\, k \; s.t. Q_j \subseteq C_k$, we notice that making the sequence proper will be completed by respecting and keeping the RIP.

The alternative method constructs successive subsets, noted as $D_1, \ldots, D_r$, such that the reversed sequence $(D_r, \ldots, D_1)$ is the sequence $(C_1^\star, \ldots, C_r^\star)$ or $(C_1, \ldots, C_p)$ sought. In other words, the constructed sequence satisfies the reversed RIP, which means the following:

$$\forall s \in \{1, \ldots, r-1\} \exists t \in \{s+1, \ldots, r\} s.t. D_s \cap (\cup_{u>s} D_u) \subseteq D_t$$

4.2. Phase 1: Constructing a Non Necessary Proper Sequence

The algorithm has exactly r steps; at each step j, it constructs a subset D_j, containing Q_j and contained in $\cup_{\ell \geq j} Q_\ell$; consequently, $\cup_{\ell > j} D_\ell = \cup_{\ell > j} Q_\ell$.

The sequence $(D_1, \ldots, D_r)$ satisfies the reversed RIP but is not necessarily proper; however, by construction, all of its elements are non-empty and distinct, and the inclusions $D_{j'} \subseteq D_j$

cannot take place if only $j' > j$.

Let us denote, for each j, $K_j = \{\alpha_j, \ldots, \alpha_{j+1} - 1\}$ (in this special case where $r = n$ therefore, for each $j \in \{1, \ldots, n\}$, $\alpha_j = j$, on a $K_j = \{j\}$).

Let us recall that the reversed RIP is automatically satisfied for the term before the last one of the sequence. For the sequence $(D_j)_{1 \leq j \leq r}$ to satisfy the reversed RIP, the algorithm saves in memory, for each j such that $1 \leq j \leq r - 2$, $S_j = D_j \cap (\cup_{\ell > j} Q_\ell)$. A sufficient condition for a subset D_k (with $k > j$) to satisfy the reversed RIP for D_j (in other words, satisfies $D_j \cap (\cup_{\ell > j} D_\ell) \subseteq D_k$) is that $S_j \subseteq D_k$. Si $S_j = \emptyset$, which means that $D_j \cap (\cup_{\ell > j} D_\ell) = \emptyset$, so the reversed RIP is trivially satisfied for D_j. Otherwise, the reversed RIP will be automatically satisfied for D_j, including in D_k, beside Q_k, S_j once $K_k \cap S_j \neq \emptyset$ (i.e., when $\alpha_k \leq \inf(S_j) < \alpha_{k+1}$, with the convention $\alpha_{r+1} = r + 1$), which will not only include in D_k the elements greater or equal to α_k contained in $\cup_{\ell \geq k} Q_\ell$.

Description of the new algorithm

Inputs a covering sequence $(Q_1, \ldots, Q_r)$ of non-empty subsets of $\{1, \ldots, n\}$;
$\alpha_i = \inf(Q_i)$; $\alpha_{r+1} = r + 1$.
The sequence $(\alpha_1, \ldots, \alpha_r)$ is strictly increasing.

Outputs sequences of subsets of I: $(D_1, \ldots, D_r)$ and $(S_1, \ldots, S_{r-2})$; for each j such that $S_j \neq \emptyset$, $s_j = \inf(S_j)$.

Initialization $D_1 = Q_1$; $S_1 = Q_1 - \{1\}$.

Current Step $j \geq 2$:

$$D_j = Q_j \cup [\cup_{k<j,\ \alpha_j \leq s_k < \alpha_{j+1}} S_k]$$

if $D_j = \emptyset, S_j = \emptyset$; if $D_j \neq \emptyset$, $S_j = D_j \cap (\cup_{k>j} Q_k)$.

End of the algorithm $j = n$.

Example 1 Given the BN in figure 1, we would like to compute $P_{i \cup pa(i)}$ (for each $i \in I$).

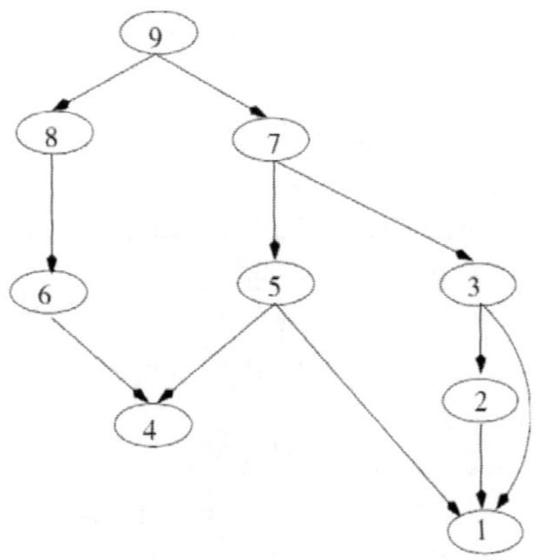

Figure 1. Example 1

We define the subsets: $Q_1 = \{1, 2, 3, 5\}, Q_2 = \{2, 3\}, Q_3 = \{3, 7\}, Q_4 = \{4, 5, 6\}, Q_5 = \{5, 7\}, Q_6 = \{6, 8\}, Q_7 = \{7, 9\}, Q_8 = \{8, 9\}, Q_9 = \{9\}$.

If we do not first clean the sequence, the algorithm will run as shown in table 1.

Table 1. Steps of the algorithm without prior cleaning.

Step i	D_i	S_i
1	$D_1 = Q_1 = \{1, 2, 3, 5\}$	$S_1 = \{2, 3, 5\}$
2	$D_2 = Q_2 \cup S_1 = \{2, 3, 5\}$	$S_2 = \{3, 5\}$
3	$D_3 = Q_3 \cup S_2 = \{3, 5, 7\}$	$S_3 = \{5, 7\}$
4	$D_4 = Q_4 = \{4, 5, 6\}$	$S_4 = \{5, 6\}$
5	$D_5 = Q_5 \cup S_3 \cup S_4 = \{5, 6, 7\}$	$S_5 = \{6, 7\}$
6	$D_6 = Q_6 \cup S_5 = \{6, 7, 8\}$	$S_6 = \{7, 8\}$
7	$D_7 = Q_7 \cup S_6 = \{7, 8, 9\}$	$S_7 = \{8, 9\}$
8	$D_8 = Q_8 \cup S_7 = \{8, 9\}$	$S_8 = \{9\}$
9	$D_9 = Q_9 \cup S_8 = \{9\}$	$S_9 = \emptyset$

We notice that the obtained sequence is not proper ($D_2 \subset D_1$, $D_8 \subset D_7$, and $D_9 \subset D_8$).

If we proceed by first making the sequence proper, we will preserve: $Q_1 = \{1, 2, 3, 5\}, Q_3 = \{3, 7\}, Q_4 = \{4, 5, 6\}, Q_5 = \{5, 7\}, Q_6 = \{6, 8\}, Q_7 = \{7, 9\}, Q_8 = \{8, 9\}$ then the algorithm will run as follows:

Table 2. Steps of the algorithm with prior cleaning.

Step j	K_j	D_j	S_j
1	$1, 2$	$D_1 = Q_1 = \{1, 2, 3, 5\}$	$S_1 = \{2, 3, 5\}$
2	3	$D_2 = Q_3 \cup S_1 = \{2, 3, 5, 7\}$	$S_2 = \{5, 7\}$
3	4	$D_3 = Q_4 = \{4, 5, 6\}$	$S_2 = \{5, 6\}$
4	5	$D_4 = Q_5 \cup S_2 = \{5, 6, 7\}$	$S_4 = \{6, 7\}$
5	6	$D_5 = Q_6 \cup S_4 = \{6, 7, 8\}$	$S_5 = \{7, 8\}$
6	7	$D_6 = Q_7 \cup S_5 = \{7, 8, 9\}$	
7	$8, 9$	$D_7 = Q_8 = \{8, 9\}$	

We notice that, in this case as well, the obtained sequence is not proper ($D_7 \subset D_6$)

4.3. Phase 2: Cleaning the Sequence Obtained in Phase 1

By inverting the $(D_1, \ldots, D_n)$ obtained by the algorithm, we obtain a covering sequence of distance elements $(C_1^\star, \ldots, C_n^\star)$ that satisfies the RIP, and for each (i, j) such that $1 \le i < j \le n$, $C_i^\star \supset C_j^\star$ is false.

If this sequence is not proper, the possibility of making it, that is removing each C_i for which it exists j ($> i$), satisfying $C_i^\star \subset C_j^\star$ is possible through the following algorithm.

Description of the cleaning phase of the algorithm

Inputs a covering sequence $(C_1^\star, \ldots, C_n^\star)$ that has the RIP and such that $C_i^\star \subset C_j^\star$ implies $i < j$.
Current Object a covering sequence $(C_1, \ldots, C_k)$ satisfying the same properties as the given sequence.
Outputs a proper covering sequence $(C_1, \ldots, C_k)$ at the end of the algorithm.
Initialization $k = n$, $(C_1, \ldots, C_k) = (C_1^\star, \ldots, C_n^\star)$
Current Step Determine whether $p = \min\{j \ ; \ \exists\, i < j \ s.t. C_i \subset C_j\}$ exists.
If p does not exist, the algorithm ends.
If p exists, consider $K = \{i < p \ ; \ C_i \subset C_p\}$ and $k = \min K$.
- If they exist, remove from the sequence the C_i such that $i \in K$ and $i \neq k$ from the sequence.
- replace C_p with C_k.
- place $C_l = C_{l-p}$ for all l greater than p.

The algorithm will eventually stop because at each step, the length of the sequence will strictly decrease.

Remark 2 Any other cleaning algorithm on $(C_1^\star, \ldots, C_n^\star)$ would lead to a set of identical subsets to $\{C_1, \ldots, C_k\}$, as it is constructed by the above-described algorithm and therefore has the JT structure. But the proposed algorithm has the advantage of getting it with an order ensuring the RIP.

Example 2 The sequence $(C_1^\star, \ldots, C_9^\star)$, found in example 1 was the following: $C_1^\star = \{9\}, C_2^\star = \{8,9\}, C_3^\star = \{7,8,9\}, C_4^\star = \{6,7,8\}, C_5^\star = \{5,6,7\}, C_6^\star = \{4,5,6\}, C_7^\star = \{3,5,7\}, C_8^\star = \{2,3,5\}, C_9^\star = \{1,2,3,5\}$, we notice that $C_1^\star \subset C_2^\star \subset C_3^\star$ and $C_8^\star \subset C_9^\star$; by application of the above algorithm, we find that the proper sequence satisfying the RIP, $(C_1, \ldots, C_6)$, matches the sequence $(C_1^\star, \ldots, C_9^\star)$: $C_1 = \{7,8,9\}, C_2 = \{6,7,8\}, C_3 = \{5,6,7\}, C_4 = \{4,5,6\}, C_5 = \{3,5,7\}, C_6 = \{1,2,3,5\}$.

5. Conclusion

In this paper, we have presented a method for constructing a proper covering sequence of cliques–a junction tree–that satisfies the running intersection property starting from a simple sequence of subsets in a Bayesian network. The proposed method is advantageous in that it provides a proper covering that has the RIP based on the computation task and without restrictions. The obtained set of cliques and separates coincide with the JTs obtained by the moralization and triangulation process, but it has the advantage of adapting to any computation task by adding links to the BN graph and therefore updates the set of subsets at the start of our algorithm. Future work will compare this method to the moralization and triangulation method on more complex examples of BNs and will also explore the algorithm's complexity and efficiency.

References

[1] Cecil H and Adnan D 1996 Inference in belief networks: A procedural guide *International Journal of Approximate Reasoning.* **15** pp 225-263
[2] Cooper F G 1990 The computational complexity of probabilistic inference using Bayesian belief networks *Artificial Intelligence* **42** pp 393-405
[3] Cooper F G and Herskovits E 1991 Determination of the entropy of a belief network is NP-hard *Knowledge Systems Lab.*
[4] Cooper F G 1987 Probabilistic inference using belief networks is NP-hard *Knowledge Systems Lab.* **42** pp 393-405
[5] Cowell R G, Dawid A P, Lauritzen S L and Spiegelhalter D J 1999 *Probabilistic Networks and Expert Systems* (Berlin: Springer)
[6] Jensen F V 1999 *An introduction to Bayesian Networks* (London: UCL Press),
[7] Geiger D, Heckerman D, King H and Meek C 2001 Stratified exponential families: Graphical models and model selection *The Annals of Statistics* **29** pp 505 - 526

IOP Conf. Series: Journal of Physics: Conf. Series **893** (2017) 012056 doi:10.1088/1742-6596/893/1/012056

[8] Lauritzen S L and Spiegelhalter D J 1988 Local computation with probabilities on graphical structures and their application to expert systems *Proceedings of the Royal Statistical Society* Series B **50** (2)
[9] Madsen A L and Jensen F V 1998 Lazy propagation in junction trees ed Gregory F. Cooper and Serafin Moral *Proceedings of the Fourteenth Conference on Uncertainty in Artificial Intelligence* (Morgan Kaufmann Publishers) pp 362 - 369
[10] Meila M and Heckerman D 2001 *An experimental comparison of several clustering and initialization methods. Machine Learning.* **42** pp 9-29
[11] Shenoy P and Shafer G 1990 Axioms for probability and belief-function propagation ed Shachter, R. et al. *Uncertainty in Artificial Intelligence* **4** (USA: Univ. California Press) pp 169-198
[12] Smyth P, Heckerman D and Jordan M 1997 Probabilistic independence networks for hidden Markov probability models *J. Neural Computation* **9** pp 227-269
[13] Suermondt H J and Cooper F G 1990 Probabilistic inference in multiply connected brief networks using loop cutsets *Journal of Approximate Reasoning* **4** pp 283 - 306
[14] Suermondt H J and Cooper F G 1991 Initialization for the method of conditioning in Bayesian belief networks *Artificial Intelligence* **50** pp 83-94
[15] Suermondt H J, Cooper F G and Heckerman D 1991 A combination of cutset conditioning with clique-tree propagation in the pathfinder system *Uncertainty in Artificial Intelligence* vol 6 ed P B Bonissone, M Henrion, L N Kanal and J F Lemmer pp 245-253
[16] Uffe K 2000 *Inference in Bayesian networks using nested junction trees. Statistics,* **19** pp 301-312
[17] Xiang Y and Jensen F V 1999 Inference in multiply sectioned Bayesian networks with extended Shafer-Shenoy and lazy propagation *Proceedings of the Fifteenth Conference on Uncertainty in Artificial Intelligence* **43** pp 680-687

IOP Conf. Series: Journal of Physics: Conf. Series **893** (2017) 012057 doi:10.1088/1742-6596/893/1/012057

Modeling and simulation of ocean wave propagation using lattice Boltzmann method

Dian Nuraiman

Graduate School of Natural Sciences and Technology, Kanazawa University, Japan
Department of Mathematics, UIN Sunan Gunung Djati Bandung, Indonesia

E-mail: dianuraiman@uinsgd.ac.id

Abstract. In this paper, we present on modeling and simulation of ocean wave propagation from the deep sea to the shoreline. This requires high computational cost for simulation with large domain. We propose to couple a 1D shallow water equations (SWE) model with a 2D incompressible Navier-Stokes equations (NSE) model in order to reduce the computational cost. The coupled model is solved using the lattice Boltzmann method (LBM) with the lattice Bhatnagar-Gross-Krook (BGK) scheme. Additionally, a special method is implemented to treat the complex behavior of free surface close to the shoreline. The result shows the coupled model can reduce computational cost significantly compared to the full NSE model.

1. Introduction

Observation of behavior of ocean wave propagation is one of the most important issues on ocean and coastal engineering research field. The big problem arise when considering large domain. The larger computational domain, the higher computational cost.

In order to reduce computational cost for large domain simulation, it is necessary to find a coupling strategy between a cheap model with a precise model. Some coupling strategies have been proposed, such as coupling a Boundary Element Model (BEM) for the potential flow equations with a Volume of Fluid (VOF) model for the NSE [1], a 1D model of Boussinesq equations with a 2D SPH model for the NSE [2], [3], and a 2D SWE model with a 3D NSE model using LBM [4].

This paper presents on modeling and simulation of ocean wave propagation from the deep sea to the shoreline. In order to save the computational cost, we propose to couple a 1D SWE model for the deep sea with a 2D incompressible NSE model close to the shoreline. We need a cheap model for the deep sea as it represents the largest part of computational domain and a precise model close to the shoreline where waves break and overturn. The coupled model is then compared to the full NSE model.

2. Models

2.1. Incompressible Navier-Stokes Equations (NSE)

The NSE are the governing equation in describing general fluid flows. The incompressible NSE consist of the continuity and momentum equation which can be written in tensor form as, respectively,

$$\frac{\partial u_j}{\partial x_j} = 0 \tag{1}$$

and

$$\frac{\partial u_i}{\partial t} + \frac{\partial (u_i u_j)}{\partial x_j} = -\frac{1}{\rho}\frac{\partial p}{\partial x_i} + \nu \frac{\partial^2 u_i}{\partial x_j^2} + F_i \tag{2}$$

The Enstein summation convention is used in the above equations.

2.2. Shallow Water Equations (SWE)

The SWE can be derived from incompressible NSE by taking depth-integration under assumption the wavelength is much larger than the corresponding depth. The SWE can be written in tensor form for the continuity equation

$$\frac{\partial h}{\partial t} + \frac{\partial (h u_j)}{\partial x_j} = 0 \tag{3}$$

and the momentum equation

$$\frac{\partial (h u_i)}{\partial t} + \frac{\partial (h u_i u_j)}{\partial x_j} = -\frac{\partial}{\partial x_i}\left(\frac{g h^2}{2}\right) + \nu \frac{\partial^2 (h u_i)}{\partial x_j^2} + F_i \tag{4}$$

3. Methods

The lattice Boltzmann method (LBM) was originated from Boltzmann's kinetic theory of gases. The basic idea is the fluid can be imagined containing a lot of particles. The LBM does not consider behavior of each particle but behavior of a collection of particles as a unit represented by a distribution function. The LBM can be used to solve both the SWE model [5] and the NSE model [4], [6].

The LBM can be derived from Boltzmann equation by taking first order discretization in discrete phase-space. Basic formulation of the lattice Boltzmann equation in descrete form with the lattice BGK model is given by

$$f_i(\mathbf{x} + \mathbf{e}_i \Delta t, t + \Delta t) = f_i(\mathbf{x}, t) - \frac{1}{\tau}(f_i(\mathbf{x}, t) - f_i^{eq}(\mathbf{x}, t)) \tag{5}$$

where f is distribution functions (DFs), f^{eq} is local equilibrium distribution function, i is direction indice, and τ is relaxation parameter.

The LBM is a mesh-based method. The computational domain is divided into Cartesian grid. There are two main steps in the LBM namely streaming and collision step. In the streaming step, the particles move to neighboring cells in direction of their lattice velocities. The streaming step can be formulated as

$$f_i^*(\mathbf{x}, t) = f_i(\mathbf{x} + \mathbf{e}_{\bar{i}} \Delta t, t) \tag{6}$$

where f_i^* denote DFs after streaming step.

In the collision step, the arriving particles at the cells collide with one another and change their velocity directions toward their equilibrium state. The collision step can be written as

$$f_i(\mathbf{x}, t + \Delta t) = f_i^*(\mathbf{x}, t) - \frac{1}{\tau}(f_i^* - f_i^{eq}) \tag{7}$$

where τ represents the fluid viscosity with $\tau = 3\nu + \frac{1}{2}$.

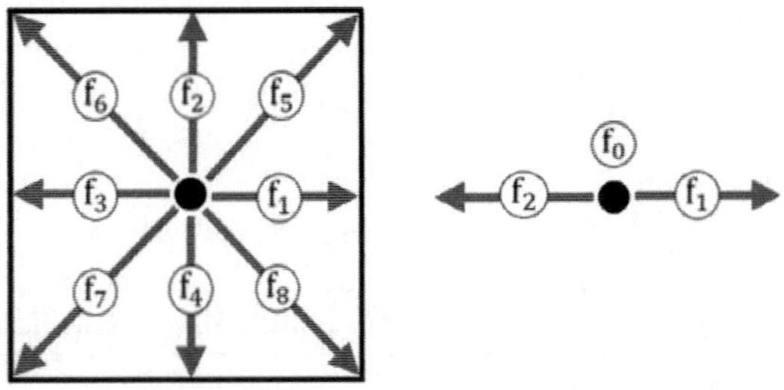

Figure 1: Lattice pattern D2Q9 (left) and D1Q3 (right)

3.1. LBM for a 2D Incompressible NSE Model

The most commonly used lattice pattern in 2D simulation is D2Q9 model which has nine lattice directions as shown in the left part of Figure 1. Macroscopic density and velocity of fluid can be written as, respectively,

$$\rho(\mathbf{x},t) = \sum_{i=0}^{8} f_i(\mathbf{x},t) \tag{8}$$

$$\mathbf{u}(\mathbf{x},t) = \sum_{i=0}^{8} \mathbf{e}_i f_i(\mathbf{x},t) \tag{9}$$

where f_i^{eq} is given by

$$f_i^{eq} = w_i \left[\rho + 3\mathbf{e}_i \cdot \mathbf{u} - \frac{3}{2}\mathbf{u}^2 + \frac{9}{2}(\mathbf{e}_i \cdot \mathbf{u})^2\right] \tag{10}$$

3.2. LBM for a 1D SWE Model

For SWE model, we use a 1D lattice pattern of D1Q3 which has three lattice directions as shown in the right part of Figure 1. Macroscopic water depth and velocity are given by, respectively,

$$h(x,t) = \sum_{i=0}^{2} g_i(x,t) \tag{11}$$

$$v(x,t) = \frac{1}{h(x,t)} \sum_{i=0}^{2} e_i g_i(x,t) \tag{12}$$

Local equilibrium distribution function g_i^{eq} is given by

$$g_i^{eq} = \begin{cases} h + w_i h(-\frac{3gh}{4c^2} - \frac{3v^2}{2c^2}) & i = 0 \\ w_i h(\frac{3gh}{2c^2} + \frac{3e_i v}{c^2} + \frac{3e_i^2 v^2}{c^4}) & i = 1,2 \end{cases} \tag{13}$$

3.3. Free Surface Treatment

The free surface treatment method is only used for the NSE model. The method was proposed by Thürey [7]. This requires two additional values to be stored for each cell, the mass m and the fluid fraction ϵ, where

$$\epsilon(\mathbf{x},t) = \frac{m(\mathbf{x},t)}{\rho(\mathbf{x},t)} \tag{14}$$

This method does not consider the gas phase as a separate fluid under assumption the viscosity difference between gas and fluid is high enough to approximate gas velocity near the surface with the surface velocity [8].

3.4. Parametrization
Given S as domain length $[m]$, r as grid resolution, and g_c as compressibility constant, simulation parameters are set as shown in Table 1.

Table 1: Simulation Parameters

Parameters	**Real Unit**	**Lattice Unit**
Cell size	$\Delta x' = \frac{S}{r}[m]$	$\Delta x = 1$
Time step	$\Delta t' = \sqrt{\frac{g_c \Delta x'}{\lvert\mathbf{g}'\rvert}}[s]$	$\Delta t = 1$
Gravitational acceleration	$\mathbf{g}'[\frac{m}{s^2}]$	$\mathbf{g} = \mathbf{g}'\frac{\Delta t'^2}{\Delta x'}$
Kinematic viscosity	$\nu'[\frac{m^2}{s}]$	$\nu = \nu' \frac{\Delta t'}{\Delta x'^2}$

4. Coupled Model
The two models are coupled by constructing a buffer zone to transfer information between two models. The coupling design as shown in Figure 2 where N_{sw} and N_{ns} are number of grid points

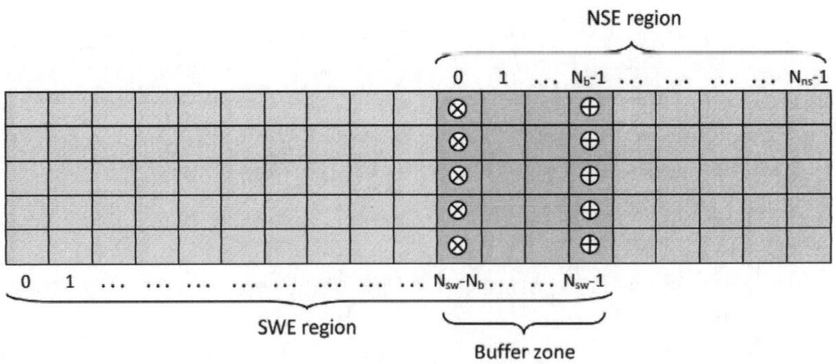

Figure 2: Coupling design

in length of the SWE region and the NSE region, respectively, and N_b is number of grid points in the buffer zone.

Both models have different characterstics. There is no depth in NSE model. Furthermore, SWE model has no velocity in vertical direction. There are two coupling procedures, SWE to NSE transfer and vice versa.

4.1. SWE to NSE transfer
The main idea is depth information from the SWE region used to define fluid cells in the NSE region by adding or removing some fluid or interface cells at the leftmost two column of NSE region. The DFs of fluid or interface cells are initialized by the SWE velocity

$$f_k = f_k^{eq}(\rho, \mathbf{u}_{ij}) \tag{15}$$

where $k = 0, 1, ..., 8$ and

$$\mathbf{u}_{ij} = (s_v v_{N_{sw}-N_b+i}, s_v w_i) \tag{16}$$

where s_v is scaling factor that was determined based on numerical experiments. Moreover, velocity in vertical direction is calculated by change of the SWE depths of the current and previous time step over time.

4.2. NSE to SWE transfer
Depth of SWE region at the right boundary is determined by depth of fluid cells of NSE region. The DFs are then initialized by taking the average velocity u^{avg} in horizontal direction over the right column given by

$$g_i = g_i^{eq}(H_{N_b-1}, u^{avg}) \tag{17}$$

where $i = 0, 1, 2$.

5. Results

Length of computational domain is 26 meters with time step for SWE model $\Delta t_{sw} = 0.001$, time step for NSE model $\Delta t_{ns} = 0.0002$, and the scaling factor for SWE velocity $s_v = 0.015$. A wall boundary condition is implemented at left and right boundary. The initial condition of wave at the fisrt half domain with zero initial velocity as shown in figure below

Figure 3: Initial condition of wave

Figure 4 shows the result at the second half domain plotted based on velocity at several times with two black vertical lines as boundaries of buffer zone. We can see how the wave is coming from the left domain, passing the buffer zone, approaching the shoreline, and going back to the deep sea. By this coupling strategy, ocean wave smoothly propagated from the deep sea to the shoreline without any reflected parasitic wave when passing the buffer zone.

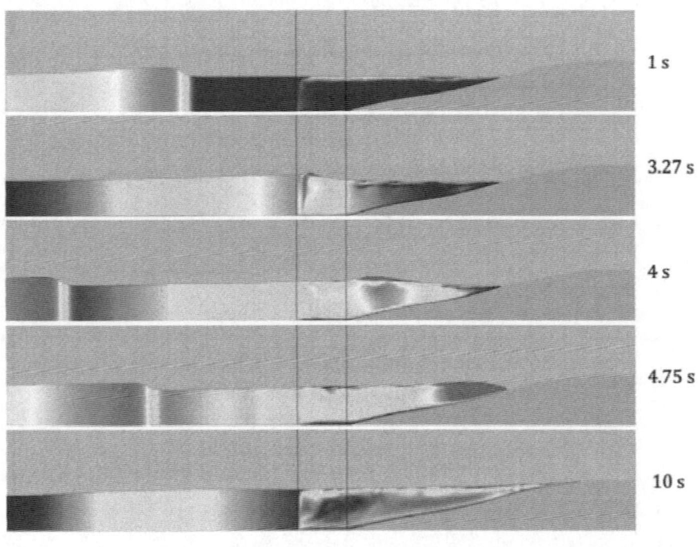

Figure 4: Result plotted based on velocity

The computation time for the coupled model was 5 minutes. It was compared with the full NSE model that requires 34 minutes to solve the same domain. Hence, the coupled model can reduce the computational cost significantly up to 85% in case of the setting in Figure 4 when compared to the full NSE model.

IOP Conf. Series: Journal of Physics: Conf. Series **893** (2017) 012057 doi:10.1088/1742-6596/893/1/012057

6. Conclusion

We have presented a coupling strategy between the SWE model and the NSE model for ocean wave propagation using the LBM. However, two points that remains are to provide a theoretical justification for the velocity scaling used in the SWE to NSE transfer and to verify the coupled model on benchmark problems.

Acknowledgment

I would like to thank Assoc. Prof. Karel Svadlenka for his help, support and suggestions in this work.

References

[1] Lachaume C, Biausser B, Grilli S T, Fraunie P and Guignard S 2003 Modeling of breaking and post-breaking waves on slopes by coupling of BEM and VOF methods *Proc. 13th Int. Offshore Polar Eng. Conf.* pp 353–359

[2] Narayanaswamy M S 2009 A hybrid Boussinesq-SPH wave propagation model with applications to forced waves in rectangular tanks *PhD Dissertation* (Maryland: The Johns Hopkins University)

[3] Kassiotis C, Ferrand M, Violeau M, Rogers B D, Standsby P K and Benoit M 2011 Coupling SPH with a 1-D Boussinesq-type wave model *6th International SPHERIC Workshop* pp 241–247

[4] Thürey N, Rüde U and Stamminger M 2006 Animation of open water phenomena with coupled shallow water and free surface simulations *Proc. of the 2006 ACM SIGGRAPH/Eurographics Symposium on Computer Animation* pp 157–164

[5] Koshimura S and Murakami K 2009 Applicability of the lattice Boltzmann method for tsunami modeling *J. Japan Soc. Civil Engrs.* **65** pp 256–260

[6] Kwak Y, Kuo J C C and Nakano A 2009 Hybrid lattice-Boltzmann/level-set method for liquid simulation and visualization, *Int. J. Comput. Sci.* **3** pp 579–592

[7] Thürey N 2007 Physically based animation of free surface flows with the lattice Boltzmann method *PhD Dissertation* (Nürnberg: University of Erlangen)

[8] Thürey N and Rüde U 2009 Stable free surfaces flows with the lattice Boltzmann method on adaptively coarsened grids *Comput. Vis. Sci.* **12** pp 247–263

IOP Conf. Series: Journal of Physics: Conf. Series **893** (2017) 012058 doi:10.1088/1742-6596/893/1/012058

Fuzzy conditional random fields for temporal data mining

Intan Nurma Yulita and Atje Setiawan Abdullah
Department of Computer Science, Faculty of Mathematics and Natural Sciences
Universitas Padjadjaran, Indonesia

E-mail: intan.nurma@unpad.ac.id, atjesetiawan@gmail.com

Abstract. Temporal data mining is one of the interesting problems in computer science and its application has been performed in a wide variety of fields. The difference between the temporal data mining and data mining is the use of variable time. Therefore, the method used must be capable of processing variables of time. Compared with other methods, conditional random field has advantages in the processing variables of time. The method is a directed graph models that has been widely applied for segmenting and labelling sequence data that appears in various domains. In this study, we proposed use of Fuzzy Logic to be applied in Conditional Random Fields to overcome the problems of uncertainty. The experiment is compared Fuzzy Conditional Random Fields, Conditional Random Fields, and Hidden Markov Models. The result showed that accuracy of Fuzzy Conditional Random Fields is the best.

1. Introduction

Several studies have introduced the method of for temporal data mining for example use artificial neural networks [1], [2], decision tree, and support vector machine. The other hand, the widely used method for sequence data is Hidden Markov Models (HMM). Research conducted by Rabiner is a preliminary study using HMM for dividing and labeling sequence data that appears in various domains such as voice recognition [3]. Since then, HMM is widely used in other pattern recognition such as speech, handwriting, gesture recognition, part-of-speech tagging, and bioinformatics. Other studies also suggest the use of HMM for the classification of motion [4], [5], hand gestures [6], and speech [7]. However, HMM has some limitations such as distribution of the conditional probability of hidden variable x only at time t, not along duration since HMM observation is a scalar (1 dimension). In order to achieve better performance than HMM, Laferty et al proposes the use of Conditional Random Fields (CRF) to divide and label the sequence data but it can overcome the limitations of the HMM [8]. CRF is a directed graph models that has been widely applied for segmenting and labeling sequence data that appears in various domains. CRF is able to combine the features of a complex sequence of observations of X which does not require the assumption of non-independence between observations in variable X. Unlike HMM, CRF accommodate observation data vectors and capable of solving the existing label is inherently biased in HMM.

2. Fuzzy Inference System

Fuzzy Inference System (FIS) is selected to implement fuzzy logic for this study. FIS have two advantages over other fuzzy logic techniques. It is better than the other in handling the linguistic concepts and also mapping the concept of non linear between input and output [9]. FIS consists of three main steps, namely:

(i) **Fuzzification**
In this step, every crisp input is converted into fuzzy input. To change it, then built a diagram of the membership function for each input.

(ii) **Inference**
In this step, every fuzzy input is converted into fuzzy output. It can be done by building a number of rules that map fuzzy input into output targets.

(iii) **Defuzzification**
In this step, conversion is done fuzzy output into a crisp output. The process is also built a diagram of membership function but it has a different purpose than fuzzification.

3. Conditional Random Fields

Input CRF is a segmentation of gesture phase with the overall features are used, while the output is a label of gesture phase. To build CRF in labeling the gesture phase, the necessary set of feature functions fi. In the CRF, each of the features has several input variables, namely:

(i) Segment of gesture phase (s)
(ii) The position i of the phase segment
(iii) Label the current phase l_i
(iv) Label l_{i-j} from the previous phases or can add l_{i+j} of the phases thereafter.

And the output is real-valued probability [10]. If the feature functions only observe a single phase with its previous phase, the function is $f_i(s, i, l_i, l_{i-1})$. As an example of the features are $f_i(s, i, l_i, l_{i-1}) = 1$ if l_i is rest and l_{i-1} is hold so that if the weights 1 associated with this $f_i$large and positive, the labeling of the phases being observed labeled rest. To build the CRF, the necessary number of feature functionality. The function of this feature is obtained through a combination of the observation of the observation before or afterwards. Large observations will be used depending on the size of the window to be used. If the window size (w) is 0 then the observation phase took only one segment without having a dependency with another segment of the mobile phase. But if the window size is 1, then shift observations will overlap with one segment before / after. In this study, experiment will test change the window size on the performance of the CRF so that it can be seen the influence of the dependence of a segment of the mobile phase to the other segments. Furthermore, the features that have been obtained are given weight. Both of these play an important role in modeling the label of each mobile phase. Modelling is indicated by a score of a label which is calculated using the following formula:

$$score(l|s) = \sum_{j=1}^{m} \sum_{i=1}^{n} \lambda_j f_j(s, i, l_i, l_{i-1}) \tag{1}$$

The first summation calculating the features j and summing the two for each position i of the segmentation of the mobile phase. Weight of features affect CRF in discovering the proper labeling of test data is provided so that the invention proper weight for each feature needs to be done. The weight of this feature is obtained through a learning process of the data used. But if the initial weights are assigned randomly then execution time to obtain the optimal weight value will be longer. The easiest way to gain weight is done by calculating all combinations or sequences for possible configurations labels and count all scores one by one through brute force. But this mechanism will require enormous computing process, therefore some ways that can be done through optimization, one of which is the algorithm Broyden-Fletcher-Goldfarb-Shanno (BFGS) is applied in this study. The final step of the process of labeling the gesture phase was

Table 1: Dataset

No	Attributes
1	lhx - Position of left hand (x coordinate)
2	lhy - Position of left hand (y coordinate)
3	lhz - Position of left hand (z coordinate)
4	rhx - Position of right hand (x coordinate)
5	rhy - Position of right hand (y coordinate)
6	rhz - Position of right hand (z coordinate)
7	hx - Position of head (x coordinate)
8	hy - Position of head (y coordinate)
9	hz - Position of head (z coordinate)
10	sx - Position of spine (x coordinate)
11	sy - Position of spine (y coordinate)
12	sz - Position of spine (z coordinate)
13	lwx - Position of left wrist (x coordinate)
14	lwy - Position of left wrist (y coordinate)
15	lwz - Position of left wrist (z coordinate)
16	rwx - Position of right wrist (x coordinate)
17	rwy - Position of right wrist (y coordinate)
18	rwz - Position of right wrist (z coordinate)

to transform it into a score of opportunities $p(l|s)$ with a range of 0 and 1 through exponential and normalization following the following formula:

$$p(l|s) = \frac{\exp[score(l|s)]}{\sum_j \exp[score(l|s)]} \tag{2}$$

4. Experiment

4.1. Data set

The data set used in this study came from School of Art, Sciences and Humanities University of Sao Paulo [11]. The data set consists of temporal segmentation of gesture. It is composed by three data recorded using Microsoft Kinect sensor. For each data has 18 features, namely: For the classification, it uses only five phases: Rest (1), Preparation (2), Stroke (3), Hold (4), and Retraction (5).

4.2. Flowchart

Three scenarios of temporal data mining in this study are described in the Figure 1.

The mechanism of this research has three scenarios, namely the implementation of HMM, CRF and Fuzzy CRF. To be able to process the data, then HMM requires vector quantization (VQ). VQ in this study is built using K-Means Clustering. While on Fuzzy CRF, it needs additional preprocessing of feature extraction to minimize number of input on the Fuzzy Inference System. The greater the input, the more rules to be built and it will improve the execution time. Some step is done before modeling is started, namely:

IOP Conf. Series: Journal of Physics: Conf. Series **893** (2017) 012058 doi:10.1088/1742-6596/893/1/012058

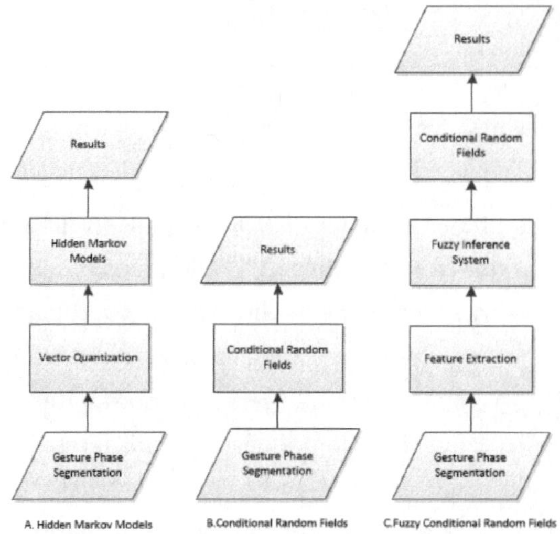

Figure 1: Architecture of proposed methods

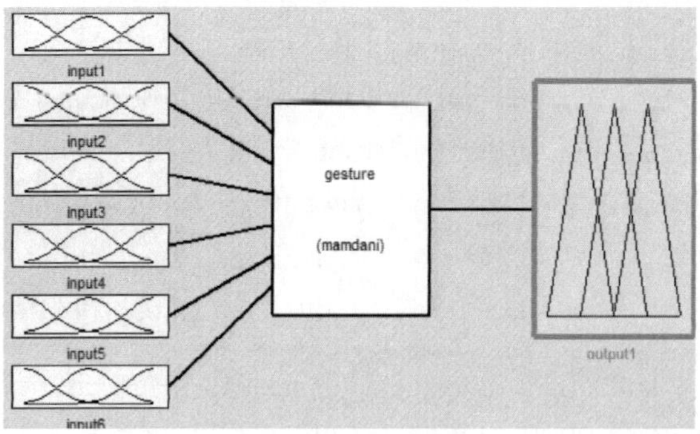

Figure 2: The Proposed Fuzzy Inference System

(i) **Preprocessing**
First scenario implements vector quantization to make all input to conform with Hidden Markov Models, and scenario 3 implements feature extraction and fuzzy inference system.

(ii) **Vector Quantization**
In order to process HMM, an observation sequence is needed. By using K-means clustering algorithm, Vector Quantization is used to construct observation sequence. The process has two steps: the formation of codebook and the determination of codebook index.

(iii) **Feature Extraction**
The process is done by replacing some feature vector with one represented feature vector by Euclidean Distance. The output of this study is six represented features.

(iv) **Fuzzy Inference System**
The system is developed by six inputs, 64 rules, and an output. The system is done by trial and error in this study. The process in this study is shown in Figure 2.

In this study, it will implements Hidden Markov Models, Conditional Random Fields, and also Fuzzy Conditional Random Fields. Hidden Markov Models is a directed graph that represents the joint probability between class and observation. But the observation of Hidden Markov Models confined to one dimension of data between two time segments. Therefore, Conditional

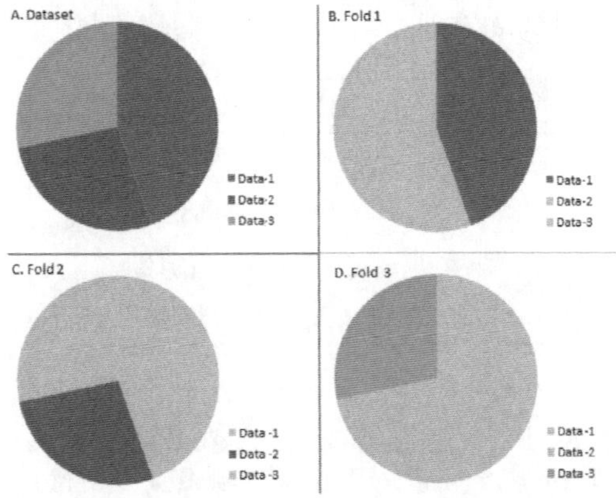

Figure 3: The cross-validation scheme

Table 2: Percentage of Accuracy

Fold	HMM	CRF	Fuzzy CRF
1	37.04	50.14	59.25
2	20.22	50.00	76.19
3	25.74	64.41	57.19
Average	27.67	54.85	64.21

Random Fields implemented to overcome it.The other hand, to handle uncertainties in gesture phase segmentation, Condition Random Fields will be combined with Fuzzy Inference Systems.

4.3. Testing
The experimental process conducted through three data to test some models produced by HMM, CRF, and Fuzzy CRF. For each data, each part of the data divided into test data and the rest into training data by cross-validation scheme. This process is illustrated in Figure 3.

5. Results
The results for three scenarios of temporal data mining are presented in Table 2 and Table 3 respectively

5.1. Accuracy
Accuracy of three methods to test data shown in Table 2. From the experiments that have been done, the percentage of accuracy with HMM, the value only ranged from 20-38, whereas using CRF ranged between 50-65, Fuzzy CRF has a range of between 57-77. It can be concluded is superior to the CRF HMM modeling based on observations due to several segments of time and data features. Accuracy is increased when combined with Fuzzy CRF because this technique is able to overcome the problem of loss of information as it happens if without using Fuzzy.

5.2. Execution time
Time execution of three methods to test data shown in Table 3. The execution time in able 3 are calculated based on only for the classification. From experiments, it is known that CRF has been combined with Fuzzy experiencing a decrease in execution time for process modeling. CRF

Table 3: Execution Time (seconds)

Fold	CRF	Fuzzy CRF
1	115.06	33.50
2	49.00	48.24
3	52.71	50.10
Average	72.26	43.95

execution time ranged from 48-116 seconds but when combined with Fuzzy Logic, the execution time becomes 33-51 seconds for each fold. This is because the number of features is reduced when it was done Fuzzy Inference System.

6. Conclusion and Future Work

From the analysis of the performance of Fuzzy Conditional Random Fields by using the data in this study, it can be concluded Fuzzy Conditional Random Fields has the highest accuracy. Also, it has the smallest execution time due to the reduction in the number of variables used for classification Since our method is an effective way for temporal Data Mining, it is highly recommended for tuning the parameter of Fuzzy Inference System by using Genetic Algorithms or Artificial Neural Networks.

Acknowledgment

The Authors thank to Rector Universitas Padjadjaran and Direktorat Riset dan Pengabdian Pada Masyarakat (DRPM) Universitas Padjadjaran. who gave funding for dissemination of this paper, and also Korean Mathematical Society (KMS) who gave funding for dissemination of this paper through travel grant.

References

[1] Priyanka Gaur 2012 Artificial Neural Networks in Data Mining *International Journal of Electronics and Computer Science Engineering* **1**(3)

[2] Sigillito V G, Wing S P, Hutton L V, Baker K B 1989 Classification of radar returns from the ionosphere using neural networks *Johns Hopkins APL Technical Digest* **10** pp 262–266

[3] Lawrence Rabiner, Biing-Hwang Juang 1993 Fundamentals Of Speech Recognition *Prentice Hall*

[4] Bobick a F, Ivanov Y a 1998 Action recognition using probabilistic parsing *In Proceedings IEEE Computer Society Conference on Computer Vision and Pattern Recognition (Cat. No.98CB36231). IEEE Comput. Soc* pp 196–202

[5] Kim T, Shakhnarovich G, Urtasun R 2010 Sparse Coding for Learning Interpretable Spatio-Temporal Primitives *NIPS* pp 1–9

[6] Elmezain M, Al-hamadi A 2008 A Hidden Markov Model-Based Isolated and Meaningful Hand Gesture Recognition *In Proceeding of World Academy of Science, Technology, Engineeering and Technology* pp 394–401

[7] Yulita I N, The H L 2012 Fuzzy Hidden Markov Models for Indonesian Speech Classification *Journal of Advanced Computational Intelligence and Intelligent Informatics (JACIII)* **16**(3)

[8] Sutton C and Mccallum A 2010 An Introduction to Conditional Random Fields *Foundations and Trends* pp 1–87

[9] Serge Guillaume 2001 Designing fuzzy inference systems from data: an interpretability-oriented review *IEEE Transactions on Fuzzy Systems* **9**(3) pp 426–443

[10] Lafferty J, Mccallum A 2001 Conditional Random Fields?: Probabilistic Models for Segmenting and Labeling Sequence Data Conditional Random Fields?: Probabilistic Models for Segmenting and (Icml) pp 282–289

[11] Madeo R C B, Wagner P K, PERES S M 2013 A Review of Temporal Aspects of Hand Gesture Analysis Applied to Discourse Analysis and Natural Conversation *International Journal of Computer Science and Information Technology* **5** pp 1–20

IOP Conf. Series: Journal of Physics: Conf. Series **893** (2017) 012059 doi:10.1088/1742-6596/893/1/012059

Data mining for clustering naming of the village at Java Island

Atje Setiawan Abdullah[1], Budi Nurani Ruchjana[2], Akik Hidayat[1], Akmal[1], and Deni Setiana[1]

[1] Department of Computer Science, Faculty of Mathematics and Natural Sciences Universitas Padjadjaran, Indonesia
[2] Department of Mathematics, Faculty of Mathematics and Natural Sciences Universitas Padjadjaran, Indonesia

E-mail: atjesetiawan@gmail.com

Abstract. Clustering of query based data mining to identify the meaning of the naming of the village in Java island, done by exploring the database village with three categories namely: prefix in the naming of the village, syllables contained in the naming of the village, and full word naming of the village which is actually used. While syllables contained in the naming of the village are classified by the behaviour of the culture and character of each province that describes the business, feelings, circumstances, places, nature, respect, plants, fruits, and animals. Sources of data used for the clustering of the naming of the village on the island of Java was obtained from Geospatial Information Agency (BIG) in the form of a complete village name data with the coordinates in six provinces in Java, which is arranged in a hierarchy of provinces, districts / cities, districts and villages. The research method using KDD (Knowledge Discovery in Database) through the process of preprocessing, data mining and postprocessing to obtain knowledge. In this study, data mining applications to facilitate the search query based on the name of the village, using Java software. While the contours of a map is processed using ArcGIS software. The results of the research can give recommendations to stakeholders such as the Department of Tourism to describe the meaning of the classification of naming the village according to the character in each province at Java island.

1. Introduction

How the ancestors in Java provides the naming of the village? Is there a difference in giving naming villages in each province in Java (West Java, Banten, Jakarta, Central Java, East Java, DIY)? How can we explain the naming of most villages by the prefix, the syllables are contained, and complete words, for every province in Java, and how its contribution to the naming of the village in Indonesia? This problem can be solved using ethnoscience approach [3].

Ethnoscience studied at culture with a scientific perspective. Ethnoscience helps to understand how people develop with different forms of knowledge and beliefs, and focuses on the ecological and historical contributions people have been given [4]. Ethnoscience is a new term and study that came into anthropological theory in the 1960s. Often referred to as "indigenous knowledge," Ethnoscience introduces a perspective based on native perceptions. It is based on a complete emic perspective, which excludes all observations, interpretations and or any personal notions belonging to the ethnographer.

The taxonomy and clustering of indigenous systems, to name a few, used to categorize plants, animals, religion and life is adapted from a linguistic analysis. The concept of Native Science is also related to the understanding the role of the environment intertwined with the meaning humans place upon their lives. Understanding the language and the native peoples linguistic system is one method to understand a native peoples system of knowledge of organization.

In the year 2016, the Ministry of Tourism Indonesia again proposed a Pesona Indonesia or Wonderful Indonesia as a country branding and slogan tourist destinations in the country [5]. As we mention before the name of villages in Indonesia, is one of interested topic to be researched, because we can use multidicipline of sciences to learn about the pattern of villages naming based on large database for many purposes. Not only for tourism, but also for supporting a toponym as the general name for any place or geographical entity. Related, more specific types of toponym include hydronym for a body of water and oronym for a mountain or hill. Place names provide the most useful geographical reference system in the world. A toponymist relies not only on maps and local histories, but interviews with local residents to determine names with established local usage. Usually the naming village can be categorized by lexical, mythos, eventual, leadership and religion.

In this research we propose the spatial data mining using Knowledge Discovery in Database (KDD) method for mapping and description of village's naming database in Java island at Indonesia via ethno-informatics approach [2]. Ethno-informatics is the discipline which it has the objective to study the basic law of information activities of culture supported by information technology. To get the good result in research of ethno-informatics, we should have relationship with other discipline, such as history, art, computer science, sociology, culture also statistics.

The special objectives of this research of visualization of villages naming for supporting tourism and indigenous culture in Indonesia can be noticed following:

(i) To get clustering of database villages naming in Java island using structure:
- prefix naming of the village
- initial syllable of villages naming
- syllables contained in the naming of the village
- complete word naming the village

In order to obtain the highest village naming information, both on the island of Java, as well as in each of the provinces in Java, West Java, Center of Java, East Java, DIY, Banten and DKI Jakarta.

(ii) To cluster database of villages naming of the village at Java island based on 16 categories that implies the following: the place, the environment, nature, feelings, herbs, vegetables, animals, poultry, fish, fruits, respect, state, color, creed, directions, and business.

(iii) To visualize database naming the village on the island of Java into map locations, to obtain picture of the position group or naming a particular village on a map of the island of Java.

(iv) To illustrate the database of villages naming at Java island in Venn diagram, so it can be shown:
- The names of villages that characterizes the island of Java and in each province in Java
- Intersection or influence the naming of the village between the provinces in Java, so it can be known to the provinces where the percentage rate of the high similarit
- Developing theorems or mathematical models associated with intersection of nearest neighbors at each province in Java island.

(v) To develop an application program based on query for villages naming based query.

(vi) To develop a data mining application for the villages naming which integrated manner, in order to facilitate the user in understanding the meaning of the villages naming of the village on the island of Java and in each province in Java.

2. Research Method

The method of this research based on KDD method. The first step, we build of the study design, utilizing the science of the latter being developed, known as data mining, which is defined to extract knowledge automatically from large databases, to obtain the patterns interesting, thus forming a knowledge [6]. The KDD method can be described as follow:

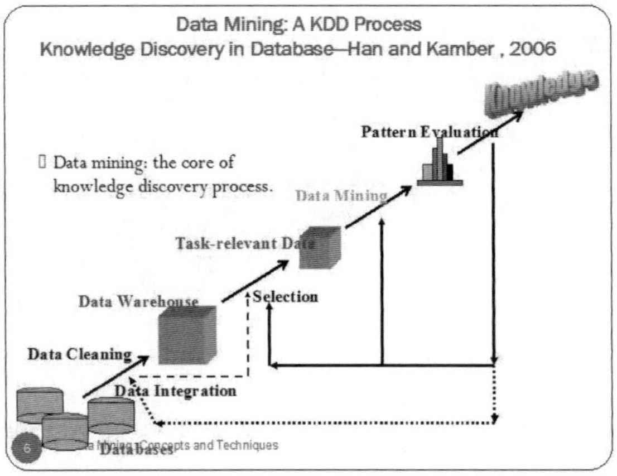

Figure 1. KDD Method in Data Mining [6]

Data mining process includes data preprocessing, data mining, and postprocessing. Data preprocessing steps including a data cleaning, data transformation, data integration and data selection, the process of data mining including the use of models to process data, while the postprocessing including processing visualization and interpretation of results, and conclusion about the result to becomes a knowledge. In this research we used large database database village from Geographic Information Bureau of Indonesia and then we build a software to create information for village's naming based on meaning of preposition, word and a whole word via descriptive statistics with supported by graphically and multivariate analysis. The research methodology for naming village can be built as Figure 2.

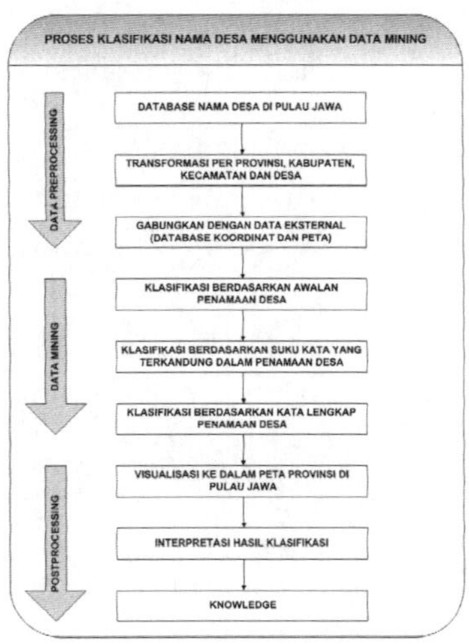

Figure 2. Research Methodology KDD Process [6]

IOP Conf. Series: Journal of Physics: Conf. Series **893** (2017) 012059 doi:10.1088/1742-6596/893/1/012059

In this research, we built Java-based application program, to do the search results of the naming village at Java island .

3. Main Result

We have the recapitulation of database of village naming at Java island as following Table 1.

Table 1. Recapitulaton of Village Database at Java Island

NO	PROVINCES	NUMBER of VILLAGE	PRECENTAGE
1	JABAR	5760	23.17
2	BANTEN	1534	6.17
3	DKI	267	1.07
4	JATENG	8355	33.60
5	JATIM	8509	34.22
6	DIY	438	1.76
	TOTAL PULAU JAWA	24863	100.00

Furthermore we built the software application for searching the whole word of village naming as Figure 3.

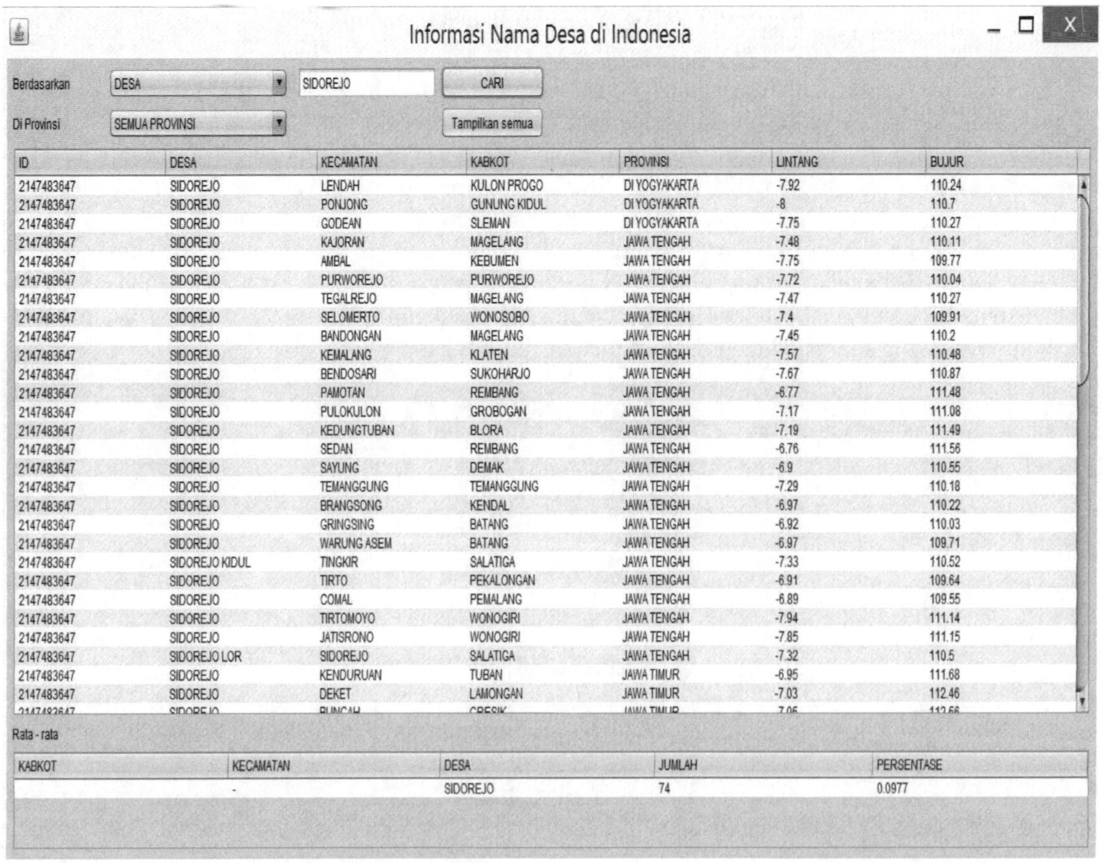

Figure 3. Application Search the Whole Name of Village Naming

In our application, we determine the prefix of village name at Java island to be Ci, Pa and Su and we got the result how many percentage of the name with three of prefix at each province at Java island. We explain the result by Figure 4.

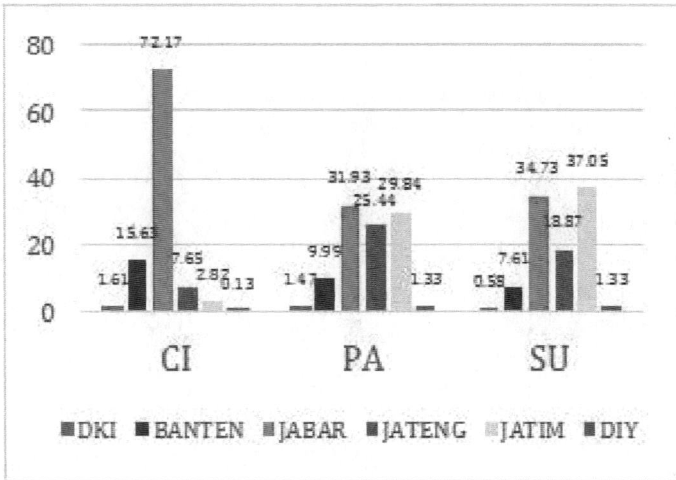

Figure 4. Example Sample Search for Prefix

We can describe that the prefix Ci means water, Pa is a tool for to do something and Su is beautiful. But, sometimes the meaning is confuse, so we need to extend the prefix to be a word and a whole word. So, we can have the table of recapitulation the village naming at Java island as following Table II and III. We divide the Table to be two sub Table, Table II and Table III. We choose 16 word from the village name, with the meaning for the place, the environment, nature, feelings, herbs, vegetables, animals, poultry, fish, fruits, respect, state, color, creed, directions, and business. Table II consist of recapitulation 16 word at Provinces: Yogyakarta, Central of java and East Java. While the Table 3 including Provinces: West java, Jakarta, Total of Java Island and National for all of informaton at Indonesia.

Table 2. Recapitulation of Word in 16 Categories at Java Island

NO	WORDS	LABEL	DIY	JATENG	JATIM
1	SARI	USAHA	26	593	375
2	KARANG	ALAM	16	379	213
3	JAYA	USAHA	1	24	18
4	MULYO/MULYA	PENGHORMATAN	21	163	110
5	KALI	ALAM	6	257	122
6	JATI	TUMBUHAN	6	143	115
7	TIMUR/WETAN	ARAH	1	124	151
8	BARAT/KULON	ARAH	1	114	151
9	LOR/UTARA/KALER	ARAH	2	126	118
10	AGUNG	PENGHORMATAN	7	74	135
11	TANJUNG	ALAM	1	59	55
12	GUNUNG	ALAM	1	71	54
13	TEGAL	ALAM	4	71	47
14	AWI/PRING/BAMBU	TUMBUHAN	5	50	31

IOP Conf. Series: Journal of Physics: Conf. Series **893** (2017) 012059 doi:10.1088/1742-6596/893/1/012059

Continue of Table III

Table 3. Recapitulation of Word in 16 Categories at Java Island

NO	WORDS	LABEL	JABAR	BANTEN	DKI	NATIONAL	JAVA ISLAND
1	SARI	USAHA	546	25	4	2298	1569
2	KARANG	ALAM	157	20	1	1062	786
3	JAYA	USAHA	376	87	3	1543	509
4	MULYO/MULYA	PENGHORMATAN	178	19	1	881	492
5	KALI	ALAM	58	4	6	590	453
6	JATI	TUMBUHAN	126	9	7	492	406
7	TIMUR/WETAN	ARAH	81	24	12	742	393
8	BARAT/KULON	ARAH	76	21	12	720	375
9	LOR/UTARA/KALER	ARAH	80	11	19	678	356
10	AGUNG	PENGHORMATAN	16	13	3	569	248
11	TANJUNG	ALAM	82	13	4	1182	214
12	GUNUNG	ALAM	64	18	3	556	211
13	TEGAL	ALAM	42	12	2	225	178
14	AWI/PRING/BAMBU	TUMBUHAN	57	9	6	361	158

Furthermore, we gave the recapitulation of village naming in the whole word for each province.

Table 4. Village Naming at Each Province using A Full Name

NO	WORD OF NAME VILLAGE	CLUSTERING	NATIONAL	JABAR	BANTEN	JATENG	DIY	DKI	JATIM	JAVA ISLAND	LABEL
1	KARANGANYAR	ALAM	108	18	0	40	0	0	19	77	JABAR,JATENG,JATIM
2	MEKARSARI	USAHA	110	43	15	2	0	0	0	60	JABAR,BANTEN,JATENG
3	SUKOREJO	PERASAAN	62	0	0	18	0	0	40	58	JATENG,JATIM
4	SIDOREJO	KEADAAN	78	0	0	23	3	0	30	56	JATENG,DIY,JATIM
5	TANJUNGSARI	ALAM	83	25	2	16	0	0	11	54	JABAR,BANTEN,JATENG,JATIM
6	BANJARSARI	TEMPAT	67	8	6	21	1	0	18	54	JABAR,BANTEN,JATENG,DIY,JATIM
7	KARANGSARI	ALAM	64	14	1	24	2	0	6	47	JABAR,BANTEN,JATENG,DIY,JATIM
8	WONOREJO	PERASAAN	60	0	0	16	0	0	29	45	JATENG,JATIM
9	MEKARJAYA	USAHA	102	37	6	0	0	0	0	43	JABAR,BANTEN
10	REJOSARI	PERASAAN	64	0	0	29	1	0	12	42	JATENG,DIY,JATIM
11	KARANGREJO	ALAM	57	0	1	16	0	0	23	40	BANTEN,JATENG,JATIM
12	SUKAMULYA	PENGHORMATAN	65	35	3	0	0	0	0	38	JABAR,BANTEN
13	SUMBERAGUNG	PENGHORMATAN	70	0	0	10	2	0	26	38	JATENG,DIY,JATIM
14	SUKAMAJU	USAHA	117	32	5	0	0	0	0	37	JABAR,BANTEN
15	GUNUNGSARI	ALAM	63	11	3	9	1	0	11	35	JABAR,BANTEN,JATENG,DIY,JATIM
16	BABAKAN	TEMPAT	33	21	4	7	0	0	1	33	JABAR,BANTEN,JATENG,JATIM
17	JATISARI	TUMBUHAN	36	10	0	10	0	0	13	33	JABAR,JATENG,JATIM
18	WONOSARI	PERASAAN	54	0	0	16	1	0	14	31	JATENG,DIY,JATIM
19	SUKASARI	PERASAAN	36	25	5	0	0	0	0	30	JABAR,BANTEN
20	NEGLASARI	TEMPAT	33	28	2	0	0	0	0	30	JABAR,BANTEN

Finally we also gave the comparison of how many data that we have for Recapitulation of a Full Name of naming village at West Java if we compare with the data in Indonesia at Table V.

Table 5. The Comparison of Full Name of Village Naming at West java Province and Indonesia (National)

NO	NAME OF VILLAGE	CLUSTERING	NATIONAL	JABAR	BANTEN	JATENG	DIY	DKI	JATIM	JAVA ISLAND	LABEL
1	KARANGANYAR	ALAM	108	18	0	40	0	0	19	77	JABAR,JATENG,JATIM
2	BABAKAN	TEMPAT	71	51	9	7	0	0	1	68	JABAR,BANTEN,JATENG,JATIM
3	MEKARSARI	USAHA	110	43	15	2	0	0	0	60	JABAR,BANTEN,JATENG
4	SUKOREJO	PERASAAN	62	0	0	18	0	0	40	58	JATENG,JATIM
5	SIDOREJO	KEADAAN	78	0	0	23	3	0	30	56	JATENG,DIY,JATIM
6	SUMBERAGUNG	PENGHORMATAN	70	0	0	10	2	0	26	38	JATENG,DIY,JATIM
7	JATISARI	TUMBUHAN	36	10	0	10	0	0	13	33	JABAR,JATENG,JATIM
8	BLIMBING	BUAH2AN	30	0	0	12	0	0	18	30	JATENG,JATIM
9	BAKUNG	BUNGA	43	2	2	4	0	0	7	15	JABAR,BANTEN,JATENG,JATIM
10	CIBODAS	WARNA	20	15	5	0	0	0	0	15	JABAR,BANTEN
11	CIBADAK	BINATANG	14	11	3	0	0	0	0	14	JABAR,BANTEN

Furthermore, we describe the result of village naming using map in Java island. So, we have the difference mapping of each criteria at Java island map. For example we can build the map of village naming of Karangsari a blue one and Wonorejo in a yellow. We can get the map that the village naming of Karangsari is at West Java-Center and also East Java, while the naming village of Wonorejo just only at Center and East Java.

IOP Conf. Series: Journal of Physics: Conf. Series 893 (2017) 012059 doi:10.1088/1742-6596/893/1/012059

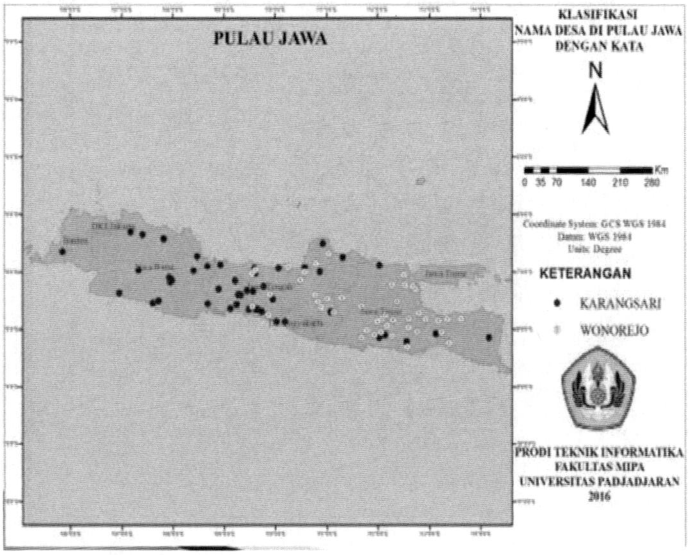

Figure 5. Map of Village Naming of Karangsari and Wonorejo at Java Island

We can continue to make a map of village naming of Mekarsari and Sukorejo and get the result as Figure 6. The map show us that the vllage naming of Mekarsari just only in West java, while the name of Sukorejo can be found at Center and East Java.

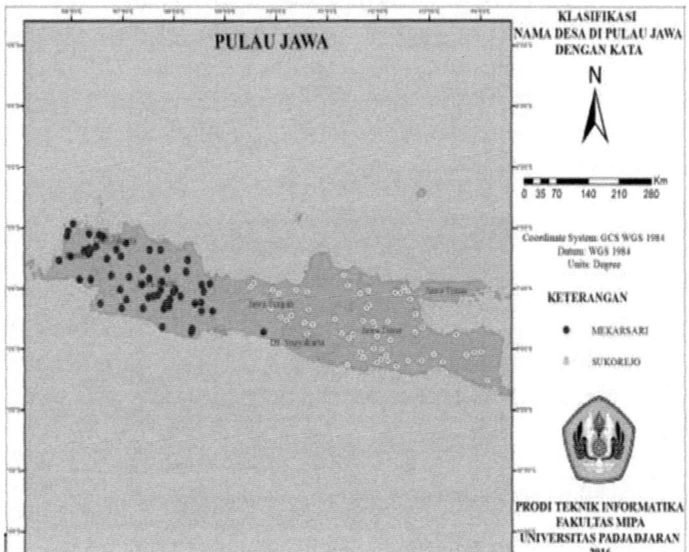

Figure 6. Map of Village Naming of Mekarsari and Sukorejo at Java Island

Furthermore to illustrate the similarity of database of villages naming at Java island we can use the Venn diagram. For example we can choose the similarity of village naming at West Jav-Center Java and East java as following Figure.

Using the Venn diagram above, we can describe the intersection of village naming between three provinces are: beringin, Bandar and Lingga. From this phenomena, finally we can continue

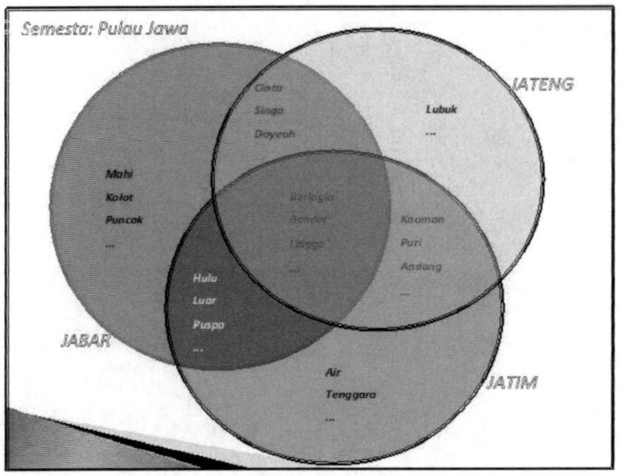

Figure 7. Venn Diagram of intersection of Villages Naming in Java Island

a mathematical work to get a theorem about the intersection of village naming in Java island.

4. Conclusion

(i) Village naming illustrate cultural behavior that is very important in daily life, past, present and future.

(ii) Community character Java can be described following criteria: has a soul strong and courageous, loving the country and the beauty of the environment, respecting local culture, having the ability of fisheries and livestock, ready to wander to another place, soulless strong and steadfast in the face of life, loves fruits local product, loved the beauty of flowers local products, can express the state under various conditions, thinking and acting well and enjoy the existing conditions, working hard to make a good situation, appreciate its leader, appreciate culture, behave religious, has a subtle feeling, love beauty, and well-behaved to anyone, loving home to live a life in peace, have a strong soul, use the plants as a supporting activity of life, be clean, and courageous, open and able to communicate with other cultures both locally and internationally.

Acknowledgment

The work reported in this paper is supported by the Academic Leadership Grant Unpad 2016, so the authors thank to Rector Universitas Padjadjaran and Directorate Research and Community Services Unpad for supporting funding of research and paper dissemination.

References

[1] Abdullah A S et al. 2015a Ethnomathematics in Cipatujah West Java, presented at LEAD Leiden Universiteit, the Netherlands

[2] Abdullah A S 2015b Data Mining for Villages Naming at West Java at National Seminar on Ethnomathematics at UHAMKA Jakarta

[3] Ahimsa-Putra H S 1985 Etnosains dan Etnometodologi: Sebuah Perbandingan dalam Masyarakat Indonesia Majalah Ilmu-ilmu Sosial Indonesia Jakarta: Lembaga Ilmu Pengetahuan Indonesia Jilid XII Nomor 2 pp 103–133

[4] Atran S 1991 Social Science Information / Sur Les Sciences Sociales. Ethnoscience Today **30**(4) pp 595–662

[5] http://pesona.indonesia.travel/overview/detail/6/indonesia-sekilas

[6] Jiawei H, Kamber M, and Jian P 2012 Data Mining Concepts and Technique, Morgan Kaufmann Publishers, USA

On the design of henon and logistic map-based random number generator

Magfirawaty[1,2], **Suryadi.M.T**[3], **Kalamullah Ramli**[1]

[1] Dept Electrical of Engineering, Universitas Indonesia, Depok 16424, Indonesia
[2] Sekolah Tinggi Sandi Negara, Bogor, Indonesia
[3] Dept of Mathemathics, Universitas Indonesia, Depok 16424, Indonesia

E-mail: magfirawaty@ui.ac.id, yadi.mt@sci.ui.ac.id, k.ramli@ee.ui.ac.id

Abstract. The key sequence is one of the main elements in the cryptosystem. True Random Number Generators (TRNG) method is one of the approaches to generating the key sequence. The randomness source of the TRNG divided into three main groups, i.e. electrical noise based, jitter based and chaos based. The chaos based utilizes a non-linear dynamic system (continuous time or discrete time) as an entropy source. In this study, a new design of TRNG based on discrete time chaotic system is proposed, which is then simulated in LabVIEW. The principle of the design consists of combining 2D and 1D chaotic systems. A mathematical model is implemented for numerical simulations. We used comparator process as a harvester method to obtain the series of random bits. Without any post processing, the proposed design generated random bit sequence with high entropy value and passed all NIST 800.22 statistical tests.

1. Introduction

In modern cryptography, the cryptosystem is a pair of algorithms that uses a key to convert plain text to ciphertext and vice versa. In cryptosystem, the algorithm is usually public and in this case the security depends on the secrecy key. Good key is a series of random bit which is generated automatically [1]. The randomness of key plays a significant role to ensures robustness of cryptosystem.

True Random Number Generator (TRNG) is a device that generates a random number which is unpredictable based on physical process and non-deterministic events [2]. TRNG consists of three main components, namely, source entropy that generates randomness, harvest process to get a randomness samplers, and post processing to minimize bias and produce random bit sequences that pass a statistical test. Furthermore, we can classify TRNG into three main classes according to entropy sources, (1) based on electrical noise, (2) jitter based, (3) chaos based.

The pioneer observation method of TRNG used the odd-even criterion with a Poisson distribution acquired by calculating random pulses from an electrical noise source [3]. Currently, the most significant concept is using ring oscillators (RO) to obtain TRNG with high throughput. RO uses the jitter approaches to generate a random signal [4, 5]. Jitter-based TRNG perform better than noise based, which is sensitive to external interference and easy to integrate with digital circuits. In addition to jitter and noise based-TRNG, recent papers used chaotic systems as a method for developing chaos based TRNG [6, 7] to increase the randomness. Chaotic is defined in a deterministic term, which is very sensitive to the changes initial conditions. TRNGs based on chaos use chaotic signals as an entropy source. Chaos-based TRNG can be applied based on continuous time (CT) and discrete time (DT). DT chaotic systems are more efficient than that of CT chaotic systems. Which it can be realized using fewer of resources, thus resulting a more efficient block of cryptosystem [6, 8–11].

In this study, the design of random number generator based on DT chaotic system is proposed. The design used a combination of two chaotic systems, i.e, Henon map and logistic map, to produce random bit sequence with high entropy value, simulated using LabVIEW. Randomness performance is assessed through entropy value and NIST 80.22 statistical test. This paper is structured as follows : Section II illustrates random number generator model by combining two-dimentional and one-dimentional chaotic systems. A threshold value is obtained using mathematical model, which is compared with the output of chaotic system to produces random bit. The randomness of the proposed method is analyzed and discussed in Section III. The last Section is a conclusion of the experiment conducted.

2. Random Number Generator Model

Discrete-time chaotic systems utilize a periodic iteration of a non-linear function, f, to accomplished the chaotic condition. Fig. 1 shows general concept of DT chaotic system which consists of chaos map $f(x)$ that depends on controlling parameters (λ) that guarantee chaotic regime of the system.

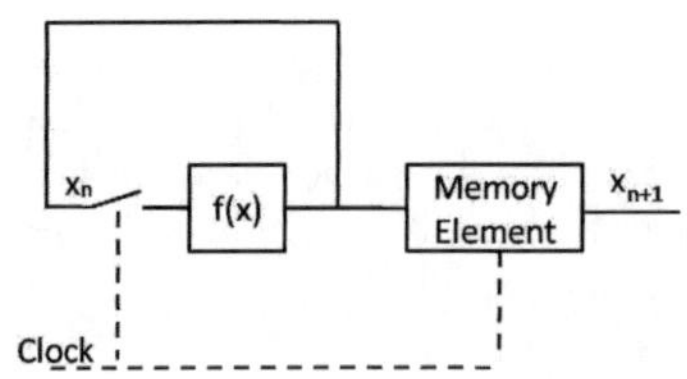

Figure 1: General concept of discrete time chaotic system [12]

$$x_{n+1} = f_{(\lambda, x_n)} \tag{1}$$

Chaotic systems are divided into several dimensions, including one- and two-dimensions. They are different due to their complexity, probability, and density of the involved function. The basic idea of our random number generator design is presented in Fig. 2. The aim of this design is to generate high entropy random bits by combining two chaotic systems, in which its main components consist of a sampling clock, sample-hold circuit, non-linear function blocks and comparator. The sample and hold block produces and forming the chaotic dynamics and the entropy core. The function blocks implement chaotic maps to increase entropy value, which is primarily limited by the Lyapunov exponent. In this study, we use Henon map and logistic map as chaotic map functions. Then, the output of last chaotic map function is compared with the threshold value to generate a bitstream.

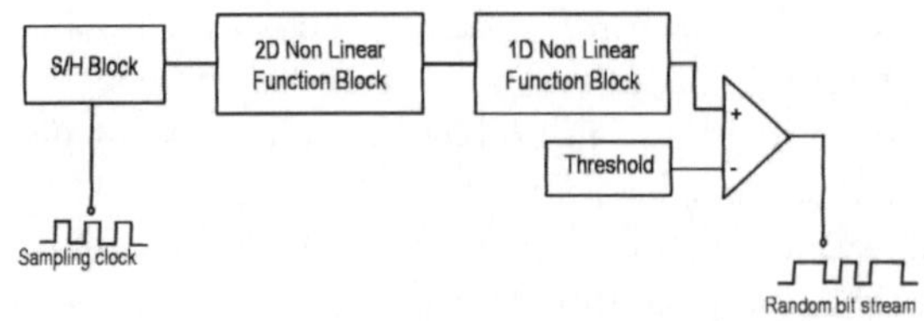

Figure 2: Proposed design

IOP Conf. Series: Journal of Physics: Conf. Series **893** (2017) 012060 doi:10.1088/1742-6596/893/1/012060

In mathematical terms, Eqn. (2) shows Henon map as a two-dimensional non-linear function block. It is a discrete time dynamical system that simplifies Lorenz system and Poincare map. The Henon equations can be expressed as

$$\begin{aligned} &(x, y) \in R^2 \\ &x_{n+1} = 1 - \alpha x_{n^2} + y_n \\ &y_{n+1} = \beta x_n \end{aligned} \tag{2}$$

in which α and β are the chaos controlling parameters of non-linear function block, where n is 0, 1, 2, etc. The initial are values symbolized by (x_0, y_0). The evolution samples of Henon map is shown in Fig. 3.

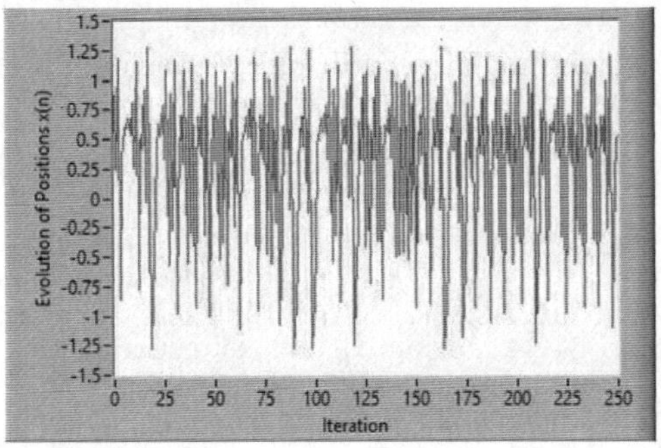

Figure 3: Evolution 250 samples of Henon map

Choosing an arbitrary initial state (which is assumed to be determined by the real time clock of hardware) is the early stages of this work, Henon map as a first non-linear function block on each iteration generates a new value relying on the disposition of chaotic samples. The chaotic behavior of Henon map is controlled by parameter values (α, β), in which $\alpha = 1.4$ and $\beta = 0.3$. For example, starting from an initial value $x_0 = 0.063752$ and $y_0 = 0.098551$, the Henon map chaotic trajectory for 8 time iterations are x_{n+1} = {0.063752, 1.092861, -0.65296, 0.730963, 0.056082, 1.214886, -1.0495, -0.17757}. Fig. 4 shows the outline of a two-dimensional sphere for the Henon map after 10000 iterations starting from initial point (0.1, 0.1). Furthermore, the values of x_{n+1} are normalized in range [0,1].

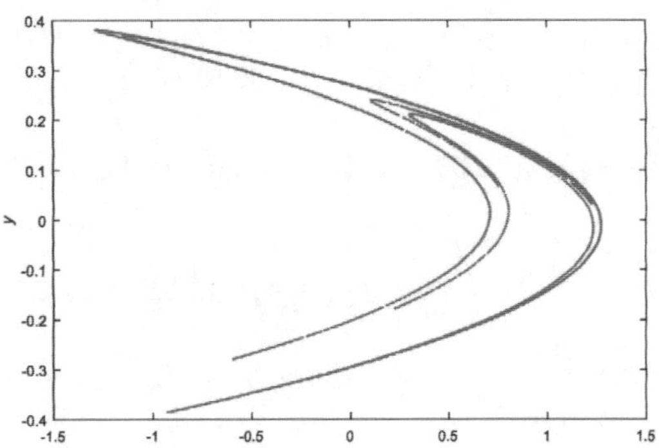

Figure 4: Henon map attractor after 10000 iterations

The second non-linear function block uses the logistic map to increase the entropy value and randomness of the output that is generated by the system. Logistic map is expressed as

$$X_{n+1} = \gamma X_n(1 - X_n) \tag{3}$$

where γ is the control parameter setting. The output value of the normalized Henon map serve as an initial value for the non-linear function block second (X_n). Then, the threshold comparator system can be expressed as

$$b_n = B(X_n) = \begin{cases} 0, & 0 \leq X_n \leq T_h \\ 1, & T_h < X_n \leq 1 \end{cases} \tag{4}$$

The randomness performance of the DT chaos based TRNG model depends on two critical parameters namely chaos controlling parameter γ and bit generation threshold T_h. We used method [7] to calculate the optimum threshold for independent bit generation. A pair of bits (b_n, b_{n+1}) generated form last function block. Their joint probability as the product of marginal probabilities, having the form of

$$\begin{aligned} P_{ij} &= P(b_n = i, b_{n+1} = j) \\ &= P(b_n = i)P(b_{n+1} = j) \ for\ i, j \in 0, 1 \end{aligned} \tag{5}$$

The probabilities of marginal bit generation can be obtained from

$$P_0 = P(0 \leq X_n \leq T_h) = \int_0^{T_h} f_X(X)dX = T_h \tag{6}$$

$$P_1 = P(T_h \leq X_n \leq 1) = 1 - P_0 = 1 - T_h \tag{7}$$

The threshold parameter is $T_h \leq \frac{2}{\gamma}$ and $X_n = \frac{2}{\gamma}$, the joint probability P_{00} as

$$\begin{aligned} P_{00} &= P(b_n = 0, b_{n+1} = 0) \\ &= P(X_n \leq T_h, X_{n+1} \leq T_h) \end{aligned} \tag{8}$$

by additivity,

$$= P(X_n \leq T_h, X_{n+1} \leq T_h, X_n \leq \frac{2}{\gamma})$$
$$+ P(X_n \leq T_h, X_{n+1} \leq T_h, X_n > \frac{2}{\gamma})$$

and by utilizing X_{n+1} from Eqn. 3

$$= P(X_n \leq T_h, \gamma X_n(1 - X_n) \leq T_h, X_n \leq \frac{2}{\gamma})$$
$$+ P(X_n \leq T_h, \gamma X_n(1 - X_n) \leq T_h, X_n > \frac{2}{\gamma})$$

$$= \int_0^{2T_h/\gamma} f_\infty(u)du = \frac{2T_h}{\gamma} \tag{9}$$

By combining P_0 as marginal probability that defined by Eqn.6 and equating to zero for $\gamma = 4$, the optimum value of T_h is expressed as

$$\begin{aligned} \Psi_{00} = P_{00} - P_0P_0 = \frac{2T_h}{\gamma} - T_h^2|_{\gamma=4} = 0 \\ \Rightarrow T_h = 0.5 \end{aligned} \tag{10}$$

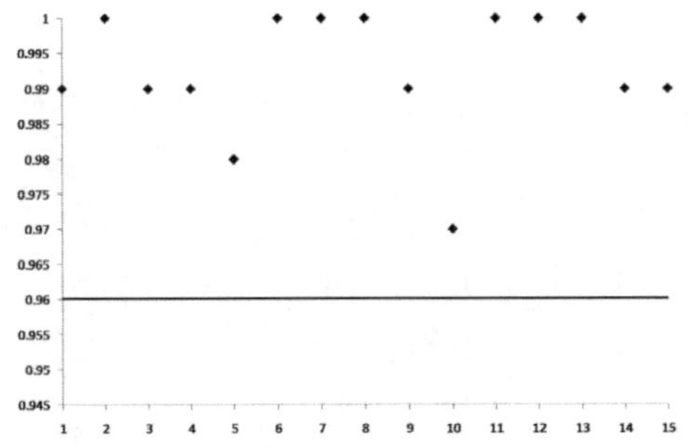

(a) The proportion of sequences that passes each statistical test

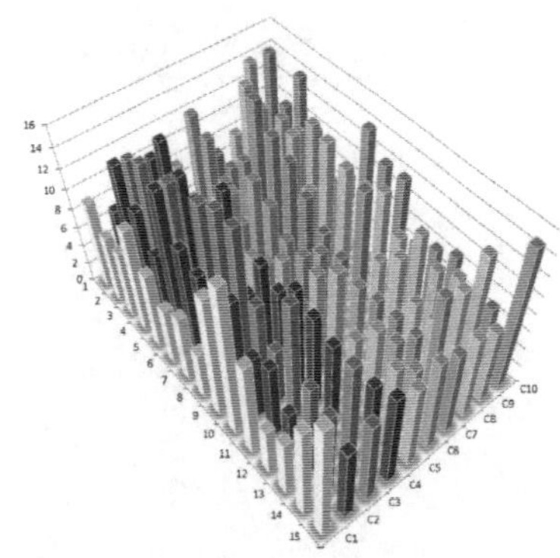

(b) The distribution of P-value for each test

Figure 5: Two approaches adopted by NIST [13]

3. Analysis and Discussion

3.1. Entropy Analysis

Theoretically, the proposed TRNG design is capable of producing a sequence which is distributed independently and unpredictable. Therefore, it is necessary to calculate the entropy value of the generated bit. The entropy in Shannon theory is a measure of uncertainty of the value of the next generated bit. According to Shannon theory, the entropy value of number series can be obtained by

$$H = -\sum_{i=0}^{1} P_i log_2 P_i \tag{11}$$

where P_i is the probability of generating a value of zero or one. Ideally, the probabilities P_0 and P_1 are equal to $\frac{1}{2}$ resulting in an entropy of 1, which is the maximum value of the entropy function. Simulation result of the proposed design obtained bit sequence, which the information entropy was approximately 0.9998. The obtained value was very close to a maximum value of the entropy function.

3.2. Statistical Analysis

The output of proposed design has been analyzed statistically. Final report file of NIST 800-22 is used to explain the randomness of bit output. Table 1 lists the NIST statistical testing results. These tests detect the non-randomness of output sequences resulted from pseudo random or random number generators. Two approaches recommended by NIST that we adopted. First, the testing of the proportion of binary output sequences that pass statistical test, check the uniformity of distribution of P-values is second approach.

The confidence interval is used to determine the acceptable proportions range. For a significance level of $\alpha = 0,01$, 100-bit sequences whith 10^6 length of each sequence, the confidence interval was equal to, 0.99 ± 0.02985. To pass each statistical test, the proportion of sequences should be above 0.96015. Fig. 5a shows the proportion value of each statistical test is greater than 0.96015. It proves with the output bit sequence is random. NIST second approach is to ensure that uniformity of P-values distribution interval range [0,1] is devided into 10 sub-intervals. P-value determine the uniformity of bit sequence. Applying a χ^2 and an additional

Table 1: NIST 800.22 Test Result

Statistical Test	Proportion	P-Value
Frequency	0.99	0.699313
Block Frequency	1	0.514124
Cumulative Sums	0.99	0.759756
Runs	0.99	0.474986
Longest Run	0.98	0.115387
Rank	1	0.534146
FFT	1	0.224821
Non Overlapping Template	1	0.474986
Overlapping Template	0.99	0.911413
Universal	0.97	0.249284
Approximate Entropy	1	0.935716
Random Excursions	1	0.602458
Random Excursions Variant	1	0.378138
Serial	0.99	0.595549
Linear Complexity	0.99	0.867692

process, we acquire a new P-value (P_T). If $P_T \geq 0.01$, then the sequences can be considered as uniformly distributed. As shown in Table 1, the value of the P_T for all statistical tests was greater than 0.01, and hence the proposed design is uniformly distributed. Furthermore, histogram can be used to visualize the distribution of p-value, as shown in the Fig. 5b.

4. Conclusion

We designed a TRNG combining two chaotic systems to generate random bit sequence with hight entropy value. We used Henon map and Logistic map as chaotic systems. We utilized a discrete time chaotic system to generate bit sequence which is distributed independently and unpredictable.The proposed design was simulated in LabVIEW. The output randomness was influenced by chaos parameter and threshold value.

The proposed design is a simple and an excellent method for generating random bits, which the information entropy was very close to a maximum value of the entropy function, and pass all NIST 800-22 statistical tests. Our design has not been implemented on hardware, and our future experiment will be performed it on FPGA board.

Acknowledgment

This research is partly funded by PITTA Grant of Universitas Indonesia, under contract number 2129/UN.R12/HKP.05.00/2016.

References

[1] Schneier B 1996 *John Wiley and Sons, NY (Oct. 1995)*
[2] Schindler W and Çetin Kaya K 2009 *About Cryptographic Engineering* (New York: Springer) ISBN 9780387718163
[3] Maddocks R, Bucci M and Luzzi R 1972 *journal of physics E: scientific instruments* **5 (6)** 542–544
[4] Sunar B, Martin W J and Stinson D R 2007 *IEEE Transactions on Computers* **56** 109–119

[5] Wold K and Tan C H 2009 *International Journal of Reconfigurable Computing* **2009**
[6] Cicek I, Pusane A E and Dundar G 2013 A novel dual entropy core true random number generator *8th International Conference on Electrical and Electronics Engineering (ELECO)* (IEEE) pp 332–335 ISBN 6050105049
[7] Cicek I, Pusane A E and Dundar G 2014 *the VLSI Journal Integration* **47** 38–47
[8] Cicek I and Dundar G 2013 A chaos based integrated jitter booster circuit for true random number generators *European Conference on Circuit Theory and Design (ECCTD)* (IEEE) pp 1–4
[9] Katz O, Ramon A and Wagner I 2008 *VLSI System, IEEE Transanctions on* **16** 1677–1686
[10] Pareschi F, Setti G and Rovatti R 2006 A fast chaos-based true random number generator for cryptographic applications *Proceedings of the 32nd European Solid-State Circuits Conference (ESSCIRC)* (Montreux: IEEE) pp 130–133 ISBN 1424403030
[11] Stojanovski T, Phil J and Kocarev L 2001 *IEEE Transactions on Circuits and Systems I: Fundamental Theory and Applications* **48** 382–385
[12] Aguilar Angulo J A, Kussener E, Barthelemy H and Duval B 2014 Discrete chaos-based Random Number Generator *IEEE Faible Tension Faible Consommation (FTFC)* (IEEE) pp 1–4
[13] Rukhin A, Soto J, Nechvatal J, Barker E, Leigh S, Levenson M, Banks D, Heckert A, Dray J and Vo S 2010

IOP Conf. Series: Journal of Physics: Conf. Series **893** (2017) 012061 doi:10.1088/1742-6596/893/1/012061

Differential evolution-simulated annealing for multiple sequence alignment

R C Addawe[1], J M Addawe[1], M R K Sueño[2], and J C Magadia[3]

[1] Department of Mathematics and Computer Science
University of the Philippines Baguio, Philippines
[2] College of Science, University of the Philippines Baguio, Philippines
[3] School of Statistics, University of the Philippines Diliman, Philippines

E-mail: {rcaddawe, jmaddawe, mksueno, jcmagadia}@up.edu.ph

Abstract. Multiple sequence alignments (MSA) are used in the analysis of molecular evolution and sequence structure relationships. In this paper, a hybrid algorithm, Differential Evolution - Simulated Annealing (DESA) is applied in optimizing multiple sequence alignments (MSAs) based on structural information, non-gaps percentage and totally conserved columns. DESA is a robust algorithm characterized by self-organization, mutation, crossover, and SA-like selection scheme of the strategy parameters. Here, the MSA problem is treated as a multi-objective optimization problem of the hybrid evolutionary algorithm, DESA. Thus, we name the algorithm as DESA-MSA. Simulated sequences and alignments were generated to evaluate the accuracy and efficiency of DESA-MSA using different indel sizes, sequence lengths, deletion rates and insertion rates. The proposed hybrid algorithm obtained acceptable solutions particularly for the MSA problem evaluated based on the three objectives.

1. Introduction

Multiple sequence alignments (MSAs) are strategies in molecular biology for homology transfer, phylogenic analyses, structural modeling, functional predictions and protein structure and interaction predictions[1, 2, 3]. The development novel experimental techniques such as the next-generation sequencing and high-throughput experiments motivated the need for MSA tools that extract biological meanings from the massive information generated by novel experimental techniques such as the next-generation sequencing and high-throughput experiments. Efficient MSA tools are capable of dealing massive information by using advanced computational approaches based on artificial intelligence and machine learning algorithms. Differential Evolution (DE) [4, 5], developed by Ken Price in 1994, is a simple population-based stochastic function minimizer which is very powerful. It basically adds the weighted difference between two population vectors to a third vector and uses a way to separate probability distributions to make a completely self-organizing algorithm. The development of DE and its improvements included hybridization of DE and its applications [6]. Simulated annealing (SA) [7, 8], on the other hand, is a probabilistic method for finding the global minimum among several local minima of an objective function. SA is based on an analogy with a physical phenomenon of the annealing process of metals [9].

In this paper, we applied the DESA algorithm [10] with the MSA problem based on structural information [11], non-gaps percentage and totally conserved columns. In the next section we briefly discuss the algorithms and multiple sequence alignment problem. The next sections present the results and conclusions.

2. Related Methodologies

In this section, DE, SA, DESA, the MSA problem, and the objective functions are summarized as the foundation to construct the proposed hybrid algorithm to solve the multiple sequence alignment problem.

2.1. Differential Evolution (DE)

Differential Evolution (DE) [4] is a robust global optimization tool for NP-hard and complex optimization problems. It has been developed with several working variants. In the *DE/rand/1/bin* variant, we minimize the objective function, $f(X) : \Re^n \to \Re$ by optimizing the values of $\mathbf{X} = (x_1, \ldots, x_D); \mathbf{X} \in \Re^D$, where $\mathbf{X}$ is a vector of D parameters subject to lower and upper constraints, $x^{(L)}$ and $x^{(U)}$, such that $x_j^{(L)} \leq x_j \leq x_j^{(U)}$, where $j = 1, 2, ..., D$. A potential solution under population i and generation G is denoted by $\mathbf{X}_{i,G}$, such that, $P_G = x_{i,G} = x_{j,i,G}$, where $i = 1, 2, \ldots, NP$. For a starting solution, let $P_G = 0 : P_0 = X_{i,j,0} = r_{i,j}(x_j^{(U)} - x_j^{(L)}) + x_j^{(L)}$ where r is a uniformly distributed random value in [0,1] for each j. The next population P_{G+1} is created from the current population P_G. A trial population for the next generation, P'_{G+1}, is $x'_{i,j,G+1} = x_{Ci,j,G} + F(x_{Ai,j,G} - x_{Bi,j,G})$ if $(r_{i,j} \leq CR) \vee (j = D_i)$; $x_{i,j,G}$ otherwise. $A_i \neq B_i \neq C_i \neq i = 1, 2, \ldots, NP$, $CR \in [0, 1]$, $F \in [0, 2]$, and $r \in [0, 1]$ is a uniformly distributed random value. Then new random values of A, B, C are assigned for each i and a new random number r is assigned for each j. D is a randomly chosen chromosome of each vector $\mathbf{X}'_{G+1}$ different from each counterpart in the previous generation, $\mathbf{X}_G$. A new random integer is assigned to D for i. Hence, from the initial population P_G and the temporary population P'_{G+1}, the next generation P_{G+1} is $X_{i,G+1} = X'_{i,G}$ if $f_{cost}(X'_{i,G+1} \leq f_{cost}(X_{i,G})$; $X_{i,G}$ otherwise.

2.2. Simulated Annealing (SA)

Simulated Annealing (SA) [8, 9] starts with a set of random values of the parameters and a temperature value T. For each variation, this set of parameters is randomly varied proportional to T, and the objective function, $f(X)$, is calculated. The change in $f(X)$ may be denoted by ΔE. If the set of parameters results in a better value of $f(X)$, the parameters are retained and forms the basis of the next variation of the potential parameter values. SA allows a deterioration in the objective function with a known probability, given by Boltzmann distribution: $P(\Delta E) = e^{(\frac{-\Delta E}{T})}$, where $P(\Delta E)$ is a probability distribution giving the probability of making an energy increasing step having magnitude ΔE. This can be done by drawing a random number in $[0, 1]$ and checking if it is less than $P(\Delta E)$; returning true, if so. Then, the value of T is decreased by an acceptably small amount and the whole set of iterations is repeated. SA surely obtains the global minimum $f(X)$. SA surely obtains the global minimum of $f(X)$, after a number of iterations and when T is sufficiently small.

2.3. Differential Evolution - Simulated Annealing (DESA)

Differential Evolution - Simulated Annealing (DESA) [10, 12] is characterized by *self-organization, mutation, crossover*, and *SA-like selection* scheme of the strategy parameters. From the current population P_G and the trial population P'_{G+1}, the population of the next generation P_{G+1} is $P_{i,G+1} = P'_{i,G}$ if $R(\Omega'_{i,G+1}) \leq R(\Omega_{i,G})$ or $e^{-[\frac{R(\Omega'_{i,G+1}) - R(\Omega_{i,G})}{T_{reduct}}]} > rand[0, 1]$; $P_{i,G}$ otherwise,

where Ω is the parameter vector and R is the fitness function. In this SA-like selection scheme of DESA, a trial vector is compared only against one individual in the current population; and that besides accepting improvements in cost, it also accepts deteriorations in cost of the objective function.

2.4. The Multiple Sequence Alignment (MSA) Problem

In a multiple sequence alignment, the idea is to maximize the number of similar residues in a given column of the alignment. An alignment can be easy or difficult depending on the degree of relationship among the sequences. More related sequences are easier to align than less related ones since insertions and deletions of gaps are greatly needed in aligning these kinds of sequence [13]. We use DESA to compute sequence alignment using the structural information as the scoring scheme. To test the performance of DESA-MSA, numerical simulations were carried out on data sets obtained from benchmark alignment data (BAliBASE) in deciding the appropriate parameters for the proposed algorithm DESA.

Given an MSA A, the score of A is the sum of the scores of each target sequence i when they are aligned with a sequence of known structure called the template. The score of all the target sequences, $Score(Target)$, normalized by the total number of contacts $|c|$ within the template is given by:

$$Score(Target) = \frac{\sum_{ij} STRIKE(c_i c_j) \times IsContact(c_i c_j)}{|c|} \tag{1}$$

with

$$IsContact(c_i c_j) = \begin{cases} 1, \; if \; c_i c_j \; are \; in \; contact \\ 0, \; else \end{cases}$$

The STRIKE score is denoted by $STRIKE(c_i c_j)$ which is the total score of the contact between c_i and c_j given by the STRIKE matrix. Therefore, the score of MSA A, where N is the number of sequences is given by:

$$Score(A) = \sum_{i}^{N} Score(Target_i) \tag{2}$$

The Non-Gaps Percentage (NGP) g is the number of gaps in the alignment, aa is the number of amino acids in the alignment; such that the total number of gaps and amino acids in the alignment equal to $g + aa$. The Percentage of Totally Conserved Columns, is denoted as $TCP = \frac{c}{l} * 100\%$, where c is the number of columns aligned with exactly the same amino acid and l is the length of the matrix.

3. Results

To find the appropriate parameters for DESA-MSA, two sets of experiments were done on five datasets obtained from BAliBASE. First, we optimized the three objective functions on STRIKE, NGP and TCP scores. Next, we maximize the STRIKE scores only to improve structural similarities between the aligned sequences. In addition, different combinations of the crossover probabilities, P_c and mutation probabilities, P_m, were investigated. The alignment with the highest STRIKE score, which gives an alignment of more quality according to the sequence structures is considered as the best alignment. The average STRIKE scores, NGP scores, TCP scores, and computer runtimes are shown in Fig. 1 to 4. A graphical comparison between the average STRIKE scores of the best alignments obtained from the two sets of simulations using different combinations of P_c and P_m is shown in Fig. 1.

The 30%-70% combination, which obtained the highest average STRIKE score of 2.2011, is just about 2% higher than the worst average STRIKE score obtained from the 20%-80% combination which is 2.1569. However, the scores from the different $P_c - P_m$ combination are not significantly different from the average STRIKE score of 2.1835, with standard deviation of

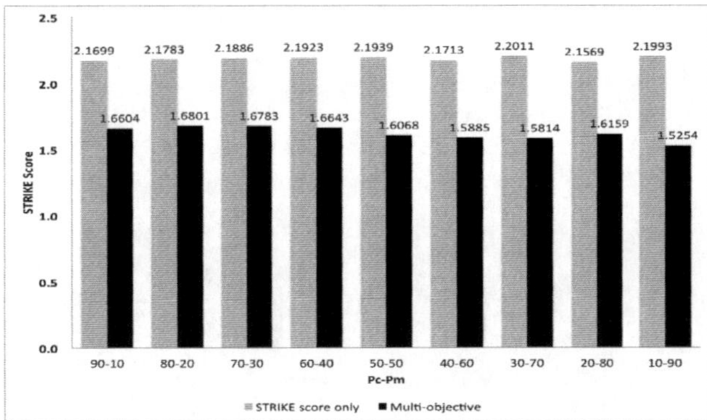

Figure 1. Average values of the STRIKE scores of the best alignments obtained from the two sets of simulations

0.152. This imply that the STRIKE score cannot be improved further by the different combination of $P_c - P_m$. On the other hand, the STRIKE scores obtained by optimizing the three objective functions on SRIKE, NGP and TCP scores had an average of 1.6224 with standard deviation 0.9238. These STRIKE scores are significantly lower than the average score of 2.1835 obtained by maximizing the SRIKE score only. However, maximizing the three objectives and using $P_c - P_m$ of 90%-10%, 80%-20%, 70%-30%, or 60%-40%, obtained significantly higher STRIKE scores with $80\% - 20\%$ combination being best. Thus, SRIKE scores may be improved considering these $P_c - P_m$ combinations. However, including TCP and NGP scores with the STRIKE scores in maximizing alignment resulted to about 25% decreased in the STRIKE scores.

For the average percentage of non-gaps (NGP) obtained from the two sets of simulations presented in Fig. 2, the 80%-20% combination yielded the highest which is 63.9912% followed by the 90%-10% combination. The NGP scores with a mean of 62.9820 and standard deviation of 0.5654 obtained by maximizing NGP, TCP and STRIKE scores are significantly higher than the NGP scores with mean of 61.1745 with a standard deviation of 0.9238, obtained by maximizing STRIKE score only. Thus, higher NGP scores may be obtained when using either 90%-10% or 80%-20% combination in optimizing the function with the NGP, TCP and STRIKE scores.

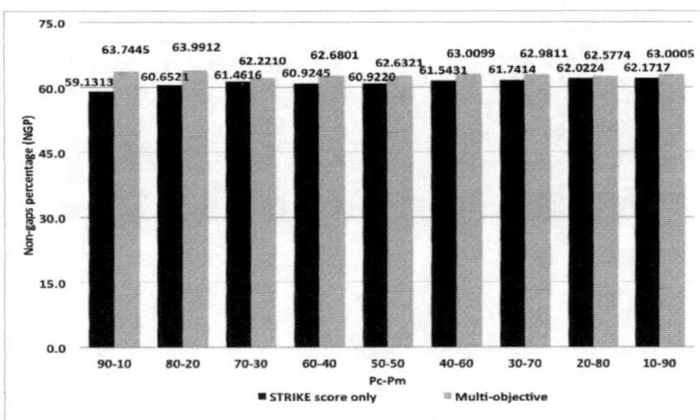

Figure 2. Average values of the NGPs of the best alignments obtained from the two sets of simulations

The average values of the percentages of totally conserved columns (TCP) obtained from the

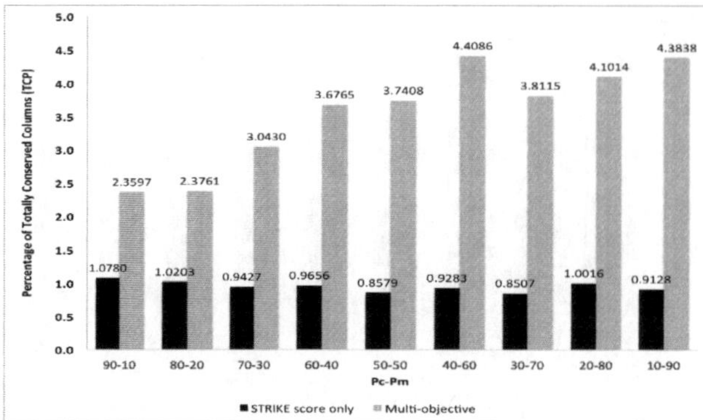

Figure 3. Average values of the TCPs of the best alignments obtained from the two sets of simulations

two sets of simulations are also presented in Fig. 3. The 40%-60% and 10%-90% combinations gave the highest average percentages with the 40%-60% combination yielding better average TCP equal to 4.4086% which only differs by 0.0248% from the 10% - 90% combination. These TCP scores using 40%-60% and 10% - 90% $P_c - P_m$ combinations are significantly higher than the rest of the combinations used.

On the average, TCP scores obtained by maximizing the three objectives with an average of 3.55446% and a standard deviation of 0.7833, is significantly higher than the TCP scores obtained by maximizing the STRIKE score only having an average of 0.9508% and a standard deviation of 0.0745. The average difference of 2.5937 on the TCP scores contributed to an improvement of about 273% on the TCP scores when the three objective functions were maximized. Hence, the optimizing the three objectives using 40%-60% or 10% - 90% combinations yielded an improved TCP scores.

For the runtime, data are presented in Fig. 4 where simulations using the 90%-10% combination had the shortest average runtime of 0.7085 minutes.

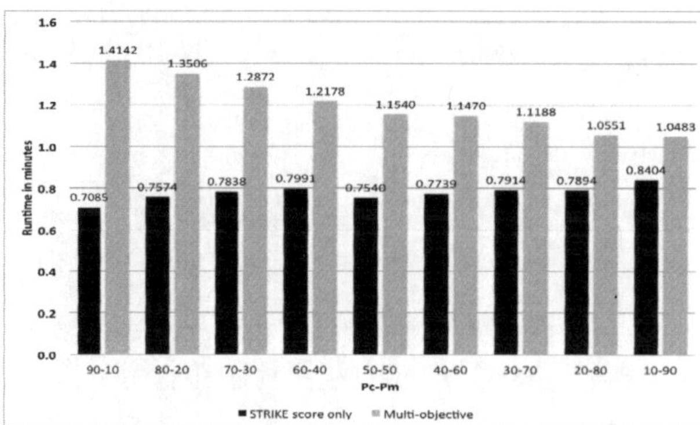

Figure 4. Average runtime (in minutes) of the best alignments obtained from the two sets of simulations

This runtime average is significantly lower than the average runtimes obtained by the rest of the $P_c - P_m$ combinations with mean of 0.7776 minutes and a standard deviation of 0.0363 minutes. On the other hand, optimizing the three objectives obtained significantly longer average runtimes. In addition, simulations maximizing only the STRIKE score are twice faster than the simulations maximizing the three objectives. Nevertheless, in maximizing the three objectives, simulations using the 20%-80% combination had the shortest average runtime of 1.0551 minutes followed by the 10% - 90% combination with an average runtime of 1.0483 minutes, with 20%-80% combination being the fastest. These two runtime averages obtained from 10%-90% and 20% - 80% are significantly lower than the average runtimes obtained by the rest of the $P_c - P_m$ combinations with mean of 1.992 minutes and a standard deviation of 0.1285. In addition, simulations maximizing only the STRIKE score are twice faster than the simulations maximizing the three objectives.

Results of the simulations obtained from DESA-MSA using multi-objective functions have shown that NGP and TCP scores may be improved by using different $P_c - P_m$ combinations. However, as the NGP and TCP scores the STRIKE score may be decreased and the average runtime is increased. Higher NGP scores may be obtained when using either 90%-10% or 80%-20% combination in optimizing the function with the NGP, TCP and STRIKE scores. The TCP scores were improved up to 273% when using $P_c - Pm$ combination of either 40%-60% or 10% - 90%. As expected, simulations in maximizing the three objectives, yielded longer runtime than the simulations using STRIKE score only. Based on the data presented, the 80%-20% combination ranked first in the average NGP score computations but obtained fairly low TCP scores. The 40%-60% combination ranked first in the average TCP score computations and third in the NGP score computations, respectively. In particular, the 40%-60% combination is more appropriate than the 80%-20%. Results show that the experiments on maximizing the STRIKE, NGP and TCP together with the 40%-60% combination obtained higher NGP and TCP values; but lower STRIKE score and longer runtime than maximizing only the STRIKE score. Though the simulations in maximizing the three objectives obtained longer runtime of about 52% longer runtime and lower STRIKE score of about 32%, it provided an NGP and TCP scores better by 3% and 273%, respectively. The 80%-20% combination is an alternative, yielding an average of about 5.5% higher NGP score, 133% higher TCP score, 32% lower SKRIKE score and 75% longer runtime.

4. Conclusions and Recommendations

In this study, we proposed Differential Evolution - Simulated Annealing (DESA), a hybrid algorithm for optimizing multiple sequence alignments (MSAs) based on three objectives: (1) structural information (2) non-gaps percentage and (3) totally conserve columns. DESA algorithm obtained acceptable and even better solutions particularly for the MSA problem evaluated based on the three objectives. DESA-MSA, a hybrid algorithm for multiple sequence alignment has been shown to be efficient in achieving optimized alignments. This paper aimed to find the best parameter by testing varying combinations of crossover and mutation probabilities. Numerical experiments were performed on five datasets obtained from the Benchmark Alignment DATABASE (BAliBASE) to decide the appropriate parameters for DESA-MSA. Results obtained from the simulations have showed that the combination 40% crossover and 60% mutation gave alignments with the best qualities in terms of sequence structures and evolutionary homologies. Thus, in this study, the best combination of of parameters is the 40% crossover and 60% mutation probabilities.

Acknowledgments

The author RCA would like to thank Japan Mathematical Society and the University of the Philippines for the conference financial support.

References

[1] Doolittle R F 1981 *Science* **214** 149
[2] Fitch W 1966 *J. Mol. Biol.* **16** 9

IOP Conf. Series: Journal of Physics: Conf. Series **893** (2017) 012061 doi:10.1088/1742-6596/893/1/012061

[3] Bacon D and Anderson W 1986 *J. Mol. Biol.* **191** 153
[4] Storn R and Price K 1997 *J. Global Optim.* **11** 341
[5] Storn R and Price K 1997 *Dr. Dobb's Journal* **22** 18
[6] Das S and Suganthan P N 2011 *IEEE Transactions on Evolutionary Computation* **15** 4
[7] Kirkpatrick S, Gelatt C D and Vecchi M P 1983 *Science* **220** 671
[8] Cerny V 1985 *J. Optim. Theory Appl.* **45** 41
[9] Bertsimas D and Tsitsiklis J 1993 *Statistical Science* **8** 10
[10] Addawe R, Adorio E, Addawe J and Magadia J 2006 *DESA: A hybrid optimization algorithm for high dimensional functions, in Proc of the 8th IASTED Int Conf on Control and Applications* (ACTA Press) pp 316–321
[11] Sueno M and Addawe J 2016 *Advanced Studies in Biology* **8** 9
[12] Addawe R and Magadia J 2014 *Differential Evolution-Simulated Annealing (DESA) algorithm for fitting autoregressive models to data, Proc of OPT-i2014 - Int Conf on Engineering and Applied Sciences Optimization* vol eds. M. Papadrakakis et al pp 434–451
[13] Florido J, Ortuno F, Pomares H, Rojas F, Urguisza J and Valenzuela O 2013 *BMC Bioinformatics* **29** 2112

IOP Conf. Series: Journal of Physics: Conf. Series **893** (2017) 012062 doi:10.1088/1742-6596/893/1/012062

Super-resolution imaging applied to moving object tracking

Galandaru Swalaganata, Dwi Ratna Sulistyaningrum, and Budi Setiyono

Mathematics Department, Institute of Technology Sepuluh Nopember, Surabaya, Indonesia

E-mail: galandaru.swalaganata@gmail.com

Abstract. Moving object tracking in a video is a method used to detect and analyze changes that occur in an object that being observed. Visual quality and the precision of the tracked target are highly wished in modern tracking system. The fact that the tracked object does not always seem clear causes the tracking result less precise. The reasons are low quality video, system noise, small object, and other factors. In order to improve the precision of the tracked object especially for small object, we propose a two step solution that integrates a super-resolution technique into tracking approach. First step is super-resolution imaging applied into frame sequences. This step was done by cropping the frame in several frame or all of frame. Second step is tracking the result of super-resolution images. Super-resolution image is a technique to obtain high-resolution images from low-resolution images. In this research single frame super-resolution technique is proposed for tracking approach. Single frame super-resolution was a kind of super-resolution that it has the advantage of fast computation time. The method used for tracking is Camshift. The advantages of Camshift was simple calculation based on HSV color that use its histogram for some condition and color of the object varies. The computational complexity and large memory requirements required for the implementation of super-resolution and tracking were reduced and the precision of the tracked target was good. Experiment showed that integrate a super-resolution imaging into tracking technique can track the object precisely with various background, shape changes of the object, and in a good light conditions.

1. Introduction

Tracking of moving objects is a way used to detect and analyze changes that occur in an object from one frame to the next frame in the video [1]. The purpose of tracking a moving object is estimated trajectory of objects in the image plane is moved in a scene. Tracking moving objects can help to detect, identify, and analyze such objects. There are three basic steps in tracking a moving object. The first step is the detection or select Region of Interest (ROI) of a moving object in a video. The second step is tracking an object that has been selected from the first step of several parts of the frame. The third step is the analysis of movement of the object to identify the object's behavior [2].

The ability to obtain visual quality and the precision of the tracked target is highly wished in modern tracking system. The fact that the tracked object does not always seem clear causes the tracking result less precise. The reasons are low quality video, system noise, small object, and other factors. By utilizing the super-resolution images in tracking process are expected to get results tracking moving objects in video more precision. Super-resolution is a technique of constructing the high resolution images from low-resolution images. There are two types of super-resolution based on the input frame, namely single frame super-resolution and multi-frame super-resolution [3]. Super-resolution single frame has advantages in terms of computation time faster than multi-frame super-resolution [4]. The process of getting high-resolution image on a single frame super-resolution is raising the image of the target so that the process does not eliminate and still contains information prior to the super-resolution [5].

Earlier research on tracking of moving objects with the Kalman filter has been done under the title Moving Object Tracking Using Kalman Filter [6]. The study was concern on objects with varied backgrounds with a target single object. Another new study with title Moving Target Tracking Based on Camshift Approach and Kalman Filter [7] successfully combines Camshift and Kalman filter. From both these studies did not use a small object. Other studies with title Detection and Tracking of Very Small Objects Low Contrast [8] succeed to detect small objects using Wavelets for a single object and Kalman filter for multiple objects. Research by Olegs Mise and Toby P. Breckonb entitled Super-Resolution Imaging Applied to Moving Targets in High Dynamic Scenes [9] successfully applying super-resolution images to obtain the results of tracking moving objects. In [9] used multi-frame super-resolution and block matching method correlation for tracking.

Studies about a small object tracking is still slightly caused by very few of the features available. Therefore, by integrated super-resolution with tracking can be obtained more features that enable tracking a small object. There is a difficulty to detect relatively small objects that will be track. The problem would be too much noise appears and objects are relatively small because it would be difficult to distinguish between the object and the noise [8]. Another difficulty was the small object is represented by a set of pixels (between 10 and 100) on a high resolution image, so it is not enough information obtained or modeled correctly [10].

In this paper discuss the result of integrated super-resolution technique into tracking approach. First step is super-resolution imaging applied into frame sequences to improve visual quality of the tracked target. This step is done by cropping the frame in several frame or all of frame. Second step is track the result of super-resolution images. The overview of super-resolution method is described in Section 2. Our main concept is introduced in Section 3. Furthermore, in Section 4 we highlight the main concepts with results shown over a variety video.

2. Single Frame Super-Resolution

Super-resolution methods generally construct a high-resolution image from single or multiple low-resolution images. There are two types of super-resolution, namely single frame super-resolution and multi-frame super-resolution [3]. In general, super-resolution consists of two phases, namely the registration of the image and image reconstruction. In process of getting high-resolution image on a single frame technique is raising the size of the image $n\times$, where n is natural number, so that in the process of super-resolution does not eliminate and still contains information prior to the super-resolution. The method which is quite popular to perform single frame super-resolution is Nearest neighbor, Bilinear, and Bicubic [5].

In this study focused on single frame super-resolution. Bicubic Interpolation [5] and Directional Bicubic Interpolation [4] were proposed for single frame super-resolution. Fig. 1 shows $2\times$ zoom process of 4×4 LR (low resolution) image by Directional Bicubic Interpolation.

3. Tracking System Enhanced by Super-Resolution

In this section discuss integrated super-resolution framework into tracking system. First step is super-resolution imaging applied into frame sequences. This step wass done by cropping the frame in several frame or all of frame in order to improve visual quality of the tracked target. Second step is tracking the result of super-resolution images. Bicubic Interpolation [5] and Directional Bicubic Interpolation [4] were proposed for super-resolution step. Every

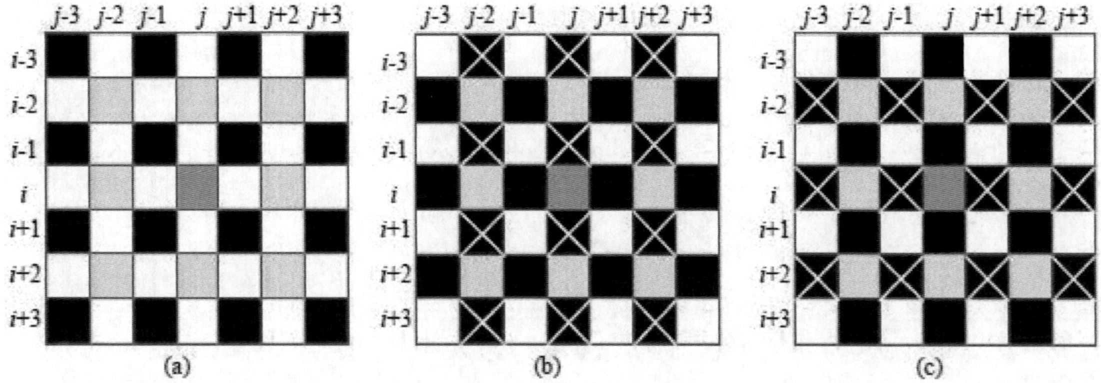

Figure 1. 2× zoom process of a LR image. The missing pixels are interpolated separately in two passes. The black pixels represent ones in the known LR image. The pixels marked with × are the interpolated ones in the first pass. The white and gray pixels represent the missing HR ones. (a) the 1st pass; (b) case 1 of the 2nd pass; (c) case 2 of the 2nd pass [4].

interpolation has their own advantages, so it will be compared for consequent after tracked. For some reason, the object changes the shape (zoom, rotate) and various color. In order to handle them, Camshift method was proposed for tracking approach. Fig. 2 shows block diagram of the tracking system integrated with super-resolution framework used in this work.

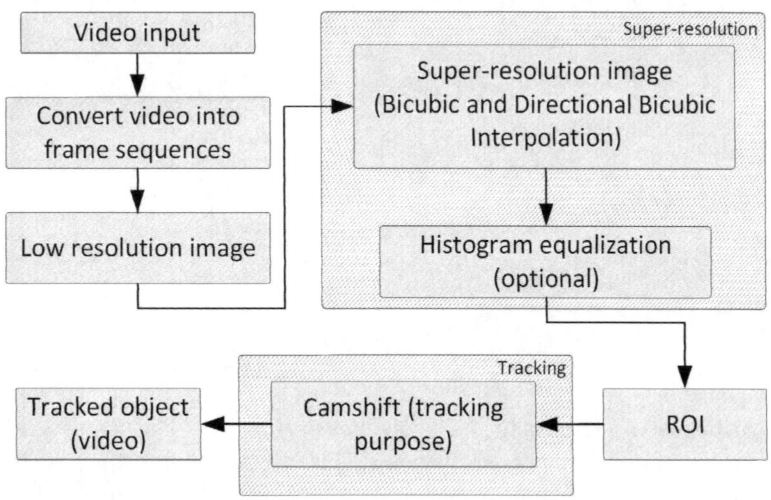

Figure 2. Tracking system with integrated super-resolution

There were two options to enhance tracking system with super-resolution, one of them was super-resolution first and then tracking the result as shown in Fig. 2. Here the chosen ROI (initialized by user) was as the first super-resolution image that being tracked. Single frame super-resolution was done by cropping the frame in several frames or all of frames. The puspose of cropping was create the search window area smaller. Therefore, it needs less computation time in searching the tracked target. An optional step was given before tracking, called histogram equalization. The purpose of equalize the histogram make a differences between forground (object) and background. It works by normalized the histogram ditribution.

3.1. Bicubic and Directional Bicubic Interpolation

Bicubic interpolation interpolates indiscriminately the missing pixels in the same direction (horizontal or vertical) and typically results in blurring, blocking, ringing or other artifacts in interpolated images. Directional Bicubic first explicitly decides the local edge direction in terms of the radio of the two orthogonal directional gradients for an interpolation pixel. If the pixel is on a strong edge, it will be interpolated using Bicubic along the edge direction; if the pixel is in a weak edge or texture region, it will be interpolated by combining the two orthogonal directional Bicubic interpolation values in direct proportion to the two orthogonal directional inverse gradients [4]. Computation time of the two interpolation is another difference. Bicubic obviously need less computation time than directional bicubic interpolation.

Bicubic interpolation is one of quite popular technique to perform single frame super-resolution [5]. Directional bicubic interpolation is new edge-directed image interpolation method based on 1-D Bicubic interpolation. Fig. 1 shows 2× zoom process of 4×4 LR (low resolution) with directional bicubic interpolation. Detailed description of the interpolation process of the method in the following [4].

Step 1: Initialization. This step expands the known $n \times n$ LR image I_l onto the grid of the $(2n-1) \times (2n-1)$ HR image I_h as shown in Fig. 1(a). The mapping relation is $I_h(2x-1, 2y-1) = I_l(x, y), x, y = 1, 2, ..., n$ where (x, y) is the pixel location.

Step 2: Interpolation in the first pass. The whole gray pixels in Fig. 1(a) are interpolated. The filled pixels are $I_h(2x, 2y), x, y = 1, 2, 3, ..., n$ (including border pixels). For every gray pixel, compute the $45°(G_1)$ and $135°(G_2)$ diagonal directional Bicubic interpolation values p_1 and p_2, the two diagonal directional gradients G_1 and G_2 using

$$G_1 = \sum_{m=3,\pm1} \sum_{n=3,\pm1} |I(i+m, j-n) - I(i+m-2, j-n+2)| \tag{1}$$

$$G_2 = \sum_{m=3,\pm1} \sum_{n=3,\pm1} |I(i+m, j+n) - I(i+m-2, j+n-2)| \tag{2}$$

Then the missing pixel p is estimated as

if $(1+G_1)/(1+G_2) > T$
  $p = p_2$;
else if $(1+G_2)/(1+G_1) > T$
  $p = p_1$;
else
  $p = (w_1p_1 + w_2p_2)/(w_1 + w_2)$
end

Step 3: Interpolation of the case 1 of the second pass. The whole gray pixels in Fig. 1(b) are interpolated. The filled pixels are $I_h(2x-1, 2y), x, y = 1, 2, 3, ..., n$ (including border pixels). For every gray pixel, compute the horizontal and vertical bicubic interpolation values p_1 and p_2, the horizontal and vertical gradients G_1 and G_2 using (1) and (2). Then the missing pixel p is estimated like Step 2.

Step 4: Interpolation of the case 2 of the second pass. The whole gray pixels in Fig. 1(c) are interpolated. The filled pixels are $I_h(2x, 2y-1), x, y = 1, 2, 3, ..., n$ (including border pixels). For every gray pixel, compute the horizontal and vertical bicubic interpolation values p_1 and p_2, the two horizontal and vertical gradients G_1 and G_2 using (1) and (2). Then the missing pixel p is estimated like Step 2.

3.2. Camshift (Continuously Adaptive Mean Shift)

The advantages of Camshift based on simple calculation based on HSV color that use its histogram for some condition and color of the object varies. Camshift method improved the searching efficiency of mean shift tracking algorithm. The coordinate of the best matching block to be the initial searching position at next continuous image is recorded [1] [7]. The procedure of Camshift algorithm is illustrated as in Fig. 3.

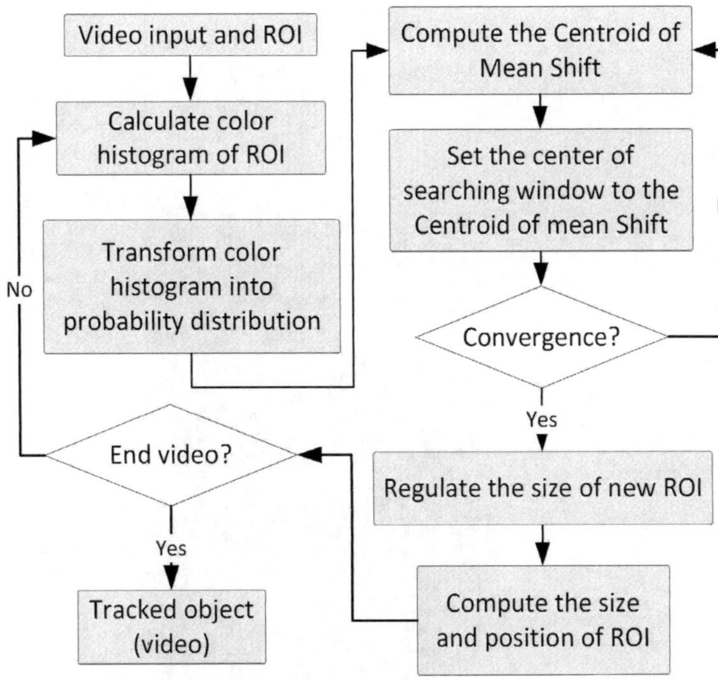

Figure 3. Camshift Method [7]

The centroid of color probability distribution image in the search window can be searched by this equation [1].

The 0th moment:

$$M_{00} = \sum_x \sum_y I(x, y) \tag{3}$$

The 1st moments for x and y :

$$M_{10} = \sum_x \sum_y xI(x, y) \tag{4}$$

$$M_{01} = \sum_x \sum_y yI(x, y) \tag{5}$$

The centroid of search window:

$$x_c = \frac{M_{10}}{M_{00}}; y_c = \frac{M_{01}}{M_{00}} \tag{6}$$

where $I(x, y)$ is a pixel color value at position (x, y) in the image and (x, y) are in the search window. M_{10} is the first moment for x, while M_{01} is the first moment for y.

Table 1. Dataset Video

	Name	Screenshot	Frame(s)
1.	Paralayang.avi	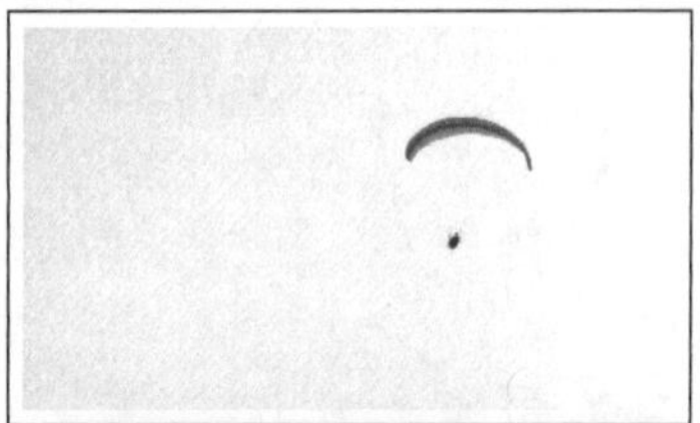	168
2.	bike.avi		151
3.	soccer.avi		76

4. Result and Discussion

In this section used Matlab application for implementation and 3 video for experiment (see Table 1). All of the video contain single moving object, small object, and various size. Experiment on this paper was conducted with extension ".avi" video which will be converted into frames sequences.

Every video in Table 1 has their own characteristics. Paralayang.avi video has size of 640×360 pixels and contains a small moving object and colour difference (between the object and background). Bike.avi video has size of 320×240 pixels and contains a small moving object and the moving camera. The last video, soccer.avi, has size of 856×480 pixels and contains more than one moving object.

4.1. Native Tracking without Super-resolution

Compares the racking results before and after the super-resolution is important to find out whether the proposed method will produce significant results. Thus, each video will tracked by Camshift method without super-resolution first.

Table 2 describes Paralayang.avi video when ROI has been initialized by user, the result of tracking a man was successfully tracked until frame 48 (28%). The object in frame 48 until end was not detected, the error was the color distribution did not read correctly. Bike.avi video succeed tracked a man in all of frames, but the rectangle for some frames was missed (55.6%). It happened because the moving camera and the object moving forward (zoom in)(see Fig.4). Bola.avi video successfully too in tracking a player in all of frames, but like previous video, the

Table 2. Tracking Result of Single Object

	Video	Native Tracking	SR* Tracking
1.	Paralayang.avi	47 frames	168 frames
2.	bike.avi	Missed ±67 frames	Missed ±12 frames
3.	soccer.avi	Missed ±23 frames	76 frames

*Bicubic and Directional Bicubic Interpolation gave same result

Figure 4. Missed rectangle in result of bike.avi tracking (frame 10, frame 112, and frame 149)

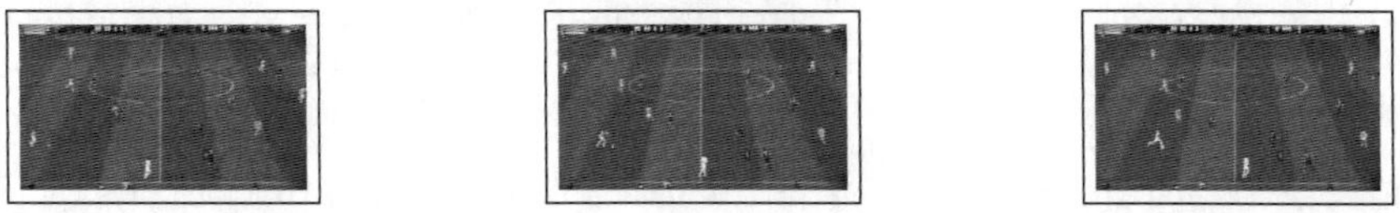

Figure 5. Missed rectangle in result of soccer.avi tracking (frame 57, frame 69, and frame 76)

rectangle for some frames was missed (70%). It was caused by a small object and the object moving too fast (see Fig.5).

4.2. Tracking Integrated with Super-resolution
Table 2 shows the tracking result of each video. As we can see, the result does not precise because of many reason depend on each video. Our proposed method proposed a solution to track the target object with integrated single frame super-resolution into tracking system. Directional Bicubic Interpolation is better than Bicubic Interpolation in high resolution image result. However, computation time Bicubic was better than Directional Bicubic. Table 2 also shows the result of tracking the target after single frame super-resolution.

Paralayang.avi and soccer.avi video can tracked successfully a single object in all frames without histogram equalization. While bike.avi video should be use histogram equalization, because the moving camera and the object moving forward (zoom in). Thus, histogram equalization is for make a differences between forground (object) and background. The rectangle can place in almost frames of bike.avi video. It missed ±12 frames of 151 frames after super-resolution.

Both interpolation, 2× Bicubic Interpolation and 2× Directional Bicubic Interpolation gave the **same result** for tracking approach. Our proposed method was resulting great impact with 100% successfully track first and third video, and 92.1% for second video. It means that integrated single frame super-resolution into tracking system can track the object precisely with various background, shape changes of the object, and good light conditions.

5. Conclution
In this paper we have presented an integration method beetwen single frame super-resolution technique and tracking system. This approach significantly improves the overall tracking system. The results show significant improvement in visual target object and the precision of object

tracking. The proposed method also works well in the various video size. Bicubic Interpolation and Directional Bicubic Interpolation gave the same result for tracking approach. The average of tracked target successfully was 51.2% for native tracking and 97.4% for our proposed method.

Acknowledgment
This research was supported by Ministry of Research, Technology and Higher Education of the Republic of Indonesia through the funding from Block Grant research. We thank our colleagues for encouraging us to present this paper on the International Conference.

References

[1] Chen X, Wu H, Li X, Luo X, and Qiu T 2012 Real-time Visual Object Tracking via Camshift-Based Robust Framework *International Journal of Fuzzy Systems* **14** p 262
[2] Yilmaz A, Javed O, and Shah M 2006 Object tracking: a survey *Acm computing surveys (CSUR)* **38** p 13
[3] Setiyono B, Hariadi M, and Purnomo M H 2012 Survey of Superresolution Using Phased Based Image Matching *Journal of Theoretical and Applied Information Technology* **43**
[4] Jing L, Zongliang G, and Xiuchang Z 2013 Directional Bicubic Interpolation-A New Method of Image Super-Resolution *Proceedings of International Conference on Multimedia Technology* (Atlantis Press) pp 470–477
[5] Raval H and Kothari A 2014 Analysis of Single Frame Super Resolution Methods *International Journal Of Engineering And Computer Science* **3** pp 9073–9076
[6] Patel A H and Thakore G D 2013 Moving Object Tracking Using Kalman Filter *International Journal of Computer Science and Mobile Computing* **2** pp 326–332
[7] Lien S, Hsia K, and Su J 2015 Moving Target Tracking Based on Camshift Approach and Kalman Filter *An International Journal Applied Mathematics and Information Sciences* **9** pp 395–401
[8] Daviesy D, Palmery P and Mirmehdiz M 1998 Detection and Tracking of Very Small Low Contrast Objects *Proceedings of the British Machine Vision Conference* (BMVA Press) pp 60.1–60.10
[9] Mise O and Breckonb T P 2013 Super-Resolution Imaging Applied to Moving Targets in High Dynamic Scenes *Emerging Technologies in Security and Defence; and Quantum Security II; and Unmanned Sensor Systems X* **8899**
[10] Lefevre s, Foissotte T and Vincent N 2003 Detection and Tracking of Small Objects by Details Removal *Proceedings of 4th European Workshop on Image Analysis for Multimedia Interactive Services* (University of London, Queen Mary) pp 137–140

IOP Conf. Series: Journal of Physics: Conf. Series **893** (2017) 012063 doi:10.1088/1742-6596/893/1/012063

Image processing based detection of lung cancer on CT scan images

Bariqi Abdillah, Alhadi Bustamam, and Devvi Sarwinda
Faculty of Mathematics and Science, Universitas Indonesia, Depok, Indonesia
E-mail: {bariqi.abdillah, alhadi, devvi}@sci.ui.ac.id

Abstract. In this paper, we implement and analyze the image processing method for detection of lung cancer. Image processing techniques are widely used in several medical problems for picture enhancement in the detection phase to support the early medical treatment. In this research we proposed a detection method of lung cancer based on image segmentation. Image segmentation is one of intermediate level in image processing. Marker control watershed and region growing approach are used to segment of CT scan image. Detection phases are followed by image enhancement using Gabor filter, image segmentation, and features extraction. From the experimental results, we found the effectiveness of our approach. The results show that the best approach for main features detection is watershed with masking method which has high accuracy and robust.

1. Introduction

Cancer is one of most dangerous disease that causes deaths [1]. Data obtained from Global Burden Cancer shows that in 2012 there are 14.1 million cases of cancer in the world, with lung cancer occupies the first position with a percentage of 13%. While the number of cancer deaths was recorded 8.2 million deaths, with lung cancer cause of death in first place with a percentage of 19% [2]. Lung cancer is a disease of abnormal cells multiplying and growing into a tumour. Cancer cells can spread out from the lungs through the blood stream or lymph fluid that surrounds the lung tissue. Generally, cancer cells often spread toward the center of the chest due to the natural flow of lymph. Metastasis occurs when cancer cells spread to other organs. The process of early detection of cancer plays an important role to prevent cancer cells from multiplying and spreading. Previous researches have been conducted for analyzing lung cancer such as using clustering method in microarray data [3], the detection of lung cancer with general image processing techniques in CT scan data with good results and accuracy [4]. In this study we proposed and evaluated additional image segmentation methods in analyzing lung cancer using image processing techniques.

Figure 1 shows a general description of lung cancer detection system that contains four basic stages. There are four stages to determine whether there is a lung cancer or not. The first phase is we get CT scan image data. The second phase, we implement image enhancement to improve quality of image. The third phase is image segmentation which is an important step in the detection of cancer. The fourth stage is feature extraction that is give us a conclusion whether there is a lung cancer or not.

2. Methodology

2.1 Image Acquisitions

The imagery used is from CT scan database. Our data is from the VIA and ELCAP Public Access

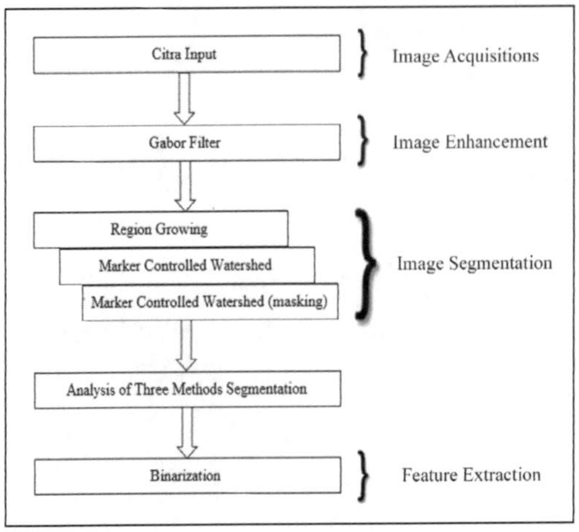

Figure 1. Image Processing for Lung Cancer Detection Stages

Database [5]. Image is in digital imaging and communications in medicine (dicom) format. There are 50 patients and each patient have about 250 pieces of images. Figure 2 shows example of CT scan image data.

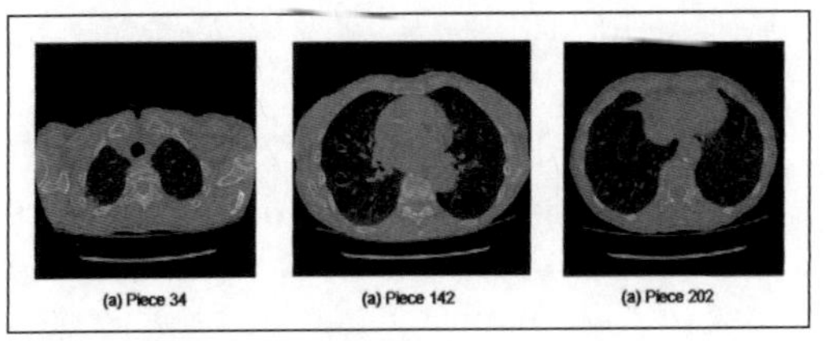

Figure 2. Example three pieces of a patient

2.2 Gabor Filter

Gabor filter named by Dennis Gabor, is a linear filter is used for edge detection. Representation of Garbor filter similar to the human visual system. In the spatial domain, 2D Gabor filter is a Gaussian filter function modulated by a sinusoidal function [4]. In the process of this cancer detection imagery used is a 2D image, so using 2D Gabor filter.

$$G(\sigma, \theta, \lambda, \psi, \gamma; x, y) = \exp^{\frac{-(x'^2+\gamma^2 y'^2)}{2\sigma^2}} \cdot cos(2\frac{x'}{\lambda} + \psi)$$

Figure 3 represents the input image and the results after enhanced by Gabor Filter.

2.3 Region Growing

Region growing is a procedure to extend the covered area by collecting pixels or sub-regions based on predetermined criteria. Basically, the approach is to initiating a set point (seed) and then expanding the area surrounding the seed which has the same properties as the seed, for example, the area that has the same range of color or gray level [6]. Figure 4 shows three stages of segmentation by region growing method.

2.4 Marker Controlled Watershed

There are two main approaches in segmentation, including edge approach and regional approach. Watershed segmentation method combining both of this approaches. Watershed method is a powerful

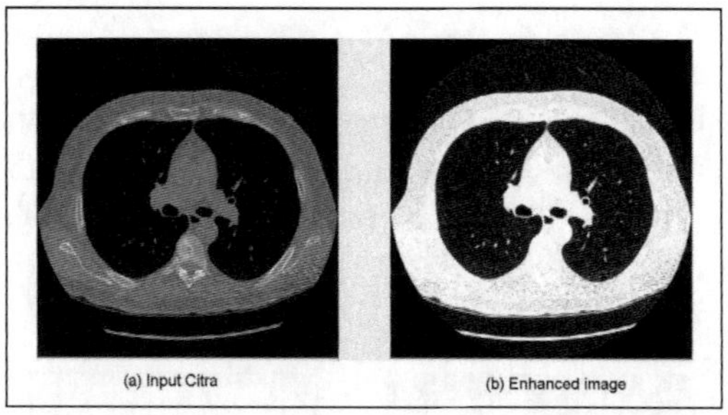

Figure 3. Image input and output by Gabor Filter

Figure 4. stages of region growing method

method to get fast segmentation results [7]. The basic idea of segmentation watershed transformation comes from geography, where an image is viewed as an earth surface (topography) with gray level as a measure surface height. Rainwater will flow and fill the lowest surface of catchment basins [6]. Figure 5 is the stages of watershed segmentation.

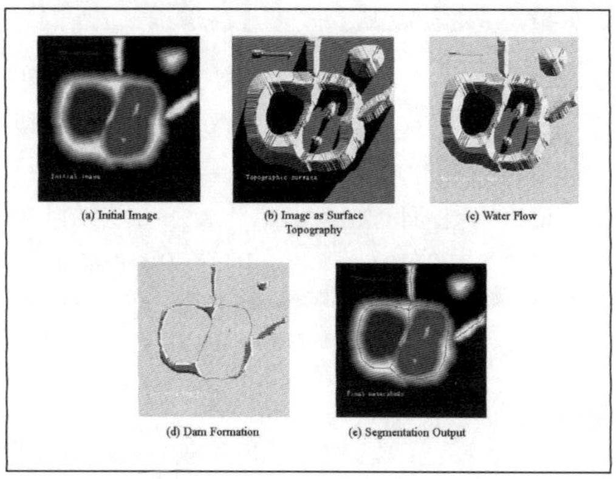

Figure 5. stages of watershed segmentation method

IOP Conf. Series: Journal of Physics: Conf. Series **893** (2017) 012063 doi:10.1088/1742-6596/893/1/012063

3. Results

3.1 Region Growing

The process to get the segmentation by region growing method is as the following; firstly, select the area that will be the target object, which are the right lung and the left lung, then, put the seed in this area. furthermore enlarge the size of the seed so that it covers all the desired areas. Figure 6 shows the result of segmentation by region growing method.

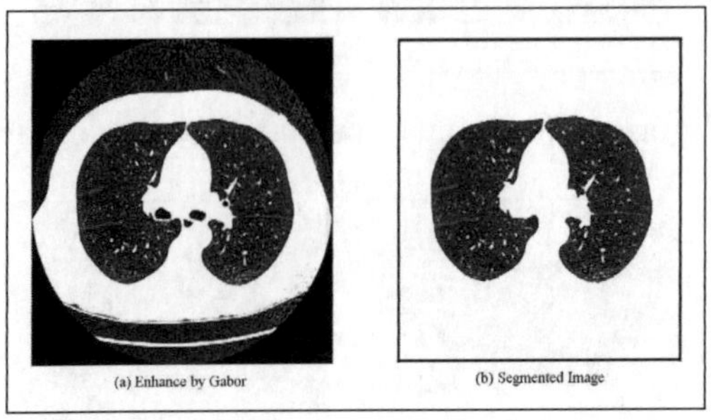

Figure 6. segmented by region growing method

3.2 Marker Controlled Watershed

To get the segmentation result of watershed method, the steps are as follows; firstly, calculate the distance gradient for edge detection, and then mark the target object by using a morphological technique called opening by reconstruction and closing by reconstruction. After marking object is revealed, other areas can be discarded. Modules and example of marker controlled watershed segmentation method is existing in Matlab software that can be directly used. Figure 7 shows the result of segmentation by using Marker-Controlled Watershed.

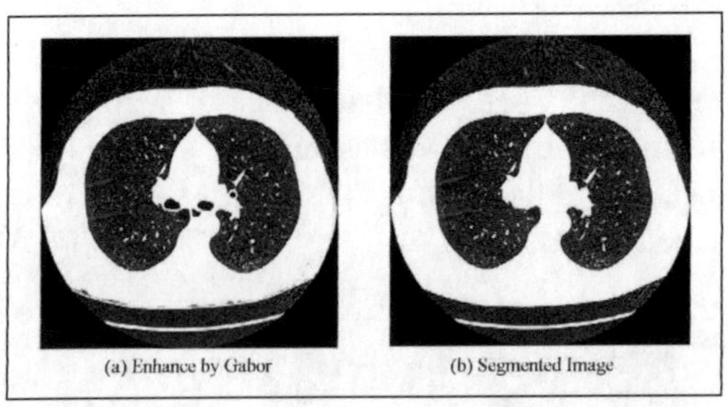

Figure 7. segmented by marker controlled watershed method

3.3 Marker Controlled Watershed with Masking

Segmentation by this method is the same process as the segmentation method at subsection 3.2 with additional masking in this process, which marks the area that contains the target object at the right and the left lung. Results of segmentation using Marker-Controlled Watershed with Masking is shown in Figure 8.

3.4 Three Methods Segmentation Analysis

The analysis of three segmentation method based on their success to find the segmentation results of the target object. Figure 6 shown two method have succesfully found the target object including (a) region growing and (c) marker controlled watershed with masking. meanwhile (b) marker controlled watershed segmentation has fail to get segmentation results, since there are background surrounding the target object.

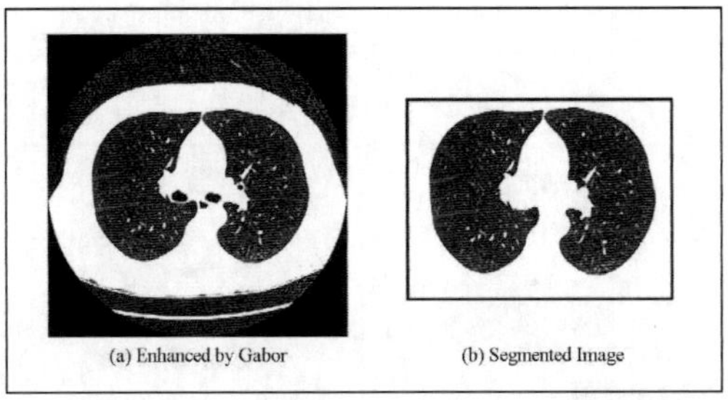

Figure 8. segmented by marker controlled watershed with masking method

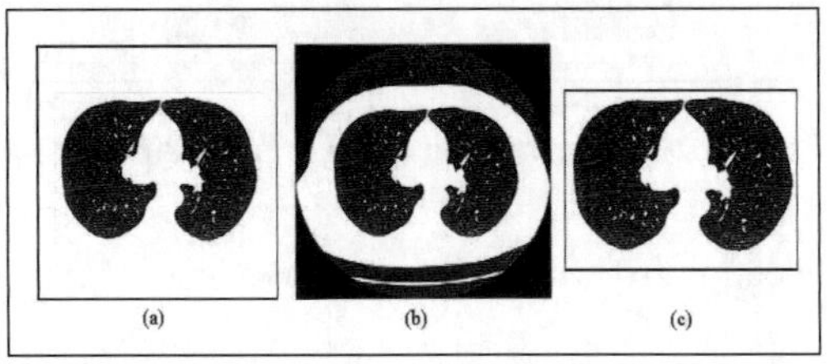

Figure 9. comparing three output segmentation method

Table I shows the comparison of running time of region growing method and marker controlled watershed segmentation with masking method. Marker controlled watershed method (b) is not included to the comparison because this method already fail to get segmentation results.

Table 1. Detection result of five sample patients.

Image	Region Growing	Marker Controlled Watershed with Masking
1	93.259 s	4.644 s
2	88.136 s	4.639 s
3	87.740 s	4.329 s
4	87.870 s	4.719 s
5	89.316 s	4.470 s

Next analysis is based on the processing time required to obtain segmentation results. Region growing takes 90 seconds on average to get the segmentation result, while the marker controlled watershed by masking is takes 5 seconds only to produce the image segmentation results. So we choose to use marker controlled watershed in this study.

3.5 Binarization

Binarization is the process of changing the color of the pixel values into two classes [8], such as black and white. After getting the quantity of black and white pixels on segmentation results, then we compared it with a threshold value to determine the the cindition of lung (normal or cancer). The threshold value is obtained from observations on normal lung. The threshold value that is used in this research is 17178,48. If the number of black pixels more than the threshold, then we conclude the lung is normal, otherwise the lung is cancer.

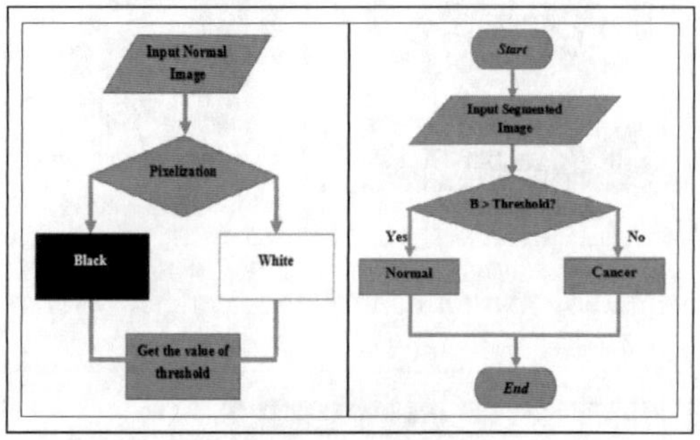

Figure 10. pixelization and binarization scheme

Table II shows result of detection from five sample patiens.

Table 2. Detection result of five sample patients.

Patient	Black pixels	conclusion
1	28504	normal
2	26661	normal
3	18673	normal
4	10399	cancer
5	13865	cancer

4. Conclusion

In this study we implement and evaluate three image segmentation methods for analyzing lung cancer, such as Region Growing, Marker Controlled Watershed, and Marker Controlled Watershed with Masking. The results show that Marker Controlled Watershed with Masking give us the best performance in term of segmentation result and running time. Therefore, we select Marker Controlled Watershed with Masking method in image segmentation stage. Furthermore, in the feature extraction stage, we use color attribute for the analysis of lung cancer using binarization. Finally, the binarization method was successfully determined condition of lung (normal or cancer) from the CT scan image.

Acknowledgment

This research is supported by HIBAH PITTA UI 2016 research grant.

References

[1] Permata T S and Bustamam A 2015 Clustering protein-protein interaction network of TP53 tumor suppressor protein using Markov clustering algorithm *International Conference on Advanced Computer Science and Information Systems (ICACSIS)* pp 221–226

[2] GLOBOCAN 2012, International Agency for Research on Cancer, World Health Organization. http://globocan.iarc.fr/

[3] Khalid R 2014 Clustering analysis of cancerous microarray data *Journal of Chemical and Pharmaceutical Research* **6**(9) pp 488-493, ISSN: 0975-7384

[4] Kajal N et al 2015 Early Detection of Lung Cancer Using Image Processing Technique: Review *International Journal of Advent Research in Computer and Electronics (IJARCE)* **2**(2), E-ISSN: 2348-5523

[5] Lung Cancer Database, Available at: https://eddie.via.cornell.edu/cgibin/datac/signon.cgi
[6] Gonzalez R C and Woods R E 2008 *Digital Image Processing Upper Saddle River* (New Jersey: Prentice Hall)
[7] Allaoui A E and Nasri M 2012 *Medical Image Segmentation by Marker Controlled Watershed and Mathematical Morphology* 1LABO MATSI, ESTO, B.P 473, University Mohammed I OUJDA, Maroko
[8] Sarwinda D and Bustamam A 2016 Detection of Alzheimers disease using advanced local binary pattern from hippocampus and whole brain of MR images *International Joint Conference on Neural Networks (IJCNN) IEEE* pp 5051–5056

IOP Conf. Series: Journal of Physics: Conf. Series **893** (2017) 012064 doi:10.1088/1742-6596/893/1/012064

Sundanese ancient manuscripts search engine using probability approach

Mira Suryani, Setiawan Hadi, Erick Paulus, Intan Nurma Yulita, and Asep K. Supriatna

University of Padjadjaran, Jalan Raya Bandung Sumedang KM21 Jatinangot 45363, INDONESIA

E-mail: mira.suryani@unpad.ac.id, setiawanhadi@unpad.ac.id, erick.paulus@unpad.ac.id, intan.nurma@unpad.ac.id, a.k.supriatna@unpad.ac.id

Abstract. Today, Information and Communication Technology (ICT) has become a regular thing for every aspect of live include cultural and heritage aspect. Sundanese ancient manuscripts as Sundanese heritage are in damage condition and also the information that containing on it. So in order to preserve the information in Sundanese ancient manuscripts and make them easier to search, a search engine has been developed. The search engine must has good computing ability. In order to get the best computation in developed search engine, three types of probabilistic approaches: Bayesian Networks Model, Divergence from Randomness with PL2 distribution, and DFR-PL2F as derivative form DFR-PL2 have been compared in this study. The three probabilistic approaches supported by index of documents and three different weighting methods: term occurrence, term frequency, and TF-IDF. The experiment involved 12 Sundanese ancient manuscripts. From 12 manuscripts there are 474 distinct terms. The developed search engine tested by 50 random queries for three types of query. The experiment results showed that for the single query and multiple query, the best searching performance given by the combination of PL2F approach and TF-IDF weighting method. The performance has been evaluated using average time responds with value about 0.08 second and Mean Average Precision (MAP) about 0.33.

1. Research Methods

There are two major phases in this study: build the collection and build Sundanese ancient retrieval system. The two activities described in the following section.

1.1. Build The Collection

The sub-phase of build the collection consists of three steps: data acquisition, transliteration, and segmentation including annotation. The detail of four steps describe below.

1.1.1. Data Acquisition The acquisition data of Sundanese ancient manuscript located in Situs Kabuyutan Ciburuy, Garut, West Java. Because of the documents are fragile, so, we canâĂŹt hold it for a long time. The image of the documents should be acquired to make it easier to be processed in the next step.

1.1.2. Transliteration After the image of the Sundanese ancient manuscript obtained, then the transliteration of the documents done by the philologist. There are three philologists read the image of Sundanese ancient manuscript, then re-write the content of manuscript in Latin letters. The results of this step are the Sundanese documents that contain the content of the Sundanese ancient manuscript.

1.1.3. Segmentation and Annotation Segmentation and annotation step done by collaboration between the researchers and the philologists. The philologists segmented and annotated the documents that obtained from the transliteration step by word manually. The obtained results, then convert to digital segmentation and annotation by the researchers using Aletheia [2]. The results of this step are segmented words, annotation in Latin letters, and the coordinates of words in the documents.

1.2. The Development of The Sundanese Ancient Manuscript Search Engine

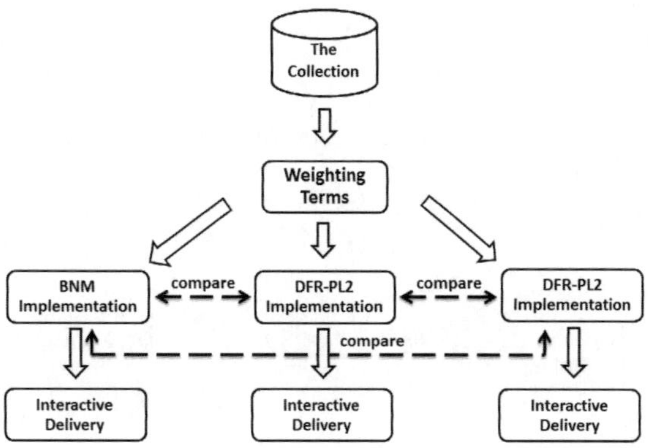

Figure 1. The development of Sundanese ancient manuscript search engine

Figure 1 shows the framework of Sundanese ancient manuscript search engine. The collection of Sundanese ancient manuscripts in Latin version processed in weight terms phase. There are three methods of weighting terms that used in this study: term occurrence (to), term frequency (tf), and term frequency âĂŞ inverse document frequency (tf-idf). After calculating the term weights, then the set of query runs in three different probabilistic methods. The three probabilistic method set with different parameter values that give based performance based best practices. The parameter values of three probabilistic methods can be seen in experiment and results part. The results of search deliver in interactive interface and the mean average precision of three methods compared.

1.3. Bayesian Network Model
Based on the reference [3], BNM is described as a directed acyclic graph (DAGs) where the nodes represent random variables. The random variables that use in this study are the documents, the terms, and inference results. The arcs in graph expressed causal relationships between variables called the queries. The conditional probabilities described the strength of this dependency relationship. The BNM is one of the probabilistic models that provide clean formalism that combined multiple sources of evidence from a document. Sources of evidence used is past queries, feedback past cycles, and distinct query formulations. For every result search that given by the system according to the query will be saved and used again in the next search process which involved the resemblance query.

Let t be a y-dimensional vector defined by $t = (t_1, t_2, ..., t_y)$ where $t_1, t_2, ..., t_y$ are binary random variables $t_i \in 0, 1$. These variables define the 2^y possible state for t. Let d_j be a binary random variable associated with a document in the collection and let q be a binary random variable associated with the user query. The inference can be calculated by:

$$P(q \wedge d_j) = \textstyle\sum_{\forall k} P(q|t) \cdot P(d_j)$$
$$(\textstyle\prod_{(\forall_i|on(i,t)=1} \cdot P(t_i|d_j) \cdot \prod_{(\forall_i|on(i,t)=0} \cdot P(t_i|d_j))$$
$$P(\overline{q \wedge d_j}) = 1 - P(q \wedge d_j) \tag{1}$$
$$where P(t_i|d_j) = 1 - P(t_i|d_j)$$

The $P(q|t)$ is term-query beliefs that can be obtained from (2). A reasonable default value for $w_q = 1$.

$$P(q|t) = \textstyle\sum_{k_i} \overline{f_{i,j}} \cdot w_q$$
$$P(q|t) = 1 - P(q|t) \tag{2}$$

The $P(d_j)$ is prior probability of observing a document d_j that can be set $\frac{1}{N}$ where N is the total number of documents in the collection.

$$For P(d_j) = 1 - \frac{1}{N} \tag{3}$$

The prior probability of the document-term beliefs can be obtained from (4) below.

$$P(t_i|d_j) = \alpha + (1 - \alpha) \cdot f_{i,j} \cdot \overline{IDF_i}$$
$$P(t_i|d_j) = 1 - P(t_i|d_j) \tag{4}$$

Where $f_{i,j}$ is normalized term frequency given by $f_{i,j} = \frac{f_{i,j}}{max f_{i,j}}$ and $\overline{IDF_i}$ is inverse document frequency variable given by (5):

$$\overline{IDF_i} = \frac{log \frac{N}{n_i}}{log N} \tag{5}$$

Based on empirical evidence, the value of α is 0.4. The final step is combining the evidence. It can be done by retrieve all the type of the past query that relevance to the new one. In order to get the ranking provided by inference network, equation (6) should be calculated.

$$P(I \wedge d_j) = \textstyle\sum_t P(I|\bar{t}) \cdot P(\bar{t}|d_j) \cdot P(d_j))$$
$$= \textstyle\sum_t P(1 - P(\overline{q}|\bar{t})) \cdot P(\overline{q_i}|\bar{t}) \cdot P(\bar{t}|\overline{d_j}) \cdot P(d_j)) \tag{6}$$

1.4. Divergence From Randomness with PL2 Distribution
Based on the reference [4], the idea of Divergence From Randomness (DFR) method is to compute the term weights by measuring the divergences between a term distribution proceeds by the random process and the actual term distributions. A word that contains a bit of information is generally scattered in the collection. We call C as a collection, then a term k_i has a probability distribution with respect to C as follows $P(t_i|C)$, then the probability of a number of other information is,$-log P(t_i|C)$. Term distribution in which a term referred to as an elite set, then the probability to a document given by $P(t_i|d_j)$. The smaller probability of occurrence of the k_i

term in the document indicates increasingly rare and important that term in representing the document.

There are many probability distribution families that can be implemented in DFR Framework. One of the distribution that can be implemented is PL2. Based on [5], the indexing weight of the query terms w_ij in the document d_j calculated with (7):

$$w_{ij} = -log_2(Prob^1_{ij}(tf)) \cdot (1 - Prob^2_{ij}) \tag{7}$$

For $Prob^1_{ij}$, can be replaced by PL2 distribution. PL2 is Poisson model with Laplace after-effect and normalisation 2. The formula of PL2 distribution can be seen in (8).

$$Prob^1_{ij} = \frac{e^-\lambda_j \cdot \lambda^t_j f_{ij}}{tf_{ij}!} where \lambda_j = \frac{to_j}{n} \tag{8}$$

The to_j indicates the number of occurrences of term t_i in the collection. The $Prob_ij^2$ can be replaced by normalizing term frequency that can be calculated with the following equation (see Eq (9)).

$$\begin{gathered} Prob^2_{ij} = \frac{tfn_{ij}}{tfn_{ij}+1}, with \\ tfn_{ij} = tf_{ij} \cdot log_2[1 + \frac{c \cdot meandl}{l_i}] \end{gathered} \tag{9}$$

The c is a constant with value 1.5, the *meandl* is the average document length, the $n =$ number of documents in the corpus, and l_i= the length (number indexing terms) of document d_j in the collection.

1.5. Divergence From Randomness with PL2F Distribution

Another probability distribution in DRF adopted from Craig, Christina, and Iadh in [1] is PL2F. PL2F is derivative from PL2. The relevance of weighting score of a document d_j for a query terms is given by (10):

$$\begin{gathered} score(d_j, q) = \textstyle\sum_{t \in q} qtw \cdot \frac{1}{tfn+1}(tfn \cdot log_2 \frac{tfn}{\lambda} + \\ (\lambda - tfn) \cdot log_2 e + 0.5 \cdot log_2(2\phi \cdot tfn) \end{gathered} \tag{10}$$

where λ given by $\frac{F}{n}$, F is the frequency of the query term t in the collection, and n is the number of document in whole collection. The query term weight qtw is given by $\frac{qtf}{qtf_{max}}$, where qtf is frequency of query term and qtf_{max} is the maximum frequency of query term among the query terms. The tfn corresponds to the normalized term frequency which is given by (11):

$$\begin{gathered} tfn = \textstyle\sum_f (w_f \cdot tf_f \cdot log_2(1 + c_f \cdot \frac{avg_{lf}}{l_f})) \\ where, (c_f > 0) \end{gathered} \tag{11}$$

where tf_f is the frequency of the term t in the field f of the document d_j; $avg_l f$ is the average length of whole documents in the collection; and l_f is the length in tokens of field f of the document d_j. The w_f and c_f are the hyper-parameter that controlled the contribution of watch field. Based on the condition of Sundanese ancient manuscript in this study, the field that use in both parameter are the body of the document with values $c = 4.10$ and $w = 1$.

Table 1. The Collection Statistics

View Point	Value
Size	100 MB
Number of Documents	12
Number of Terms	823
Number of Distinct Terms	474
Mean of docs	68.58
St Dev.	15.56
Median	73.5
Max	87
Min	29
Topics	Ramayana story

2. Experiments and Results

The collection of Sundanese terms obtained from Sundanese ancient manuscript describe in Table 1.

When the weighting term process done, each terms has their own weight, then the set of query randomly picked to run in the three probabilistic approaches. The parameter values for three different probabilistic methods shown in Table 2.

Table 2. The Parameter Settings

Method	Paramater	Value
BNM	w	1.00
	α	0.40
DFR-PL2	c	1.50
	$meandl$	68.58
DFR-PL2F	c	4.10
	w	1.00
	$meandl$	68.58

The set of queries in the proposed system. The results of search shown in interactive interface on the browser. The example of search results can be seen in Figure 2.

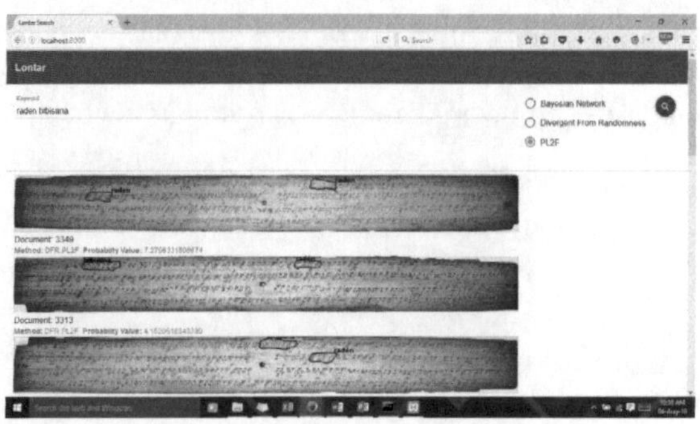

Figure 2. Beta version of Sundanese ancient manuscript search engine

When the proposed system tested by 50 queries, the query latency can be calculated. The different average time responds from three probabilistic shown that the best average time responds given by DFR-PL2F and the lowest respond given by BNM. The ratio of query latency of three methods is 16:1:1 respectively to BNM, PL2, and PL2F. The complete results of average time respond that indicates the query latency can be seen in Table 3.

Table 3. The Average Time Responses of Query Search

Method	Average Time Response (sec)
BNM	1.30 s
DFR-PL2	0.09 s
DFR-PL2F	0.08 s

Based on an experiment, there is no difference of mean average precision for the three probabilistic methods. The 50 queries that run in the system gave the MAP value about 0.34. The difference of mean average precision can be seen when the proposed system tested by two terms of query type. Even though three probabilistic methods have a derivation of MAP value, but the best performance given by DFR-PL2F with 0.22 for 50 queries. Figure 3 shows the MAP value for the two terms query that applied in the proposed systems.

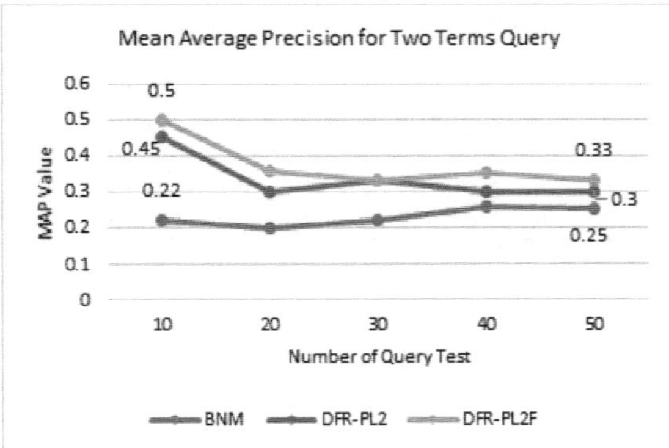

Figure 3. The comparison of mean average precision for two terms query

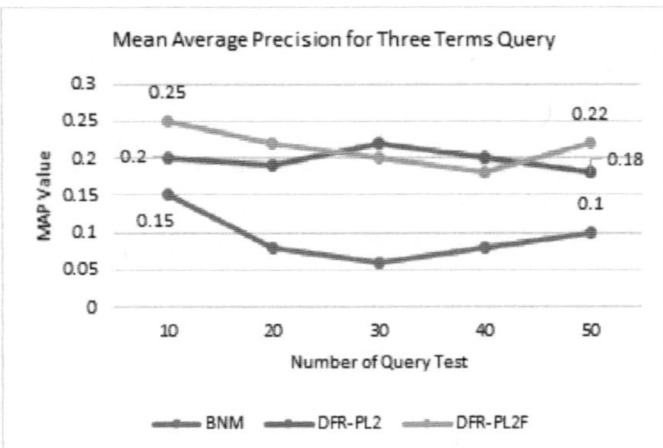

Figure 4. The comparison of mean average precision for three terms query

When the proposed system tested by three terms query type, there are significantly different with the MAP value. The best performance given by the DFR-PL2F method with MAP value is about 0.33 and the lowest performance given by the BNM method with the MAP value is about 0.25. If we looked at precisely at the Figure 3 and 4, even though the BNM method gives the lowest performance, but there are escalation results of the MAP value when tested from 10 queries until 50 queries per batch.

3. Conclusions and Future Works

Based on the experiment results, the best query latency performed when using the PL2F method with average time 0.08 second. The MAP evaluation shown that the best performance of the proposed system is given by the PL2F method, especially for 2 and 3 terms query. Even though the PL2F gives the best performance in the latest sum of the collection, the BNM method shown the better progress value of the MAP than the PL2 and the PL2F method but still the lowest. There are many recommendations that should be done in the future, such as create stop word list in Sundanese language, adding the collection, apply the variance of Sundanese stemming methods, and develop other probability methods for results search improvement.

Acknowledgement

The authors would like to thank to DRPMI Universitas Padjadjaran, the Faculty of Nat. Science and Mathematics, the Misnistry of Research, Technology, and Higher Education that gives The International Collaboration and Publication grant in 2016, the philologist from Sundanese Centre Studies of Universitas Padjadjaran, the Situs Kabuyutan Ciburuy Garut, and the L3i Lab of University La Rochelle France as a collaboration partner.

References

[1] Craig M, Crishtina L, and Iadh O 2007 Terrier Takes on The Non-English Web *in SIGIR âĂŹ07 iNEWS07 Workshop, Amsterdam*

[2] Clausner C, Pletschacher S, and Antonacopoulos A 2011 Aletheia - An Advanced Document Layout and Text Ground-Truthing System for Production Environments *in International Conference on Document Analysis and Recognition, Beijing*

[3] Omri M N, Amira K, and Kamel G, 2012 A New Information Retrieval Model Based on Possibilistic Bayesian Networks *in International Conference on Computer Related Knowledge, Sousse*

[4] Ricardo B Y and Berthie R N 2011 Modern Information Retrieval *London: Pearson Addison Wesley*

[5] Jacques S, Ljiljana D, and Mitra A 2011 Information Retrieval with Hindi, Bengali, and Marathi Language: Evaluation and Analysis*in Third International Workshop FIRE, Bombay* .

Robust optimization modelling with applications to industry and environmental problems

Diah Chaerani, Stanley P Dewanto, Eman Lesmana

Department of Mathematics, Faculty of Mathematics and Natural Sciences
Universitas Padjadjaran, Indonesia

E-mail: d.chaerani@unpad.ac.id, stanley@unpad.ac.id, eman.lesmana@unpad.ac.id

Abstract.
Robust Optimization (RO) modeling is one of the existing methodology for handling data uncertainty in optimization problem. The main challenge in this RO methodology is how and when we can reformulate the robust counterpart of uncertain problems as a computationally tractable optimization problem or at least approximate the robust counterpart by a tractable problem. Due to its definition the robust counterpart highly depends on how we choose the uncertainty set. As a consequence we can meet this challenge only if this set is chosen in a suitable way. The development on RO grows fast, since 2004, a new approach of RO called Adjustable Robust Optimization (ARO) is introduced to handle uncertain problems when the decision variables must be decided as a "wait and see" decision variables. Different than the classic Robust Optimization (RO) that models decision variables as "here and now". In ARO, the uncertain problems can be considered as a multistage decision problem, thus decision variables involved are now become the wait and see decision variables. In this paper we present the applications of both RO and ARO. We present briefly all results to strengthen the importance of RO and ARO in many real life problems.

1. Introduction: Optimization under Uncertainty

Optimization under uncertainty refers to the branch of optimization where there are uncertainties involved in the data. The importance of this class of problem is in many models the uncertainty is ignored and a representative nominal value of the data is used. However, the decision variables cannot be perfectly forecasted but instead should be considered random or uncertain.

The general formulation of uncertain optimization problem as presented in the following term.

$$\min_{x,t}\{t \ : \ f_0(x,\zeta) - t \le 0, f_i(x,\zeta) \in \mathcal{K}_i, i = 1,\ldots,m\} \tag{1}$$

where $(x,t) \in \mathbf{R}^{(n+1)}$ is the vector of decision variables, $\zeta \in \mathbf{R}^d$ is the vector of problem's data, $f_0(x,\zeta) \ : \ \mathbf{R}^n \times \mathbf{R}^d \to \mathbf{R}$, $f_i(x,\zeta) \ : \ \mathbf{R}^n \times \mathbf{R}^d \to \mathbf{R}^{k_i}, i \le i \le m$, are given function and $\mathcal{K}_i \subset \mathbf{R}^{k_i}$ are given nonempty set. This means that the data vector ζ is not known exactly at the time when the solution has to be determined.

2. Robust Optimization

In this section we present a brief discussion about the theory of Robust Optimization (RO) as proposed by [1] and discussed in [7], [8], also discussed in [4]. Let $c \in R^n, b \in R^m$ and

$A \in R^m \times R^n$ be the parameters of the canonical linear programming problem

$$\min_x \left\{ c^T x \; : \; Ax \leq b \right\} \qquad \text{(LP)}$$

The main idea of Robust Optimization is we are interested in a version of (LP) problem in which the data (c, A, b) are uncertain, but are known to reside in an uncertain set U. In this case the (LP) is not a single deterministic problem, but a family of problems, one for each realisation $(c, A, b) \in U$:

$$\min_{x \in R^n} \left\{ c^T x \; : \; Ax \leq b \right\}_{(c,A,b) \in U} \qquad \text{(ULP)}.$$

We assume that the decision environment is such that

(i) Decision is offline: the entire decision vector x is to be fixed prior to knowing which value the actual parameters take "here-and-now-decision".

(ii) In the dynamic (ULP) case only part of the variables $(x_1, \ldots, x_n)$ need to be determined off-line. The rest may be determined ofter some of the uncertain parameters become known ("wait and see" decision).

(iii) The prior information on the data (c, A, b) is crude and is captured by a compact uncertainty set U.

(iv) The inequality constraints $Ax \leq b$ are hard constraints, i.e., they must all be satisfied whenever the uncertain parameters reside in U.

Under the above conditions the robust optimization approach converts the uncertain family of the problems (ULP) into the following single deterministic problem, which we call *Robust Counterpart*:

$$\pi^* = \min_{x \in R^n} \left\{ c^T x \; : \; Ax \leq b, (c, A, b) \in U \right\} \qquad \text{(RC)}.$$

A vector x^* is the called a *robust optimal solution* if for all realizations $(c, A, b) \in U$, x^* is feasible and the value of objective function is guaranteed to be at most π^*. Problem (RC) can be written equivalently as a problem with a linear certain objective function and only uncertain constraints as follows:

$$\min t \tag{2}$$
$$s.t \;\; c^T x - t \leq 0, \tag{3}$$
$$a_i^T x - b_i \leq 0, i = 1, \ldots, m, \forall (c, A, b) \in U \tag{4}$$

2.1. The Treatment of Robust Linear Optimization (RLO)

We cite from [7] that for the treatment of Robust Linear Optimization (RLO) we will assume without loss of generality that,

(i) Firstly, the objective $c^T x$ is certain. If there is uncertainty in the objective, we can reformulate the problem such that this uncertainty appears in a constraint see (2).

(ii) Secondly, the right-hand-side b is certain. If b is uncertain, we can introduce an extra variable x_{n+1} and change the problem into

$$min \; c^T x \tag{5}$$
$$s.t \;\; a_i^T x - b_i x_{n+1} \leq 0, \tag{6}$$
$$x_{n+1} = 1, i = \ldots, m, \forall (A, b) \in U. \tag{7}$$

This assumption is made to make the notation for the uncertainty region and the resulting robust counterpart formulations easier.

(iii) Third, the robustness with respect to U can be formulated constraint-wise and the last, the uncertainty set U is closed and convex.

Notes that because this inequality on x is of the semiinfinite type (it should be satisfied for an infinite number of values of the parameters A and b), it cannot directly be efficiently solved by standard optimization methods. The challenge is to find for which types of uncertainty sets problem (RC) can be reformulated into a tractable optimization problem.

2.2. Types of uncertainty sets
The challenge is to find U for which types of uncertainty sets problem can be reformulated into a tractable optimization problem. Since the robustness with respect to U can be formulated constraint-wise, thus we can reformulate each constraint which involves the uncertain data. Refers to [7], consider the canonical robust semi-infinite constraint

$$a^T x - b \leq 0, \forall (a, b) \in U. \tag{8}$$

Here a is a vector in R^n and b which are the general representatives of a_i and b_i.

Similarly, U stands for U_i. Its convenient to describe the uncertainty parameters a and b and the uncertainty set U in terms of a primitive factor $\zeta \in R^L$. Namely,

$$a = \bar{a} + Q\zeta, b = \bar{b} + q^T \zeta \tag{9}$$

where $\bar{a} \in R^n, Q \in R^{n\times L}, \bar{b} \in R$ and $q \in R^L$ and

$$U = \left\{ \begin{pmatrix} a = \bar{a} + Q\zeta \\ b = \bar{b} + q^T \zeta \end{pmatrix} : \zeta \in Z \right\} \tag{10}$$

where $Z \subset R^L$ is the uncertainty set for the primitive factors. The fixed vector $\bar{a}$ and the scalar $\bar{b}$ will thereafter be called nominal. In the view of representation (10) the alternative formulation of (8). Replacing a and b by their expressions in ζ, we have

$$(\bar{a}^T x - \bar{b}) + (Q^T x - q)^T \zeta \leq 0, \forall \zeta \in Z. \tag{11}$$

2.2.1. Case 1: Using Interval or Box Uncertainty Set In general, a tractable robust counterpart formulation for RLO of (8) with interval (or box) uncertainty regions can be stated as the following formulation.

$$(\bar{a} + P\zeta)^T x \leq b, \forall \zeta \; : \; \|\zeta\|_\infty \leq \mu. \tag{12}$$

This problem is a semi-infinite programming problem.

2.2.2. Case 2: Using Ellipsoidal Uncertainty Set Refers to [7], in the case of ellipsoidal uncertainty, the robust counterpart of (8) becomes

$$\begin{aligned} &(\bar{a} + P\zeta)^T x \leq b, \forall \zeta \; : \; \|\zeta\|_2 \leq \mu. \\ \equiv\; &\bar{a}^T x + \max_{\zeta \,:\, \|\zeta\|_2 \leq \mu} (P^T x)^T \zeta \leq b. \end{aligned} \tag{13}$$

It can easily verify that

$$\max_{\zeta \,:\, \|\zeta\|_2 \leq \mu} (P^T x)^T = \mu \left\| P^T x \right\|_2. \tag{14}$$

IOP Conf. Series: Journal of Physics: Conf. Series **893** (2017) 012065 doi:10.1088/1742-6596/893/1/012065

Hence x satisfies (13) if and only if

$$\bar{a}^T x + \mu \left\| P^T x \right\|_2 \leq b. \tag{15}$$

Now the constraint does not have the semi-infinite structure as the original one in (13). This is a conic quadratic programming problem. Again as in the case of box uncertainty, in the final robust counterpart an extra safety term $\mu \left\| P^T x \right\|_2$ is added to account for uncertainty and it depends on the value of x.

2.3. Case 3: Using Polyhedral Uncertainty Set

In case of the uncertainty set is polyhedral, i.e., a system of linear inequalities

$$(\bar{a} + P\zeta)^T x \leq b, \forall \zeta \, : \, d - D\zeta \geq 0. \tag{16}$$

This is equivalent with

$$\bar{a}^T x + \max_{\zeta \, : \, d - D\zeta \geq 0} (P^T x)\zeta \leq b. \tag{17}$$

For this formulation, the robust counterpart cannot be obtained explicitly. Thus, refers to [7] the robust counterpart formulation can be obtained via its dual formulation.

Consider that by the duality theory we have that

$$\begin{aligned} & \max \left\{ (P^T x)\zeta \, : \, d - D\zeta \geq 0 \right\} \\ = & \min \left\{ d^T y \, : \, D^T y = P^T x, y \geq 0 \right\}. \end{aligned} \tag{18}$$

Hence x satisfies (17) if and only if x satisfy

$$\bar{a}^T x + \min_y \left\{ d^T y \, : \, D^T y = P^T x, y \geq 0 \right\} \leq b. \tag{19}$$

Notice that if this constraint is satisfied for a feasible y, i.e., that satisfies $D^T y = P^T x, y \geq 0$ then the constraint is certainly satisfied for the minimum over y. Hence we can delete the min. Thus, x satisfies (17) if and only if there exists a y such that (x, y) satisfies

$$\bar{a}^T x + d^T y \leq b, D^T y = P^T x, y \geq 0. \tag{20}$$

This result is very tractable since it is a system of linear (in) equalities.

3. Adjustable Robust Optimization

The development on RO grows fast, since 2004, a new approach of RO called Adjustable Robust Optimization (ARO) is introduced to handle uncertain problems when the decision variables must be decided as a "wait and see" decision variables. Different than the classic Robust Optimization (RO) that models decision variables as "here and now". In ARO, the uncertain problems can be considered as a multistage decision problem, thus decision variables involved are now become the wait and see decision variables. Adjustable robust counterpart handles uncertain problem when the variable has to be decided after the uncertainty parameter values (or part of it) has been revealed. In other words, in such case we can postpone part of decision until we have obtained more of full information on (some of the) uncertain parameters. Considering the Adjustable Robust Counterpart of one constraint of a multi-stage problem:

$$(\bar{a} + P\zeta)^T x + d^T y \leq b, \forall \zeta \in Z. \tag{21}$$

where $x \in R^{n_1}$ is non-adjustable, $y \in R^{n_2}$ is adjustable. This situation is called a fixed recourse situation, since the coefficients for the adjustable variable y are certain.

(i) The optimization variable $y = y(\zeta)$ called a decision rule is in fact a vector function. It has been shown that in general solving such problem is NP-hard. Therefore, $y(\zeta)$ is restricted to a given class of function.

(ii) However it appears that the linear decision rule already shows good performance in practice.

4. Applications of RO and ARO

Some results of the author on developing model for handling uncertain data, has been published in [2], [3], [4], [5], [6], [9],[12],[11] and [13].

In [2], a discussion on Robust Shortest Path Problems by means of Robust Linear Optimization is presented. Modeling Robust Design Problems via Conic Optimization is discussed in [3]. We discussed Handling Optimization under Uncertainty Problem Using Robust Counterpart Methodology in [4].

Some application topics in environmental issues such as vaccination and water supply is discussed. We have the results of developing the Robust Optimization to the Uncertain Vaccination Strategy Problem (see [5]). A recent results about Determining Robust Counterpart of Spatial Optimization Model for Water Supply Allocation can be seen in [6].

To engineering application we focus in developing the robust optimization model based Model Predictive Control using box uncertainty set(see [9]). Also a topic on robust electrical network topology design by conic optimization is discuss in [12].

For application of ARO, we have work on the problem of determining the adjustable robust counterpart of water distribution problem in the case of Waduk Jatiluhur (see [13]) and also the adjustable robust counterpart for transportation problem with discount factor (see [11]).

5. Conclusion

The fast development on RO has brought us to a new era of ARO. Some applications has been discuss in many papers. The future research will be the ARO version with binary integer variables .

Acknowledgment

This research is funded by Competency Research Grant from The Indonesian Ministry of Research and Higher Education for the year of 2016 and also partly funded by TWAS Individual Research Grant 2016 to the first author.

References

[1] Ben-Tal A and Nemirovskii A 2002 Robust optimization methodology and applications *Mathematical Programming* **92**(3 Ser. B) pp 453-480

[2] Chaerani D, Roos C and Aman A 2004 The robust shortest path problems by means of robust linear optimization, ed H Flueren, D den Hertog and P Kort, *Operations Research Proceedings* pp 335-342

[3] Chaerani D and C Roos 2006 Modeling robust design problems via conic optimization *Proceedings of Operations Research* (Berlin, Heidelberg, New York: Springer Verlag)

[4] Chaerani D and Roos C 2013 Handling optimization under uncertainty problem using robust counterpart methodology *Jurnal Teknik Industri* **15**(2) pp 111-118

[5] Chaerani D, Anggriani N and Firdaniza 2014 On the robust optimization to the uncertain vaccination strategy problem *American Institute of Physics (AIP) Conf. Proc.* **1587** pp 34-37

[6] Chaerani D, Ruchjana B N, Setiawan A, Rejito J, Dewanto S P, Dharmawan I A and Rosandi Y 2016 Determining robust counterpart of spatial optimization model for water supply allocation *AIP Conf. Proc.* **1716** (020002)

[7] Den Hertog D, Ben-Tal A and Brekelmans 2015 *Practical robust optimization* (Lecture Notes LNMB Course)

[8] Gorissen B L, Yanikoglu I, and Hertog D D 2015 *A practical guide to robust optimization.* **Omega** (53) pp 124-137

[9] Joelianto E, Rismayasari D and Chaerani D 2009 The robust optimization based Model Predictive Control using box uncertainty set *IEEE Xplore* pp 1-5

[10] Murray A T, Gober P, Anselin L, Rey S J, Sampson D, Padegimas P D and Liu Y 2012 *Spatial optimization models for water supply allocation* (Water resources management vol 26) pp 2243-57 (Berlin: Springer)
[11] Nurnaini R, Chaerani D and Sukono 2016 Adjustable robust counterpart for transportation problem with discount factor and external buying *SNIPS Proceedings ITB*
[12] Roos C, Bai Y and Chaerani D 2012 Robust electrical network topology design by conic optimization *A special issue of SQU Journal for Science* **75** (1) pp 125-146
[13] Yuniar P D 2015 Determining the optimal solution of water distribution system using adjustable robust counterpart approach *thesis* (Bandung: Departement of Mathematics Universitas Padjadjaran)

IOP Conf. Series: Journal of Physics: Conf. Series **893** (2017) 012066 doi:10.1088/1742-6596/893/1/012066

Shortest path problem on a grid network with unordered intermediate points

Veekeong Saw, Amirah Rahman, and Wen Eng Ong

School of Mathematical Sciences, Universiti Sains Malaysia, 11800 Penang, Malaysia

E-mail: kentvk_90@hotmail.com, amirahr@usm.my, weneng@usm.my

Abstract. We consider a shortest path problem with single cost factor on a grid network with unordered intermediate points. A two stage heuristic algorithm is proposed to find a feasible solution path within a reasonable amount of time. To evaluate the performance of the proposed algorithm, computational experiments are performed on grid maps of varying size and number of intermediate points. Preliminary results for the problem are reported. Numerical comparisons against brute forcing show that the proposed algorithm consistently yields solutions that are within 10% of the optimal solution and uses significantly less computation time.

1. Introduction

Shortest path problem (SPP) is a classical optimization problem in network theory. The objective of SPP is to find a path from a starting point to a destination point in a network, such that the resource consumption is minimized. There are a handful of well-known algorithms used to find optimal solution for SPP, such as Dijkstra algorithm, A* algorithm, Bellman-Ford algorithm and Floyd-Warshall algorithm, depends on the nature of SPP itself ([4, 7, 1]). Different variations of SPP has been applied in various fields such as polar image segmentation, texture analysis, vehicle routing and military operations ([2, 5, 8]). SPP can be represented in various metric representations such as grid network, visibility graphs and Voronoi diagram. The metric representation that is widely used in path planning is grid network. For example, in military operations, the path planning for unmanned combat vehicles was represented in grid map ([3, 6]).

In this paper, we consider the SPP with multiple unordered intermediate points (SPP-UIP) on a grid network, where the objective is to find the minimum cost Hamilton path with unique starting and ending points. Note that the order of the intermediate points are not given in advance and it is to be determined during the process of searching for optimal path. To date, there has been no apparent research on shortest path problem that involves grid maps and unordered intermediate points as in our research. A two phase heuristic is proposed as a solution method for the SPP-UIP.

2. Problem Description

In this work, the grid network which is represented in an $N \times N$ grid, was modeled as a graph $G = (V, E)$, where V is the set of vertices, which represents all cells in the grid and E is the set of edges which represents the link between two adjacent cells in the grid. Associated with each vertex is the cost value. We assume that the grid is 8-connected, meaning that any cell in the grid can move to any one of the eight adjacent cells (Figure 1), except the cells on the border of the grid. A path between two vertices can be represented by a sequence of cells. Figure 2 shows an example of a shortest path between two cell, $(1, 1)$ and $(14, 8)$, on a 15×15 grid. The total cost of a path is calculated by summing up the cost of all the traversed vertices.

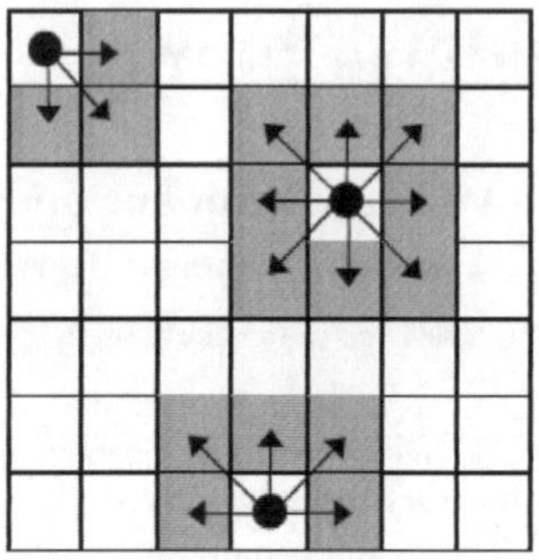

Figure 1. Accessible directions from a cell in different positions. ([6])

	0	1	2	3	4	5	6	7	8	9	10	11	12	13	14
0	6	1	6	5	3	2	5	0	5	6	0	5	6	6	0
1	1	6	0	4	4	2	2	3	6	5	6	5	5	6	1
2	1	5	1	2	0	4	5	0	4	3	3	4	4	2	5
3	2	0	4	1	1	4	1	0	4	3	0	1	4	5	1
4	3	4	0	1	1	6	3	3	0	3	6	5	0	2	3
5	6	5	0	5	6	1	2	4	4	3	3	3	5	1	2
6	4	1	0	4	0	6	4	3	4	0	3	0	5	6	3
7	4	5	3	5	0	2	3	2	4	5	5	2	5	1	6
8	4	4	5	2	4	1	2	6	6	5	1	2	1	3	1
9	0	6	0	3	4	1	2	5	2	0	5	2	0	1	3
10	3	5	3	3	5	6	4	6	2	3	1	3	2	6	2
11	1	1	2	4	4	1	1	1	2	4	4	6	1	5	5
12	0	2	4	4	0	1	5	1	4	5	4	4	1	6	6
13	6	2	2	0	1	1	0	0	0	6	6	6	0	6	0
14	1	4	6	3	2	5	3	3	6	0	3	5	5	4	0

Figure 2. Shortest path from $(1, 1)$ to $(14, 8)$ with a total cost of 24.

A path between any two vertices is known as a partial path. A path that links the starting point to the destination point, passing through all intermediate points once is known as a complete path. A complete path is given as a concatenation of partial paths. The total cost of a path is known as the cumulative cost of the path.

Before proceeding to the mathematical formulation of SPP-UIP, a list of notation used throughout this paper is shown as follows.

i, j, k, w	Vertices
s	Starting point
t	Destination point
n	Number of intermediate points to be visited
v_l	Intermediate points l, $l \in \{1, 2, \cdots, n\}$
Λ	Set of vertices need to be visited, $\Lambda = \{s, v_1, v_2, \ldots, v_n, t\}$
$c(i)$	Cost value associated with vertex i
$\pi(i,j)$	Path from vertex i to vertex j
$C(i,j)$	Cumulative cost of the path $\pi(i,j)$, $C(i,j) = \sum_{i \in \pi(i,j)} c(i)$
$Suc(i)$	Set of successors of vertex i, $Suc(i) = \{j \mid (i,j) \in E\}$
$Pre(i)$	Set of predecessors of vertex i, $Pre(i) = \{j \mid (j,i) \in E\}$
S	Set of vertices k of which the $C(P_i, k)$ is not obtained yet, where $P_i \in \Lambda$
S^*	Set of vertices k of which the $C(P_i, k)$ is obtained, $S^* = V - S$
$Prev(w)$	Previous vertex of vertex w in shortest partial path
x_i	$x_i = 1$ if vertex i is included in path, otherwise $x_i = 0$

The integer linear programming formulation of SPP-UIP is as follows.

$$\text{Minimize} \sum_{i \in V} x_i c(i) \tag{1}$$

s.t.

$$\sum_{j \in Suc(i)} x_j = 1, \quad \forall i \in \Lambda \backslash \{t\} \tag{2}$$

$$\sum_{j \in Pre(i)} x_j = 1, \quad \forall i \in \Lambda \backslash \{s\} \tag{3}$$

$$\sum_{j \in Suc(i)} x_j - \sum_{j \in Pre(i)} x_j = 0, \quad \forall i \in V \backslash \{s, t\} \tag{4}$$

$$\sum_{i \in S} \sum_{j \in Suc(i)} x_j \leq |U| - 1, \quad \forall U \subset V, U \neq \emptyset \tag{5}$$

$$x_i \in \{0, 1\}, \quad \forall i \in V$$

Equation (1) represents the objective function to minimize the total cost of the complete path. Equation (2) and (3) ensure that there can only be one arrival and one departure from any point except for the starting point (no arrival) and destination point (no departure) respectively. Equation (4) is the flow conservation constraint. Inequality (5) prevents the solution path from forming subtours.

3. Two Phase Heuristic

We present a two phase heuristic to solve the SPP-UIP. In Phase 1, we use the Dijkstra algorithm to generate the shortest partial path between every pair of points in Λ, and represent it in incidence matrix form. Then, in Phase 2, we choose the best combination of partial paths that form the complete path with minimum cost. There are many ways to search for such combinations. In this study, we compare the repetitive nearest neighbor algorithm (RNNA) with the brute force method (BF). A detailed description for each phase is presented as follows.

3.1. Phase 1: Generate the complete graph
In Phase 1, we compute the shortest partial path for all pairs points in Λ by applying Dijkstra algorithm $\frac{(n+2)(n+1)}{2}$ times. Note that in typical SPP, the cost value is associated with edges of the graph and the cumulated path cost is calculated by adding all the cost of path edge associated to the complete path. However, in SPP-UIP the cost value is associated with vertices of the graph. Hence, while applying the Dijkstra algorithm, the path cost is calculated by adding the cost associated to the vertices instead of edges. The details of algorithm is given in Algorithm 1.

Algorithm 1 Dijkstra algorithm to find shortest distance for all pairs of input points

Require: $\Lambda = \{s, v_1, v_2, \ldots, v_n, t\}$ and re-indexing it as $\{P_1, P_2, \ldots, P_{n+2}\}$
  for all two vertices $P_i, P_j \in \Lambda$, $(i \neq j, i > j)$ **do**
    $S \leftarrow V, S^* \leftarrow \emptyset, C(P_i, P_i) \leftarrow c(P_i), C(P_i, k) \leftarrow \infty$ for $k \in V \backslash \{P_i\}$
    while $P_j \notin S^*$ **do**
      Select vertex with minimum $C(P_i, k)$ among $k \in S$ and name it as i^*
      $S \leftarrow S - \{i^*\}, S^* \leftarrow S^* \cup \{i^*\}$
      for all $w \in Suc(i^*)$ **do**
        if $C(P_i, w) > C(P_i, i^*) + c(w)$ **then**
          $C(P_i, w) \leftarrow C(P_i, i^*) + c(w)$
          $Prev(w) \leftarrow i^*$
        end if
      end for
    end while
  end for

The shortest path between P_i and P_j can be traced by tracking from $Prev(P_j)$ until reaching $Prev(w) = P_i$, and rewriting the whole sequence in reverse order. After executing Algorithm 1, we have a subgraph $G' = (V', E')$, which is a complete graph made up of the paths linking all pairs of given points in Λ. Note that G' can be represented as an $(n+2) \times (n+2)$ incidence matrix $I_{G'} = [a_{i,j}]_{(n+2)\times(n+2)}$.

3.2. Phase 2: Search for the best path
In Phase 1, we obtain a complete graph G' which contains the cost of all pairs of points in Λ. Our next task is to search for the best Hamilton path from s to t. Note that this is similar to the well known travelling salesman problem (TSP), except that in TSP, the path starts and ends at the same point while SPP-UIP have unique starting and ending points. The nearest neighbor algorithm (NNA) is a greedy but simple algorithm that solves the TSP. It selects the immediate best (nearest) point to visit next, until all have been visited. Hence NNA executes quickly but does not guarantee the optimality of solution all the time. To increase the likelihood of getting good solution, we apply the repetitive nearest neighbor algorithm (RNNA), where the NNA is repeated $n+2$ times by starting at each point in Λ once. Finally, the shortest path among the $n+2$ solution paths is selected. The procedure is described in Algorithm 2.

In the beginning of Algorithm 2, we model the problem into traveling salesman problem by changing $C(s,t)$ in G' into $-M$, where M is a very big integer that exceed all the entries in $I_{G'}$. Using this new problem set, we select one point in Λ as a beginning point. Then we repeatedly select the next point in Λ with the least cumulative cost until all the points in Λ are visited, and

Algorithm 2 RNNA to find best shortest complete path

Require: Re-indexed Λ and $I_{G'} = [a_{i,j}]_{(n+2)\times(n+2)}$
  $a_{1,n+2} = a_{n+2,1} = -M$, where $M > \max\limits_{a_{i,j}\in I_{G'}} a_{i,j}$
  for all point $P_i \in \Lambda$ **do**
    $\Lambda \leftarrow \Lambda - \{P_i\}$, append P_i in $\pi(P_i, P_i)$
    while $\Lambda \neq \emptyset$ **do**
      Let i^* be last node in $\pi(P_i, P_i)$
      Select $k = \min\limits_{j\in\Lambda} a_{i^*,j}$
      Append j in $\pi(P_i, P_i)$, $\Lambda \leftarrow \Lambda - \{j\}$,
    end while
    Append P_i in $\pi(P_i, P_i)$ to join last node in $\pi(P_i, P_i)$ to P_i
    $C(P_i, P_i) = \sum\limits_{i\in\Lambda} c(P_i) + \sum \text{selected } k + M$
  end for
  Select path with minimum $C(P_i, P_i), P_i \in \Lambda$.
  Detach path s, t from the selected path and rewrite the path with s as starting point and t as ending point.

finally return to the first point. This will form a loop. The entire process is repeated for $n + 2$ times, and finally the trials which yields the least cumulative cost was selected. The selected path was rewritten and we will get a sequence of points which yields the least cost complete path.

4. Computational Experiments

For evaluation purposes, computational experiments were conducted on grid networks of different sizes (N = 30, 50, 75, 100) with various number of intermediate points. The cost of each cell in the grid was randomly assigned with numbers ranging from 0 to 6. The position of the starting point, destination point and the intermediate points are provided by the user. To ensure that all points are far enough apart, we set that no point is within one cell radius of another.

In all cases, Phase 1 is performed to obtain G' and $I_{G'}$, then followed by Phase 2 where the best solution path is searched by RNNA. As a benchmark for solution quality measurement, Brute Force (BF) search is also performed by making a list of all possible Hamilton paths, then calculate the cost of each Hamilton path by adding up the cost of its edges. Finally, the Hamilton path with the least total cost is the optimal path. All algorithms were coded in C++ and tests were done on a PC with a 2.5GHz processor and 4GB of RAM. Computation time recorded are average time from 30 trials for each instance and the maximum computation time allowed for each instance is 300 seconds. The performance of RNNA and BF are tabulated in Table 1 and Table 2.

Table 1 shows the performance of RNNA and BF in searching for a solution path. In general, the computation time for both algorithms increase as the map size and number of intermediate points increase. Note that by using BF, the computation time starts to exceeded the time limit when $n = 11$. This is because as the number of intermediate point increases from n to $n + 1$, BF uses $n + 1$ times as many computation time as previous one. On the other hand, RNNA always obtains a solution in under a second. We also observe that BF works well on problems with fewer than 11 intermediate points.

Table 2 shows the quality of the solution found by using BF and RNNA. From the table we can see that the solution found by both BF and RNNA are almost the same when the number of intermediate points is small. But as the number of intermediate point increases,

Table 1. Average computation time (s) of BF and RNNA in path searching

Int pts	30× 30		50× 50		75× 75		100× 100	
	BF	RNNA	BF	RNNA	BF	RNNA	BF	RNNA
5	0.0082	0.1401	0.0088	0.1441	0.0141	0.2324	0.0094	0.1554
7	0.0551	0.1034	0.1248	0.2444	0.0494	0.1051	0.0815	0.1488
9	4.7841	0.1821	5.4248	0.1976	3.4975	0.1420	4.5268	0.1870
11	403.8979	0.1424	400.1723	0.1366	402.1891	0.1680	512.4457	0.2458

Table 2. The solution quality of RNNA compared to BF

Int pts	30× 30		50× 50		75× 75		100× 100	
	Diff count[a]	% Diff Avg[b]	Diff count[a]	% Diff Avg[b]	Diff count[a]	% Diff Avg[b]	Diff count[a]	% Diff Avg[b]
5	9	2.2618	9	1.6250	7	0.7375	10	2.0422
7	14	1.5994	12	1.6971	19	2.8215	19	2.6532
9	19	2.9797	21	2.8959	19	2.4916	16	2.2945
11	19	3.4323	21	3.0108	25	5.3175	23	4.2385

[a] Number of trials (out of 30 for each problem) where both BF and RNNA obtain different $C(s,t)$.
[b] Average percentage difference of solution cost obtained by RNNA compared to BF.

the cost of solution path found by RNNA starts to deviate from solution found by BF. This is because RNNA is a greedy algorithm. It always chooses the nearest point at the moment without considering the effect afterwards, hence resulting differences with the optimal solution path found by BF.

To observe how well can RNNA perform with increasing problem size, we conducted another experiment by increasing the number of intermediate points gradually. The computation time in Phase 1 and Phase 2 are recorded. The result are shown in Table 3.

Table 3. Average computation time (s) of RNNA

Int pts	30× 30			50× 50			75× 75			100× 100		
	Dijkstra	RNNA	Total	Dijkstra	RNNA	Total	Dijkstra	RNNA	Total	Dijkstra	RNNA	Total
5	0.3126	0.1401	0.4527	0.7700	0.1441	0.9141	3.4081	0.2324	3.6405	5.6362	0.1554	5.7916
10	0.4525	0.1359	0.5884	1.4486	0.1626	1.6112	4.9650	0.1600	5.1250	13.3273	0.1499	13.4772
15	0.8427	0.2206	1.0633	2.6621	0.2435	2.9056	9.8358	0.2476	10.0834	26.3822	0.2515	26.6337
20	1.6473	0.3767	2.0240	4.3760	0.4180	4.7940	15.8896	0.3949	16.2845	43.5400	0.3465	43.8865
25	2.2728	0.5722	2.8450	6.4552	0.5538	7.0090	22.7885	0.5369	23.3254	63.6984	0.4972	64.1956
30	3.3194	0.8923	4.2117	9.1396	0.7403	9.8799	31.0910	0.6597	31.7507	96.6226	0.7311	97.3537
35	4.5330	1.0010	5.5340	12.7168	0.9275	13.6443	40.0883	0.7880	40.8763	159.1426	1.0444	160.1870
40	6.1171	1.3000	7.4171	16.2596	1.1103	17.3699	55.3650	1.0130	56.3780	219.6215	1.3499	220.9714

From Table 3, we observe that the time used in Phase 1 increases with map size and the number of intermediate points. However, the computation time for Phase 2 is within 1 second for all problem instances. This shows that RNNA has stable performance even when both map size and number of intermediate points are increased.

5. Conclusion

In this paper, we study the shortest path problem with multiple unordered intermediate points (SPP-UIP) on a grid network, and present a two phase heuristic algorithm to find the solution path in SPP-UIP. In Phase 1, we use the Dijkstra algorithm to generate the shortest paths between every pair of points that need to be visited. Then, in Phase 2, we choose the best combination of partial paths that form the best complete path using RNNA. We conduct computational experiments on randomly generated grid networks of different sizes with different number of intermediate points. Experimental results show that despite deviation in solution found by RNNA from optimal solution, RNNA uses much lesser time to solve SPP-UIP disregarding the map size and the number of intermediate points. Therefore, we conclude that RNNA can be applied to solve SPP-UIP with large map size and large number of intermediate points.

This research can be extended in several ways. For example, we can consider SPP-UIP with multiple cost factors, as many real life optimization problems are multi-objective based. Besides that, we can also consider cases where the cost values in the grid network changes over the time, which occurs in situations such as rescue path planning in disaster area, where the cost value function are rarely constant. We may also consider the cases where the intermediate points are in clustered form. Such cases can be applied in route planning where we need to visit multiple sites in a cluster region before departing to another region.

Acknowledgments

The authors would like to acknowledge the funding by Universiti Sains Malaysia Short Term Research Grant 304/PMATHS/6313066.

References

[1] Alexander S 2002 Combinatorial Optimization: polyhedra and efficiency *Springer Science & Business Media* **24**

[2] Appleton B and Sun C 2003 Circular shortest paths by branch and bound *Pattern Recognition* **36** (11) pp 2513-20

[3] Bae K Y, Kim Y D and Han J H 2015 Finding a risk-constrained shortest path for an unmanned combat vehicle *Computers & Industrial Engineering* **80** pp 245-253

[4] Cormen T H 2009 *Introduction to Algorithms*, 3rd ed (USA: MIT Press)

[5] De Mesquita Sá Junior J J, Backes A R and Cortez P C 2013 Texture analysis and classification using shortest paths in graphs *Pattern Recognition Letters* **34** (11) pp 1314-19

[6] Han D H, Kim Y D and Lee J Y 2014 Multiple-criterion shortest path algorithms for global path planning of unmanned combat vehicles *Computers & Industrial Engineering* **71** pp 57-69

[7] Hart P E, Nilsson N J and Raphael B 1968 A formal basis for the heuristic determination of minimum cost paths *IEEE Transactions on System Science and Cybernetics* **4** (2) pp 100-107

[8] Kuby M, Araz O M, Palmer M and Capar I 2014 An efficient online mapping tool for finding the shortest feasible path for alternative-fuel vehicles *International Journal of Hydrogen Energy* **39** (32) pp 18433-39

IOP Conf. Series: Journal of Physics: Conf. Series **893** (2017) 012067 doi:10.1088/1742-6596/893/1/012067

Optimal strategy for supplier selection problem integrated with optimal control problem of single product inventory system with piecewise holding cost

Sutrisno, Widowati, and R H Tjahjana

Department of Mathematics, Diponegoro University, Semarang, Indonesia

E-mail: widowati_math@undip.ac.id

Abstract. In this paper, we formulate a hybrid mathematical model of supplier selection problem integrated with inventory control problem of a single product inventory system with piecewise holding cost. This model will be formulated in a piecewise affine (PWA) form that can be converted into mixed logical dynamic (MLD) form. By using this MLD model, we solve the supplier selection problem and control this inventory system so that the stock level tracks a desired level as the reference trajectory as closed as possible with minimal total cost. We use model predictive control for hybrid system to solve the problem. From the numerical experiment results, the optimal supplier was selected at each time period and the evolution of the stock level tracks the desired level well.

1. Introduction

Management on supply chain tries to minimize the operational cost and improve the profit. It focuses in relationships in order to maximize outcome for all parties in the chain [1]. Supply chain network consisting of all parties that are manufacturer, suppliers, carriers or transporters, warehouses, retailers and customers [2]. The optimal strategy in supply chain management is made to minimize the operational cost with some service level constraint and determine the optimal quantity, optimal locations and optimal time of a product to be produced and distributed [3]. Supply chain contains inventory system that has to be optimized so that it gives the minimal holding cost. Inventory problem can appear in the several forms like raw material inventory, work-in-process inventory and finished product inventory and each of them needs to be controlled in order to minimize the holding cost [3]. Most of researchers were developed a mathematical model to optimize supply chain management. The basic mathematical model in a linear state space equation for periodic review inventory system was developed in [4] and the optimal strategy was determined by using linear quadratic control method. Another approach was used a particle swarm optimization to optimize the basic inventory system considering discount [5].

Supply chain management is also contains supplier selection problem. Supplier selection problem is appeared if we have several supplier alternatives to supply the product and we have to decide how many the product that will be purchased from each supplier. Several approaches were used to find the optimal strategy for supplier selection problem solving. The most approach is mathematical model [6]. In control theory, there is a mathematical model called hybrid

IOP Conf. Series: Journal of Physics: Conf. Series **893** (2017) 012067 doi:10.1088/1742-6596/893/1/012067

dynamical system where the state value is depends on the current location (or event) which is called as piecewise-affine (PWA) model [7]. The PWA model can be converted into mixed logical dynamical system (MLD) by writing the PWA model in hybrid systems description language (HYSDEL) and the equivalent MLD model can be generated by using mld function in hybrid toolbox for MATLAB [8, 9]. For an inventory system with piecewise holding cost where the holding cost is different for several stock level intervals, [10] was formulated a hybrid mathematical model in piecewise-affine (PWA) model and controlled the stock level to track some desired level by using model predictive control for hybrid system. Predictive control is an optimal control method that can be used to solve a trajectory tracking control problem. The basic MPC for non-hybrid system was developed in [11] and the MPC for hybrid system was developed in [12, 13]. This optimal control method solves the problem i.e. determine the optimal input by defining an objective function and optimizing the corresponding optimization by using some optimization method.

In this paper, we integrate the supplier selection problem with inventory control problem of a single product inventory system with piecewise holding cost by formulating a PWA model and determine the optimal strategy by using model predictive control for hybrid system. We formulate the integrated supplier selection problem and inventory control problem in a PWA model, convert it into MLD model and determine the optimal strategy by using predictive control for hybrid system.

2. Mathematical Model

Let $u_s(k)$ be the amount of arriving shipment of the product from supplier s that arrived at time/review period k and n_p be the lead time delay. Then, the dynamic of stock level of the product $y(k) \geq 0$ for any $k \geq 0$ can be stated as follows

$$\begin{aligned} y(k) =& u_1(0) + u_2(0) + \cdots + u_S(0) \\ &+ u_1(1) + u_2(1) + \cdots + u_S(1) \\ &+ \cdots + u_1(k-n_p-1) + u_2(k-n_p-1) \\ &+ \cdots + u_m(k-n_p-1) - d(0) \\ &- d(1) - \cdots - d(k-1) \\ =& \sum_{j=0}^{k-n_p-1} \sum_{s=1}^{S} u_s(j) - \sum_{j=0}^{k-1} d(j). \end{aligned} \tag{1}$$

where S denotes the number of the suppliers and $d(k)$ denotes the demand of the product at review period k. We develop the model for inventory system with piecewise holding cost (or holding cost considering discount) as follows. Let $q_j, j = 1, 2, ..., m$ denotes the holding cost where

$$\text{holding cost} = \begin{cases} q_1, \text{ if } \hat{y}_0 = 0 \leq y(k) \leq \hat{y}_1 \\ q_2, \text{ if } \hat{y}_1 < y(k) \leq \hat{y}_2 \\ \cdots \\ q_m, \text{ if } \hat{y}_{m-1} < y(k) \leq \hat{y}_m. \end{cases} \tag{2}$$

Let $x_1(k) = y(k)$, $x_i(k) = q_j u(k - n_p + i)$ for $i = 2, 3, ..., n_p + 1$ and

$$\begin{cases} x_1(k+1) = x_1(k) + x_2(k) - q_j d(k) \\ x_2(k+1) = x_3(k) \\ x_3(k+1) = x_4(k) \\ \vdots \\ x_{(n-1)}(k+1) = x_n(k) \\ x_n(k+1) = q_j u(k). \end{cases}$$

where $n = n_p+1$, $x(k) = [x_1(k), x_2(k), ..., x_n(k)]'$, then the dynamic of stock level of the product can be written as a PWA model as follows

$$x(k+1) = \begin{cases} Ax(k) + B_1 u(k) + D_1 d(k) \text{ if } \hat{y}_0 \le y(k) \le \hat{y}_1 \\ Ax(k) + B_2 u(k) + D_2 d(k) \text{ if } \hat{y}_1 < y(k) \le \hat{y}_2 \\ \vdots \\ Ax(k) + B_m u(k) \\ +D_m d(k) \text{ if } \hat{y}_{m-1} < y(k) \le \hat{y}_m, \end{cases} \tag{3}$$

$$y(k) = \begin{cases} \frac{1}{q_1} x(k), \text{ if } \hat{y}_0 < y(k) \le \hat{y}_1 \\ \vdots \\ \frac{1}{q_j} x(k), \text{ if } \hat{y}_{j-1} < y(k) \le \hat{y}_j \\ \vdots \\ \frac{1}{q_m} x(k), \text{ if } \hat{y}_{m-1} < y(k) \le \hat{y}_m. \end{cases} \tag{4}$$

where $A = \begin{bmatrix} 1 & 1 & 0 & \cdots & 0 \\ 0 & 0 & 1 & \cdots & 0 \\ 0 & 0 & 0 & \cdots & \vdots \\ \vdots & \vdots & \vdots & \ddots & 1 \\ 0 & 0 & 0 & \cdots & 0 \end{bmatrix}$, $B_j = \begin{bmatrix} 0 \\ 0 \\ \vdots \\ q_j \end{bmatrix}$

and $D = \begin{bmatrix} -q_j \\ 0 \\ 0 \\ \vdots \\ 0 \end{bmatrix}$.

This PWA model can be controlled by using predictive control for hybrid system by converting it into equivalent MLD model then determining the optimal input by using mixed-integer quadratic programming.

3. Numerical Experiment

Suppose that there are four suppliers namely s_1, s_2, s_3 and s_4 to supply a product where the purchasing cost per unit product for each review period 1, 2, 3, ..., 80 from these suppliers is given by Table 1. The maximum capacity of each supplier to supply the product is given by Table 2.

Table 1. Purchasing cost per unit ($)

Time period (k)	s_1	s_2	s_3	s_4
1, 2, ..., 20	9	9.5	10	10.5
21, 22, 23, ..., 40	11	9	8	8
41, 42, 43, ..., 60	10	11	12	8
61, 62, 63, ..., 80	9	12	11	10

Table 2. Supplier capacity (unit)

Time period (k)	Supplier capacity			
	s_1	s_2	s_3	s_4
1, 2, 3, ..., 80	150	200	250	300

Suppose that the holding cost per unit is \$25 if the product volume is 0 to 100 units and it will be \$20 if the product volume is more than 100 units, hence $q_1 = \$25$ and $q_2 = \$20$, the maximum capacity of the storage is 300 units and the lead time delay $n_p = 2$. We solved this problem in Windows 8 with AMD A6 2.7 GHz of processor and 4 GB of Memory by using software MATLAB R2013a with hybrid system toolbox. By writing PWA model (3)-(4) in HYSDEL (available in Appendix 1) and using *mld* function in hybrid system toolbox, we convert the PWA model (3)-(4) into equivalent MLD model by adding auxiliary states δ and z in the form of

$$x(k+1) = Ax(k) + B_1u(k) + B_2\delta(k) + B_3z(k) \tag{5}$$
$$y(k) = Cx(k) + D_1u(k) + D_2\delta(k) + D_3z(k) \tag{6}$$
$$E_2\delta(k) + E_3z(k) \leq E_1u(k) + E_4x(k) + E_5 \tag{7}$$

where the values of the matrices $A, B_1, B_2, B_3, C, D_1, D_2, D_3, E_1, E_2, E_3, E_4$ and E_5 are available in the Appendix 2. Let H_p be the length of the horizon prediction for controller designing by using model predictive control for hybrid system given by [9, 12], the objective function for this optimal control problem is defined as the gain of the actual stock level from the desired stock level with weighting matrix Q_y and the input cost with weighting matrix Q_u and this optimal control problem is defined as follows

$$\min_{[u,\delta,z]_0^{H_p-1}} \sum_{k=0}^{H_p-1} \left[\|y(k) - y_d\|_{Q_y}^2 + \|u(k)\|_{Q_u}^2 \right] \tag{8}$$

subject to :

$$\begin{cases} x(k+1) = Ax(k) + B_1u(k) + B_2\delta(k) + B_3z(k) \\ y(k) = Cx(k) + D_1u(k) + D_2\delta(k) + D_3z(k) \\ -E_4x(k) - E_1u(k) + E_2\delta(k) + E_3z(k) \leq E_5 \\ u_{\min} \leq u(k) \leq u_{\max} \\ x_{\min} \leq x(k) \leq x_{\max} \\ y_{\min} \leq y(k) \leq y_{\max} \end{cases}$$

where $k = 0, 1, ..., H_p - 1$, notations $v_{\min}$ and $v_{\max}$ mean the lower bound and upper bound for v respectively and the notation $\|v\|_Q^2$ means v^TQv. The value of the desired stock level y_d can be found in the simulation result. The demand for this simulation is assumed to be constant that is $d(k) = 20$ for all $k > 0$. Assume that the initial product stock level is 70 units and the other parameters are given by Table 3 and Table 4.

Table 3. Parameter value

Parameter	k	Q_y	H_p	u_{min}	u_{max}	y_{min}	y_{max}
Value	any	1	10	$\begin{bmatrix}0\\0\\0\\0\end{bmatrix}$	$\begin{bmatrix}150\\200\\250\\300\end{bmatrix}$	0	300

Table 4. Weighting matrices for input

Time period (k)	Q_u
1,2,...,20	$diag(9,9.5,10,10.5)$
21,22,...,40	$diag(11,9,8,8)$
41,42,...,60	$diag(10,11,12,8)$
61,62,...,80	$diag(9,12,11,10)$

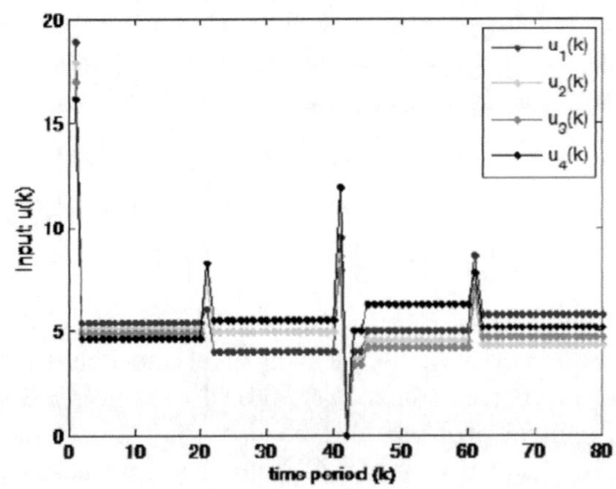

Figure 1. Optimal input $u(k)$

By solving (8) using mixed-integer quadratic programming, we have the simulation results shown by Fig. 1-2. The optimal input is shown by Fig. 1. The actual stock level $y(k)$ and the desired stock level $y_d(k)$ are shown by Fig. 2. Fig. 2 shows the optimal input u i.e. the amount of the arriving shipment of the product for each time or review period k. From this figure, it can be seen that all suppliers were selected to supply the product with different volume. This optimal input shown by Fig. 1 is applied to the system to generate the output i.e. the actual stock level $y(k)$ shown by Fig. 2. Fig. 2 shows the actual stock level and the desired stock level decided by decision maker. At time period 1 to 5, the product stock level was fluctuated due to the lead time delay is 2 periods and it has to satisfy the demand at time period 1 and 2. For time periods 5 to 80, the product stock level is following the desired level well although after the desired level is changed it needs 2 to 3 time periods to follow the desired level.

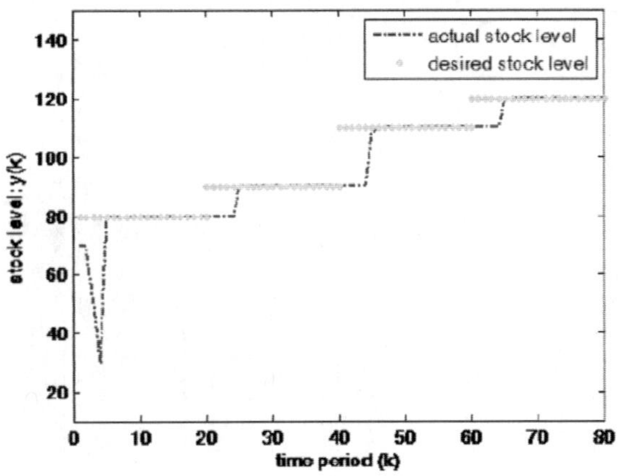

Figure 2. Output: actual stock level $y(k)$ and desired stock level y_d

4. Conclusion and Future Research

In this paper, the integrated supplier selection problem and tracking control problem of a single product inventory system with piecewise holding cost was considered. The problem was solved by formulating a piecewise affine model and determining the optimal amount of the product shipment from the optimal supplier by using model predictive control method for hybrid system. From the numerical experiment result, it can be conclude that the supplier selection problem was solved i.e. the optimal product volume from each supplier was determined and the stock level was controlled and followed the reference level well. In the future work, we will develop the model for multi-product inventory system and nonstationary demand. Furthermore, we will develop the model in deterministic and probabilistic environments.

Acknowledgment
The authors would like to thank Universitas Diponegoro for financial support through Research Professorship Program, DIPA PNBP Universitas Diponegoro 2016.

Appendix A. Code in HYSDEL for PWA model (3)-(4)

```
SYSTEM pwa_hybrid_inventory {
INTERFACE {
STATE { REAL x1 [0,4000];
        REAL x2 [0,4000];
        REAL x3 [0,4000];
        REAL y1 [0,200]; }
INPUT { REAL u1 [0,150];
        REAL u2 [0,200];
        REAL u3 [0,250];
        REAL u4 [0,300]; }
OUTPUT{ REAL y; }
PARAMETER { REAL y_hat_1, d, q1, q2; } }
IMPLEMENTATION {
AUX { REAL z1,z2,z3,z4;  BOOL d1; }
AD  { d1 = y1 >= y_hat_1; }
DA  { z1 = {IF d1 THEN  x1 + x2 - q2*d  ELSE  x1 + x2 - q1*d};
      z2 = {IF d1 THEN  x3  ELSE  x3 };
```

```
    z3 = {IF d1 THEN  q2*u1+q2*u2+q2*u3+q2*u4  ELSE   q1*u1+q1*u2+q1*u3+q1*u4 };
    z4 = {IF d1 THEN  x1/q2  ELSE  x1/q1 };  }
  CONTINUOUS { x1 = z1;   x2 = z2;  x3 = z3;  y1 = z4; }
  OUTPUT { y = y1;  } } }
```

Appendix 2. Matrices for system (5)-(7)
$A = zeros(4,4)$, $B_1 = zeros(4,4)$, $B_2 = zeros(4,1)$, $B_3 = eyes(4,4)$, $C = \begin{bmatrix} 0 & 0 & 0 & 1 \end{bmatrix}$, $D_1 = zeros(1,4)$, $D_2 = 0$, $D_3 = zeros(1,4)$, $E_5^T = [100, 0, 8500, 7500, 500, -500, 4000, 4000, 0, 0, 18000, 22500, 0, 0, 200, 160, 0, 0]$, $E_2^T = [-100, 100, 8100, 7900, -7900, -8100, 4000, 4000, -4000, -4000, 18000, 22500, -22500, -18000, 200, 160, -160, -200]$,

$$E_3 = \begin{bmatrix} 0 & 0 & 0 & 0 \\ 0 & 0 & 0 & 0 \\ -1 & 0 & 0 & 0 \\ 1 & 0 & 0 & 0 \\ -1 & 0 & 0 & 0 \\ 1 & 0 & 0 & 0 \\ 0 & -1 & 0 & 0 \\ 0 & 1 & 0 & 0 \\ 0 & -1 & 0 & 0 \\ 0 & 1 & 0 & 0 \\ 0 & 0 & -1 & 0 \\ 0 & 0 & 1 & 0 \\ 0 & 0 & -1 & 0 \\ 0 & 0 & 1 & 0 \\ 0 & 0 & 0 & -1 \\ 0 & 0 & 0 & 1 \\ 0 & 0 & 0 & -1 \\ 0 & 0 & 0 & 0 \end{bmatrix}, E_4 = \begin{bmatrix} 0 & 0 & 0 & -1 \\ 0 & 0 & 0 & 1 \\ -1 & -1 & 0 & 0 \\ 1 & 1 & 0 & 0 \\ -1 & -1 & 0 & 0 \\ 1 & 1 & 0 & 0 \\ 0 & 0 & -1 & 0 \\ 0 & 0 & 1 & 0 \\ 0 & 0 & -1 & 0 \\ 0 & 0 & 1 & 0 \\ 0 & 0 & 0 & 0 \\ 0 & 0 & 0 & 0 \\ 0 & 0 & 0 & 0 \\ 0 & 0 & 0 & 0 \\ -0.5 & 0 & 0 & 0 \\ 0.05 & 0 & 0 & 0 \\ -0.04 & 0 & 0 & 0 \\ 0.04 & 0 & 0 & 0 \end{bmatrix}, E_1 = \begin{bmatrix} 0 & 0 & 0 & 0 \\ 0 & 0 & 0 & 0 \\ 0 & 0 & 0 & 0 \\ 0 & 0 & 0 & 0 \\ 0 & 0 & 0 & 0 \\ 0 & 0 & 0 & 0 \\ 0 & 0 & 0 & 0 \\ 0 & 0 & 0 & 0 \\ 0 & 0 & 0 & 0 \\ 0 & 0 & 0 & 0 \\ -20 & -20 & -20 & -20 \\ 20 & 20 & 20 & 20 \\ -25 & -25 & -25 & -25 \\ 25 & 25 & 25 & 25 \\ 0 & 0 & 0 & 0 \\ 0 & 0 & 0 & 0 \\ 0 & 0 & 0 & 0 \\ 0 & 0 & 0 & 0 \end{bmatrix}.$$

References

[1] Christopher M 2011 *Logistics and supply chain management* (Great Britain: Pearson Education Limited)
[2] Chopra S and Meindl P 2007 *Supply chain management, strategy, planning, and operation* (New Jersey: Prentice Hall)
[3] Levi D S 2000 *Designing and managing the supply chain* (USA: McGraw-Hill Companies)
[4] Ignaciuk P and Bartoszewicz A 2010 Linear-quadratic optimal control strategy for periodic-review perishable systems *IEEE Trans. on Con. Sys. Tech.* **20** 1400
[5] Mousavi S M Niaki S Bahreininejad A and Musa S N 2014 Multi-Item multiperiodic inventory control problem with variable demand and discounts: A particle swarm optimization algorithm *The Sci. World Journal* **2014** 1
[6] Kara S S 2011 Supplier selection with an integrated methodology in unknown environment *Exp. Sys. with App.* **38** 2133
[7] Branicky M *Introduction to hybrid systems* (USA: Department of Electrical Engineering and Computer Science, Case Western Reserve University)
[8] Bemporad A 2004 Efficient conversion of mixed logical dynamical systems into an equivalent piecewise affine form *IEEE Trans. Automat. Control* **49** 832
[9] Bemporad A 2012 *Hybrid toolbox, users guide*
[10] Sutrisno, Widowati and Munawwaroh D A 2015 Hybrid mathematical model of inventory system with piecewise holding cost and its optimal strategy *IEEE International conference on advanced mechatronics, intelligent manufacture, and industrial automation* 29
[11] Maciejowski J M 2001 *Predictive control with constraints* (USA: Prentice Hall)
[12] Borrelli F, Bemporad A and Morari M 2011 *Predictive control for linear and hybrid systems*
[13] Lazar M, Heemels W, Weiland W and Bemporad A 2006 Stabilizing model predictive control of hybrid systems *IEEE Trans. Automat. Control* **51** 1813

IOP Conf. Series: Journal of Physics: Conf. Series **893** (2017) 012068 doi:10.1088/1742-6596/893/1/012068

Multimodal optimization by using hybrid of artificial bee colony algorithm and BFGS algorithm

S Anam

Department of Mathematics, Brawijaya University, Malang, Indonesia

E-mail: syaiful@ub.ac.id

Abstract. Optimization has become one of the important fields in Mathematics. Many problems in engineering and science can be formulated into optimization problems. They maybe have many local optima. The optimization problem with many local optima, known as multimodal optimization problem, is how to find the global solution. Several metaheuristic methods have been proposed to solve multimodal optimization problems such as Particle Swarm Optimization (PSO), Genetics Algorithm (GA), Artificial Bee Colony (ABC) algorithm, etc. The performance of the ABC algorithm is better than or similar to those of other population-based algorithms with the advantage of employing a fewer control parameters. The ABC algorithm also has the advantages of strong robustness, fast convergence and high flexibility. However, it has the disadvantages premature convergence in the later search period. The accuracy of the optimal value cannot meet the requirements sometimes. Broyden–Fletcher–Goldfarb–Shanno (BFGS) algorithm is a good iterative method for finding a local optimum. Compared with other local optimization methods, the BFGS algorithm is better. Based on the advantages of the ABC algorithm and the BFGS algorithm, this paper proposes a hybrid of the artificial bee colony algorithm and the BFGS algorithm to solve the multimodal optimization problem. The first step is that the ABC algorithm is run to find a point. In the second step is that the point obtained by the first step is used as an initial point of BFGS algorithm. The results show that the hybrid method can overcome from the basic ABC algorithm problems for almost all test function. However, if the shape of function is flat, the proposed method cannot work well.

1. Introduction

Optimization has become one of the important fields in Mathematics. Many problems can be formulated into optimization problems. The optimization problems play an important role in industry and science fields. For example, optimization has been applied in industrial wastewater [4] and in tomotherapy for prostate cancer patients [9].

The optimization problem maybe has many local optima. The optimization problem with many local optima, known as multimodal optimization problem, is how to find a global solution from several optimum points. Several metaheuristic methods have been proposed to solve multimodal optimization problems such as Particle Swarm Optimization (PSO), Genetics Algorithm (GA), Artificial Bee Colony (ABC) algorithm, etc [2].

The ABC algorithm is one of the popular evolutionary algorithms that proposed by Karaboga in 2005 [5]. The ABC algorithm was successfully applied in many applications such as job shop scheduling. The ABC algorithm has several advantages compared with the other population-based algorithms. The first advantage is that it is easy to be used, high flexible and tardiness [7].

The second advantage is that it employs a fewer control parameters than the other algorithms such as Particle Swarm Optimization and Genetic Algorithm [6]. The third advantage is that the ABC Algorithm hybridizes easily with the other optimization algorithms [8]. The forth advantage is that the ABC Algorithm is strong robustness and fast convergence. The ABC algorithm also can handle stochastic cost objective function [3]. However, the ABC algorithm has the disadvantages that it easily meets premature convergence in the later search period. The optimum value cannot be found sometimes [10].

Quasi-Newton method is one of the good methods for solving the minimization problem. The size of the gradient change is constructed well enough to produce super linear convergence. This method has better convergence than gradient descent method. Broyden–Fletcher–Goldfarb–Shanno (BFGS) algorithm is one of type Quasi Newton methods which is a good iterative method for finding a local optimum. This method has also been proven to be better than gradient descent when applied to the neural network [1] .

Based on the advantages of the ABC algorithm and the BFGS algorithm, this paper proposes a hybrid of the artificial bee colony algorithm and the BFGS algorithm to solve the multimodal optimization problem. The principle of a hybrid method utilizes the strengths and reduces the weaknesses possessed by each method, so it can optimize function successfully. The performance of the proposed method is evaluated by using 5 test functions.

2. The ABC Algorithm

The ABC algorithm is one of the popular evolutionary algorithms that is proposed by Karaboga in 2005 [5]. It is very simple, clear and adaptable. The ABC algorithm is inspired by the behavior of bee colonies in search for food sources. The search cycle of ABC algorithm consists of three rules. The first rule is that the employee bees are sent to a food source and the nectar quality is evaluated. The second rule is that onlookers choose the food sources after obtaining information from employee bees and calculating the nectar quality. The last rule is that determining the scout bees and sending them onto possible food sources. The ABC procedure could be represented in **Algorithm 1**.

Algorithm 1 : Schematic pseudo code of ABC procedure.

(i) Initialize the ABC and problem parameters
(ii) Initialize the Food Source Memory (FSM)
(iii) Repeat
 (a) Send the employee bees to the food sources.
 (b) Send the onlookers to select a food source.
 (c) Send the scouts to search possible new food.
 (d) Memorize the best food source.
 Until (termination criterion are met)

2.1. Initialization of the ABC and problem parameter.

In the optimization problem, if function $f(\mathbf{x})$ will be minimized and $\mathbf{x}$ is decision variable, then the optimization problem is formulated as follows:

$$\min_{\mathbf{x}\in\mathbf{X}} f(\mathbf{x}) \tag{1}$$

If $\mathbf{x}$ has N dimensions then $\mathbf{x}$ can be written as follows $\mathbf{x} = \{x_1, x_2, ..., x_N\}$ where $x_i \in [LB_i, UB_i]$. LB_i and UB_i are the lower and upper bound values for the variable x_i. The parameters of the ABC algorithm are the following:

(i) Population size (SN) is the number of food sources (candidate of solutions) in the population. SN is equal to the number of employee bees or onlooker bees.

(ii) Maximum Cycle Number (MCN) is representation of the maximum number of generations.

(iii) Limit is used to diversify the search, to determine the number of allowable generations for which each non improved food source is to be abandoned.

2.2. *Initialization of the Food Source Memory (FSM)*

A food source is a candidate solution to the optimization problem. Food Source Memory (FSM) is a memory allocation consisting of food resources and may be determined by SN in equation (2).

$$FSM = \begin{bmatrix} x_1(1) & x_1(2) & ... & x_1(N) \\ x_2(1) & x_2(2) & ... & x_2(N) \\ ... & ... & ... & ... \\ x_N(1) & x_N(2) & ... & x_N(N) \end{bmatrix} \tag{2}$$

with each vector generated from:

$$\begin{array}{l} x_j(i) = LB_i \pm (UB_i - LB_i) \times r \\ \forall j \in \{1, 2, ..., SN\} \forall i \in \{1, 2, ..., N\} \end{array} \tag{3}$$

Note: $r \sim (0, 1)$ is uniform random number between 0 and 1.

2.3. *Sending of the employee bees to the food sources*

In this step, each employee bee is assigned to its food source and in turn, a new one is generated from its neighbouring solution, using equation (4).

$$\begin{array}{l} x'_j(i) = x_j(i) \pm (x_j(i) - x_k(i)) \times r \\ \forall k \in \{1, 2, ..., SN\} k \neq j \text{ and } r \sim (0, 1) \end{array} \tag{4}$$

In this step, the quality nectars on food sources ($x_j(i)$ and $x'_j(i)$) also are evaluated. If the quality nectar of $x'_j(i)$ is better than the quality nectar of $x_j(i)$ then $x_j(i)$ will be replaced by $x'_j(i)$.

2.4. *Sending of the onlookers to select a food source*

The onlooker bee has the same number of food sources as the employee. In this step the probability of each source of food from the previous step is calculated food sources are chosen by bees onlooker by using Roulette Wheel Selection.

2.5. *Sending of the scouts to search possible new food*

Scout bees will randomly searching for new food sources to replace the abandoned food source by using equation (3). The abandoned food source is a food source that can not be updated on a particular experiment. These food sources will be determined by the parameter limit [10].

3. The BFGS Algorithm

Quasi-Newton method is an alternative method that can be used for solving optimization problems. This method was successfully used to minimize errors on artificial neural networks [1]. Quasi-Newton method is method that can be used if the calculation of the Hessian matrix is difficult or time consuming. This method has a rapid convergence when compared with the method of gradient descent. One of Quasi Newton method types is the BFGS method. The BFGS algorithm can be seen in **Algorithm 2**.

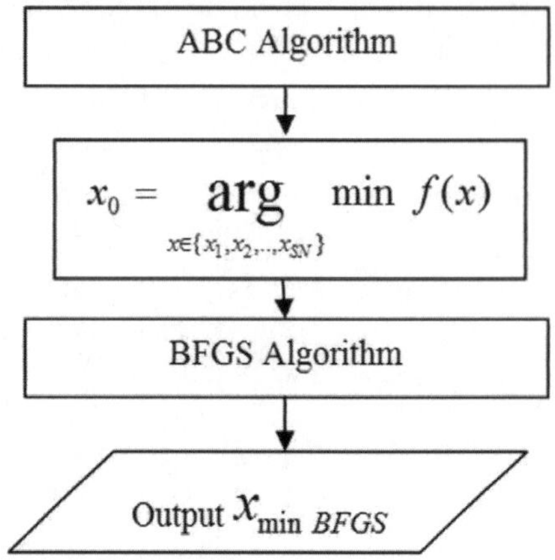

Figure 1. Flowchart of the proposed method.

Algorithm 2 : BFGS algorithm method

(i) Input $\mathbf{x_0}, \epsilon$ (stopping criteria) and k_{max}.
(ii) Set initial iteration $k = 0$ and $\mathbf{B} = \mathbf{I}$ where $\mathbf{I}$ is an indentity matrix.
(iii) Calculate $f(\mathbf{x}_k)$
(iv) While ($\|\nabla(\mathbf{x}_k)\|| > \epsilon$) or ($k <= k_{max}$) do
  (a) Calculate $\mathbf{d}_k = -\mathbf{B}_k \nabla f(\mathbf{x}_k)$, where $\mathbf{d}_k$ generating direction.
  (b) Select α that is able to minimize $f(\mathbf{x}_k + \alpha_k \mathbf{d}_k)$
  (c) $\mathbf{x}_{k+1} = \mathbf{x}_k + \alpha_k \mathbf{d}_k$
  (d) Calculate $\mathbf{B}_{k+1}$ by using equation (5).

$$\mathbf{B}_{k+1} = \mathbf{B}_k - \frac{\mathbf{B}_k\mathbf{s}_k(\mathbf{B}_k\mathbf{s}_k)^T}{\mathbf{s}_k^T\mathbf{B}_k\mathbf{s}_k} + \frac{\mathbf{y}_k\mathbf{y}_k^T}{\mathbf{y}_k^T\mathbf{s}_k} + \phi_k[\mathbf{s}_k^T\mathbf{B}_k\mathbf{s}_k]\mathbf{v}_k\mathbf{v}_k^T \tag{5}$$

  where $\phi_k \in [0,1], \mathbf{s}_k = \mathbf{x}_{k+1} - \mathbf{x}_k$ and
  $\mathbf{v}_k = \frac{\mathbf{y}_k}{\mathbf{y}_k^T\mathbf{s}_k} - \frac{\mathbf{B}_k\mathbf{s}_k}{\mathbf{s}_k^T\mathbf{B}_k\mathbf{s}_k}$.
  (e) $k = k + 1$
End while

4. Proposed Method

The ABC Algorithm is a global optimization technique that is well known, inspired by the behavior of bee colonies. The ABC algorithm has advantages of memory, local search and solution improvement mechanism over the other meta-heuristic algorithms. The scout bee in the ABC algorithm controls the exploration process to find the area of global solution point. The onlooker bee and employee bee controls the exploitation process to find global solution point in the area which found by the scout bee. However, the ABC algorithm does not guarantee to give the optimal solution and often has premature convergence. To overcome this problem, in this paper proposes a hybrid of the ABC algorithm and the BFGS algorithm.

Hybridization schemes of ABC algorithm and BFGS algorithm as shown in Figure 1. In the proposed method, the best solution of the ABC algorithm is used as an initial value for the

BFGS method. Figure 2 illustrates how to the proposed method works. $\mathbf{x}_0$ is the best solution of the ABC algorithm which is represented by a yellow point. It is used to be an initial point of the BFGS algorithm. Furthermore, the BFGS method will find the solution that is represented by a green point. The point is very close to the optimal solution that represented by a red point.

5. Experimental Results and Discussion

The numerical solution of the proposed method is evaluated by using a computer system with the following specifications: processor: Intel(R) Core i5 3317U CPU 1.7G.Hz, memory: 4 GB, and operating system: Windows 7 Home Basic, Service Pack 1. The hybrid of the ABC algorithm and the BFGS algorithm is implemented using software Matlab 2008.

The proposed method is compared with the basic ABC algorithm to evaluate the performace of the proposed method. In order to evaluate the performance of the proposed method, several benchmark functions are used for evaluating. The benchmark functions are the following:

(i) **Schaffer function**
Schaffer function is defined by equation (6). It is also multimodal function. $\mathbf{x}$ is in the interval of [-100, 100]. The global optimum value is 0 and the optimum solution is $\mathbf{x}_{opt} = (0, 0, \ldots, 0)$.

$$f_1(\mathbf{x}) = 0.5 + \frac{sin^2\left(\sqrt{x_1^2 + x_2^2}\right) - 0.5}{(1 + 0.001(x_1^2 + x_2^2))^2} \quad (6)$$

(ii) **Sphere function**
Sphere function is defined by equation (7). It is unimodal function, convex and continous. $\mathbf{x}$ is in the interval of [-100, 100]. The global optimum value is 0 and the optimum solution is $\mathbf{x}_{opt} = (0, 0, \ldots, 0)$.

$$f_2(\mathbf{x}) = \sum_{i=1}^{n} x_i^2 \quad (7)$$

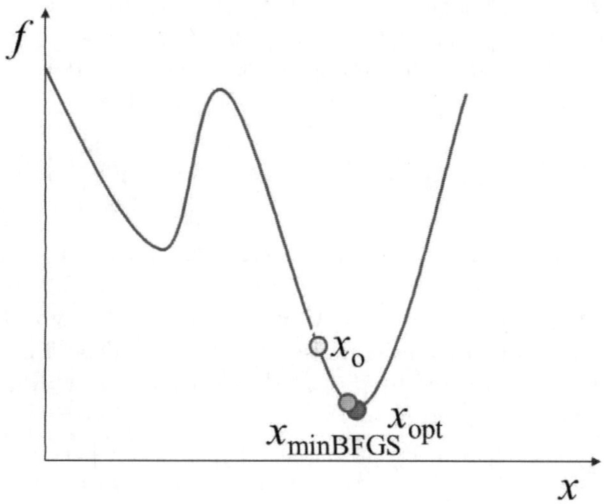

Figure 2. Ilustration of the proposed method.

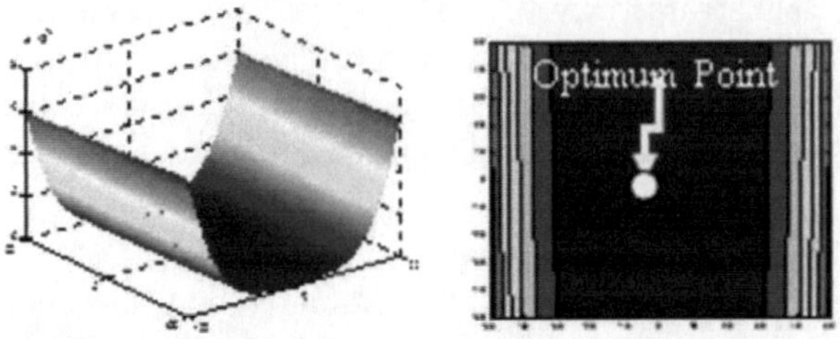

(a) (b)

Figure 3. (a) Rosenbrock function view. (a) Rosenbrock function contour.

(iii) **Griewank function**
Griewank function is generally defined by equation (8). $\mathbf{x}$ is in the interval of [-200,200]. The global minimum value for Griewank function is 0 and the corresponding with global optimum solution $\mathbf{x}_{opt} = (100, 100, \ldots, 100)$. Because the number of local optima increases with the dimentionality, Griewank function is strongly multimodal function.

$$f_3(\mathbf{x}) = \frac{1}{4000}\left(\sum_{i=1}^{n}(x_i - 100)^2\right) - \left(\prod_{i=1}^{n} cos\left(\frac{x_i - 100}{\sqrt{i}}\right)\right) + 1 \tag{8}$$

(iv) **Rastrigin function**
Rastrigin function is defined by equation (9). It is multimodal function. The difficult part about finding local optimal solution is that an optimization easily can be trapped in local optimum on its way toward global optimum. $\mathbf{x}$ is in the interval of [-5.12, 5.12]. The global optimum value is 0 and the optimum solution is $\mathbf{x}_{opt} = (0, 0, \ldots, 0)$.

$$f_4(\mathbf{x}) = \sum_{i=1}^{n}\left(x_i^2 - 10cos(2\pi x_i) + 10\right) \tag{9}$$

(v) **Rosenbrock function**
Rosenbrock function is defined by equation (10). It is also multimodal function. $\mathbf{x}$ is in the interval of [-50, 50]. The global optimum value is 0 and the optimum solution is $\mathbf{x}_{opt} = (1, 1, \ldots, 1)$. The global optimum is inside a long, narrow, parabolic-shaped flat valley. Since it is difficult to converge to the global optimum.

$$f_5(\mathbf{x}) = \sum_{i=1}^{n-1} 100(x_{i+1} - x_i^2)^2 + (x_i - 1)^2 \tag{10}$$

Each of the experiment was repeated 5 times with different random seeds. The mean and the deviation standard of the function value, and the mean and the deviation standard of the computational time obtained by the basic ABC algorithm and the hybrid of the ABC algorithm and BFGS algorithm for under the same condition are given in Table 1 and Table 2, respectively. The objective function value obtained by proposed method is better than the objective function value obtained by the basic ABC algorithm for almost all test function. However, the proposed method cannot works well for Rosenbrock function. The BFGS algorithm cannot work well because the Rosenbrock function shapes is very flat as shown in Figure 3 (a) and the global optimum of the Rosenbrock function is inside a long, narrow and parabolic-shaped flat valley which is shown by the Rosenbrock function contours in Figure 3 (b). The BFGS algorithm is failed to find a true step size that is able to minimize $f(\mathbf{x}_k + \alpha_k \mathbf{d}_k)$. The computational time of the proposed method is better than the basic ABC algorithm as shown in Table 2.

Table 1. The objective function value and optimum point obtained by the basic ABC algorithm and the proposed method.

f	D	ABC Algorithm Mean (Std)	Proposed Method Mean (Std)
f_1	2	5.3×10^{-3} (4.8×10^{-3})	$\mathbf{3.9 \times 10^{-3}}$ ($\mathbf{5.0 \times 10^{-3}}$)
f_2	2	5.1×10^{-17} (2.2×10^{-17})	$\mathbf{1.77 \times 10^{-36}}$ ($\mathbf{7.7 \times 10^{-37}}$)
	4	1.47×10^{-5} (4.25×10^{-5})	$\mathbf{3.2 \times 10^{-35}}$ (1.3×10^{-35})
f_3	2	2.2×10^{-3} (3.6×10^{-3})	$\mathbf{1.5 \times 10^{-3}}$ ($\mathbf{3.1 \times 10^{-3}}$)
	4	3.48×10^{-2} (2.18×10^{-2})	$\mathbf{1.48 \times 10^{-2}}$ ($\mathbf{6.0 \times 10^{-3}}$)
f_4	2	0 (0)	0 (0)
	4	4.05×10^{-1} (1.28)	$\mathbf{7.11 \times 10^{-15}}$ ($\mathbf{2.3 \times 10^{-15}}$)
f_5	2	$\mathbf{4.4 \times 10^{-3}}$ ($\mathbf{6.4 \times 10^{-3}}$)	9.49×10^{-1} (1.301)
	4	$\mathbf{5.86 \times 10^{-1}}$ ($\mathbf{8.97 \times 10^{-1}}$)	7.38×10^{2} (1.44×10^{-3})

Table 2. The computational time of the basic ABC algorithm and the proposed method corresponding optimum point search in Tabel 1.

f	D	ABC Algorithm		Proposed Method	
		Mean	Std	Mean	Std
f_1	2	11.7836	0.8247	**9.5275**	**0.6212**
f_2	2	10.4157	0.7202	**9.3355**	**0.3464**
	4	10.4156	0.7037	**9.6526**	**0.3244**
f_3	2	12.1211	1.0269	**9.8939**	**0.2869**
	4	12.0705	1.0471	**10.0757**	**0.3712**
f_4	2	12.0267	0.4039	**10.207**	**0.4329**
	4	12.0263	0.3856	**10.5047**	**0.4673**
f_5	2	12.9441	1.6708	**10.0831**	**0.6389**
	4	12.1936	0.8299	**10.0019**	**0.6749**

6. Conclusion

We have proposed a method for multimodal function optimization by using a hybrid of the ABC algorithm and the BFGS method. The results show that the hybrid method can overcome from the basic ABC algorithm problems for almost all test function. The numerical results show the objective function value produced by the proposed method is very closed to global minumum value for almost all test functions. However, if the shape of function is flat, the proposed method cannot work well.

References

[1] Anam S, Yanti I and Widodo A 2010 *Jurnal Teknika* **2** p 43

[2] Bahesti Z and Shamsuddin S M H 2013 *International Journal Advanced Soft Computing Application* **5** p 1
[3] Bolaji A L, Khader A T, Al-Betar M A and Awadallah M A 2013 *Journal of Theoritical and Applied Information Technology* **47** p 434
[4] Feng X and Chu K H 2004 *Process Safety and Environmental Protection* **8** p 249
[5] Karaboga D 2005 *Technical Report Erciyes University Engineering Faculty, Computer Engineering Department*
[6] Karaboga D and Akay B 2009 *International Journal of Applied Mathematics and Computation* **214** p 108
[7] Rao R S, Narasimhama S V L and Ramalingaraju M 2008 *International Journal of Electrical Computer, Energetic, Electronic, and Communication Engineering* **2** p 116
[8] Singh 2009 *International Journal of Applied Soft Computing* **9** p 625
[9] Skórska M and Piotrowskia T 2013 *Physica Medica* **29** p 273
[10] Yan G and Li C 2011 *Journal of Computational Information Systems* **7** p 3309